Introduction to ELECTRICAL POWER *and* POWER ELECTRONICS

Introduction to ELECTRICAL POWER *and* POWER ELECTRONICS

MUKUND R. PATEL

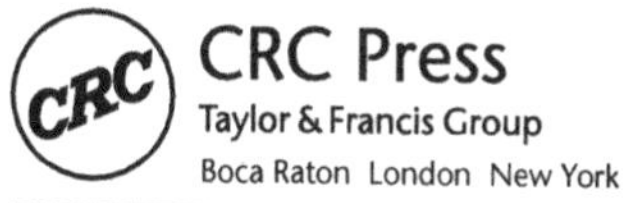

CRC Press is an imprint of the
Taylor & Francis Group, an **informa** business

CRC Press
Taylor & Francis Group
6000 Broken Sound Parkway NW, Suite 300
Boca Raton, FL 33487-2742

First issued in paperback 2017

Version Date: 20120611

ISBN 13: 978-1-138-07625-9 (pbk)
ISBN 13: 978-1-4665-5660-7 (hbk)

Library of Congress Cataloging-in-Publication Data

Patel, Mukund R., 1942-
Introduction to electrical power and power electronics / by Mukund R. Patel.
p. cm.
Includes bibliographical references and index.
ISBN 978-1-4665-5660-7
1. Power electronics. I. Title.

TK7881.15.P375 2013
621.31'7--dc23 2012021707

Visit the Taylor & Francis Web site at
http://www.taylorandfrancis.com

and the CRC Press Web site at
http://www.crcpress.com

to…

Sarla,

my friend and my wife,

and to her late parents,

Kantaben and Shantilal Patel,

for inspiring her to be an electrical engineer

in 1950s rural India with no electricity.

Contents

PART A Power Generation, Distribution, and Utilization

PART B Power Electronics and Motor Drives

Preface

The United States, Canada, the United Kingdom, and many other countries are presently experiencing a shortage of power engineers. The shortage is expected to get worse in the United States, where about 45% of the utility power engineers will become eligible to retire before 2020. They would require about 7000 new power engineering graduates to replace them and about an equal number in the supporting industries. At the same time, many countries are making huge investments in building new power plants and power grids that will require even more power engineers to serve the industry. It is for this trend that university students are now once again getting attracted to the electrical power programs.

No other technology has brought a greater change in the electrical power industry—and still holding the potential of bringing future improvements—than power electronics. The power electronics equipment prices have declined to about 1/10 since the late 1980s, fueling a rapid growth in their applications throughout the power industry. Large cruise ships' power plant ratings approach 100 MW_e since they use electric propulsion with power electronics in abundance. The navies of the world have started changing from mechanical propulsion to electric propulsion with power electronics for the numerous benefits they offer. With high-power combat weapons onboard, they will need even larger power plants in size comparable to those on land.

Until now, there has been no single book available that covered the entire scope of electrical power and power electronics systems. Traditional books on power systems focused on high voltage transmission, in which only a handful of power engineers get involved. A vast majority of power engineers work in cities, towns, large factories, steel mills, commercial and institutional buildings, refineries, data centers, railways, shipping ports, commercial and navy ships, etc. They mostly deal with power taken from the utility company or self-generated mostly for running motors and other loads via transformers, cables, and protection devices such as fuses and circuit breakers. They also routinely deal with batteries of various electrochemistries, power electronics converters of all kinds, variable-frequency motor drives, etc. This book is the first comprehensive volume of its kind that focuses on all of these topics, which are directly relevant to most power engineers on a daily basis. In addition, engineers working in the renewable energy technologies would also find this book useful since the electrical power system for extracting energy from wind and ocean currents is the exact mirror image of the variable-frequency motors driving pumps, fans, and ship propellers.

It is hoped that the book is a timely addition to the literature and a one-volume resource for students of electrical power and power electronics at various universities around the world and for a range of industry professionals working as electrical, mechanical, and chemical engineers, managers, and supervisors—actually, anyone who uses or deals with electrical power.

Acknowledgments

A book of this nature incorporating all aspects of electrical power and power electronics requires help from many sources. I have been extremely fortunate to receive full support from many organizations and individuals in the field. They not only encouraged me to write the book on this timely subject but also provided valuable suggestions and comments during the development of the book.

At the U.S. Merchant Marine Academy, Kings Point, NY, I am grateful to Dr. David Palmer, engineering department head, and Dr. Shashi Kumar, academic dean, for supporting the research work that led to my writing this book. I have benefited from many students at the Academy, both the undergraduate seniors and the graduate students, who contributed to my learning by pointed questions and discussions based on their professional experience.

My special gratitude goes to Dr. Robert Degeneff of Utility Systems Technology Inc., Niskayuna, NY, a colleague from the General Electric Company and a retired professor of electrical power engineering at the Rensselaer Polytechnic Institute, Troy, NY, and Dr. Anurag Srivastava of Washington State University, Pullman. Both professors reviewed the book proposal and provided valuable comments.

My grandchildren Sevina, Naiya, Dhruv, and Rayna and my wife Sarla cheerfully contributed good time they would have otherwise spent with me.

I heartily acknowledge the valuable support and encouragement I received from all.

Mukund R. Patel
Yardley, Pennsylvania

The Author

Mukund R. Patel, PhD, PE, is a professor of engineering at the U.S. Merchant Marine Academy in Kings Point, NY. He has over 45 years of hands-on involvement in research, development, and design of state-of-the-art electrical power equipment and systems. He has served as a principal engineer at the General Electric Company in Valley Forge, PA; fellow engineer at the Westinghouse Research and Development Center in Pittsburgh, PA; senior staff engineer at the Lockheed Martin Corporation in Princeton, NJ; development manager at Bharat Bijlee (Siemens) Limited, Bombay, India; and 3M McKnight Distinguished Visiting Professor at the University of Minnesota, Duluth.

Dr. Patel obtained his PhD degree in electric power engineering from the Rensselaer Polytechnic Institute, Troy, NY; MS degree in engineering management from the University of Pittsburgh; ME degree in electrical machine design from Gujarat University; and BE degree from Sardar University, India. He is a fellow of the Institution of Mechanical Engineers (U.K.), associate fellow of the American Institute of Aeronautics and Astronautics, senior life member of the IEEE, registered professional engineer in Pennsylvania, chartered mechanical engineer in the United Kingdom, and an elected member of Eta Kappa Nu, Tau Beta Pi, Sigma Xi, and Omega Rho.

Dr. Patel is an associate editor of *Solar Energy*, the journal of the International Solar Energy Society, and a member of the review panels for the government-funded research projects on renewable energy in the state of California and the emirate of Qatar. He has authored five books, three of which have been translated into Chinese, and major chapters in two international handbooks. He has taught 3-day courses to practicing engineers in the electrical power industry for over 15 years, has presented and published over 50 papers at national and international conferences and journals, holds several patents, and has earned NASA recognition for exceptional contribution to the power system design for the Upper Atmosphere Research Satellite. The high-voltage, high-frequency cable he developed for the International Space Station was nominated by NASA for an Industrial Research-100 award in 1987. He can be reached at patelm@usmma.edu or patelm30@gmail.com.

About This Book

This book has evolved from the author's 30 years of work experience at General Electric, Lockheed Martin, and Westinghouse Electric, and 15 years of teaching electrical power at the U.S. Merchant Marine Academy at Kings Point, NY. The book has 17 chapters, divided into two parts:

- *Part A*: Power Generation, Distribution, and Utilization
- *Part B*: Power Electronics and Motor Drives

Part A of the book focuses on all aspects of electrical power that most power engineers get involved in during their professional careers. It covers the generation and distribution of power to load equipment such as motors via step-down transformers, cables, circuit breakers, relays, and fuses. Some engineers working on standalone power plants get involved with generators. All these topics are covered in Chapters 1 through 9.

Chapter 10 covers various ways of designing and operating power systems for economic utilization of power and basic methods of quantifying profitable energy-saving opportunities. The regenerative braking that converts the kinetic energy of a moving mass into electricity—as is done in hybrid automobiles—and energy-saving benefits of variable speed motor drives are dealt with in detail in Chapter 10.

The battery is always an integral part of power system design for providing emergency power to essential and critical loads and power to dc control circuits and computers. It is covered in Chapter 11 in detail not found in traditional power system books.

Part B of the book starts with power electronics switches in Chapter 12. The next three chapters cover dc–dc converters, ac–dc–ac converters, and frequency converters used in variable-frequency motor drives. Chapter 16 discusses the quality-of-power issues in modern power systems with many power electronics loads of relatively large size.

Chapter 17 covers power converter cooling by air and also by water, which presents important interdisciplinary design topics relevant to power engineers.

The unique features of this book are as follows:

- This is the only book available that covers all aspects of electrical power and power electronics under one cover that will otherwise require several book.
- Many examples, exercise problems, and concept questions appear at the end of each chapter.
- Long visible suffixes in equations make their use fluent to save time.

Both international and British systems of units are used in the book to present data, as they came from various sources. An extensive conversion table connecting the two systems of units is therefore given in the next section for ready reference.

The book can be used for a one-semester course for electrical majors in four-year degree programs or for two-quarter courses split between power systems and power electronics as needed. For professional engineers working in the industry, the book would be ideal for a refresher course or as a single-source reference book for all topics on electrical power and power electronics.

Systems of Units and Conversion Factors

Both the international units (SI or MKS system) and British units are used in this book. The table below relates the international units with the British units commonly used in the United States.

Category	Value in SI Unit =	Factor Below ×	Value in English Unit
Length	m	0.3048	ft.
	mm	25.4	in.
	μm	25.4	mil
	km	1.6093	mi.
	km	1.852	nautical mi.
Area	m^2	0.0929	$ft.^2$
	μm^2	506.7	circular mil
Volume	L (dm^3)	28.3168	$ft.^3$
	L	0.01639	$in.^3$
	cm^3	16.3871	$in.^3$
	m^3/s	0.02831	$ft.^3/h$
	L	3.7853	gal. (U.S.)
	L/s	0.06309	gal./min
Mass	kg	0.45359	pound mass
	kg	14.5939	slug mass
Density	kg/m^3	16.020	lb. mass/$ft.^3$
	kg/cm^3	0.02768	lb. mass/$in.^3$
Force	N	4.4482	lb. force
Pressure	kPa	6.8948	lb./$in.^2$ (psi)
	kPa	100.0	bar
	kPa	101.325	std atm (760 torr)
	kPa	0.13284	1mm Hg at 20°C
Torque	N m	1.3558	lb.-force ft.
Power	W	1.3558	ft. lb./s
	W	745.7	hp
Energy	J	1.3558	ft. lb.-force
	kJ	1.0551	Btu international
	kWh	3412	Btu international
	MJ	2.6845	hp h
	MJ	105.506	therm
Temperature	°C	(°F − 32)·5/9	°F
	°K	(°F + 459.67) × 5/9	°R
Heat	W	0.2931	Btu (international)/h
	kW	3.517	Ton refrigeration

	W/m^2	3.1546	Btu/(ft.^{2} h)
	W/(m^2 °C)	5.6783	Btu/(ft.^{2} h. °F)
	MJ/(m^3 °C)	0.0671	Btu/(ft.^{3} °F)
	W/(m °C)	0.1442	Btu in./(ft.^{2} h °F)
	W/(m °C)	1.7304	Btu ft./(ft.^{2} h °F)
	J/kg	2.326	Btu/lb.
	MJ/m^3	0.0373	Btu/ft.^{3}
	J/(kg °C)	4.1868	Btu/(lb. °F)
Velocity	m/s	0.3048	ft./s
	m/s	0.44704	mi./h
Knot	m/s	0.51446	knot
Magnetics	Wb	10^{-8}	line
	Wb/m^2 (T)	0.0155	kiloline/in.^{2}

Prefixes to Units

μ	Micro	10^{-6}	m	mili	10^{-3}
k	Kilo	10^3	M	mega	10^6
G	Giga	10^9	T	tera	10^{12}

Other Conversions

1 nautical mi. = 1.15081 mi.
1 bar pressure = 14.50 psi = 100 kPa
= 29.53 in Hg = 10.20 m water = 33.46 ft. water
1 cal (CGS unit) = 4.1868 J
1 kg cal (SI unit) = 4.1868 kJ
1 hp = 550 ft.-lb./s
1 T magnetic flux density = 1 Wb/m^2 = 10,000 G (lines/cm^2)
Absolute zero temperature = 273.16°C = 459.67°F
Acceleration due to earth gravity = 9.806 7 m/s^2 (32.173 5 ft./s^2)
Permeability of free space $\mu_o = 4\pi \times 10^{-7}$ henry/m
Permittivity of free space $\varepsilon_o = 8.85 \times 10^{-12}$ F/m

Energy Content of Fuels

1 tip of match-stick = 1 Btu (heats 1 lb. water by 1°F)
1 therm = 100,000 Btu (105.5 MJ = 29.3 kWh)
1 Quad = 10^{15} Btu

1 $ft.^3$ of natural gas = 1000 Btu (1055 kJ)
1 gal. of LP gas = 95,000 Btu
1 gal. of gasoline = 125,000 Btu
1 gal. of no. 2 oil = 140,000 Btu
1 gal. of oil (U.S.) = 42 kWh
1 barrel = 42 gal. (U.S.)
1 barrel of refined oil = 6×10^6 Btu
1 barrel of crude oil = 5.1×10^6 Btu
1 ton of coal = 25×10^6 Btu
1 cord of wood = 30×10^6 Btu
1 million Btu = 90 lb. coal, or 8 gal. gasoline, or 11 gal. propane
1 quad (10^{15} Btu) = 45 million tons of coal, or 10^{12} $ft.^3$ of natural gas, or 170 million barrels of oil
1 lb. of hydrogen = 52,000 Btu = 15.24 kWh primary energy
World's total primary energy demand in 2010 was about 1 quad (10^{15} Btu) per day = 110×10^{16} J/day = 305×10^{12} kWh/day

---

About 10,000 Btu of primary thermal energy input at the power generating plant produces 1 kWh of electrical energy at the user's outlet.

Part A

Power Generation, Distribution, and Utilization

Part A of the book focuses on all aspects of electrical power that most power engineers get involved in during their professional careers. A vast majority of power engineers work on distribution and utilization of electrical power in cities and towns, factories and oil refineries, commercial and institutional buildings, electric trains, shipping ports, electric cruise ships, etc. Their everyday task is related to the distribution of power to load equipment such as motors and process heaters via step-down transformers and cables. The system protection is an integral part of their work, which requires circuit breakers, relays, and fuses. Some engineers working on cogenerating power plants get involved with generators. All these topics are covered in Chapters 1 through 9. Only a small percentage of power engineers deal with high-voltage power transmission, which is therefore excluded from this book.

Chapter 10 covers various ways of economic utilization of power for getting the required work done with the minimum expense of electrical energy. It includes the regenerative braking that recovers the kinetic energy from a moving mass and converts it into electricity, as is done in hybrid automobiles. The motors use almost 60% of all electrical energy in the world; hence, the energy-saving benefits of variable speed motor drives are also covered briefly in this chapter and then again in greater detail in Chapter 15.

The battery is covered in Chapter 11 in such detail that is not found in traditional power system books. It forms an integral part of the power system design for providing (1) backup power to essential and critical loads in case of the main power failure and (2) power to the low voltage dc control circuits and computers. Moreover, the battery performance depends greatly on many variables in a nonlinear manner, as covered in the chapter.

1 AC Power Fundamentals

Thomas Edison's Pearl Street low-voltage direct current (dc) power generating station opened in 1882 to serve the New York City market. Its low-voltage power had to be utilized in the vicinity of the generating station to keep the conductor I^2R loss to an economically viable level. Many more dc power stations were built by Edison and his competitors in the neighborhood of local users. Then came Westinghouse's alternating current (ac) power systems, with large steam and hydro power plants built where economical, such as on Niagara Falls. The high-voltage ac power was brought to the load centers, which was stepped down by transformers before feeding to the end users. The ac system proved to be much more economical and flexible and soon drove away the dc competitors. New Yorkers paid an inflation-adjusted price of about $5/kWh for dc power in 1890 versus the average of about $0.12/kWh we pay for ac power today in the United States. With the transformer and Nicola Tesla's induction motor for which Westinghouse acquired the patent, ac soon became universally adopted for electric power all around the world.

The fundamentals of power flow in ac circuits—usually covered in an undergraduate course in electrical engineering—are reviewed in this chapter. A clear understanding of these fundamentals will prepare the student for the chapters that follow and is also essential for working in the electrical power field.

1.1 CURRENT VOLTAGE POWER AND ENERGY

The basis of electricity is an electric charge (measured in coulombs) moving between two points at different electrical potentials, either absorbing or releasing the energy along its way. The basic electrical entities in power engineering are defined below with their generally used symbols and units.

- Current I (amperes) = flow rate of electrical charge (1 A = 1 C/s)
- Voltage V (volts) = electrical potential difference, that is, energy absorbed or released per coulomb of charge moving from one point to another (1 V = 1 J/C)
- Power P (watts) = rate of energy flow (1 W = 1 J/s)

$$\text{Therefore, } P = \frac{\text{Joules}}{\text{Second}} = \frac{\text{Joules}}{\text{Coulomb}} \times \frac{\text{Coulombs}}{\text{Second}} = V \times I \text{ Watts} \tag{1.1}$$

With time-varying voltage and current, the instantaneous power $p(t) = v(t) \times i(t)$ and energy = power × time duration. With time-varying power, the energy used between 0 and T seconds is given by the integral (i.e., area under the power versus time curve)

$$\text{Energy} = \int_0^T v(t) \cdot i(t)\,\mathrm{d}t \quad \text{watt seconds (joules).} \tag{1.2}$$

Inversely, the power is the time differential of energy, that is,

$$\text{Power} = \frac{\mathrm{d}}{\mathrm{d}t}(\text{Energy}). \tag{1.3}$$

Example 1.1

The electrical potential of point A is 200 V higher than that of point B, and 30 C of charge per minute flows from A to B. Determine the current and power flow from A to B and the energy transferred in 1 min.

Solution:

Current flow from point A to B = 30 ÷ 60 = 0.50 A

Power flow from A to B = Voltage × Current = 200 × 0.5 = 100 W

Energy transferred in 60 s = 100 × 60 = 6000 W s (J).

It is worth repeating here that 1 W = 1 J/s or 1 J = 1 W s. The commercial unit of electrical energy is kilowatt-hour (kWh); 1 kWh = 1000 W × 3600 s = 3,600,000 W s = 3.6 MJ. The utility bill is based on kilowatt-hour electrical energy used over the billing period. An average urban U.S. customer using 1000 kWh per month pays about $0.15/kWh, out of which about 40% is for generation, 5% for transmission, 35% for distribution, and 20% for account service and the utility company's profit.

For example, if a heater consumes 1500 W for 4 h and 750 W for 8 h every day in a winter month of 30 days, then the electrical bill at 15 cents/kWh tariff will be (1.5 × 4 + 0.750 × 8) kWh/day × 30 days/month × $ 0.15/kWh = $54 for that month.

1.2 ALTERNATING CURRENT

AC is used all over the world for electrical power. It varies sinusoidally with time t as

$$i(t) = I_m \sin \omega t \text{ or } i(t) = I_m \cos \omega t. \tag{1.4}$$

Although its representation by either the sine or cosine function is called sinusoidal, the cosine function shown in Figure 1.1 is more common, where I_m is the maximum value, or amplitude, or peak value of the sinusoid (in amperes); $\omega = 2\pi f$ is the angular speed (called angular frequency) of the alternations (in radians per second); T is the period of repetition (in seconds per cycle); and $f = 1/T = \omega/2\pi$ is the frequency [in cycles per second (called hertz)].

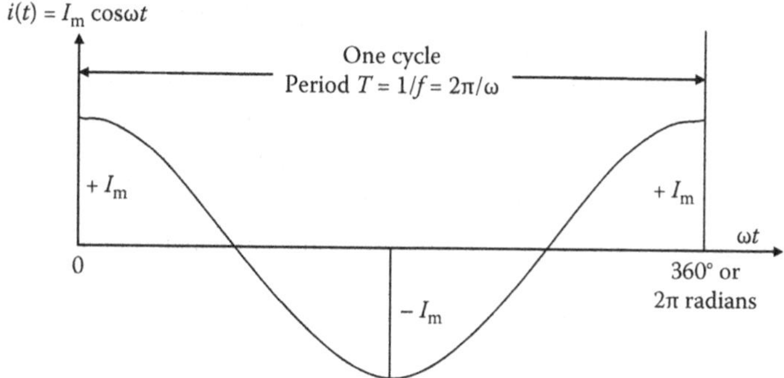

FIGURE 1.1 Sinusoidal ac over one cycle represented by cosine function.

In sine or cosine representation, the ac current completes one cycle of alternation in $\omega t = 360°$ or 2π radians. For this reason, one cycle of ac is customarily displayed with respect to ωt and not with time t. As an example, for sinusoidal current $i(t) = 170 \cos 377t$,

- Peak value (amplitude) $I_m = 170$ A (front number)
- Angular frequency $\omega = 377$ rad/sec (number in front of t)
- Numerical frequency $f = \omega/2\pi$ cycles/sec (Hz)
- Period of repetition $T = 1/f = 2\pi/\omega$ sec/cycle.

We recall that the dc needs only one number to specify its value, but we see in Equation 1.4 that the ac needs three numbers to specify its value at any instant of time t, namely, the peak value I_m, the angular frequency ω, and time t. This makes the mathematics in ac circuits complex.

1.2.1 RMS Value and Average Power

Although ac varies sinusoidally with time with no real fixed value, we often speak in terms of a fixed ac value, for example, ac current of 10 A or ac voltage of 120 V. We see below what a fixed number in ac means, taking an example of resistor carrying current $i(t)$. The power absorbed by the resistor at any time t is given by $p(t) = i(t)^2R$, which varies in time as a square function of the current. It is always positive even when the current is negative during one-half cycle. Therefore, the average of i^2 is always a positive nonzero value, although the average of sinusoidal current is always zero. To find the average power, we must therefore use (average of i^2), not (average of $i)^2$, carefully noticing the placement of the parentheses in each case. We also note that since i_{avg} is always zero, $(i_{avg})^2$ is also always zero. However, since i^2 is always positive in both positive and negative half cycles, $(i^2)_{avg}$ is never zero, unless $i(t) = 0$ at all times. Therefore, $P_{avg} = \{i(t)^2\}_{avg} \times R$. The square root of $\{i(t)^2\}_{avg}$ is called the root-mean-squared (rms) value of the current $i(t)$.

In all practical applications, it is the average power that matters. For example, we are mostly interested in how much a room heater, a fan motor, or a pump motor produces at the end of an hour or any other time duration. Since ac repeats every cycle, the average power over one cycle is the same as that over 1 min or 1 h or 1 day, as long as the power is *on*. Therefore, the effective value of the current for determining the average power is the square root of (average of i^2) over one cycle, that is, $I_{\text{eff}} = \sqrt{\text{Mean of } i^2} = I_{\text{root.mean.squared}} = I_{\text{rms}}$. It is the equivalent dc value that would result in the same *average power*. For any wave shape in general, that is, cosine, sine, square, rectangular, triangular, etc.,

$$I_{\text{rms}} = \sqrt{\frac{\int_0^T i^2(t) \cdot dt}{T}}. \tag{1.5}$$

For a sinusoidal current, $I_{\text{rms}} = \sqrt{\frac{\int_0^T (I_{pk} \cos \omega t)^2 \cdot dt}{T}} = \frac{I_{pk}}{\sqrt{2}}$,

and similarly, for a sinusoidal voltage, $V_{\text{rms}} = \sqrt{\frac{\int_0^T (V_{pk} \cos \omega t)^2 \cdot dt}{T}} = \frac{V_{pk}}{\sqrt{2}}$. (1.6)

The divisor $\sqrt{2}$ above is for the sinusoidal ac only. It is different for different wave shapes. Using the basic calculus of finding the rms value, the student is encouraged to derive the divisor $\sqrt{2}$ for a sinusoidal wave, 1.0 for a rectangular wave, and $\sqrt{3}$ for a triangular wave.

1.2.2 Polarity Marking in AC

Although ac circuit terminals alternate their + and – polarities every one-half cycle, we still mark them with + and – polarities, as if they were of fixed polarities, as in dc. Such polarity marking in ac has the following meanings:

- With multiple voltage sources (generators and transformers) in parallel, all + marks are positive at the same time, and all – marks are negative at the same time. This information is needed to connect multiple voltage sources in parallel to share a large load.
- With multiple voltage sources or loads in series, the (– +) and (– +) sequence indicates an additive voltage pair, whereas (– +) and (+ –) sequence indicates a subtractive voltage pair.
- In a single-voltage-source circuit, the + terminal is usually connected to the load, and the – terminal is connected to the ground.

To eliminate such technical contradiction in using the + and – marks, the modern polarity marking is sometimes done with dots • for positive terminals and no mark for the negative terminals.

1.3 AC PHASOR

The ac current $i(t) = I_m \cos \omega t$ can be represented by an arm of length I_m rotating at an angular speed (frequency) ω rad/sec as shown in Figure 1.2. The actual instantaneous value of $i(t)$ at any time t is $I_m \cos \omega t$, which is the projection on the reference axis. It alternates between $+I_m$ and $-I_m$ peaks, going through zero twice at $\omega t = 90°$ and 270° every cycle. Since the actual value of $i(t)$ depends on the phase angle of the rotating arm, the rotating arm is called the *phasor*. Decades ago, it used to be called vector since the algebra dealing with it works like the vector algebra.

The ac voltage and current phasors, in general, can have phase difference between their peaks, that is, their peaks can occur at different instants of times. Voltage V and current I phasors shown in Figure 1.3a have the phase difference of angle θ, with the current peak lagging the voltage peak by angle θ. Two phasors of the same frequency with phase difference θ between their peaks will also have the same phase difference θ between their zeros. The wavy hat sign ~ on V and I signifies the sinusoidal variations with respect to time.

In the electrical power industry, since the voltage is given by the generator or the utility company, the power engineer always takes the voltage as the reference phasor and then designates the current as leading or lagging the voltage. In all practical power circuits, the current lags the voltage, so we say that the current lags in most practical power circuits.

Drawing a neat sine wave by hand is difficult. Power engineers generally circumvent this difficulty by drawing the phasor diagram as shown in Figure 1.3b, where the V and I phasors are rotating at the same angular speed ω, keeping their phase difference θ fixed. The actual instantaneous values of both V and I are their respective projections on the reference axis at any given instant of time.

The phasor diagram can be drawn using the arm length equal to the peak value or the rms value. Since the power engineer is always interested in the average

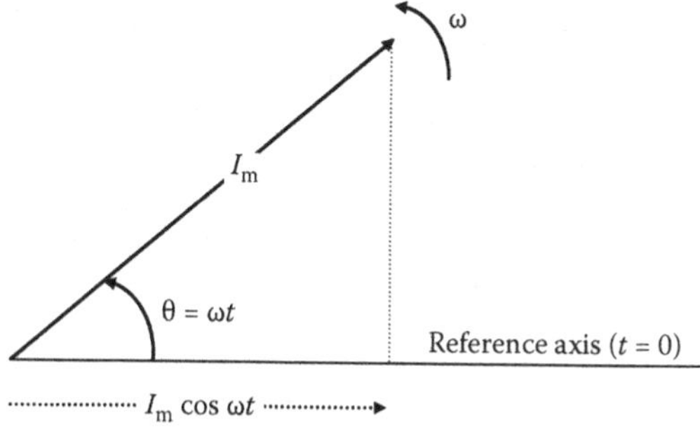

FIGURE 1.2 Rotating phasor $\tilde{I}$ representing ac.

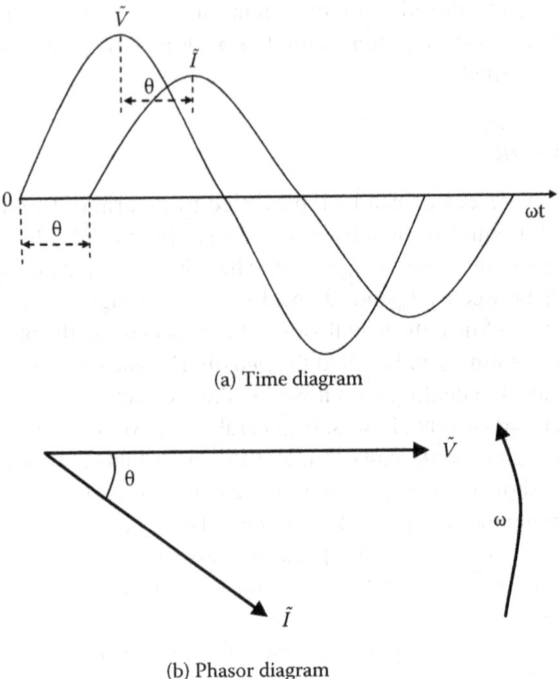

FIGURE 1.3 Two sinusoidal phasors out of phase by angle θ.

power, he or she always draws the phasor diagram using the rms values. The rms values are customarily implied in the power field, and we will do the same in this book as well.

For average power, we recognize that the voltage and current that are out of phase would produce less average power than those in phase with their peaks occurring at the same time. If V and I are in phase (i.e., when $\theta = 0$), their product $P = V \times I$ is always positive even when both V and I are negative during one-half cycle. However, when $\theta \neq 0$, the instantaneous power is negative when either V or I is negative and the other is positive. If positive power means the power flowing from the source to the load, then the negative power means the power flowing backward to the source from the energy stored in the load inductance or capacitance. The average power in such a case is always less than the maximum power V and I would produce if they were in phase. The average power of voltage V and current I lagging the voltage by phase angle θ is given by the time-average of $p(t) = v(t) \cdot i(t)$ over one cycle of period T, that is,

$$P_{avg} = \frac{1}{T}\int_0^T V_{pk}\cos(\omega t)\cdot I_{pk}\cos(\omega t - \theta)\,\mathrm{d}t = \frac{V_{pk}I_{pk}}{2}\cos\theta = V_{rms}I_{rms}\cos\theta. \quad (1.7)$$

The voltage and current, if in phase with $\theta = 0°$, would produce the maximum possible average power equal to $V_{rms} \times I_{rms}$. When not in phase, they produce less average power. The power reduction factor cos θ is called the power factor (pf). Obviously, pf = 1.0 (unity) when $\theta = 0°$, and pf = 0 when $\theta = 90°$.

Example 1.2

A circuit element has sinusoidal voltage of 300 cos 314*t* volts across its terminals and draws 80 cos(314*t* – 25°) amperes. Determine the average power delivered to the element.

Solution:

The current is lagging the voltage by 25°, so the pf of this element is cos 25°. For the average power, we use rms values and pf, that is,

$$P_{avg} = (300 \div \sqrt{2}) \times (80 \div \sqrt{2}) \times \cos 25° = 10{,}876 \text{ W}.$$

1.3.1 Operator *j* for 90° Phase Shift

An uppercase letter with a wavy hat sign (e.g., $\tilde{I} = I\angle\theta$) in this book represents a sinusoidally varying ac phasor with rms magnitude I and a phase difference of angle θ with respect to a reference sinusoidal wave (usually the voltage). If $\tilde{A}$ in Figure 1.4 represents any voltage or current phasor, then another phasor $\tilde{B}$ that is of the same magnitude as $\tilde{A}$ but with 90° phase shift in the positive (counterclockwise) direction can be written in long hand as $\tilde{B} = \tilde{A}$ with added phase shift of +90°, or $\tilde{B} = \tilde{A}\angle 90° = \tilde{A}j$ or $j\tilde{A}$ in short hand where the *operator j* represents the phase shift of $\tilde{A}$ by +90° in the positive (counterclockwise) direction.

Shifting $\tilde{B}$ further by + 90°, we get another phasor $\tilde{C} = j\tilde{B} = j(j\tilde{A}) = j^2\tilde{A}$. We graphically see in Figure 1.4 that $\tilde{C} = j^2\tilde{A} = -\tilde{A}$ and therefore deduce that $j^2 = -1$ or $j = \sqrt{-1}$. Thus, j is an imaginary number generally denoted by letter i in the

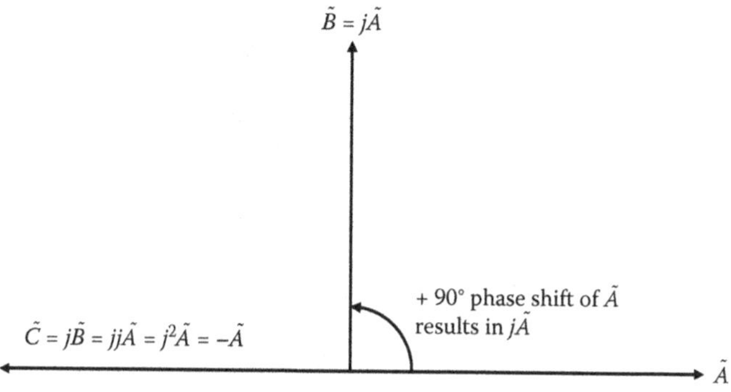

FIGURE 1.4 Operator *j* representing 90° phase shift in positive (counterclockwise) direction.

mathematics of complex numbers (we use letter j to avoid confusion with current i in electrical circuits).

Thus, in mathematical operations,

$$j = \sqrt{-1} \text{ represents } +90° \text{ phase shift.} \tag{1.8}$$

From $j^2 = -1$ or $1 = -j^2$, we get $\frac{1}{j} = -j$, which is a useful relation to remember.

1.3.2 Three Ways of Writing a Phasor

Alternative ways of writing a phasor $\tilde{I}$ are depicted in Figure 1.5.

In polar form (also called θ-form), we write phasor $\tilde{I} = I_m \angle\theta$, where I_m = magnitude of the phasor (can be peak, but rms magnitude is used in this book, as is customary in power engineering), and θ = phase angle of the phasor. Here, the hat sign ~ signifies a sinusoidal phasor. In routine use, we often drop the hat sign ~ and write $I = 3\angle{-20°}$, meaning an ac current of 3 A rms value lagging the voltage by 20°.

In rectangular form (also called j-form), we write $\tilde{I} = x + jy = I_m \cos\theta + jI_m \sin\theta$, where x and y are the phasor's rectangular components on the real and imaginary axes, respectively.

In exponential form (also called e-form), we use Euler's trigonometric identity $e^{j\theta} = \cos\theta + j\sin\theta$ to write the phasor in yet another form, that is, $\tilde{I} = I_m \cos\theta + jI_m \sin\theta = I_m \times (\cos\theta + j\sin\theta) = I_m \cdot e^{j\theta}$.

The three alternative ways of representing a phasor are then summarized below:

$$\tilde{I} = I_m \angle\theta = I_m \cos\theta + jI_m \sin\theta = I_m e^{j\theta}, \text{ where } \theta = \omega t. \tag{1.9}$$

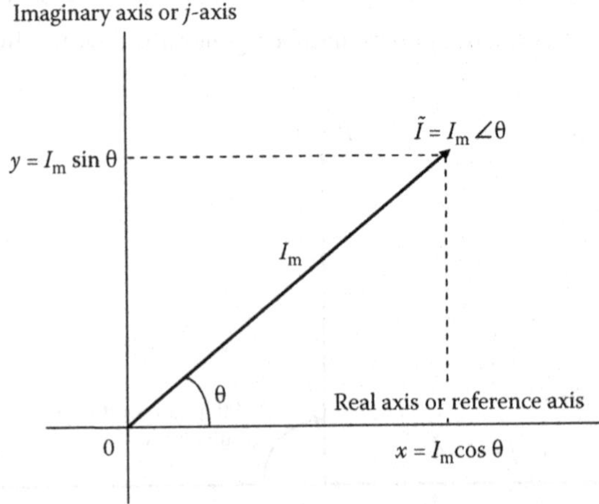

FIGURE 1.5 Polar and rectangular components of phasor $\tilde{I}$.

It is important to take a note here that two phasors $\tilde{A} = A_m\angle\theta_1 = x_1 + jy_1$ and $\tilde{B} = B_m\angle\theta_2 = x_2 + jy_2$ are equal if, and only if, both their magnitudes and phase angles are equal, that is, $A_m = B_m$ and $\theta_1 = \theta_2$, or $x_1 = x_2$ and $y_1 = y_2$, that is, when both their real and imaginary components are individually equal.

1.3.3 Phasor Form Conversion

The form we choose to represent various phasors depends on the algebraic operation required on a given set of phasors. Certain algebraic operations require the phasors in certain form, as seen in the next section. Therefore, converting the phasor from one form to another is often necessary and is done using the trigonometry of Figure 1.5.

Polar to rectangular (θ to j) conversion: Consider a phasor given in θ-form, that is, $\tilde{I} = I_m\angle\theta$, where I_m and θ are known. To convert it in *j*-form, we write $\tilde{I}$ using the rectangular components, that is,

$$\tilde{I} = x + jy = I_m \cos\theta + jI_m \sin\theta \text{ therefore } x = I_m \cos\theta \text{ and } y = I_m \sin\theta. \quad (1.10)$$

Rectangular to polar (j to θ) conversion: Consider a phasor given in *j*-form, that is, $\tilde{I} = x + jy$ where *x* and *y* are known. To convert it in θ-form, we write $\tilde{I}$ using the polar components, that is,

$$\tilde{I} = I_m\angle\theta \text{ where } I_m = \sqrt{x^2 + y^2} \text{ and } \theta = \tan^{-1}\left(\frac{y}{x}\right). \quad (1.11)$$

1.4 PHASOR ALGEBRA REVIEW

AC power engineers are routinely required to perform six basic mathematical operations on phasors, namely, to add, subtract, multiply, divide, differentiate, and integrate phasors. These operations are normally covered in books on algebra of complex numbers and also on ac circuits. This section is a brief summary of such operations.

Consider two phasors $\tilde{A}$ and $\tilde{B}$ given by

$$\tilde{A} = A_m\angle\theta_1 = x_1 + jy_1 = A_m e^{j\theta 1} \text{ and } \tilde{B} = B_m\angle\theta_2 = x_2 + jy_2 = B_m e^{j\theta 2}.$$

We add or subtract two phasors in the rectangular form, that is,

$$\tilde{A} + \tilde{B} = (x_1 + jy_1) + (x_2 + jy_2) = (x_1 + x_2) + j(y_1 + y_2). \quad (1.12)$$

$$\tilde{A} - \tilde{B} = (x_1 + jy_1) - (x_2 + jy_2) = (x_1 - x_2) + j(y_1 - y_2). \quad (1.13)$$

We multiply or divide two phasors in the polar and exponential forms, that is,

$$\tilde{A}\cdot\tilde{B} = A_m\angle\theta_1 \cdot B_m\angle\theta_2 = A_m e^{j\theta_1}\cdot B_m e^{j\theta_2} = A_m B_m e^{j(\theta_1+\theta_2)} = A_m B_m\angle(\theta_1+\theta_2) \tag{1.14}$$

$$\frac{\tilde{A}}{\tilde{B}} = \frac{A_m\angle\theta_1}{B_m\angle\theta_2} = \frac{A_m e^{j\theta_1}}{B_m e^{j\theta_2}} = \frac{A_m}{B_m}e^{j(\theta_1-\theta_2)} = \frac{A_m}{B_m}\angle(\theta_1-\theta_2). \tag{1.15}$$

We differentiate or integrate phasor $\tilde{A}$ with respect to time t in the exponential form in the time domain, that is, $\tilde{A} = A_m\angle\theta = A_m e^{j\theta} = A_m e^{j\omega t}$. Then,

$$\frac{d\tilde{A}}{dt} = \frac{d}{dt}A_m e^{j\omega t} = j\omega A_m e^{j\omega t} = j\omega\tilde{A} \tag{1.16}$$

$$\int \tilde{A}\,dt = \int A_m e^{j\omega t}\,dt = A_m\frac{e^{j\omega t}}{j\omega} = \frac{\tilde{A}}{j\omega}. \tag{1.17}$$

The summary of above phasor operations in words follows:

- To add two phasors, add their x and y components separately.
- To subtract two phasors, subtract their x and y components separately.
- To multiply two phasors, multiply their magnitudes and add their angles.
- To divide two phasors, divide their magnitudes and subtract their angles.
- To differentiate a phasor, multiple it by $j\omega$, that is, $d/dt = j\omega$.
- To integrate a phasor, divide it by $j\omega$, that is, $\int dt = 1/j\omega = -j\omega$.

Tip-to-tail method is the graphical method of adding or subtracting two phasors as illustrated in Figure 1.6. To add $\tilde{A}$ and $\tilde{B}$ in Figure 1.6 (top), first draw $\tilde{A}$. Then,

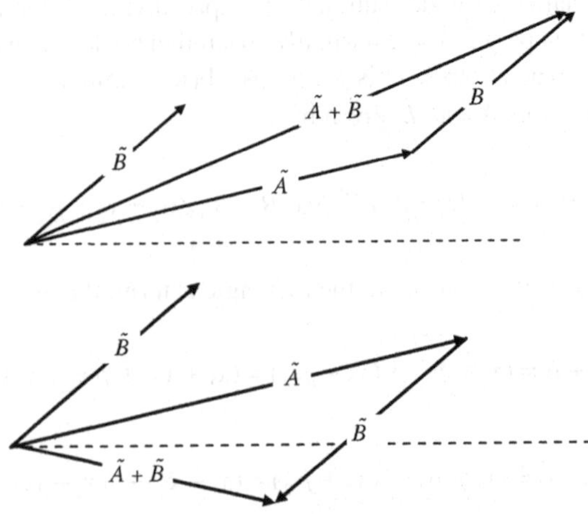

FIGURE 1.6 Tip-to-tail method of adding and subtracting two phasors.

at the tip of $\tilde{A}$, place the tail of $\tilde{B}$ and draw $\tilde{B}$. The end point from the origin is then $\tilde{A} + \tilde{B}$. To subtract $\tilde{B}$ from $\tilde{A}$ in Figure 1.6 (bottom), first draw $\tilde{A}$. Then, at the tip of $\tilde{A}$, place the tail of $\tilde{B}$ and draw $-\tilde{B}$ (i.e., $\tilde{B}$ in the negative direction.) The end point from the origin is then $\tilde{A} - \tilde{B}$.

Example 1.3

Given two phasors, $\tilde{A} = 60\angle 30°$ and $\tilde{B} = 40\angle 60°$, determine (1) $\tilde{A} + \tilde{B}$, (2) $\tilde{A} - \tilde{B}$, (3) $\tilde{A} \times \tilde{B}$, and (4) $\tilde{A}/\tilde{B}$. Express each result in j-form and also in θ-form.

General note:

For simplicity in writing, all angles we write following $\angle$ signs in this book are in degrees, whether or not expressly shown with superscripts (°).

Solution:

For adding and subtracting $\tilde{A}$ and $\tilde{B}$, we must express phasors in j-form, that is,

$$\tilde{A} = 60\ (\cos 30° + j\sin 30°) = 51.96 + j30, \text{ and } \tilde{B} = 40\ (\cos 60° + j\sin 60°) = 20 + j34.64$$

Then,

$$\tilde{A} + \tilde{B} = (51.96 + 20) + j(30 + 34.64) = 71.96 + j64.64 = 96.73\angle 41.93°$$

$$\tilde{A} - \tilde{B} = (51.96 - 20) + j(30 - 34.64) = 31.96 - j4.64 = 32.30\angle -8.26°.$$

For multiplying and dividing, we must use $\tilde{A}$ and $\tilde{B}$ in θ-form as given, that is,

$$\tilde{A} \times \tilde{B} = 60\angle 30° \times 40\angle 60° = 60 \times 40\angle 30 + 60 = 2400\angle 90°$$
$$= 2400\ (\cos 90 + j\sin 90) = 0 + j2400$$

$$\tilde{A}/\tilde{B} = 60\angle 30° \div 40\angle 60° = (60 \div 40)\angle 30 - 60 = 1.5\angle -30°$$
$$= 1.5\ (\cos 30 - \sin 30) = 1.3 - j0.75$$

Example 1.4

Determine $\dfrac{2}{j1.5}$ in θ-form.

Solution:

All ac voltages, currents, impedances, and powers are phasors, that is, complex numbers having $x + jy$ components or the rms magnitude and phase angle θ. When we write 2 in ac, it really means $x = 2$ and $y = 0$, or magnitude 2 and $\theta = 0°$. Also, when we write $j1.5$, it really means $x = 0$ and $y = 1.5$, or magnitude 1.5 and $\theta = 90°$.

The algebra of dividing two phasors requires first converting them in θ-form and then dividing the two as follows:

$$\frac{2}{j1.5} = \frac{2 + j0}{0 + j1.5} = \frac{2\angle 0}{1.5\angle 90} = \frac{2}{1.5}\angle 0 - 90 = 1.33\angle -90°.$$

Recognizing that a real number alone always means $\theta = 0°$, and an imaginary number alone always means $\theta = 90°$, we can skip the formal j-form to θ-form conversion, and write directly as

$$\frac{2}{j1.5} = \frac{2\angle 0}{1.5\angle 90} = \frac{2}{1.5}\angle 0 - 90 = 1.33\angle -90°$$

or, recalling that $1/j = -j = \angle -90°$, we can simplify the algebra further as

$$\frac{2}{j1.5} = -j\frac{2}{1.5} = -j1.33 = 1.33\angle -90°.$$

Understanding all of the above three ways of arriving at the same results adds fluency in the complex algebra of phasors in ac power circuits, which is required for professional electrical power engineers.

1.5 SINGLE-PHASE AC POWER CIRCUIT

In analyzing any power circuit (ac or dc), we use the flowing basic circuit laws:

Kirchhoff's voltage law (KVL): In any closed loop, phasor sum of all source voltages = phasor sum of all voltage drops in load elements. It applies in every loop, covering every segment along the loop, in ac or dc. In a voltage driven circuit—usually the case in power engineering—it is KVL that determines the current, that is, the circuit draws current that will satisfy KVL in every closed loop of the circuit.

Ohms law: It basically gives the voltage drop across two terminals of R, L, or C element as summarized in Table 1.1. It is local; it applies only between two terminals of R, L, or C (but not of a source).

Kirchhoff's current law (KCL): At any junction node (point or a closed box), phasor sum of all currents going inward = phasor sum of all currents going outward. The KCL is even more local than Ohm's law; it applies only at a junction node.

TABLE 1.1
Voltage Drop, Current, and Energy Storage Relations in *R, L*, and *C* Elements in AC or DC Circuits

Parameter between Two Terminals of the Element	Resistance *R* of Wire (ohm)	Inductance *L* of Coil (henry)	Capacitance *C* between Two Conductors (farad)
Voltage drop in the element in dc or ac (Ohm's law or its equivalent)	$v = R \times i$	$v = L\dfrac{di}{dt}$	$v = \dfrac{\text{Charge } Q}{C} = \dfrac{\int i \cdot dt}{C}$ or $i = C\dfrac{dv}{dt}$
Voltage drop in steady state dc (i.e., when $di/dt = 0$ and $dv/dt = 0$)	$V = R \cdot I$	$V_{drop} = 0$ (i.e., coil in steady dc behaves like an ideal wire shorting the coil)	$I_{cap} = 0$ after fully charged (i.e., capacitor in steady dc behaves like open circuit)
Energy stored in the element	None (dissipates energy in heat, which leaves the circuit)	$1/2\ LI^2$ (stores energy in magnetic flux through the coil)	$1/2\ C\ V^2$ (stores energy in electrical charge on the capacitor conductors)
Energy conservation	Dissipates the absorbed energy into heat	Tends to hold on stored energy by retaining its current *I*	Tends to hold on stored energy by retaining its voltage *V*

1.5.1 Series *R–L–C* Circuit

The basic power circuit is made of one or more of the three basic load elements, namely, the resistor *R*, inductor *L*, and capacitor *C*, powered generally by a voltage source (e.g., the utility grid). Since the *R*, *L*, and *C* elements are distinctly different in units and also in behavior, they cannot be combined with each other using any series-parallel combination formulas until they are converted into their respective ohm values with phase shifts.

Consider a series *R–L–C* circuit of Figure 1.7 excited by a sinusoidal voltage source $\tilde{V}_s$. Since the driving voltage in this circuit is sinusoidal, the current must also be sinusoidal of the same frequency, say, $\tilde{I} = I\angle\theta$ where I = rms magnitude (note that we now drop the suffix m from I_m and imply that *I* without the hat sign means the rms magnitude).

In the series loop of Figure 1.7, we write KVL using Ohm's law relations given in Table 1.1 and Equations 1.16 and 1.17:

$$\tilde{V}_s = \tilde{V}_R + \tilde{V}_L + \tilde{V}_c = R\tilde{I} + L\frac{d\tilde{I}}{dt} + \frac{\int \tilde{I}\,dt}{C} = R\tilde{I} + Lj\omega\tilde{I} + \frac{\tilde{I}}{Cj\omega}.$$

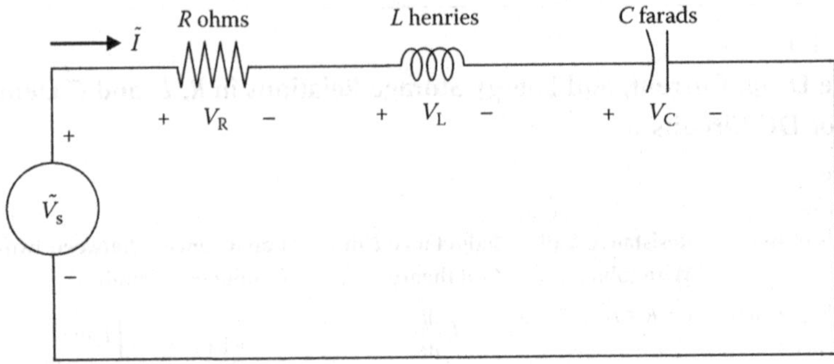

FIGURE 1.7 Series *R–L–C* circuit powered by sinusoidal source voltage $\tilde{V}_s$.

Recalling that $j = -1/j$, we rewrite the above equation as

$$\tilde{V}_s = R\tilde{I} + j\omega L\tilde{I} - \frac{j}{\omega C}\tilde{I} = R\tilde{I} + j\left(\omega L - \frac{j}{\omega C}\right)\tilde{I} \tag{1.18}$$

where $j = \sqrt{-1} = \angle 90° = 90°$ phase shift in the positive direction. We note here the following phasor relations in the *R*, *L*, and *C* elements:

- $\tilde{V}_R = R\tilde{I}$, that is, V_R is in phase with *I* with $\theta = 0°$, meaning that a pure *R* always draws current at pf = 1.0 (unity), absorbing average power that is not zero in *R*.
- $\tilde{V}_L = j\omega L\tilde{I}$, that is, V_L leads *I* by 90°, or *I* lags V_L by 90°, meaning that a pure *L* always draws current at pf = 0 lagging, giving zero average power absorbed in *L*.
- $\tilde{V}_C = -j\dfrac{1}{\omega C}\tilde{I}$, that is, V_c lags *I* by 90° or *I* leads V_c by 90°, meaning that a pure *C* always draws current at pf = 0 leading, again giving zero average power absorbed in *C*.

The voltmeter measures only the rms magnitude without the phase angle. Therefore, three voltmeter readings across *R*, *L*, and *C* (i.e., V_R, V_L, and V_C) will not add up to the total source voltage magnitude V_s. We must add the three load voltages taking their phase differences into account (called the phasor sum) to obtain the total source voltage. Figure 1.8 shows the phasor sum using the tip-to-tail method, where

$$\tilde{V}_s = \tilde{V}_R + \tilde{V}_L + \tilde{V}_c = \tilde{V}_s\angle\theta. \tag{1.19}$$

We next write Equation 1.18 as

$$\tilde{V}_s = \left\{R + j\left(\omega L - \frac{1}{\omega C}\right)\right\} \times \tilde{I} = \tilde{Z} \times \tilde{I} \tag{1.20}$$

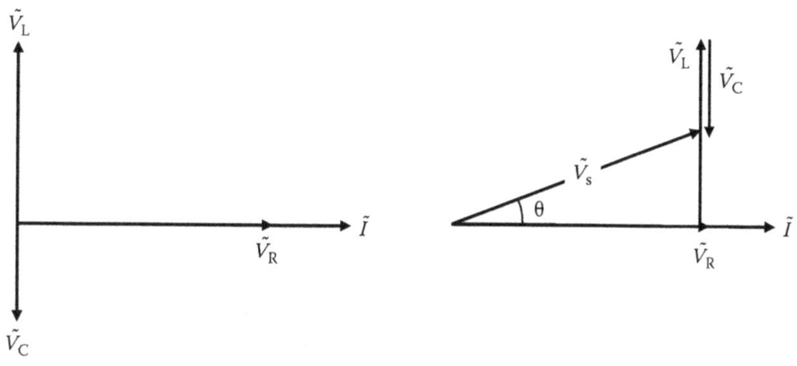

(a) Individual phasors $\tilde{I}$, $\tilde{V}_R$, $\tilde{V}_L$, and $\tilde{V}_C$ (b) Tip-to-trail addition $\tilde{V}_R + \tilde{V}_L + \tilde{V}_C = \tilde{V}_s$

FIGURE 1.8 Phasors $\tilde{V}_R$, $\tilde{V}_L$, and $\tilde{V}_C$ and their phasor sum $\tilde{V}_s$ in series *R–L–C* circuit.

where $\tilde{Z} = R + j\left(\omega L - \dfrac{1}{\omega C}\right) = R + jX$ ohms.

The *R* and *X* both have units of ohms and so does $\tilde{Z}$. Therefore, Equation 1.20 gives ac voltage as a product of ampere and ohm, as in dc. Equation 1.20 is Ohm's law in ac, where $\tilde{Z}$ is called the total circuit *impedance*. The $\tilde{Z}$ is generally written as $\tilde{Z} = Z\angle\theta = R + jX$, where $X = (\omega L - 1/\omega C)$ is called the *reactance*, which comes from *L* and *C*. Since the reactance has two parts, we write $X = X_L - X_C$, where

$$X_L = \omega L = \text{reactance from inductance (ohms)}$$

$$X_C = 1/(\omega C) = \text{reactance from capacitance (ohms).}$$

The contributions of *L* and *C* in the total reactance *X* are subtractive, that is, they tend to neutralize each other. As a result, *X* can be zero in some circuits even if *L* and *C* are not individually zero.

The KVL in a closed loop in any power circuit with multiple voltage sources and impedances in series can be expressed in a simple form as follows:

$$\tilde{I}_{\text{loop}} = \frac{\text{phasor sum of all source voltages}}{\text{phasor sum of all impedances}} = \frac{\sum \tilde{V}_{\text{source}}}{\sum \tilde{Z}\angle\theta}. \tag{1.21}$$

The current lags the voltage if the phase angle θ of all load impedance combined is positive (i.e., when the loop is more inductive than capacitive). Thus, the overall pf of a circuit is determined by the nature of the load. The pf has nothing to do with the source voltage.

Example 1.5

A one-loop circuit has source voltage of 120 V and load impedance of $10\angle{+84.3°}$ Ω. Determine the current and average power delivered to the load. Can you guess about the nature of the load—could it be a heater, capacitor, motor, etc.?

Solution:

The 120 V with no angle stated implies that it is the reference phasor at 0°. It is also implied in power engineering that it is the rms value. The current phasor in this loop will then be

$$\tilde{I}_{loop} = \frac{\tilde{V}_{loop}}{\tilde{Z}_{loop}} = \frac{120\angle 0°}{10\angle 84.3°} = \left(\frac{120}{10}\right)\angle 0° - 84.3° = 12\angle{-84.3°}\text{ A.}$$

The 12 A is the rms value since we used 120 V_{rms} to derive the current, which lags the voltage by 84.3°, giving the pf = cos (84.3°) = 0.10 lagging. The average power delivered by the source and absorbed by the load is given by $P_{avg} = V_{rms} I_{rms} \cos\theta =$ 120 × 12 × 0.10 = 144 W (versus 1440 W 120 V and 12 A could produce if they were in phase with θ = 0°). This is an example of extremely poor pf.

As for the nature of the load, since it draws current with a large lag angle close to 90°, it must have a large inductive reactance with small resistance. Therefore, it is most likely to be a large coil.

Example 1.6

Determine the average real power drawn by a coil having inductance of 10 mH and resistance of 1.5 Ω connected to a 120-V, 60-Hz source.

Solution:

For a 60-Hz source, ω = 2π 60 = 377 rad/s, hence the coil impedance

$$\tilde{Z} = 1.5 + j377 \times 0.010 = 1.5 + j3.77\ \Omega = 4.057\angle 68.3°\ \Omega.$$

Taking 120 V as the reference phasor, the current drawn by the coil is

$$\tilde{I} = 120\angle 0° \div 4.057\angle 68.3° = 29.58\angle{-68.3°}\text{ A,}$$

which lags the voltage by 68.3°

therefore, real power P = 120 × 29.58 × cos 68.3° = 1312.45 W.

Alternatively, the average real power is absorbed only by the circuit resistance, and the power absorbed by the inductance or capacitance always averages out to zero over one cycle. Thus, the average real power can also be derived simply from the following:

$$\text{real power } P = I^2R = 29.58^2 \times 1.5 = 1312.45 \text{ W},$$

Example 1.7

Convert the time-domain circuit shown in figure (a) below in the phasor domain with impedances in ohms. Then, determine the current phasor $\tilde{I}$ and the average power.

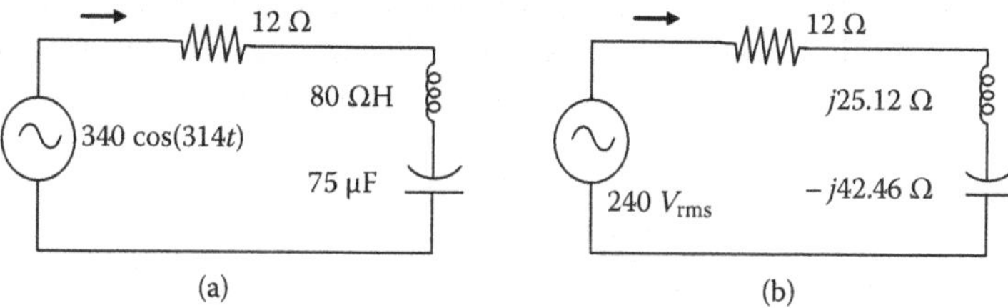

Solution:

First, the voltage phasor's rms value = 340 ÷ √2 = 240∠0° V (reference for angle).
Angular frequency $\omega = 314$, which gives $f = \omega/2\pi = 50$ Hz.
The loop impedance of series R–L–C circuit in figure (a) is

$$Z = 12 + j\left(314 \times 0.080 - \frac{1}{314 \times 75 \times 10^{-6}}\right)$$
$$= 12 + j(25.12 - 42.46) = 12 - j17.34 = 21.09\angle{-55.3}\ \Omega.$$

The phasor-domain circuit is shown in figure (b), from which we derive the circuit current

$$I = \frac{V}{Z} = \frac{240\angle 0}{21.09\angle{-55.3}} = \left(\frac{240}{21.09}\right)\angle 0 - (-55.3) = 11.38\angle{+55.3^\circ}\text{ A}.$$

The 11.38-A current phasor leads the voltage by 55.3°, meaning that the capacitance dominates the inductance in this circuit.

$$\text{Average power } P = 240 \times 11.38 \times \cos 55.3^\circ = 1555 \text{ W}.$$

1.5.2 Impedance Triangle

As seen in Equation 1.20, we write the total circuit impedance in the form $Z = R + jX$, where the operator $j = \sqrt{-1}$ in algebra or 90° phase shift in the phasor diagram.

For engineers and managers not fully versed with the operator j, the reason for assigning the inductive reactance X with 90° phase shift from R can be simply explained as follows in view of the circuit shown Figure 1.9a. The voltage drop in R is IR, which always subtracts from the source voltage. The voltage drop in the inductor, however, is equal to $L \times di/dt$. This drop is positive for one-half cycle of rising current from point A to B under a sinusoidal current as in Figure 1.9b when the inductor is absorbing the energy that causes the voltage to drop. It is negative for the other one-half cycle of falling current from point B to C when the inductor is supplying the energy that adds into the circuit voltage. The negative voltage drop means it subtracts from the positive voltage drop in the resistance. Therefore, for one-half cycle of rising current, the total voltage drop is $IR + IX$ as in Figure 1.9c, whereas for the following one-half cycle of falling current, the net voltage drop is $IR - IX$ as in Figure 1.9d. We graphically show the effective rms value of

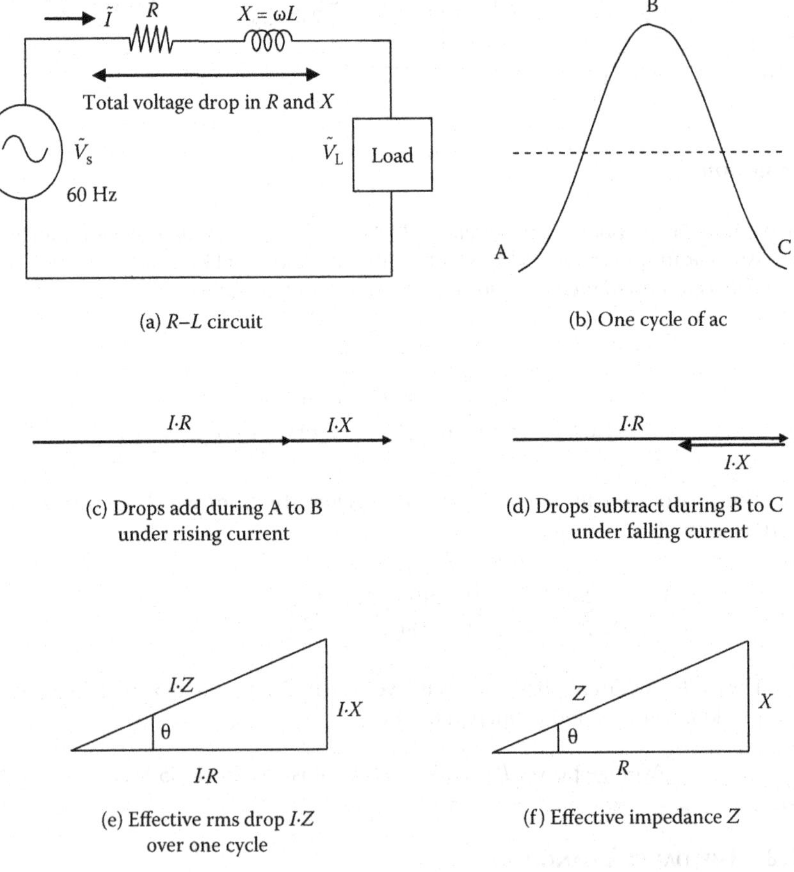

FIGURE 1.9 Simple explanation of voltage drop in R–L circuit and impedance triangle.

the voltage drop over one cycle as in Figure 1.9e not positive nor negative, but in an in-between position with 90° phase shift. Analytically, the effective rms value of the total voltage drop is

$$\text{Total } V_{\text{drop.rms}} = \sqrt{\frac{(IR+IX)^2+(IR-IX)^2}{2}} = \sqrt{(IR)^2+(IX)^2} = I\sqrt{R^2+X^2} = I\cdot Z$$

where $Z=\sqrt{R^2+X^2}$, which has the right-angle triangle relation as shown in Figure 1.9f, which is known as the impedance triangle. Therefore, the circuit impedance, in general, is expressed as a phasor $\tilde{Z}=R+jX$, as drawn in Figure 1.9f using the tip-to-tail method, where cos θ is the circuit pf.

1.5.3 Circuit Laws and Theorems

All circuit laws and theorems learned in dc apply in the same manner in ac as well, except that all numbers in ac are phasors (complex numbers), each with magnitude and phase angle, making the ac algebra complex. All dc formulas remain valid in ac if we replace *R* in dc with $\tilde{Z}$ in ac. For example, if the circuit has more than one impedance connected in a series–parallel combination, then the total equivalent $\tilde{Z}$ can be determined by using the corresponding dc formula after replacing *R* with $Z\angle\theta$ and doing the phasor algebra of the resulting complex numbers, that is,

$$\tilde{Z}_{\text{series}} = \tilde{Z}_1+\tilde{Z}_2+\ldots. \text{ and } Z_{\text{parallel}} = \frac{1}{\frac{1}{Z_1}+\frac{1}{Z_2}+\ldots.}. \tag{1.22}$$

Example 1.8

A 120-V, 60-Hz, single-phase source shown below powers a 0.10-mF capacitor in parallel with 20-mH coil that has 10-Ω winding resistance. Determine the current, pf, and power delivered by the source. (Note: Practical capacitors have negligible resistance, but the inductors have significant resistance due to long wires needed to wind numerous turns.)

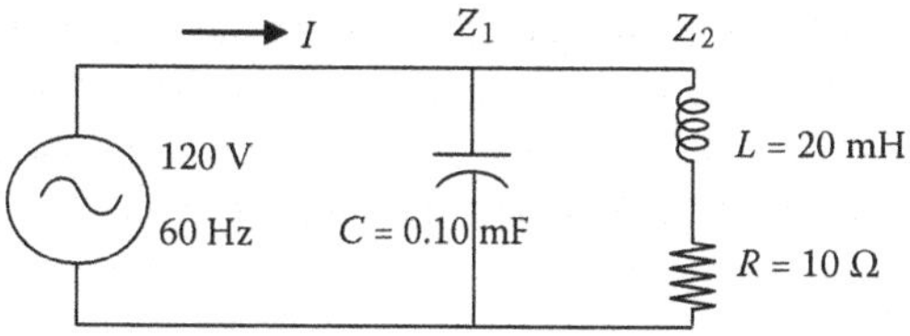

Solution:

We first determine the circuit impedances (all angles are in °):

capacitor impedance $Z_1 = -j/\omega C = -j/(2\pi 60 \times 0.10 \times 10^{-3}) = -j26.52\ \Omega$
$= 26.52\angle{-90}°\ \Omega$

coil impedance $Z_2 = R + j\omega L = 10 + j2\pi 60 \times 20 \times 10^{-3} = 10 + j7.54\ \Omega$
$= 12.52\angle 37°\ \Omega$.

Z_1 and Z_2 are in parallel, so they give

$$Z_{\text{Total}} = \frac{1}{\dfrac{1}{Z_1} + \dfrac{1}{Z_2}} = \frac{Z_1 \cdot Z_2}{Z_1 + Z_2}.$$

Notice that, the second term $Z_1\ Z_2 \div (Z_1 + Z_2)$ is simple to write and easy to remember as the product divided by the sum, although it is valid only for two impedances in parallel (it does not extend to three impedances in parallel).

$$\text{Therefore, } Z_{\text{Total}} = \frac{26.52\angle{-90} \times 12.52\angle 37}{-j26.52 + 10 + 7.54} = \frac{332\angle{-53}}{21.45\angle{-62.2}} = 15.48\angle 9.2\ \Omega$$

$$I = \frac{120V}{15.48\angle 9.2} = 7.75\angle{-9.2}\ \text{A}.$$

Power factor of the circuit = cos 9.2° = 0.987 lagging

$$P_{avg} = 120\ \text{V} \times 7.75\ \text{A} \times 0.987 = 918\ \text{W}.$$

The nodal analysis, mesh analysis, superposition theorem, and Thevenin and Norton equivalent source models are all valid in ac as they are in dc. The maximum power transfer theorem, however, has a slight change. For maximum power transfer in ac, we must have $\tilde{Z}_{\text{Load}} = \tilde{Z}^*_{\text{Th}}$, where Z_{Th} = Thevenin source impedance at the load points, and superscript * denotes the complex conjugate defined below.

For phasor impedance $\tilde{Z}_{\text{Th}} = Z_{\text{Th}}\angle\theta = R_{\text{Th}} + jX_{\text{Th}}$, its complex conjugate $\tilde{Z}^*_{\text{Th}} = Z_{\text{Th}}\angle{-\theta} = R_{\text{Th}} - jX_{\text{Th}}$, which is obtained by flipping the sign of j-part in the rectangular form or by flipping the sign of angle θ in the polar form. Thus, the complex conjugate of a phasor is its mirror image in the real axis as shown in Figure 1.10. With load impedance $\tilde{Z}_{\text{Load}} = \tilde{Z}^*_{\text{Th}}$ (called the *load matching*), the maximum power that can be transferred to the load remains the same as in dc, that is,

$$P_{\text{max}} = \frac{V_{\text{Th(rms)}}}{4R_{\text{Th}}}. \tag{1.23}$$

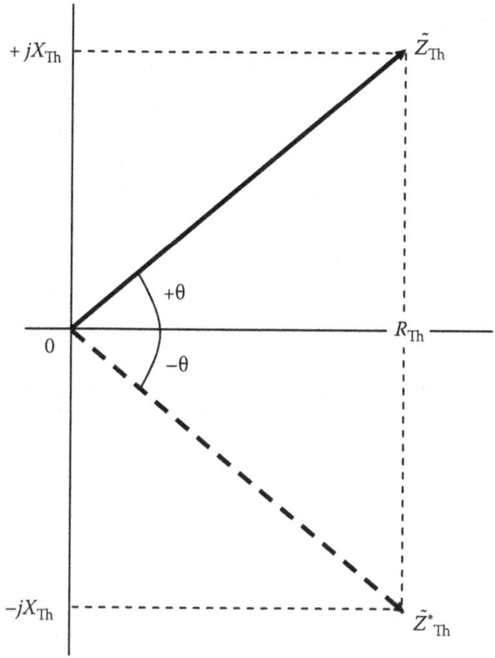

FIGURE 1.10 Complex conjugate of phasor is its own mirror image in reference axis.

1.6 AC POWER IN COMPLEX FORM

For a general form of ac power delivered from a source or absorbed in a load, we consider $V\angle\theta_v$ and $I\angle\theta_i$ phasors shown in Figure 1.11. The ac power is given by the product of the voltage and the complex conjugate of the current phasor. Therefore, the power is also a phasor, a complex number, hence the name

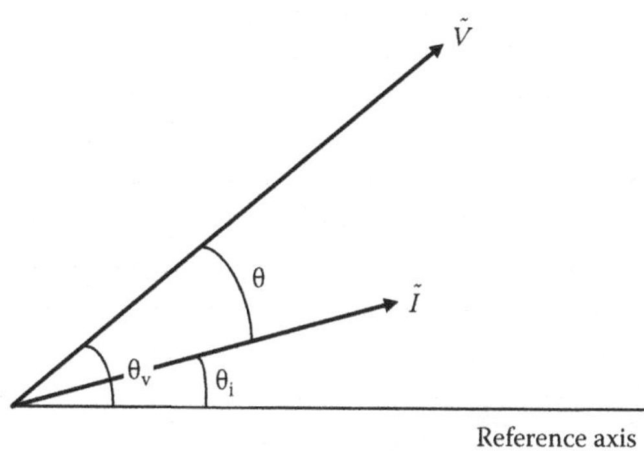

FIGURE 1.11 Product of $\tilde{V}$ and $\tilde{I}^*$ gives complex power in ac.

complex power. For average power, we use the rms values along with their phase angles, that is,

$$\text{complex power } \tilde{S} = V_{rms}\angle\theta_v \cdot \{I_{rms}\angle\theta_i\}^* \tag{1.24}$$

where * denotes the complex conjugate, that is, the mirror image of $\tilde{I}$ on the reference axis.

$$\text{Therefore, } \tilde{S} = V_{rms}\angle\theta_v \cdot I_{rms}\angle{-\theta_i} = V_{rms} \cdot I_{rms}\angle(\theta_v - \theta_i) = V_{rms} \cdot I_{rms}\angle\theta \tag{1.25}$$

where $\theta = \theta_v - \theta_i$ = phase difference between $\tilde{V}$ and $\tilde{I}$ phasors. In retrospect, if $\tilde{I}$ were used instead of $\tilde{I}^*$ in the power product, θ would be equal to $\theta_v + \theta_i$, giving an incorrect pf.

In *j*-form, Equation 1.25 for the complex power becomes

$$\tilde{S} = V_{rms} \cdot (I_{rms}\cos\theta + jI_{rms}\sin\theta) = V_{rms} \cdot I_{rms}\cos\theta + jV_{rms} \cdot I_{rms}\sin\theta = P + jQ \tag{1.26}$$

where the first term is the power on the real axis, called real power $P = V_{rms} \times I_{rms} \cos\theta$, which has the unit of watts, or kilowatts, or megawatts, depending on the power system size. The second term is the power on the imaginary axis, going in and out of the reactance *X*, hence called the reactive power $Q = V_{rms} \times I_{rms} \sin\theta$, which has the unit of volt-ampere reactive (VAR, or kVAR, or MVAR). The pf cos θ of the load has no units. The complex power $\tilde{S}$, being the product of volts and amperes, has the unit of volt-ampere (VA, or kVA, or MVA).

Therefore, complex power $S\angle\theta = P + jQ$ is a phasor, with *Q* out of phase by +90° from *P* as shown by the *power triangle* in Figure 1.12. The trigonometry of power triangle gives

$$S = \sqrt{P^2 + Q^2}$$

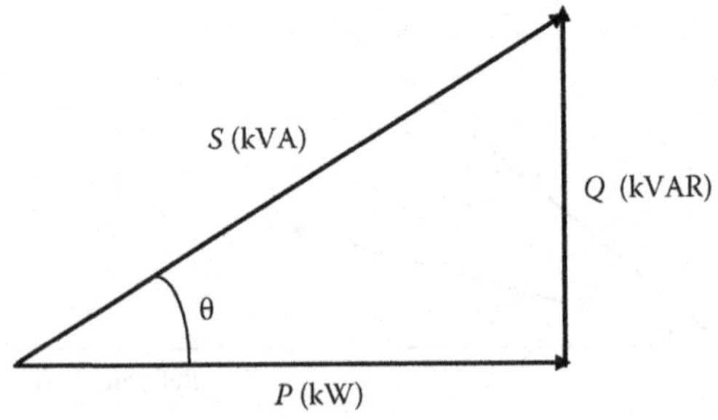

FIGURE 1.12 Power triangle of real power *P*, reactive power *Q*, and apparent power *S*.

or

$$kVA = \sqrt{kW^2 + kVAR^2}$$

and

$$\text{pf} = \frac{P}{S} = \frac{\text{kW}}{\text{kVA}}$$

or

$$\text{kW} = \text{KVA} \cdot \text{pf}. \tag{1.27}$$

It is noteworthy that the lagging pf of the load circuit produces positive Q and vice versa. Also, in complex power $\tilde{S} = S\angle\theta = V_{rms} \cdot I_{rms}\angle\theta$, the $S = V_{rms} \cdot I_{rms}$ without θ is known as the apparent power. With θ, $\tilde{S}$ is known as the complex power. Both $\tilde{S}$ and S have units of volt-amperes or kilovolt-amperes, or megavolt-amperes.

The real power P (watts) delivered by the generator requires fuel (real source of power) to drive the generator engine. The reactive power Q (VARs) on the imaginary axis is the power going in and out of L and C—charging and discharging L and C every half cycle—with zero average power. It does not require fuel to drive the generator engine, although it loads the generator voltage and current capabilities. Therefore, VAR adds in the capital cost of power equipment but does not add in the fuel cost for running the prime mover.

The ac power absorbed in load impedance $\tilde{Z}$ can also be derived as follows. The complex power absorbed in $\tilde{Z}$ is given by (voltage across Z) × (complex conjugate of I in Z), that is,

$$\tilde{S} = \tilde{V} \times \tilde{I}^* = (\tilde{I}\tilde{Z}) \times \tilde{I}^* = I^2\tilde{Z} = I^2(R + jX) = I^2R + jI^2X = P + jQ,$$

where

real power $$P = I^2R,$$

reactive power $$Q = I^2X,$$

and

apparent power $$S = I^2 \cdot Z = \sqrt{P^2 + Q^2} \tag{1.28}$$

which is the same power triangle relation as in Figure 1.12.

We take note here, that in impedance $\tilde{Z} = R + jX$, the real power P is absorbed only in R, and the reactive power Q is absorbed only in X.

Example 1.9

Determine the current and power in the following circuit, and draw the phasor and time diagrams.

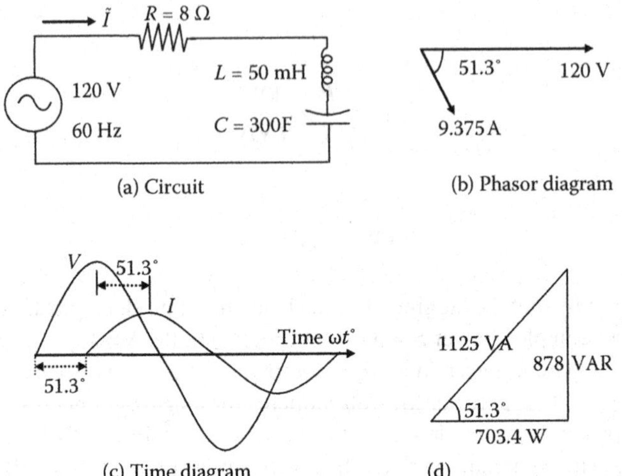

(a) Circuit (b) Phasor diagram (c) Time diagram (d)

Solution:

We first convert the *R*–*L*–*C* values into the total impedance $\tilde{Z}$ in ohms:

$$\tilde{Z} = R + j\omega L - j1/\omega C,$$

where

$$\omega = 2\pi f = 2\pi \cdot 60 = 377 \text{ rad/s}$$

$$\text{Therefore, } \tilde{Z} = 8 + j377 \times 0.050 - j1/\left(377 \times 300 \cdot 10^{-6}\right) = 8 + j10\ \Omega$$
$$= \sqrt{\left(8^2 + 10^2\right)} \angle \tan^{-}(10/8)° = 12.8\angle 51.3°.$$

In a one-loop circuit, current $I = \tilde{V}_{\text{Loop}} \div \tilde{Z}_{\text{Loop}} = 120\angle 0°$ (reference) $\div (12.8\angle 51.3)$, which gives $\tilde{I}$ = (120 ÷ 12.8)∠0−51.3° = 9.375∠−51.3° A.

The current lags the reference voltage by 51.3° as shown in phasor diagram (b) and time diagram (c).

Since 120 V is the implied rms value, 9.375 A is also an rms value. The complex power in any element (source or load) is $\tilde{S} = \tilde{V} \times \tilde{I}^*$ volt-amperes, where $\tilde{V}$ = voltage across and $\tilde{I}^*$ = complex conjugate of current through the element.

Thus, in this case,

$$\tilde{S}_{\text{source}} = 120\angle 0° \times 9.375\angle{+51.3°} = (120 \times 9.375)\angle(0 + 51.3°) = 1125\angle 51.3°$$
$$= 1125(\cos 51.3° + j\sin 51.3°) = 703.4 + j878 \text{ VA}.$$

Apparent power drawn from the source $S = 1125$ VA.

Real power $P = 703.4$ W (absorbed in R and leaving the circuit as heat).

Reactive power $Q = 878.0$ VAR (charging and discharging L and C, but remaining in the circuit).

The power triangle is shown in (d). The following is another way to determine the ac powers:

Apparent power = $I^2Z = 9.375^2 \times 12.8 = 1125$ VA; real power = $I^2R = 9.375^2 \times 8 = 703.4$ W; and reactive power = $I^2X = 9.375^2 \times 10 = 878$ VAR, all matching with the above values.

Example 1.10

Two parallel loads draw power from a 480-V source as shown in the figure below, where Load-1 draws 15 kW and 10 kVAR lagging, and Load-2 draws 7 kW and 5 kVAR leading. Determine the combined pf and the total kVA drawn from the source.

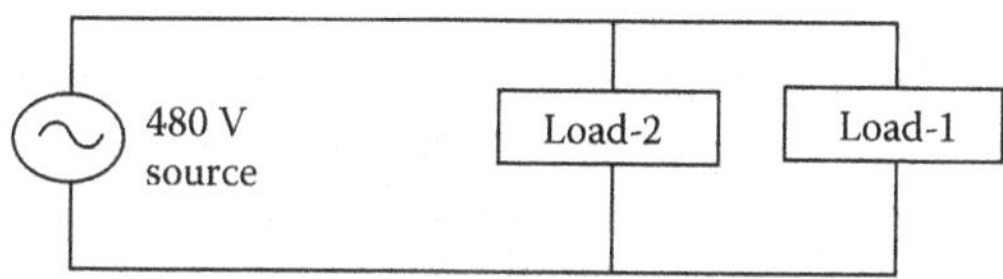

Solution:

In power engineering, the lagging kVAR has, a + sign, and the leading kVAR has, a – sign. The parallel loads add up to make the total, and their real and reactive powers are added individually to make the total real and reactive powers.

$$\text{Therefore, } \tilde{S}_{\text{Total}} = \tilde{S}_1 + \tilde{S}_2 = (15 + j10) + (7 - j5) = 22 + j5 = 22.56\angle{+12.8^\circ}.$$

Total kVA drawn from the source = 22.56, and the combined pf = cos 12.8° = 0.975 lagging since the combined power has + 5 kVAR, with the + sign indicating the lagging pf (or the + sign with 12.8° in total complex power $\tilde{S}$ comes from the inductive load, which draws lagging current).

1.7 REACTIVE POWER

If the current lags the voltage by 90° as shown in Figure 1.13—as in a pure inductor—the instantaneous power shown by a heavy curve is positive in the first one-half cycle and negative in the second one-half cycle. The power is first positive (from source to inductor) and then negative (from inductor to source). Therefore, the average power in an inductor over one full cycle is always zero, although wires are occupied with current at all times, leaving less room for the real power to flow from the source to the load. Similarly, the average power is also zero over one full cycle in a pure capacitor, which draws current leading the voltage by 90°.

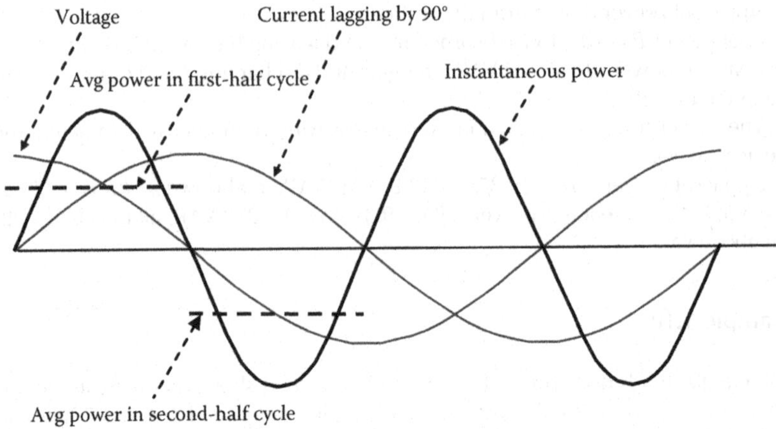

FIGURE 1.13 Voltage, current, and instantaneous and average powers in inductor.

On the other hand, if the current and voltage are in phase—as in a pure resistor—the power is always positive even when both the voltage and the current are negative in one-half cycle. This results in the average power having the same positive value in both one-half cycles and also over one full cycle.

1.8 THREE-PHASE AC POWER SYSTEM

The ac power is now universally adopted all over the world because of its easy conversion from one voltage level to another using energy-efficient power transformers. From a large ac power system (power grid), large power users are catered to at high voltage, medium power users at medium voltage, and low power users at safe low voltage of 120 or 240 V.

In a single-phase (1-ph) power circuit, the instantaneous power varies between the maximum and minimum values every cycle, with the average power positive. In a balanced three-phase (3-ph) power circuit, when the instantaneous power in one phase decreases, the power in remaining phases increases, making the sum of power in all three phases a steady constant value with no time variations. For having such a smooth power flow, the three-phase ac system has been universally adopted around the world. Since all three phases are balanced and identical, the balanced three-phase power circuit is generally shown by a one-line diagram and analyzed on a single-phase (per-phase) basis. Then, average three-phase power = 3 × average single-phase power.

The three-phase source or load can be connected in Y or Δ. The basic voltage and current relations between the line-to-neutral (called phase) values and the line-to-line (called line) values are reviewed below.

1.8.1 Balanced Y- and Δ-Connected Systems

In a balanced three-phase system, the magnitudes of line voltage V_L, line current I_L, and three-phase power in terms of the magnitudes of phase voltage V_{ph} and phase current I_{ph} are derived as follows, where the term phase voltage (V_{ph}) also means line-to-neutral voltage (V_{LN}), and the term line voltage (V_L) also means line-to-line voltage (V_{LL}).

The Y-connected system shown in Figure 1.14a gives the voltage of line a from line b, $\tilde{V}_{ab} = \tilde{V}_{an} - \tilde{V}_{bn}$, which is shown in Figure 1.14b by a heavy solid line. The geometry would give the line-to-line voltage magnitude $V_{LL} = \sqrt{3}\ V_{ph}$. We note that the line voltage $\tilde{V}_{ab}$ leads the phase voltage $\tilde{V}_{an}$ by 30°.

The line current is obviously the same as the phase current, i.e., $\tilde{I}_L = \tilde{I}_{ph}$.

Three-phase power $P_{3\text{-ph}} = 3 \times P_{1\text{-ph}} = 3 \times V_{ph} \times I_{ph} \times \text{pf} = 3 \times (V_{LL}/\sqrt{3}) \times I_L \times \text{pf}$.

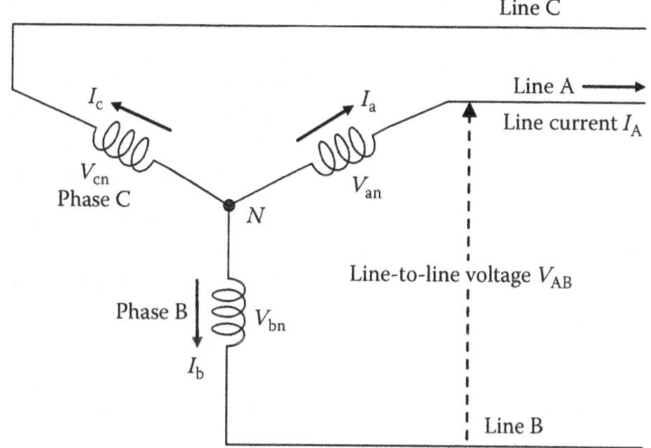

(a) Phase and line voltage and current definitions

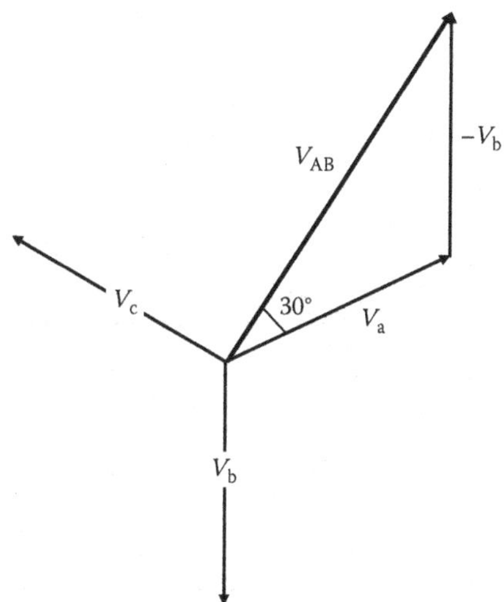

(b) Line-to-line voltage phasor $V_{AB} = V_a - V_b$

FIGURE 1.14 Phase and line voltages in balanced Y-connected source or load.

Therefore, $P_{3\text{-ph}} = \sqrt{3}\, V_{LL} I_L$ pf watts and $S_{3\text{-ph}} = \sqrt{3}\, V_{LL} I_L$ volt-amperes. (1.29)

The Δ-connected system shown in Figure 1.15a gives the line current $\tilde{I}_A = \tilde{I}_a - \tilde{I}_c$, which is shown by a heavy line in Figure 1.15b, the geometry of which leads to $I_L = \sqrt{3}\, I_{ph}$.

The line-to-line voltage here is obviously the same as the phase voltages, that is, $V_{LL} = V_{ph}$.

Three-phase power $P_{3\text{-ph}} = 3 \times P_{1\phi} = 3 \times V_{ph} \times I_{ph} \times \text{pf} = 3 \times V_{LL} \times (I_L/\sqrt{3}) \times \text{pf}$.

Therefore, $P_{3\text{-ph}} = \sqrt{3}\, V_{LL} I_L$ pf watts and $S_{3\text{-ph}} = \sqrt{3}\, V_{LL} I_L$ volt-amperes. (1.30)

Note that the expression for three-phase power comes out to be the same in both *Y* and Δ.

In a three-phase, four-wire, *Y*-connected system, the neutral current is the sum of three line currents. In a balanced *Y* system, the current in neutral wire (if used) is always zero because it is the phasor sum of three equal line currents out of phase by 120° from each other. For this reason, a balanced three-phase system performs exactly the same with or without the neutral wire, except in unbalanced load condition.

In a three-phase, Δ-connected system, no wire exists for the return currents, so the phasor sum of all line currents is forced to be zero regardless of balanced or

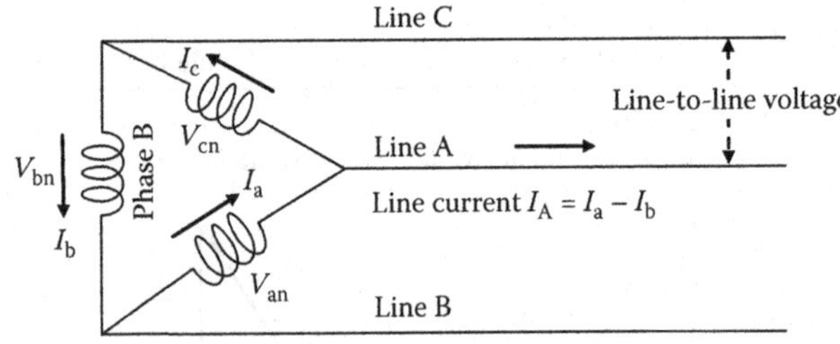

(a) Phase and line voltage and current definitions

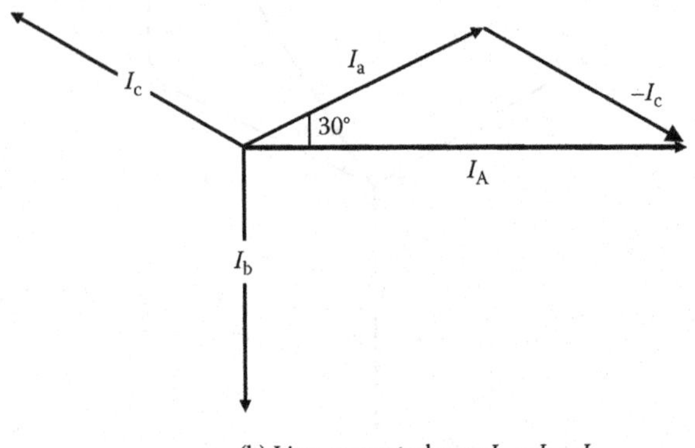

(b) Line-current phasor $I_A = I_a - I_b$

FIGURE 1.15 Phase and line currents in balanced Δ-connected source or load.

unbalanced load. With an unbalanced three-phase load, however, this imposed zero on the return current produces unbalanced line-to-line voltages.

Example 1.11

A Y-connected generator has the line-to-neutral voltage phasor $\tilde{V}_{an} = 100\angle 0$. Determine the line-to-line voltage phasors $\tilde{V}_{ab}$, $\tilde{V}_{bc}$, and $\tilde{V}_{ca}$ with their phase angles with respect to $\tilde{V}_{an}$.

Solution:

The phasor relations between the line and phase voltages are shown in the figure below, the trigonometry of which gives the following:

$$\tilde{V}_{ab} = \sqrt{3} \times 100\angle{+30°},$$

$$\tilde{V}_{bc} = \sqrt{3} \times 100\angle{-90°},$$

and

$$\tilde{V}_{ca} = \sqrt{3} \times 100\angle{-210°} \text{ or } \sqrt{3} \times 100\angle{+150°}.$$

It means that with reference to the phase voltage $\tilde{V}_{an}$, the line voltage $\tilde{V}_{ab}$ leads $\tilde{V}_{an}$ by 30°, $\tilde{V}_{bc}$ lags $\tilde{V}_{an}$ by 90°, and $\tilde{V}_{ca}$ lags $\tilde{V}_{an}$ by 210° or leads $\tilde{V}_{an}$ by 150°.

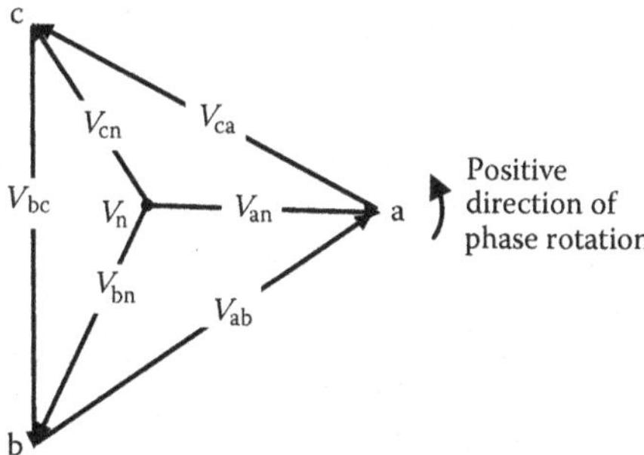

1.8.2 Y–Δ Equivalent Impedance Conversion

Almost all three-phase generators are Y-connected, but some transformers and loads may be Δ-connected. Even so, we always analyze the power system assuming all equipment to be Y-connected since it is much easier to determine the phase values in each phase of the generator and then convert the results into three-phase values. For this purpose, if a load is Δ-connected, we first convert it into an equivalent

Y-connected load, which would absorb the same total power. The equivalence can be derived from the rigorous circuit analysis, which would lead to the following final results in view of Figure 1.16 (all *Z*'s are complex phasors in the equations below).

From Δ to *Y* conversion, that is, from given Z_{ab}, Z_{bc}, and Z_{ca}, we find

$$Z_{an} = \frac{Z_{ab}Z_{ca}}{Z_{ab}+Z_{bc}+Z_{ca}} \quad Z_{bn} = \frac{Z_{ab}Z_{bc}}{Z_{ab}+Z_{bc}+Z_{ca}} \quad Z_{cn} = \frac{Z_{bc}Z_{ca}}{Z_{ab}+Z_{bc}+Z_{ca}}. \tag{1.31}$$

From *Y* to Δ conversion, that is, from given Z_{an}, Z_{bn}, and Z_{cn}, we find

$$Z_{ab} = \frac{Z_{an}Z_{bn}+Z_{bn}Z_{cn}+Z_{cn}Z_{an}}{Z_{cn}} \quad Z_{bc} = \frac{Z_{an}Z_{bn}+Z_{bn}Z_{cn}+Z_{cn}Z_{an}}{Z_{an}} \text{ and}$$
$$Z_{ca} = \frac{Z_{an}Z_{bn}+Z_{bn}Z_{cn}+Z_{cn}Z_{an}}{Z_{bn}} \tag{1.32}$$

Equations 1.31 and 1.32 hold true for three-phase balanced or unbalanced load impedances. However, for balanced three-phase *Y* or Δ loads, where load impedances in each phase are equal (i.e., both *R* and *X* in each phase are equal), they get simplified to

$$Z_{LN} = 1/3\ Z_{LL} \quad \text{and} \quad Z_{LL} = 3\ Z_{LN} \tag{1.33}$$

where $Z_{LN} = Z_{an} = Z_{bn} = Z_{cn}$ and $Z_{LL} = Z_{ab} = Z_{bc} = Z_{ca}$.

Equation 1.33 can also be deduced as follows. Since ac power absorbed in a load impedance is given by $S = I^2 Z = V^2/Z$, we have $Z = V^2/S$. Therefore, the equivalent *Z* in ohms for the same power (required for equivalency) must change as a V^2 ratio. For converting Δ into equivalent *Y*, since V_{LN} in $Y = (1/\sqrt{3})\ V_{LL}$ in Δ, the equivalent *Y* value $Z_{LN} = (1/\sqrt{3})^2\ Z_{LL} = 1/3$ of the Z_{LL} value in Δ. Similarly, for converting *Y* into equivalent Δ, the equivalent Δ value $Z_{LL} = (\sqrt{3})^2 = 3 \times Z_{LN}$ value in *Y*.

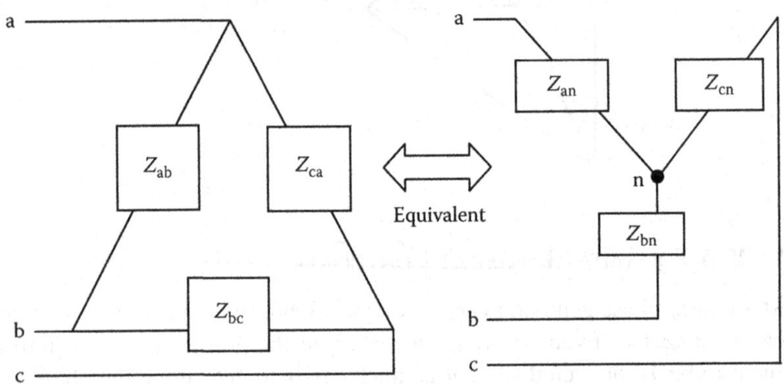

FIGURE 1.16 Δ to *Y* or *Y* to Δ equivalent impedance conversions.

After transforming the entire system into an equivalent Y, and then combining series–parallel impedances into one equivalent total impedance per phase, we can easily determine the following:

$$\text{Phase current} = \frac{\text{Voltage per phase}}{\text{Total impedance per phase}}. \tag{1.34}$$

Example 1.12

A balanced Δ-connected load-1 has $Z_{LL1} = 6 + j9\ \Omega$ in each phase of Δ. It is connected in parallel with a balanced Y-connected load-2 with $Z_{LN2} = 2 + j3\ \Omega$ in each phase of Y. If these two parallel loads are powered by a three-phase, Y-connected generator with a 480-V line voltage, determine the line current at the generator terminals and the real power delivered by the generator.

Solution:

We first convert Δ-connected Z_{LL1} into equivalent Y-connected Z_{LN1}, which is 1/3 Z_{LL1}, that is, $Z_{LN1} = 1/3\ (6 + j9) = 2 + j3\ \Omega$/ph. The Y-connected Z_{LN2} is in parallel with Z_{LN1}. Therefore,

$$Z_{LN.Total} = \frac{1}{\dfrac{1}{Z_{LN1}} + \dfrac{1}{Z_{LN2}}} = \frac{Z_{LN1}Z_{LN2}}{Z_{LN1} + Z_{LN2}} = 1 + j1.5$$

$$= 1.8\angle 56.3^\circ\ \Omega/\text{ph (detailed algebra is left as an exercise for the students).}$$

The Y-connected generator phase voltage = 480 ÷ √3 = 277.1 V

$$\text{Therefore, generator line current} = \text{generator phase current} = \frac{277.1\angle 0\,(\text{Ref.})}{1.8\angle 56.3}$$

$$= 153.9\angle -56.3\ \text{A}.$$

Real power per phase, $P = V \times I \times \cos\theta = 277.1 \times 153.9 \times \cos 56.3^\circ = 23{,}672$ W/phase.

Alternatively, using total circuit resistance $R = 1\ \Omega$ in $Z_{LN.Total}$, we get

$P = I^2R = 153.9^2 \times 1 = 23{,}685$ W/phase (checks within rounding error).

PROBLEMS

1. The electrical potential of point 1 is 3000 V higher than that of point 2. If 200 C of charge per minute flows from point 1 to 2, determine the current and power flow from point 1 to 2 and the energy transferred in 15 min.
2. A sinusoidal voltage of 900 cos 377t volts applied across a circuit element draws 250 cos(377t – 30°) amperes. Determine the average power absorbed in the element.

3. For two phasors $\tilde{A} = 85\angle 40°$ and $\tilde{B} = 700\angle 50°$, find (i) $\tilde{A}+\tilde{B}$, (ii) $\tilde{A}-\tilde{B}$, (iii) $\tilde{A}\times\tilde{B}$, and (iv) $\tilde{A}/\tilde{B}$. Express each result in j-form and also in θ-form.
4. If phasor $\tilde{A} = \dfrac{8.5}{j6.5}$, express it in j-form, θ-form, and e-form.
5. A source voltage of 240 V_{rms} is applied to $30\angle 30°$ Ω load impedance. Find the average current, rms current, peak current, and average power delivered to the load.
6. A coil having inductance of 30 mH and resistance of 3 Ω is connected to a 120-V, 60-Hz source. Determine the kilowatt-hour energy drawn by the coil in an 8-h period.
7. The figure below is in time domain. Express it in phasor domain with impedances in ohms, and determine the current phasor and the average power.

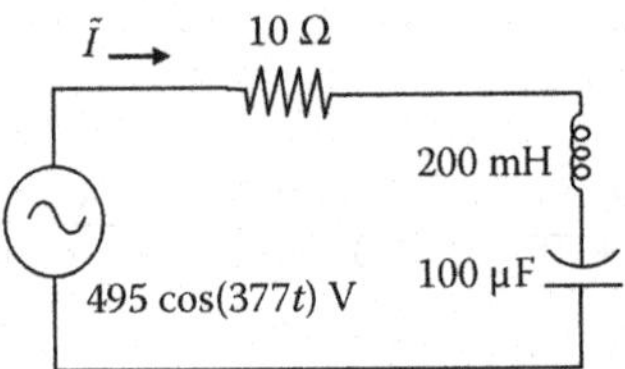

8. A 240-V, 50-Hz, single-phase voltage is applied to the circuit shown in the figure below. Determine the current, pf, and power delivered by the source.

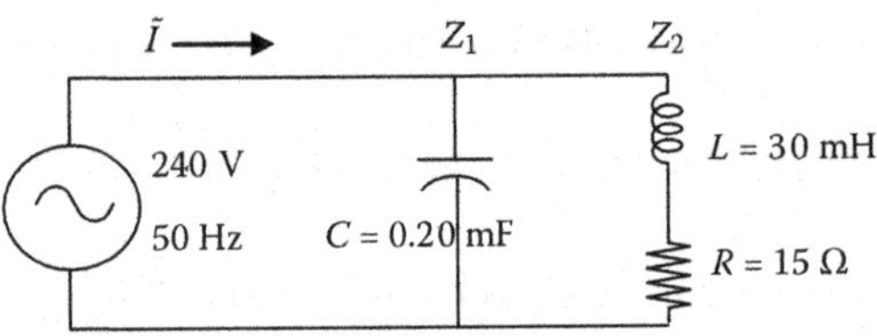

9. For the circuit shown below, determine the current and power and draw the phasor and time diagrams.

Ĩ R = 6 Ω
L = 150 mH
240 V
60 Hz
C = 600 μF

10. Three parallel loads draw power from a 480-V source. Load-1 is 25 kW and 15 kVAR lagging, load-2 is 10 kW and 8 kVAR leading, and load-3 is 18 kW and 10 kVAR lagging. Determine the combined power factor and the total kVA drawn from the source.
11. A Δ-connected transformer has the line-to-line voltage phasor $\tilde{V}_{ab} = 480\angle 0$. Determine the line-to-neutral voltage phasors $\tilde{V}_{an}$, $\tilde{V}_{bn}$, and $\tilde{V}_{cn}$ in magnitudes and phase angles with respect to $\tilde{V}_{ab}$.

12. A balanced Y-connected load-1 has $Z_{LN1} = 4 + j1\ \Omega$ in each phase of Y. It is connected in parallel with a balanced Δ-connected load-2 with $Z_{LL2} = 12 + j6\ \Omega$ in each phase of Δ. If these two parallel loads are powered by a three-phase, Y-connected generator with a 480-V line voltage, determine the line current at the generator terminals and the real power delivered by the generator.
13. Determine the average and rms values of a square wave current of +10 A for $t = 0$ to 2.5 ms and −10 A for $t = 2.5$ to 5 ms.
14. Determine the average and rms values of a rectangular wave current of +5 A for 0 to 2 ms and −5 A for $t = 2$ to 6 ms.
15. Powered by a 120-V rms voltage source, three impedances are connected in parallel: (i) $10 + j20\ \Omega$, (ii) $5 + j20\ \Omega$, and (iii) $10 + j30\ \Omega$. Determine the real power drawn by each impedance. Without making numerical calculations, how would you have identified the impedance that will draw the highest power?
16. A 120-V source is connected to a load impedance of $31.8 + j42.4\ \Omega$. Determine (a) the load current, (b) complex power, (c) real power, (d) reactive power, and (e) apparent power. Be sure to include with each answer the phase angle and the unit as applicable.
17. In the circuit shown below, determine $\tilde{I}$, $\tilde{I}_1$, and $\tilde{I}_2$, and draw the phasor diagram, taking 120 V as the reference phasor.

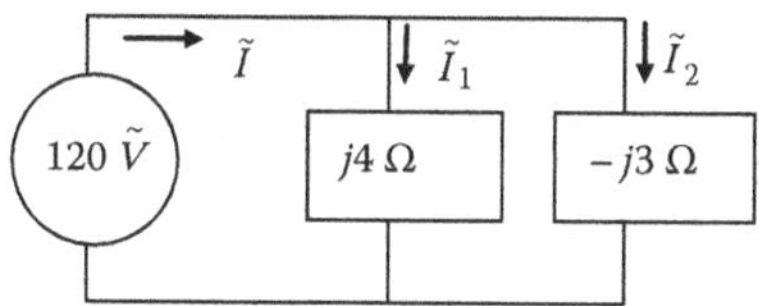

18. A three-phase, 440-V, 60-Hz load draws 30-kW real power and 40 kVAR reactive power from the source: (i) determine the line current, complex power, and apparent power drawn from the source; (ii) draw the power triangle; and (iii) if the reactive power is reduced to 10 kVAR, keeping the same real power, determine the % change in the prime mover's fuel consumption rate.
19. A 440-V, three-phase, Y-connected generator delivers power to a Δ-connected load with $6 + j9\ \Omega$ in each phase of Δ. Determine the line current and three-phase power to the load.
20. A three-phase Δ-connected load has $4.5 + j6.3\ \Omega$ impedance in each phase of Δ. For easy circuit calculations with other parallel loads connected in Y, the Δ-load needs to be converted into an equivalent Y. Determine the impedance Z in each phase of the equivalent Y.
21. The circuit in the figure below is shown in time domain. First, convert it into the phasor domain showing the voltage and impedance as phasors, and then find the total impedance and current drawn from the source.

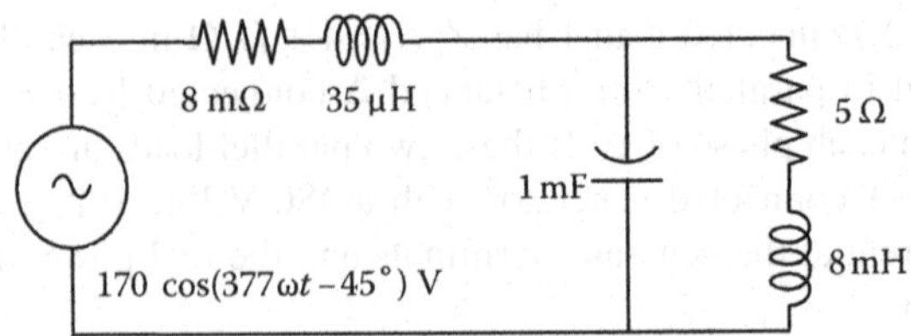

22. Prepare an EXCEL spreadsheet as shown in the table below for one cycle of 60-Hz ac in the increments of 15°. With $\theta = 0°$, plot the voltage, current, and power versus ωt using the CHART function in EXCEL. Find the average values of columns 2 through 5. Then, verify that (a) the square root of Avg4 is equal to the rms value $I_{peak}/\sqrt{2}$ and (b) average power Avg5 is equal to $V_{rms} I_{rms} \cos\theta$. Make sure to convert ωt degrees into radians for the EXCEL computations if necessary. Make and plot the second spreadsheet with $\theta = 30°$ and the third sheet with $\theta = 90°$. Examining your numerical answers and the charts, draw your conclusions as to the effect of lag angle θ on the average power.

$\omega t°$	$v = 100 \sin \omega t$	$i = 150 \sin(\omega t - \theta)$	i^2	$v * i$
0				
15				
.				
360°				
Avg	Avg2	Avg3	Avg4	Avg5

QUESTIONS

1. Why is the ac universally adopted for the electrical power grid over dc?
2. Why is the three-phase power system universally adopted over a single-phase ac power system?
3. Explain the terms phasor, phase difference, and operator j.
4. For the impedance boxes Z_1, Z_2, and Z_3 shown below, which of the R, L, and C element(s) and their relative magnitudes, can you expect to see if you open the boxes? Answer for each box individually.

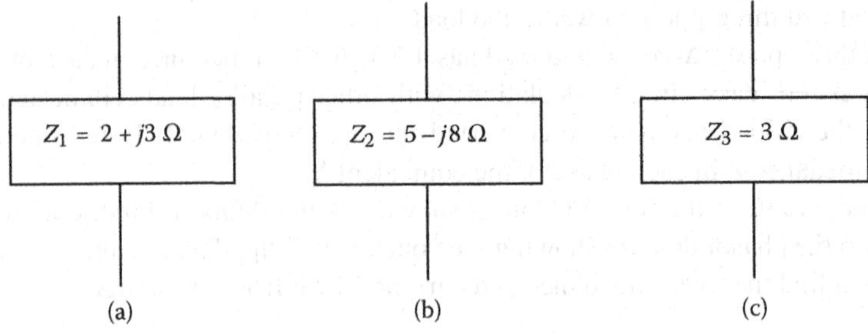

5. How do the basic circuit laws and theorems in dc and ac differ?
6. If a 50-Hz impedance box of value $10\angle{-30°}\ \Omega$ designed to work on a 240-V, 50-Hz source in Europe is connected to a 240-V, 60-Hz source in the United States, how would the current magnitude and phase angle change at higher frequency (will they increase or decrease?)
7. What is the complex conjugate of a current, and why is the conjugate (and not the straight) current used to determine the complex power in ac circuits?
8. If the real power from the source to the load delivers real work (pumps water, ventilates air, heats space, etc.), what does the reactive power do?
9. Why should the plant power engineer use power at or near a unity power factor as much as possible?
10. Draw a set of balanced three-phase currents in a *Y*-connected system with neutral wire, and graphically, using the tip-tail method, find their sum $\tilde{I}_a + \tilde{I}_b + \tilde{I}_c = \tilde{I}_n$ returning via the neutral wire.
11. In a balanced three-phase *Y*-connected power circuit, why does the circuit behave exactly the same with or without the neutral wire? What purpose does the neutral wire, where provided, serve in a three-phase four-wire system?

FURTHER READING

Alexander, C.K. and Sadiku, M.N.O. 2012. *Fundamentals of Electric Circuits*. Boston: McGraw Hill.

Dorf, R.C. and Svoboda, J.A. 1999. *Introduction to Electric Circuits*. New York: John Wiley & Sons.

Rizzoni, G. 2007. *Principles and Applications of Electrical Engineering*. New York: McGraw Hill.

2 Common Aspects of Power Equipment

All electrical machines have two sets of coils; one set (field coil in generator and motor and primary coil in transformer) produces the working magnetic flux φ, and the other set (armature coil in generator and motor and secondary coil in transformer) reacts to flux φ and produces an electromotive force (voltage) in the generator and transformer and a mechanical force (torque) in the motor. Both sets of coils are embedded in magnetic steel to produce the required flux with a minimum magnetizing (field excitation) current. The electromagnetic interaction between two sets of coils is governed by the following basic laws of physics.

2.1 FARADAY'S LAW AND COIL VOLTAGE EQUATION

A coil of N turns linking magnetic flux φ coming from an independent external source as shown in Figure 2.1a induces an internal voltage equal to $-\mathrm{d}\varphi/\mathrm{d}t$ in every turn of the coil. With N turns in series, it adds up to the terminal voltage

$$v = -N \times \mathrm{d}\varphi/\mathrm{d}t. \tag{2.1}$$

The – sign is due to Lenz, which indicates that the polarity of induced voltage and the resulting current in the coil (if in a closed loop) is such as to oppose $\mathrm{d}\varphi/\mathrm{d}t$, that is, to oppose any change (increase or decrease) in flux φ. Stated differently, the coil current direction will be such as to produce an internal reaction flux φ_i to counter the flux coming from the external source as shown in Figure 2.1b.

Constant flux linkage theorem: A corollary to Faraday's law when applied to a coil with shorted terminals as shown in Figure 2.1b, where the terminal voltage v = 0, leads to $-N\cdot\mathrm{d}\varphi/\mathrm{d}t = 0$ or flux φ = constant, that is, the flux linking any shorted coil remains constant. To do this, the shorted coil produces an internally circulating current and the resulting flux just sufficient to cancel out the change in incoming flux such that the net flux through the coil remains constant (zero or any other value). Some examples of the working of this theorem are as follows.

- Losing radio or cell phone signals in a long road tunnel or a concrete building with reinforcing steel grid of numerous shorted metal rings (called Faraday rings) shown in Figure 2.1c
- Faraday cage shown in Figure 2.1d that shields the inside from electromagnetic signals coming from the outside

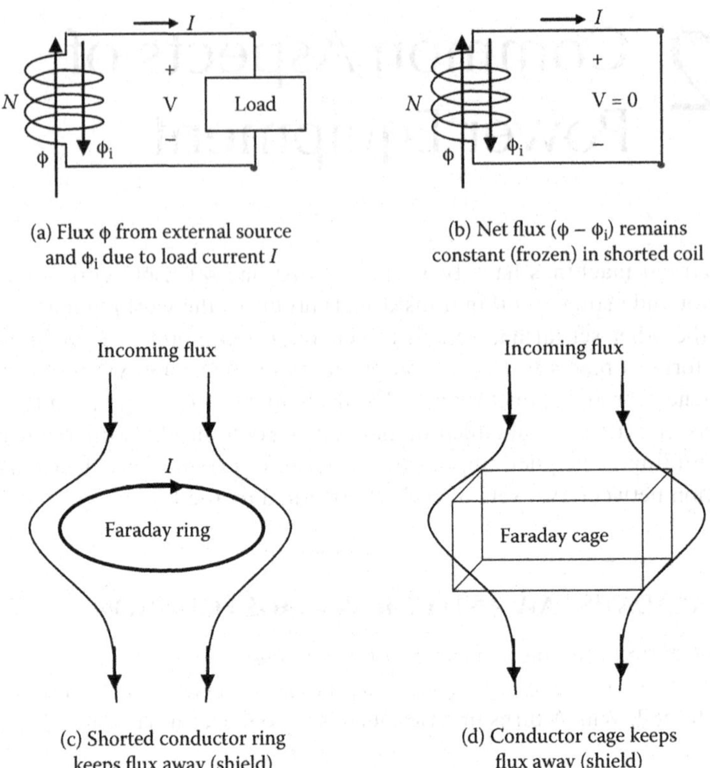

FIGURE 2.1 Faraday's law and constant flux theorem in shorted coil.

- Superconducting coil for energy storage where the coil with zero resistance retains its flux and the associated magnetic energy forever

Basic voltage equation: In any electrical machine, either rotating (generator and motor) or stationary (transformer), if a coil of N turns is subjected to an alternating flux $\varphi = \varphi_m \cos \omega t$, where φ_m = maximum value (peak amplitude) of the magnetic flux in weber, and $\omega = 2\pi f$ = angular frequency in radians per second, then a voltage is induced in the coil that is given by

$$v = -N\frac{d\phi}{dt} = -N\frac{d\{\phi_m \cos \omega t\}}{dt} = \{N\varphi_m\omega\}\sin \omega t = \{V_m\}\sin \omega t$$

where the voltage amplitude $V_m = N\varphi_m\omega = 2\pi f N \varphi_m$

$$\text{Therefore, } V_{rms} = V_m/\sqrt{2} = (2\pi/\sqrt{2})\, f\, N\, \varphi_m = 4.444\, fN\varphi_m \text{ volts.} \tag{2.2}$$

Example 2.1

Determine the rms value of voltage induced in a 30-turn coil on a 20 × 20 cm cross-section core with peak flux density of 1.65 T alternating at 50 Hz.

Solution:

Tesla is the unit of flux density B in webers per square meter, from which we get the peak flux in the core = 1.65 T × (0.20 m × 0.20 m) = 0.066 Wb. Then, using Equation 2.2, we have

$$V_{rms} = 4.444 \times 50 \times 30 \times 0.066 = 440 \text{ V}.$$

2.2 MECHANICAL FORCE AND TORQUE

Two parallel conductors with currents I_1 and I_2 in the *opposite* direction as shown in Figure 2.2 produce a *repulsive* force

$$F = \frac{5.4KI_1I_2}{S_{\text{inches}}} \cdot 10^{-7} \text{ lbf/foot length} \quad (2.3)$$

where K = conductor shape factor ($K = 1$ for round conductors and < 1 for rectangular bars), and S = center-to-center spacing between the conductors (in inches). In SI units with S in centimeters, it becomes

$$F = \frac{200KI_1I_2}{S_{\text{cm}}} \cdot 10^{-7} \text{ Newtons/meter length.} \quad (2.4)$$

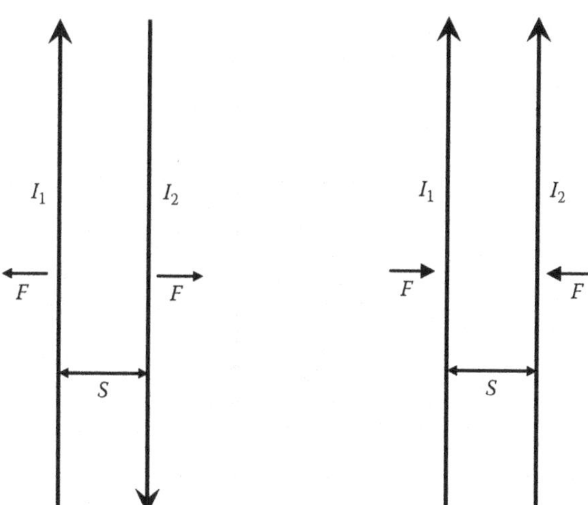

FIGURE 2.2 Mechanical force between two parallel bus bars.

If we use proper polarities of I_1 and I_2 in Equations 2.3 and 2.4, then positive force would mean attractive force, and negative force would mean repulsive force. The conductors carrying currents I_1 and I_2 in the same direction produce attractive force, and I_1 and I_2 in the opposite direction—as in single-phase and dc bus bars—produce repulsive force.

Example 2.2

In a single-phase switchboard, two parallel rectangular bus bars 1.1 in. apart at center lines experience 20,000 A_{rms} current when a load gets short-circuit fault. Determine the peak mechanical force per foot length of the bus bars. Assume the bus bar shape factor $K = 0.8$. State whether this force is attractive or repulsive.

Solution:

In single-phase power circuits, the currents in two parallel bus bars are equal and opposite. Therefore, we must use peak currents $I_1 = -I_2 = \sqrt{2} \times 20{,}000$ in Equation 2.3, which gives the peak force

$$F = \frac{5.4 \times 0.8 \times \left(\sqrt{2} \times 20{,}000\right) \times \left(-\sqrt{2} \times 20{,}000\right)}{1.1} \times 10^{-7} = -314.2 \text{ lb./ft.}$$

The – sign indicates repulsive force. The bars must be braced to avoid mechanical damage due to bending stresses or deflections between supports.

A coil with current I placed in a magnetic flux φ of another coil (Figure 2.3) produces a mechanical torque equal to

$$\text{torque} = KI\varphi \sin\theta \tag{2.5}$$

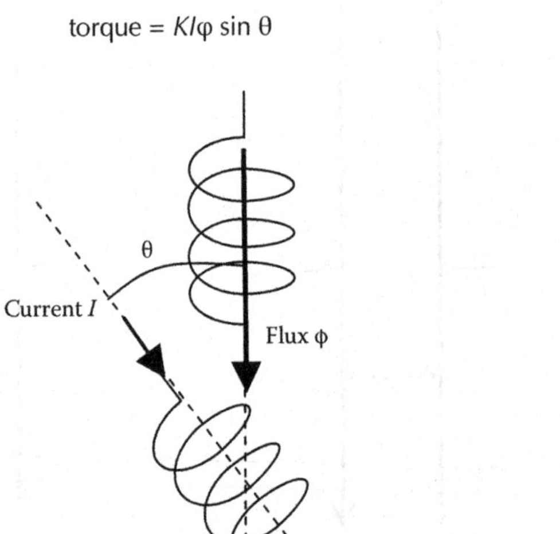

FIGURE 2.3 Mechanical torque between two coils.

where K = constant of the geometry, and θ = angle between the magnetic center lines of two coils.

Two coils with currents I_1 and I_2 and the maximum mutual inductance M_o, which occurs when $\theta = 0$, produce a torque with magnitude given by

$$\text{torque} = M_o I_1 I_2 \sin\theta. \tag{2.6}$$

2.3 ELECTRICAL EQUIVALENT OF NEWTON'S THIRD LAW

If the rotor of a motor is held fixed (blocked from rotating), and the stator is allowed to rotate freely, the "stator" would rotate at the same speed as the rotor would have in the opposite direction. This is due to the reaction torque on the stator. It can be viewed as the electrical equivalent of Newton's third law: *the action and reaction are equal and opposite*, both of which can be electrical or mechanical. If the generator stator coil is pushing the electrical current outward, a back torque (reaction) is produced on the rotor coil, which opposes the rotor from turning forward. The prime mover must overcome this back torque by using fuel to maintain a constant speed. We feel such back torque when turning a megger for testing the insulation resistance. In the motor operation, if the rotor coil is pushing a mechanical torque outward, a back voltage (reaction) is produced in the stator coil, which opposes the current from going inward. This back voltage must be overcome by the applied source voltage at the stator terminals in order to maintain constant torque.

The mechanical analogy of the back voltage in the motor and the back torque in the generator is the back pressure in water pipes and air compressors. The pump must overcome the back pressure in order to push the fluid forward in the desired direction, that is, inward to the compressor.

Thus, there is a back torque in the rotor of the electrical generator, and there is a back voltage in the stator of the motor. Both the back torque and the back voltage are the reactions that must be overcome by the primary power source to keep the machine delivering steady power outward. Stated differently, there is always a motor reaction in a generator, and a generator reaction in a motor.

2.4 POWER LOSSES IN ELECTRICAL MACHINES

Three types of power losses occur in all electrical machines. They can be expressed as follows, where K = proportionality constant (different in different equations):

1. Ohmic loss in conductor = I^2R or $K{\cdot}P^2$ if delivering power P at a constant voltage.
2. Magnetic loss in iron core = $K\varphi_m^2$ or KV^2 since V and the magnetic flux φ_m are linearly related in the unsaturated region of the machine operation as per Equation 2.2.
3. Mechanical loss in bearing friction and aerodynamic windage depending on the speed of rotation.

Regarding the magnetic loss, the alternating magnetic flux produces two types of power loss in the ferromagnetic iron core: (1) hysteresis loss $P_{hyst} = K_h f B_m^n$ due to hysteresis (magnetic friction) and (2) eddy current loss $P_{eddy} = K_e f^2 B_m^2 t^2$ due to internally circulating (eddy) currents in the core body. Here, K_e and K_h = proportionality constants, f = frequency, B_m = maximum (peak) flux density, t = thickness of the core laminations perpendicular to the flux flow, and n = Steinmetz constant (typically 1.6 to 1.8 for magnetic steel used in electrical machines).

Since P_{eddy} depends on the core lamination thickness squared, the magnetic core is made of thin sheets (laminations) to minimize the eddy current loss. Widely used laminations are 9, 11, 13, 19, and 23 mils thick (1 mil = 1/1000 in. = 25.4 μm). In a given machine, using 9-mil-thick laminations would result in about one-fourth the eddy loss compared with that with 19-mil-thick laminations. The hysteresis loss, however, does not change with lamination thickness.

The total power loss in the magnetic core is then $P_{core} = P_{hyst} + P_{eddy}$, which leads to

$$P_{core} = K_h f B_m^n + K_e f^2 B_m^2 t^2 \text{ watts.} \tag{2.7}$$

Vendors of the magnetic core laminations provide the curves of total power loss in watts per kilogram of core weight at various frequencies, flux densities, and lamination thickness for the grades of core steel they supply. The electrical machine design engineer uses such vendor-supplied curves.

2.5 MAXIMUM EFFICIENCY OPERATING POINT

The losses in any power equipment can also be classified broadly in three categories:

- Voltage-related loss, which varies approximately with the flux density squared or applied voltage squared. Since most power equipment operates at a constant voltage and constant frequency, this loss is a *fixed loss* at all load currents (from zero to full load).
- Current-related loss in diode-type elements, varying with the load current or power output.
- Current^2-related loss in the conductor resistance, varying with the square of the load current or square of the power output in a constant-voltage equipment.

At a constant-voltage operation, therefore, the total loss can be expressed as a function of the load power output P_o as follows:

$$\text{total loss} = \text{fixed loss } K_o + \text{linearly varying loss } K_1 \cdot P_o + \text{square variable loss } K_2 P_o^2.$$

Efficiency η is the ratio of output to input powers, that is,

$$\eta = \frac{\text{power output}}{\text{power input}} = \frac{P_o}{P_o + \text{total loss}} = \frac{P_o}{P_o + K_0 + K_1 P_o + K_2 P_o^2}. \tag{2.8}$$

The efficiency varies with load power as per Equation 2.8, which is depicted in Figure 2.4. Due to the fixed loss, the efficiency is zero at zero load and is low at light loads. The efficiency peaks at some load point and then decreases due to high square-variable loss at high loads. We, therefore, deduce that there has to be a load point where the efficiency has the maximum value. To find that point, we find the derivative of η with respect to P_o and equate to zero, that is, $d\eta/dP_o = 0$. Solving the resulting equation leads to $K_o = K_2 P_o^2$. That is, the equipment efficiency is maximum when the fixed loss equals the square variable loss.

It is noteworthy that the linearly varying loss in diode-type devices does not matter in determining the maximum efficiency point. It just degrades the efficiency in a fixed ratio, that is, $\eta = P_o/(P_o + K_1 \cdot P_o) = 1/(1 + K_1)$, which does not vary with load.

For equipment like transformers, motors, and generators, the fixed loss comes from the magnetic loss in the core operating at constant voltage (constant flux amplitude), and the square-variable loss comes from the I^2R loss in winding conductor resistance. Thus, in electrical machines, an important conclusion is stated below.

At maximum efficiency point,

$$\text{fixed core loss} = \text{square-variable conductor loss.} \tag{2.9}$$

For saving energy, the equipment should be operated at the load P_{op} in Figure 2.4 at which the efficiency is maximum. This load point is usually at 75%–80% of the full rated load. Large power users often specify the maximum efficiency point at the power level at which the equipment is likely to operate most of the time, and specify the rated load equal to the maximum continuous load capability needed during the equipment service life.

Example 2.3

An electrical machine has a no-load loss of 10 kW and a full-load loss of 26 kW when delivering 500 kW. Determine its (1) full-load efficiency and (2) maximum efficiency with the corresponding load point in percentage of the full load and in kilowatts.

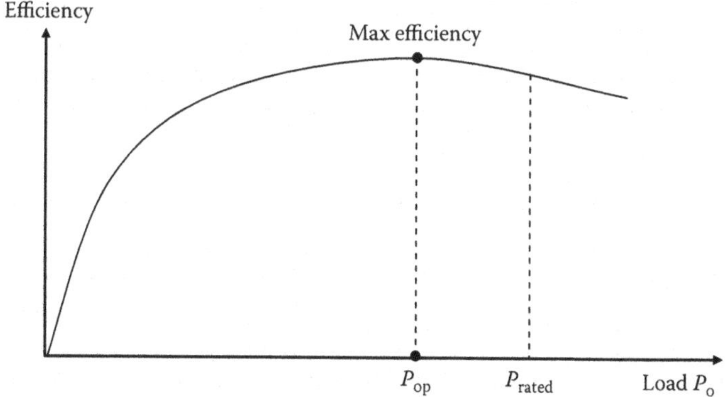

FIGURE 2.4 Efficiency versus load power output and maximum efficiency point.

Solution:

Full-load efficiency = 500 ÷ (500 + 26) = 0.9506.

The no-load power loss of 10 kW remains constant as long as the machine remains connected to the lines, regardless of its level of loading. Therefore, the I^2R power loss in the conductor at 100% load = 26 − 10 = 16 kW, which varies with load squared. As per Equation 2.9, the efficiency is maximum when conductor loss is equal to fixed loss, that is, when 10 = 16 × load^2. This gives load = (10 ÷ 16)$^{1/2}$ = 0.79 or 79% of the full load, or 500 × 0.79 = 395 kW.

Maximum efficiency at 79% load = 395 ÷ (395 + 10 + 10) = 0.9518.

This is better than 0.9506 at full load. Although this is a small difference, it can add up to a good energy saving over the year if the machine is large and is working long hours every day of the year.

2.6 THEVENIN EQUIVALENT SOURCE MODEL

Any electrical power source (generator, transformer, battery, fuel cell, etc.) or, for that matter, any complex electrical network can be modeled by an internal source voltage V_s (also called Thevenin voltage V_{Th}) in series with an internal source impedance Z_s (also called Thevenin impedance Z_{Th}) as shown in Figure 2.5. If no load were connected, the terminal voltage would be the same as the source voltage. However, when the source is loaded as depicted in Figure 2.6a, the terminal voltage at the load point is given by the source voltage less the voltage drop in Z_s under the load current I, that is,

$$\tilde{V}_T = \tilde{V}_S - \tilde{I} \times \tilde{Z}_s. \tag{2.10}$$

The terminal voltage droop line with increasing load current is shown in Figure 2.6b.

The Thevenin equivalent source parameters $\tilde{V}_s$ and $\tilde{Z}_s$ can be found by calculations or by test. Referencing to Figure 2.6 again, if we measure the voltage at no load and voltage drop $\tilde{I} \times Z_s$ at some load current, then it is easy to see that

$$\tilde{V}_s = \tilde{V}_T \text{ at no load, and } \tilde{Z}_s = \tilde{V}_{drop} \div \tilde{I} = \left(\tilde{V}_{T\text{ at no-load}} - \tilde{V}_{T\text{ at load I}}\right) \div \tilde{I}. \tag{2.11}$$

Since Equation 2.11 is a phasor relation in ac, mere voltmeter and ammeter readings are not enough to determine $\tilde{V}_s$ and $\tilde{Z}_s$. Some means of accounting for the phase angles between the voltage and current (e.g., oscilloscope) may be needed to determine $\tilde{V}_s$ and $\tilde{Z}_s$. In dc circuits, however, the voltmeter and ammeter alone can establish the Thevenin source parameters V_s and R_s.

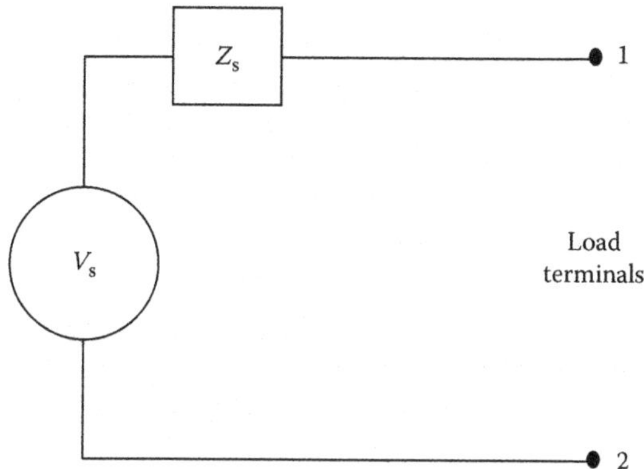

FIGURE 2.5 Thevenin equivalent source model of any electrical power source.

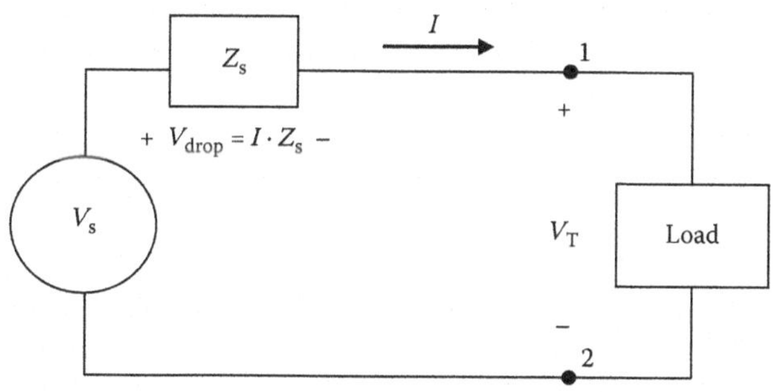

(a) Thevenin equivalent source with load current

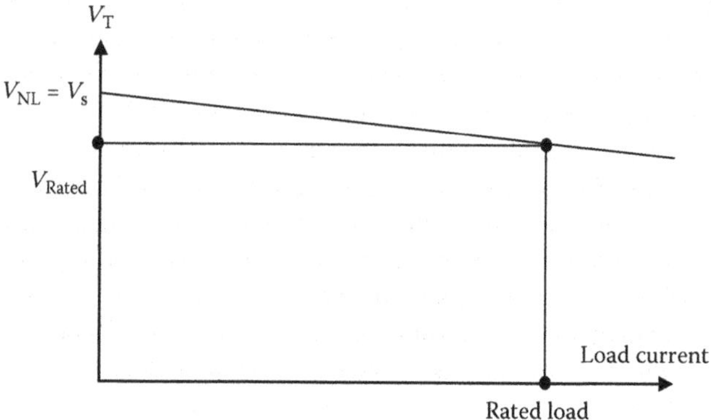

(b) Terminal voltage droops with load

FIGURE 2.6 Drooping voltage with increasing load current at output terminals.

Example 2.4

If a dc source with open circuit voltage of 240 V drops to 220 V under a 25-A load, determine its Thevenin parameters, that is, the source voltage V_s and the source resistance R_s.

Solution:

The open circuit voltage is the Thevenin source voltage, that is, $V_s = 240$ V.
Since the voltage drop of 240 – 220 = 20 V at 25-A load current must be in the internal source resistance, the Thevenin source resistance $R_s = 20\text{ V} \div 25\text{ A} = 0.8\ \Omega$.

2.7 VOLTAGE DROP AND REGULATION

The rated voltage of an electrical power source is defined as the terminal voltage at full rated load current. The output terminal voltage rises with decreasing load current, rising to its maximum value at zero load (no load) as indicated by Equation 2.10 and depicted in Figure 2.6b. The % voltage regulation of a power source is defined as

$$\%\text{ voltage regulation} = \frac{V_{\text{no load}} - V_{\text{rated load}}}{V_{\text{rated load}}} \times 100 \tag{2.12}$$

Voltage regulation is a common term used in describing the voltage change that a load connected to the transformer or generator terminals, or at the end of a long cable, would see. It indicates the voltage variation in the terminal voltage of the source from fully loaded to fully unloaded conditions. All loads should be designed to withstand this voltage variation. Alternatively, if a constant voltage is desired at the load terminals, the source voltage must be regulated by automatic feedback control in the percentage range equal to the % voltage regulation. Therefore, the voltage regulation defines the load voltage deviation without the voltage regulator or the voltage range over which the regulator must control the source voltage.

For the most generic analysis that is repeatedly used in many chapters in this book, we can represent the total series impedance $Z = R + jX$ ohms between the source and the load as shown in Figure 2.7a. The total Z can be due to one or more impedances of the generator, transformer, and cable. The load draws current I at lagging pf cos θ at the load terminal voltage V_{load}, which must be equal to source voltage V_{source} minus the voltage drop $I \times Z$ in the series impedance. Therefore, the source voltage, which is maintained constant, must be equal to

$$\tilde{V}_{\text{source}} = \tilde{V}_{\text{load}} + \tilde{I} \times (R + jX). \tag{2.13}$$

The phasor diagram of Equation 2.13 is shown in Figure 2.7b, the trigonometry of which gives the exact magnitude of the source voltage for any pf = cos θ:

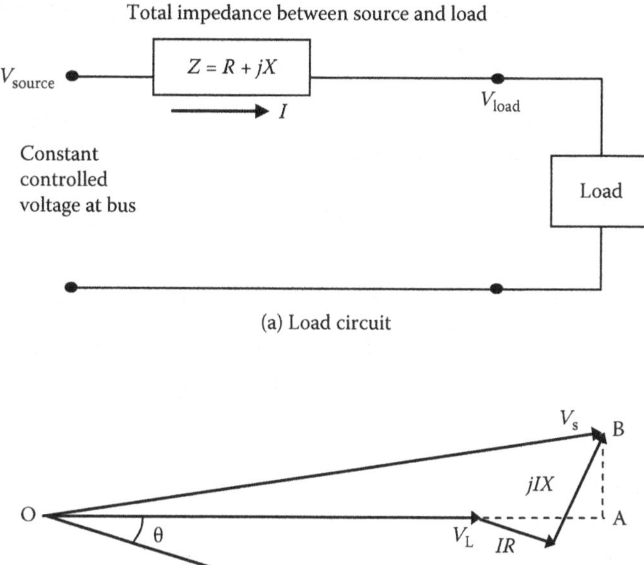

FIGURE 2.7 Voltage drop in series impedance and voltage regulation.

$$V_s^2 = OA^2 + AB^2 = \left(V_L + I \times R\cos\theta + I \times X\sin\theta\right)^2 + (I \times X\cos\theta - I \times R\sin\theta)^2. \quad (2.14)$$

For typical pf around 0.85, we can approximately write the source voltage V_s = OB = OA and, subsequently, the voltage drop magnitude:

$$V_{drop} = V_s - V_L = OA - V_L = I \times R \cos\theta + I \times X \sin\theta = I \times (R \cos\theta + X \sin\theta). \quad (2.15)$$

We can take the following as the effective impedance between the source and the load:

$$Z_{eff} = R \cos\theta + X \sin\theta \quad (2.16)$$

which, when multiplied with current *I*, gives the voltage drop by simply applying Ohm's law:

$$V_{drop} = I \times Z_{eff}. \quad (2.17)$$

All calculations are done using volts, amperes, and ohms per phase and taking θ positive for lagging pf and negative for leading pf. Equation 2.16 is easy to remember: *R* (real part of *Z*) × cos θ that gives real power + *X* (reactive part of *Z*) × sin θ that gives reactive power. Also, of course, the voltage drop is linearly proportional to the current and series impedances, as in Ohm's law. Therefore, in practical power circuits, we can explicitly write Equation 2.18 below in terms of the pf:

$$V_{drop} = I\left\{R \cdot pf + X \cdot \sqrt{1 - pf^2}\right\} \text{for all practical power factors.} \tag{2.18}$$

$$\text{Then, \% voltage regulation} = (V_{drop}/V_L) \times 100\%. \tag{2.19}$$

For $R \ll X$, Equation 2.18 is valid for a wide range of pf ranging from 0 lagging to 0.9 leading, except that the algebraic sign before the reactance X is + for lagging pf and – for leading pf. Although this approximation is fairly accurate for most practical use, we must remember that the exact voltage drop is given by V_s from Equation 2.14 minus V_L.

Example 2.5

A three-phase, 480-V, 5000-kVA generator connects to many feeders, one of which takes single-phase power to two equipment at the other end (see figure below) via a cable and transformer with combined $Z = 0.03 + j0.07\ \Omega$. The voltage at point A at the receiving end rises and falls as a large single-phase, 300-kVA load varies over time. The small 1-kVA load remains online continuously and must accommodate the voltage variations at point A with a 300-kVA load fully on or fully off. Determine the voltage range that the 1-kVA load will see, over which it must perform as intended.

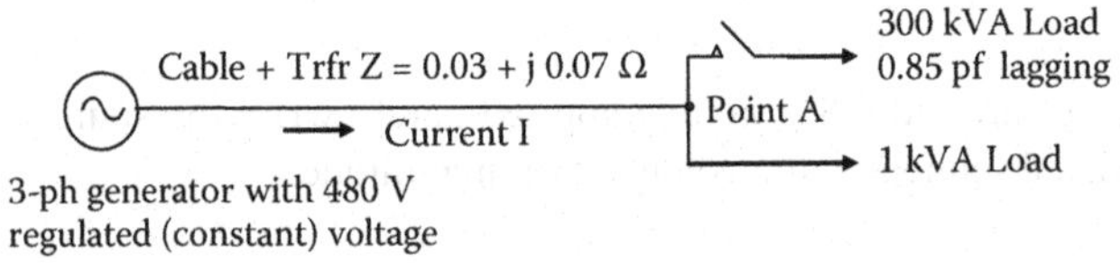

Solution:

The 300-kVA load current I = 300 kVA × 1000 ÷ (√3 × 480) = 361 A/ph.

Assuming 0.85 pf lagging (i.e., cos θ = 0.85 and sin θ = 0.5268), Equation 2.15 gives the voltage drop in the cable:

V_{drop} = 361 (0.03 × 0.85 + 0.07 × 0.5268) = 22.5 V/ph = 22.5 √3 = 39 V_{LL}.

Therefore, voltage regulation = 39 ÷ 480 = 0.081 or 8.1% of 480 V (a rather high percentage).

Therefore, voltage at point A when 300-kVA load is fully on = 480 – 39 = 441 V_{LL}.

Voltage at point A when 300-kVA load is fully off = 480 V_{LL} (same as the generator voltage since there is negligible current and hence negligible voltage drop in the cable).

In order to supply all loads, small and large, at a reasonable voltage range, the voltage variation at point A must be kept below 5% from full load to no load by proper design of the system.

2.8 LOAD SHARING AMONG SOURCES

Two or more electrical power sources are often connected in parallel for increasing the load capability or reliability of the system. In the case of two sources, both sources must meet the following conditions before they can be connected in parallel to share the load:

- Voltage magnitudes must be equal within a few percent.
- Voltage polarities must be the same (+ connected to + and – connected to –).
- Phase sequence of three-phase sources must be the same.
- Frequencies of ac sources must be equal (else, voltages will get out of phase).

2.8.1 Static Sources in Parallel

The load shared by two static sources—such as two transformers, batteries, fuel cells, etc.—is determined by their terminal voltage droop lines shown in Figure 2.8. With two static sources 1 and 2 sharing load in parallel, their individual terminal voltages are given by

$$V_{T1} = V_{s1} - K_1 \cdot P_1 \text{ and } V_{T2} = V_{s2} - K_2 \cdot P_2 \tag{2.20}$$

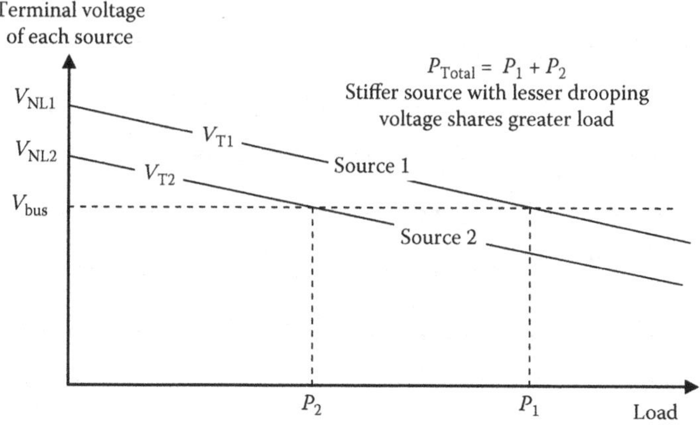

FIGURE 2.8 Voltage droop lines for two static power sources sharing load in parallel.

where K_1 and K_2 = voltage droop rates = $\Delta V/\Delta P$ in volts per kilowatt, and V_{s1} and V_{s2} are the internal voltages of sources 1 and 2, respectively, that will be equal to their voltages at zero load.

By parallel connection at a common bus, we force both sources to share the load such that they have the same terminal voltage equal to the bus voltage, that is, $V_{T1} = V_{T2} = V_{bus}$.

$$\text{Therefore, } V_{s1} - K_1 \cdot P_1 = V_{s2} - K_2 \cdot P_2 = V_{bus}. \tag{2.21}$$

We must also have the total load,

$$P_{Total} = P_1 + P_2. \tag{2.22}$$

The parameters V_{s1}, K_1, V_{s2}, and K_2 are the source parameters given or known to us from the manufacturer's technical data sheets. Therefore, solving Equations 2.21 and 2.22 simultaneously gives P_1 and P_2 for a given P_{Total}.

Graphically, using the droop lines of sources 1 and 2 shown in Figure 2.8, the load shared by two individual sources can be determined by trial and error. In that process, we find P_1 and P_2 at different values of V_{bus} until their sum $P_1 + P_2$ adds up to P_{Total}. As an example, for V_{bus} shown by the dotted line in Figure 2.8, the load shared by Source 1 is P_1 and that by Source 2 is P_2. If P_1 and P_2 add up to P_{Total}, the dotted line is indeed the V_{bus} value. Otherwise, we try another value of V_{bus} until P_1 and P_2 add up to P_{Total}. When the fun of trial and error fades, we can use the graphical method shown in Figure 2.9 to determine the load sharing in one step. In this method, the two voltage droop lines with vertical axes V_{s1} and V_{s2} are drawn apart by P_{Total} on the horizontal load axis, and the droop lines are drawn in the opposite directions. Then, the point of their intersection gives the individual load P_1 and P_2 and also the bus voltage V_{bus} in one step.

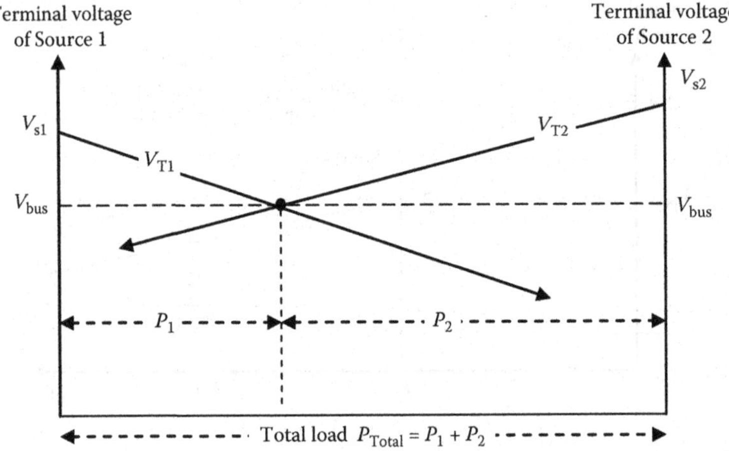

FIGURE 2.9 One-step method of determining load shared by two static sources in parallel.

Example 2.6

Battery-1 and Battery-2 have their terminal voltage versus load current droop lines shown in the following figure. Determine analytically the load current shared by each battery and the battery bus voltage if they share a total load current of (1) 900 A and (2) 600 A. In case of a 600-A total load, verify your answers by one-step method with Battery-2 line redrawn as needed.

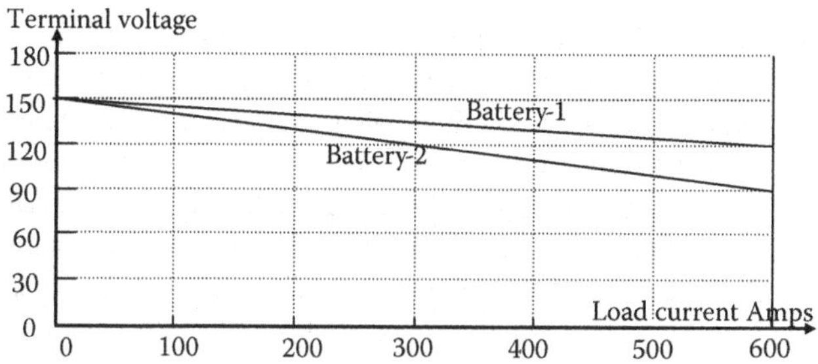

Solution:

We first determine the droop line equations from the above figure. Battery-1 voltage drops from 150 V at zero load to 120 V at 600 A, which gives

$$V_{T1} = 150 - \frac{150-120}{600} I_1 = 150 - 0.05 I_1.$$

Similarly, we have

$$V_{T2} = 150 - \frac{150-90}{600} I_2 = 150 - 0.10 I_2.$$

1. For a 900-A total load at a common bus voltage V_{bus}, we write

$$V_{T1} = 150 - 0.05\, I_1 = V_{T2} = 150 - 0.10\, I_2 = V_{bus} \quad \text{(a)}$$

and the total load current $I_1 + I_2 = 900$ A. (b)

The simultaneous solution of (a) and (b) gives

$I_1 = 600$ A, $I_2 = 300$ A, and $V_{bus} = 120$ V.

2. For a 600-A total current, we repeat the above procedure, which leads to

$I_1 = 400$ A, $I_2 = 200$ A, and $V_{bus} = 130$ V.

As the load current decreases from 900 to 600 A, we note that the battery bus voltage rises from 120 to 130 V, as expected.

These answers can also be derived graphically in one step by plotting the Battery-2 line backward starting from the right-hand side and drooping as we increase its load current from 0 to 600 A toward the left-hand side. This is left as an exercise.

2.8.2 Load Adjustment

The load sharing can be adjusted by raising or lowering the voltage droop line of one or both sources working in parallel. This is generally done by changing the no-load voltage, which, in turn, raises or lowers the terminal voltage versus load line while maintaining the droop rate. In transformers, this is easily done by changing the taps on the primary or secondary side. In batteries and fuel cells, it is done by adding or subtracting one or more cells in one stack. For two sources in parallel to share equal load, it is important to have their no-load voltages and the droop rates equal or as close as possible.

For two dc generators in parallel, the above analysis for static sources also applies. For two motors in parallel sharing a torque load on common shaft, the analysis presented here applies equally well after substituting the voltages in Equation 2.21 with individual motor speeds versus horsepower load. The horsepower sharing between two locomotive engines hauling a long train can also be determined by using their respective speed versus horsepower droop lines.

The load sharing among ac generators is complex, as both the real and reactive powers are shared. It is covered in Chapter 3.

2.9 POWER RATING OF EQUIPMENT

The following apply for all equipment (electrical or mechanical) commonly found in electrical power plants.

The electrical power ratings of generators and transformers are stated in kilovolt-amperes or megavolt-amperes (not in the real power kilowatts or megawatts). These machines can deliver rated voltage and rated current, but the real power delivered by the machine depends on the power factor, which is primarily determined by the nature of the load connected. We often hear the electrical power plant capacity in megawatts since it refers to the mechanical output rating of the prime mover, which is the dominant investment in the power plant. Moreover, the megawatt rating also determines the plant's fuel consumption rate, such as the number of coal cars coming daily in the thermal power plant. That is why the megawatt rating draws more attention than the megavolt-ampere rating.

- Nameplate rating refers to the *full load output,* and input = output + internal losses.
- Efficiency = output power/input power, which varies with the output power.
- Cable rating is stated in terms of the voltage and the current carrying capacity (ampacity).

- The mechanical power rating of motor or engine is stated in horsepower in British units or kilowatts in SI units.
- Equipment rating is limited by the operating temperature limit on insulation in electrical equipment and the cooling method used in the design.

The service factor (SF), indicated on the equipment nameplate, is the equipment's ability to carry temporary overload without thermal or mechanical failure or measurable reduction in life. Industry standards require all electrical equipment to have the SF of at least 1.15 typically for 2 h. It means that the equipment can carry 15% overload for 2 h. However, the SF is often misinterpreted as the factor by which we can continuously overload the motor by 15%. If it continuously overloads by 15%, the equipment may run hotter than the design temperature by more than 10°C, reducing the service life to less than one-half, which is a significant penalty to pay.

2.9.1 Temperature Rise under Load

The power rating of the electrical equipment is limited by the temperature rise above the ambient air that cools the equipment. Industry standard ambient air temperature is 40°C. It can be higher, up to 65°C, in special places like the boiler room uptake. The maximum permissible operating temperature depends on the class of insulation used in the equipment. For example, the temperature rise of 80°C is permitted in equipment with class B insulation. In 40°C standard ambient air, it would have the operating temperature of 80°C + 40°C = 120°C. For some insulations, such as Nomex, the allowable operating temperature can be up to 200°C.

The temperature rise is determined by the cooling medium and the surface area available for dissipating the internal power loss that heats the equipment body. Air-cooled equipment dissipates the internal heat mostly by convection and radiation, and negligibly by conduction. For electrical equipment normally operating around 100°C, the temperature rise above ambient air is approximately given by

$$\Delta T_{^\circ C} = K \times (\text{watts})^{0.8} \tag{2.23}$$

where watts = power loss to be dissipated, and K = constant for a given body.

2.9.2 Service Life under Overload

During an overload, the power loss in the conductor rises with the load squared, and the temperature also rises according to Equation 2.23. Overloading above the rated temperature limit reduces the equipment life to about one-half for every 10°C rise above the design temperature. Mathematically,

$$\text{actual life} = \frac{\text{rated life}}{2^{\left(\Delta T/10\right)}} \tag{2.24}$$

where $\Delta T = \Delta T_a - \Delta T_R$ = temperature rise above the rated design temperature rise; ΔT_a = actual temperature rise above ambient air at the operating load; and ΔT_R = design temperature rise above ambient air at the rated load.

This 10°C rule for half-life is based on typical properties of insulations normally used in electrical machines. Experience on other electrical components, such as power electronics devices, suggests a different rule, such as every 7°C rise for reducing the life to one-half.

Example 2.7

A cable is designed for 30-year life with 50°C rise in 40°C ambient air at rated load. If it is continuously overloaded by 15%, determine its new expected life.

Solution:

The power loss in a cable is due only to resistance; hence, it varies as the load current squared. Denoting the rated and overload conditions by suffixes 1 and 2, respectively, the power loss at 15% overload is given by $W_2 = 1.15^2 \times W_1 = 1.3225\ W_1$. With the same heat dissipation area, Equation 2.23 in the ratio gives

$$\frac{\Delta T_2}{\Delta T_1} = \left(\frac{W_2}{W_1}\right)^{0.8} = 1.3225^{0.8} = 1.25\,.$$

Since $\Delta T_1 = 50°C$, we have $\Delta T_2 = 1.25 \times 50 = 62.5°C$ and $\Delta T_2 - \Delta T_1 = 62.5 - 50 = 12.5°C$.

The cable, therefore, operates 12.5°C hotter than the design temperature. If this continues until the end, its life gets reduced by the 10°C rule in Equation 2.24, that is,

$$\text{life with continous overload} = \frac{30}{2^{\left(\frac{12.5}{10}\right)}} = 12.61\ \text{years}.$$

This is a significant reduction of life, which illustrates that continuous overload, even by 10%–15%, is not economical. That is why 15% overload on electrical equipment is limited for an hour or two and only on occasion.

2.10 TEMPERATURE EFFECT ON RESISTANCE

The conductor resistance (usually in ohms) rises with the operating temperature. The conductor resistance R_2 at new temperature T_2 can be derived from known resistance R_1 at T_1 as follows:

$$\text{for conductor, } R_2 = R_1 \times \{1 + \alpha\,(T_2 - T_1)\} \tag{2.25}$$

where α = temperature coefficient of conductor resistance = 0.0039/°C for both copper and aluminum of electrical grade.

The insulation resistance (usually in megaohms) drops at higher temperature. For the equipment to function at rated voltage, the insulation must be sound with certain minimum resistance value measured by megger, applied between the coil conductor and ground. The industry standards require the insulation resistance, adjusted to 40°C if measured at a different temperature, to be greater than the minimum recommended value R_{min}. The IEEE Standard 43-2000 suggests the following R_{min} values at 40°C for different equipment, where kV_{LL} = line-to-line rated voltage of the equipment in kilovolts:

$$R_{min} = kV_{LL} + 1 \text{ M}\Omega \text{ for rotating machines built before 1970}$$
$$R_{min} = 5 \text{ M}\Omega \text{ for new rotating machines rated below 1 } kV_{LL}$$
$$R_{min} = 100 \text{ M}\Omega \text{ for new rotating machines rated above 1 } kV_{LL}$$
$$R_{min} = 2\ kV_{LL} \text{ M}\Omega \text{ for all transformer coils.}$$

If the insulation resistance were measured at temperature T, then its value adjusted at 40°C must be compared with the R_{min} value that is required at the standard ambient temperature of 40°C. The adjusted value is given by

$$R_{40^\circ C} = K_T \times R_T \tag{2.26}$$

where R_T = insulation resistance measured at T°C, and K_T = temperature correction factor for insulation resistance. The industry data suggest that factor K_T doubles for every 10°C rise in insulation temperature above the standard ambient temperature of 40°C, that is,

$$K_T = 2^{\left(\frac{T-40}{10}\right)} \text{ where } T = \text{ actual temperature of insulation in degree Celsius.} \tag{2.27}$$

Example 2.8

The stator winding insulation resistance of a 100-hp, 460-V motor was measured to be 10 MΩ when sitting idle in normal room temperature of 20°C. Determine whether this motor is worthy of connecting to the line as per the IEEE Standard.

Solution:

First, for this motor to be worthy of connecting to the line voltage, the winding insulation must have R_{min} = 5 MΩ at 40°C. The measured value of 10 MΩ at 20°C must be first adjusted to 40°C before comparing with R_{min} and making judgment about its service worthiness. Using Equation 2.27, the temperature correction factor

$$K_T = 2^{\frac{20-40}{10}} = 2^{-2} = 0.25, \text{ which gives } R_{40^\circ C} = 0.25 \times 10 = 2.5\ \Omega.$$

Since this is less than the 5-MΩ minimum required in service, the motor winding insulation does not meet the industry standard. The stator winding must be rewound and varnish-impregnated before connecting to the lines.

PROBLEMS

1. Determine the rms value of voltage induced in an 80-turn coil on a 30 × 25 cm cross-section core with peak flux density 1.7 T alternating at 60 Hz.
2. Two parallel single-phase rectangular bus bars 2 cm apart at center lines experience 30,000 A_{rms} current when a load gets short-circuit fault. Determine the peak mechanical force per meter length of the bus bars. Assume the bus bar shape factor $K = 0.85$.
3. A transformer has no-load loss of 30 kW and full-load loss of 90 kW when delivering 3-MW power. Determine its (1) full load efficiency and (2) maximum efficiency with the corresponding loading level in percentage of the full load and in kilowatts.
4. A dc source with open circuit voltage of 120 V drops to 105 V under a 15-A load. Determine its Thevenin parameters, that is, the source voltage V_s and the source resistance R_s.
5. From a three-phase, 480-V generator switchboard, a feeder takes three-phase power to two equipments at the other end—similar to the figure in Example 2.5—via cable and transformer with the combined series impedance $Z = 0.04 + j0.09$ Ω/phase. The voltage at the receiving end point A rises and falls as a large single-phase, 500-kVA load at point A varies over time. The small 1-kVA load, also connected at point A, remains online continuously and must accommodate the voltage variations at point A with 500-kVA load fully on or fully off. Determine the voltage range that the 1-kVA load will see, over which it must perform within its specifications.
6. Two batteries' terminal voltage versus load current droop lines are shown in the figure below. Determine analytically the load current shared by each battery and the battery bus voltage if they share a total load current of (1) 500 A and (2) 800 A. In case of a 500-A total load, verify your answers by one-step method with Battery-2 line redrawn backward as needed.

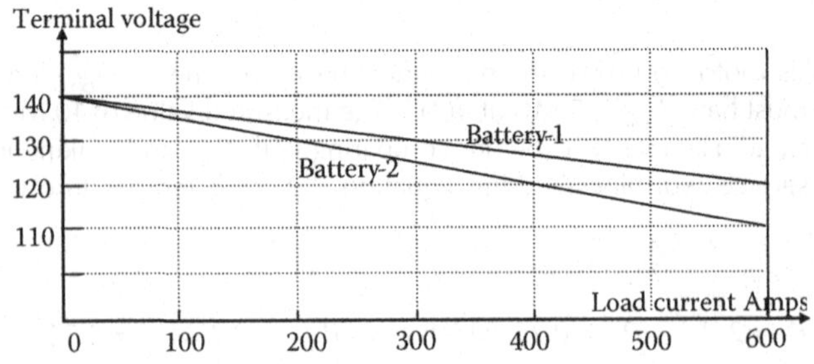

7. An electrical equipment is designed for 25-year life with 70°C rise in 40°C ambient air at a rated load. If it is continuously overloaded by 10%, determine its new expected life.
8. A 500-hp, 460-V motor winding insulation resistance was measured to be 15 MΩ when sitting idle in normal room temperature of 25°C. Determine whether this motor is worthy of continuing service as per the IEEE Standard.
9. A 75-kVA, single-phase, 60-Hz transformer primary coil with 100 turns is connected to a 265-V source. Determine the peak flux in the core. If the magnetomotive force of 500 ampere-turns is needed to establish this flux, determine the core excitation current.
10. A feeder transformer delivers full load current at its output voltage of 460 V. When the load is removed, the output voltage rises to 480 V. Determine the voltage regulation of this transformer.
11. A generator winding delivering rated load of 1000 kVA rises to 50°C above 40°C ambient air, making its operating temperature 40 + 50 = 90°C. If the generator is overloaded by 30%, determine its operating temperature. If always operated at 30% overload, determine the generator life if the rated design life is 25 years.
12. A 4160-V generator phase coil insulation resistance is 20 MΩ measured at 70°C soon after it was tripped. Compare this with the minimum required by the industry standard, and state whether this generator is good to continue in operation.
13. A single-phase, 500-kVA, 440/120-V transformer feeds service loads via bus duct with two bus bars, each 1/2 × 3 in. in cross section and spaced ½ in. apart. An electrical system study has concluded that the worst-case fault current in the bus is 30,000 A at the first peak. Determine the peak mechanical force between the bus bars that will cause defection and bending stress between the supports. Assume the bus bar shape factor $K = 0.7$.
14. A power electronics component designed to last 25 years under rated load is always operated 5°C above the design temperature limit. Determine its expected life if 8°C rule for one-half life is found to be applicable to electronics components.
15. A 1000-kVA, single-phase transformer operating at 85% power factor lagging at a full rated load has conductor power loss of 10 kW and core loss of 5 kW. Determine its maximum possible efficiency and the corresponding load level as percentage of the rated load.

QUESTIONS

1. Identify at least two electrical products (other than the transformer, motor, generator, and those identified in Section 2.1) that work on Faraday's law of electromagnetic induction.
2. Using Faraday's law, explain why we lose radio and cell phone communications while going through a long tunnel made of steel-reinforced concrete structure.

3. Using Faraday's law, explain the purpose of the front door screen in the microwave oven.
4. How would the magnetic loss in a 60-Hz motor built in the United States change when operated in Europe at 50 Hz, and vice versa?
5. State in one sentence when the electrical equipment operates at the maximum efficiency.
6. How would you determine the Thevenin equivalent source parameters V_s and Z_s of the entire power grid bringing power to the wall outlet at your workplace? Answer separately assuming that (1) the source reactance is negligible, that is, $Z_s = R_s$ (a pure resistance), and (2) the source reactance is not negligible, that is, $Z_s = R_s + j X_s$.
7. You have two 120-V batteries working in parallel, one old and one new, each with 60 cells in series. You wish the new battery to share greater load. What would you do?
8. In a long locomotive going uphill with two engines pulling, how would you determine their horsepower load-sharing to meet the total traction load?
9. At what rate does the electrical equipment's service life degrade at higher operating temperature? Identify specific reasons of the life degradation.

FURTHER READING

Bosela, T.R. 1997. *Introduction to Electrical Power System Technology*. Upper Saddle River: Prentice Hall.

Hubert, C.I. 2002. *Electrical Machines*. Upper Saddle River: Prentice Hall.

Wildi, T. 2002. *Electrical Machines, Drives, and Power Systems*. Upper Saddle River: Prentice Hall.

3 AC Generator

Nearly all electrical power in the world is generated by the three-phase synchronous generator, which is also known as the ac generator or the alternator. It consists of three stationary coils (called the stator, armature, or phase coils), which are physically separated in space by 120° from each other, and a rotor with dc coil that produces the dc magnetic field. Both the stator and rotor coils are individually embedded in ferromagnetic cores with an air gap that is consistent with the electrical and mechanical design requirements. Figure 3.1 is a simplified cross section of a three-phase generator with three stator coils and salient poles on the rotor.

A thermodynamic prime mover drives the rotor, which generates voltage in each phase of the three identical stator coils. The three phase voltages are equal in magnitude but 120° out of phase in time (or in ωt, to be precise). The stator coils are usually connected in three-phase *Y*. That way, the conductors in the stator slots need to have insulation to the ground for only $1/\sqrt{3} = 0.577$ or 57.7% of the line voltage; hence, more conductor can be packed in the slots. In the conventional generator, the dc excitation field current comes from a small separate exciter via slip rings and carbon brushes.

One mechanical revolution of a two-pole rotor generates one electrical cycle (360 electrical degrees) in the stator coil voltage. In a four-pole rotor, one mechanical revolution generates 4/2 = 2 electrical cycles (2 × 360 electrical degrees). A rotor with *P* number of poles driven at *n* rpm generates *P*/2 × (*n*/60) electrical cycles per second. Therefore, the generator frequency is given by

$$f = \frac{P}{2} \cdot \frac{n}{60} = \frac{n \cdot p}{120} \text{ cycles/second (hertz).} \tag{3.1}$$

The prime mover must drive the generator at a constant speed to generate power at a constant frequency (60 Hz in the United States and 50-Hz in Europe). We also note here for later use that with a *P*-pole rotor,

$$\text{one mechanical degree} = P/2 \text{ electrical degrees.} \tag{3.2}$$

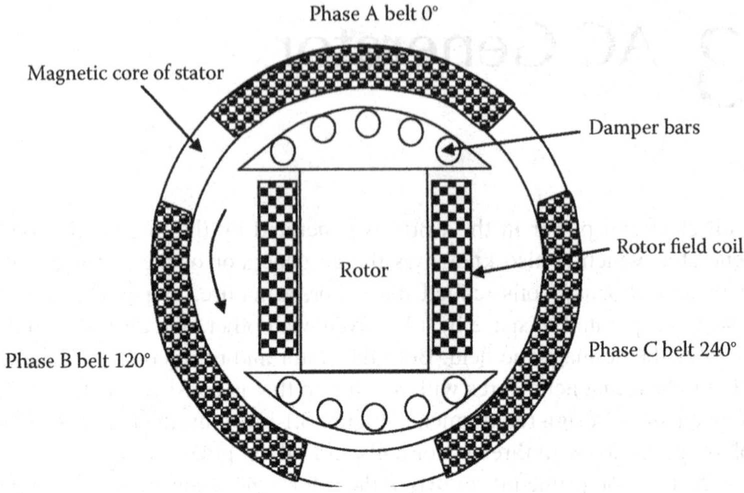

FIGURE 3.1 Three-phase synchronous generator cross section with salient pole.

3.1 TERMINAL PERFORMANCE

The performance of a three-phase generator at the terminals is based on the three-phase power fundamentals covered in Chapter 1. The generator typically supplies power to more than one load, some of which may be connected in Y and some in Δ. In such cases, the Δ-connected load impedance is usually converted into its equivalent Y-connected impedance equal to one-third of the impedance value in Δ and combined in parallel with other Y-connected loads. All calculations are then done on a per-phase basis. This is illustrated in Examples 3.1 and 3.2.

Example 3.1

A three-phase, 460-V, Y-connected generator is powering a 500-kW balance three-phase Δ-connected load at 0.85 pf lagging. Determine (1) the line-to-line voltage and line current, (2) the generator phase voltage and phase current, and (3) the load phase voltage and phase current.

Solution:

As customarily implied, the generator voltage of 460 V is line to line. The line current is then derived from the three-phase power (Equation 1.29), that is, $P = 500 \times 1000 = \sqrt{3} \times 460 \times I_L \times 0.85$, which gives $I_L = 738.3$ A.

Referring to Figure 1.13, Y-connected generator phase voltage = 460 V ÷ $\sqrt{3}$ = 265.6 V, and phase current = line current = 738.3 A.

Referring to Figure 1.14, Δ-connected load phase voltage = line-to-line voltage = 460 V, and phase current = 738.3 ÷ $\sqrt{3}$ = 426.3 A.

Example 3.2

A balanced Y-connected load with $2 + j3\ \Omega$/ph (L-N) is connected in parallel with a balanced Δ-connected load with $9 + j12\ \Omega$/ph (L-L). Determine the combined total equivalent Y-connected impedance value per phase. If these loads were powered by a Y-connected 480-V_{LL} generator, determine the current drawn from the generator lines.

Solution:

We first convert the balanced Δ-connected $9 + j12\ \Omega$ into the equivalent Y-connected load, which is $1/3\ (9 + j12) = 3 + j4\ \Omega$ per phase. This is in parallel with another Y-connected load with $2 + j3\ \Omega$ per phase. We now combine the two parallel Y-connected loads to obtain

$$Z_{\text{Total}} = \frac{1}{\dfrac{1}{3+j4}+\dfrac{1}{2+j3}} = \frac{(2+j3)(3+j4)}{(2+j3)+(3+j4)} = \frac{-6+j17}{5+j7}$$

$$= \frac{18.03\angle(180-70.56)}{8.60\angle 54.46} = 2.1\angle 55 = 1.2 + j1.72\ \Omega/\text{phase}.$$

$$\text{Therefore, generator phase current (also line current)} = \frac{480\big/\sqrt{3}}{2.1\angle 55^\circ} = 132\angle -55^\circ\ \text{A}.$$

3.2 ELECTRICAL MODEL

The equivalent electrical model of the ac generator is shown in Figure 3.2, where

E_f = emf induced inside the armature coil due to rotor field;
R_a = resistance of the armature coil (usually negligible);

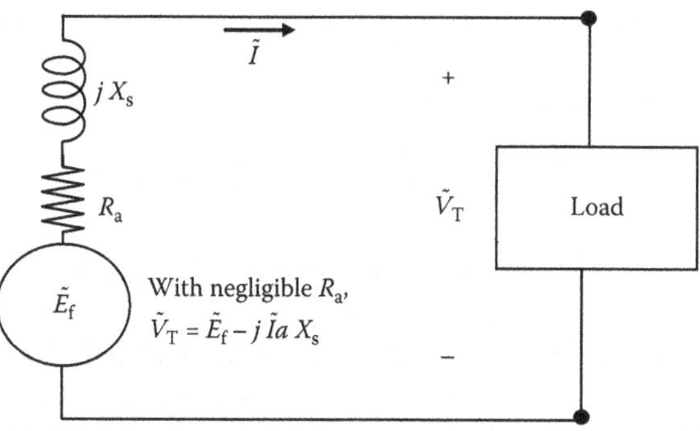

FIGURE 3.2 AC generator equivalent circuit model with load.

X_s = synchronous reactance of the armature coil;
V_T = voltage at the output terminals; and
I = load current (usually lagging the terminal voltage).

If n = rotor speed (in revolutions per minute), ϕ_p = flux per pole, I_f = field current, and K_{m1} and K_{m2} are the machine constants, then in absence of the magnetic saturation,

$$E_f = K_{m1} n \phi_p = K_{m2} n I_f. \tag{3.3}$$

The terminal voltage is equal to E_f induced in the phase coil minus the voltage drop in R_a and X_s. In the phasor sense, it is given by

$$\tilde{V}_T = \tilde{E}_f - \tilde{I}\left(R_a + jX_s\right) = \tilde{E}_f - j\tilde{I}X_s \tag{3.4}$$

The value of R_a, being generally much smaller than X_s, is often ignored to simplify the analysis, as done in the right-hand side of Equation 3.4.

3.3 ELECTRICAL POWER OUTPUT

The performance analysis of a three-phase machine is always conducted on a per-phase basis since three phases are identical except the 120° phase difference. If the generator is supplying a load current I lagging the terminal voltage V_T by phase angle θ, the phasor diagram of Equation 3.4, all with per-phase values, is shown in Figure 3.3, which ignores the armature resistance R_a.

The power generated by each phase $P = V_T\, I \cos\theta$. Using the trigonometry of the phasor diagram, $I\, X_s \cos\theta = E_f \sin\delta$ or $I \cos\theta = E_f \sin\delta / X_s$, which leads to

$$P = V_T\,(E_f \sin\delta / X_s)$$

or

$$P = \frac{E_f V_T}{X_s} \sin\delta = P_{max} \sin\delta \tag{3.5}$$

where

$$P_{max} = \frac{E_f V_T}{X_s} \text{ watts/phase.} \tag{3.6}$$

Since the power output depends on angle δ between V_T and E_f, δ is known as the power angle (measured in electrical degrees).

The three-phase stator currents produce a magnetic field that rotates exactly at the same speed as the rotor field (called the synchronous speed), and δ is the physical angle between the magnetic center lines of the rotor field and the stator field, with the rotor field leading the stator field by angle δ.

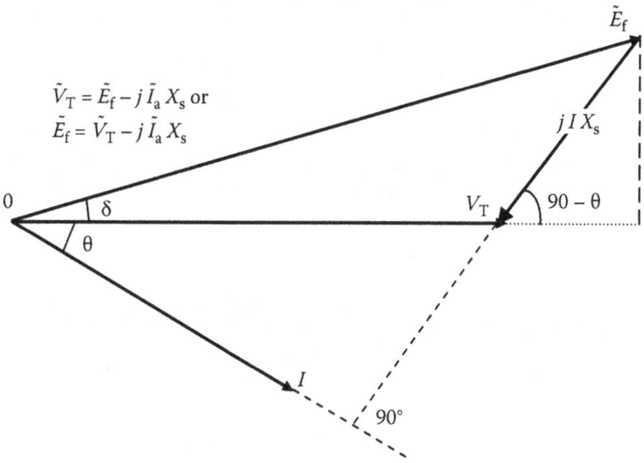

FIGURE 3.3 Phasor diagram of ac generator with negligible armature resistance.

The output power versus power angle relation as per Equation 3.5 is a half-sine curve, shown in Figure 3.4. An increase in output power results in an increase in power angle only up to P_{max} at $\delta = 90°$. Beyond this limit, the rotor and stator fields would no longer follow each other in a magnetic lock step and would step out of the synchronous mode of operation, that is, the machine would become unstable and unable to produce steady power. Therefore, P_{max} is called the steady-state stability limit (or the pull-out power) of the machine. It occurs at $\delta = 90°$ and is equal to

$$P_{max} = \frac{E_f V_T}{X_s}.$$

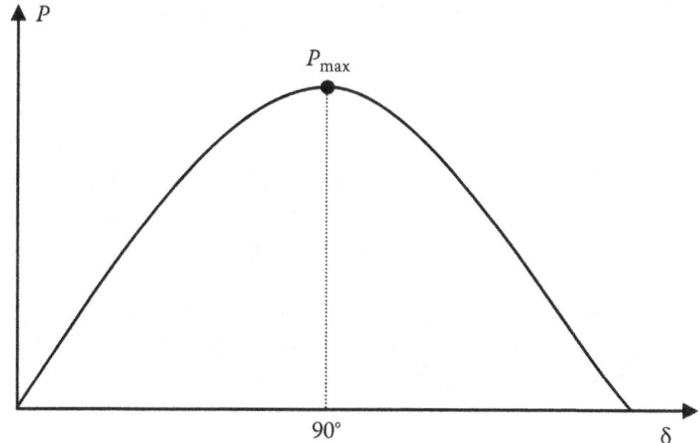

FIGURE 3.4 AC generator power output versus power angle δ.

Example 3.3

A 30-MVA, 11-kV, 60-Hz, three-phase, Y-connected synchronous generator has $X_s = 2\ \Omega$/ph and negligible R_a. Determine its power angle when delivering rated power at unity power factor.

Solution:

On a per-phase basis, $V_T = 11$ kV ÷ √3 = 6.351 kV and 1/3 × 30,000 kVA = 6.351 kV × I_a.

Therefore, armature current $I_a = 1575$ A. From the phasor diagram shown below at unity pf, we get

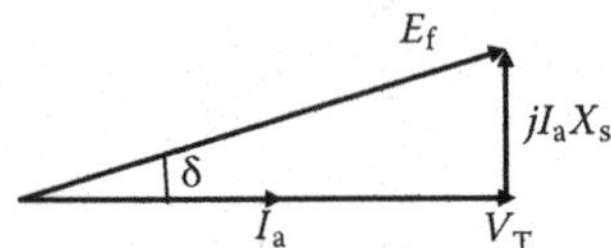

$$E_f = \sqrt{6351^2 + (1575 \times 2)^2} = 7089\ \text{V} = 7.089\ \text{kV and}$$

$$P_{max} = \frac{6.351 \times 7.089}{2} = 22.51\ \text{MW}.$$

The power delivered by the generator $P = 1/3 \times 30$ MVA × 1.0 pf = 10 MW/phase.

The power angle is then derived from Equation 3.5:

$$P_{gen} = 10 = 22.51 \sin\delta, \text{ which gives } \delta = 26.4°.$$

Example 3.4

A three-phase, Y-connected synchronous generator is rated 10 MVA, 11 kV. Its resistance is negligible, and the synchronous reactance is 1.5 Ω per phase. Determine the generator voltage E_g when delivering rated megavolt-amperes at 0.85 power factor lagging.

Solution:

E_g is an alternative symbol used in some books and national standards for the field excitation voltage E_f. With that note, we make per-phase calculations below.

For Y-connection, $V_T = 11{,}000 \div \sqrt{3} = 6351$ V/ph, which we take as the reference phasor.

Armature current at 0.85 power factor lagging, $\tilde{I}_a = 10 \times 10^6 \div (\sqrt{3} \times 11000) =$ 525 A at $\angle -\cos^{-1}0.85 = 31.8°$ lagging, that is, $I_a = 52 \angle -31.8°$ A.

The generated voltage is then derived from

$$\tilde{E}_g = \tilde{V}_T + \tilde{I}_a \times jX_s = 6351\angle 0° + 525\angle -31.8° \times 1.5\angle 90°$$
$$= 6351\angle 0° + 787.5\angle 58.2° = 6766 + j669.3 = 6799\angle 5.65°\ \text{V}.$$

Thus, the power angle δ is 5.65° at the rated load and 0.85 pf lagging. The positive sign indicates that the generated voltage leads the terminal voltage, that is, the rotor magnetic field leads the stator magnetic field by angle δ. In the synchronous motor, the opposite is true. In rotating machines, the flux of the coil, which is the primary source of energy, leads the other coil's flux by a mechanical angle equal to ($\delta_{electrical}$ ÷ number of pole pairs).

The electromechanical torque required from the prime mover to generate electrical power in all three phases is equal to (3 × power per phase) divided by the mechanical angular speed of rotor, that is,

$$T_{em} = \frac{3E_f V_T}{\omega_m Xs} \sin\delta \text{ Newton meter} \tag{3.7}$$

where $\omega_m = 2\pi n/60$ mechanical rad/s. The prime mover output kilowatt or horsepower rating is then given by

$$\text{Prime mover kW} = \frac{T_{em(N\cdot m)} \cdot \omega_{m(rad/sec)}}{746 \cdot \text{generator efficiency}}$$

or

$$\text{hp} = \frac{T_{em(lb\cdot ft)} \cdot \text{speed}_{rpm}}{5252 \cdot \text{generator efficiency}}. \tag{3.8}$$

Example 3.5

A 5-MVA, 60-Hz, 6.6-kV, four-pole, 95% efficient round rotor synchronous generator has the synchronous reactance of 7.0 Ω/phase and negligible armature resistance. When operating at unity power factor, determine (1) the maximum power it can deliverer under steady state (no step load changes), (2) the power angle (the rotor lead angle), and (3) back torque on the rotor in Newton meters and in pound feet.

Solution:

1. $V_{LN} = 6600 \div \sqrt{3} = 3810.5$ V and $I_L = 1/3 \times 5 \times 10^6 \div 3810.5 = 437.4$ A in phase with voltage at unity pf. From the phase diagram at unity pf similar to the Example 3.3 figure, we get

$$E_f = \{3810.5^2 + (437.4 \times 7.0)^2\}^{1/2} = 4888.2 \text{ V}.$$

2. $P_{max\cdot 3ph} = 3P_{max\cdot 1ph} = 3 \times \dfrac{4888.2 \times 3810.5}{7.0} = 7{,}982{,}785 \text{ W} = 7.983 \text{ MW}.$

Using Equation 3.5 with power in three-phase megawatts on both sides,

$$5 = 7.983 \sin \delta, \text{ which gives } \delta = 38.8°.$$

3. Back torque T on rotor is derived from power = $T \times \omega_{mech}$, where ω_{mech} = $2\pi \times 1800$ rpm ÷ 60 = 188.5 mechanical rad/s.

$$\text{Therefore, } T = 5 \times 10^6 \text{ W} \div 188.5 = 26{,}526 \text{ Newton m.}$$

In British units, using Equation 3.8, hp = 5000 kW ÷ 0.746 = $T \times 1800 \div 5252$, which gives back torque T = 19,556 lb/ft.

Alternatively, we can use the unit conversion table in the front of the book, which relates

$$T_{N\cdot m} = 1.3558 \times T_{lb/ft}, \text{ or } T_{lb\cdot ft} = 26{,}526 \div 1.3558$$
$$= 19{,}565 \text{ lb/ft (within rounding error).}$$

3.3.1 Field Excitation Effect

As per Equation 3.3, increasing the field current I_f increases E_f. One can visualize in Figure 3.3 that higher (longer) E_f, in turn, increases the terminal voltage V_T or the power factor angle θ, increasing the lagging reactive power output (kVAR) of the generator. The generator can supply more lagging kVARs to the load until I_f reaches its permissible limit. On the reverse, decreasing I_f decreases both E_f and the lagging kVAR output of the generator. The underexcitation can be adjusted such that $E_f = V_T$, when the machines would operate at unity power factor delivering the maximum real kilowatt power to the load. However, this can exceed the prime mover capability, which is typically rated to drive the generator at a lagging power factor of 0.9. Further underexcitation can even make the generator operate at a leading power factor absorbing kVAR from the system. One downside of the underexcitation is lower E_f and lower steady-state power limit P_{max} as per Equation 3.5 and, in turn, a lower transient stability limit, which will be discussed later.

If the load draws lagging kVAR, it must be supplied from the generator, or else, the system cannot operate in a stable mode. The excitation of the machine in normal operation is adjusted such that it maintains the required terminal voltage at around 0.9 pf lagging to match with that of the load.

Example 3.6

One of six generators on a large cruise ship was repaired and brought back online by synchronizing with other five generators, which in a group can be assumed to make an infinite bus for the incoming generator. The incoming generator has the synchronous reactance X_s = 1.5 pu, V_T = 1.0 pu, and the field current I_f = 1000 A at the time of synchronization. After synchronization, while keeping the field current unchanged, the steam valves are adjusted such that it delivers 0.2 pu of its rated power.

1. Determine the armature current in per unit (pu) value when delivering 20% power with rated $I_f = 1000$ A.
2. If the field current I_f is now increased by 60% while keeping the steam input the same (i.e., the power output unchanged), determine the new armature current in pu value.

Solution:

This example is posed in pu values of power, voltage, current, and reactance. The pu system is discussed in Chapter 5. Therefore, it is recommended to read this example after covering Chapter 5 to appreciate the advantage of using the pu system not only in the transformer but also overall in power engineering studies. Until then, we just say here that 1 *per unit* = 100 *percent*, that is, 1.0 pu = 100% of the rated value.

1. At synchronization, the incoming generator has no load, that is, no internal voltage drop, so it must have $E_f = V_T = 1.0$ pu $\angle 0°$ (say the reference phasor). Then,

$$P = \frac{E_f V_T}{Xs}\sin\delta = \frac{1\times 1}{1.5}\sin\delta = 0.20 \text{ pu},$$

which gives $\delta = 17.46°$, and $\tilde{E}_f = 1.0\angle 17.46°$

$$\text{Therefore, } I_a = \frac{E_f - V_T}{jXs} = \frac{1\angle 17.46° - 1\angle 0°}{j1.5} = 0.2\angle 8.73° \text{ pu.}$$

2. The field current I_f is now increased by 60%, so the new E_f magnitude is now 1.6 × old value, which was 1.0:

$$P = \frac{E_f V_T}{Xs}\sin\delta = \frac{1.6\times 1}{1.5}\sin\delta = 0.20 \text{ pu},$$

which gives $\delta = 10.8°$ and $\tilde{E}_f = 1.6\angle 10.8°$ pu

$$\text{Therefore, } I_a = \frac{E_f - V_T}{jXs} = \frac{1.6\angle 10.8° - 1\angle 0°}{j1.5} = 0.43\angle -62.3° \text{ pu.}$$

We note here that by increasing the field current while delivering the same output power, the power angle δ decreased, the armature current increased, and the power factor decreased from 8.73° leading to 62.3° lagging with respect to $\tilde{V}_T$.

3.3.2 Power Capability Limits

As described above, the excitation current not only affects the terminal voltage; it also affects the operating power limit of the machine. Under normal excitation of the field coil, the permissible armature current limits the power output. Underexcitation

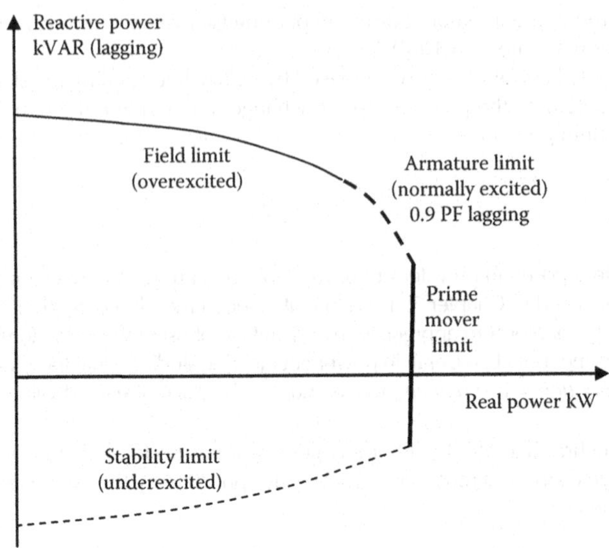

FIGURE 3.5 AC generator power capability curve with four limiting boundaries.

of the field makes the machine operate near unity power factor capable of delivering more real power, but that would overload the prime mover, as most prime movers are rated to drive the generator operating around 0.9 pf lagging. An overexcited machine can deliver more lagging kVARs until it reaches the heating limit of the field coil. All these limits jointly determine the generator's actual real power generating capability as shown in Figure 3.5, with limiting boundaries coming from different performance limitations.

3.3.3 Round and Salient Pole Rotors

Large ac generators are made of solid magnetic steel rotor of cylindrical shape, whereas small- and medium-size generators are made with laminated salient poles (protruding poles) as shown in Figure 3.6. The salient pole construction provides more physical space for the rotor winding and also produces additional torque (called reluctance torque) that increases the machine power output at small power angle δ. The flux in the main magnetic axis (called direct axis or d-axis) has lower-reluctance path (less air gap) than the flux in the quadrature axis (called q-axis). For this reason, the protruding salient poles have a natural tendency to align with the stator field axis under ferromagnetic attraction even in the absence of the rotor current. This requires additional torque from the prime mover, twice in one electrical cycle, to drive the rotor away from the said natural alignment. The total torque input and hence the power output of the generator have a $\sin(2\delta)$ component superimposed on the $\sin \delta$ component, as shown in Figure 3.7.

The power output analysis of the salient pole generator requires two-axis theory using Park transformation. It resolves the total machine magnetics in the d- and

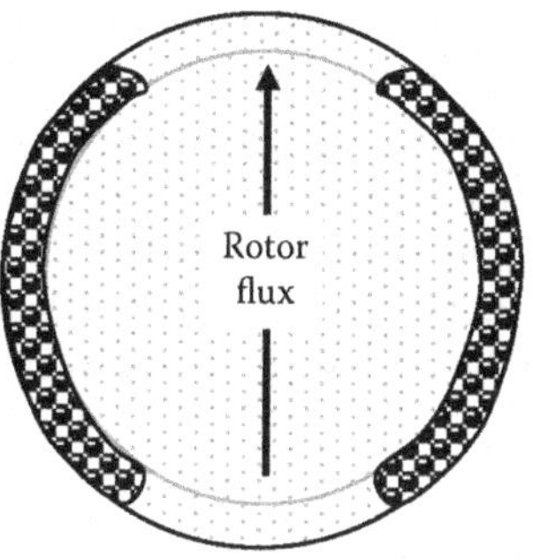

(a) Round rotor has field coil in slots on rotor periphery

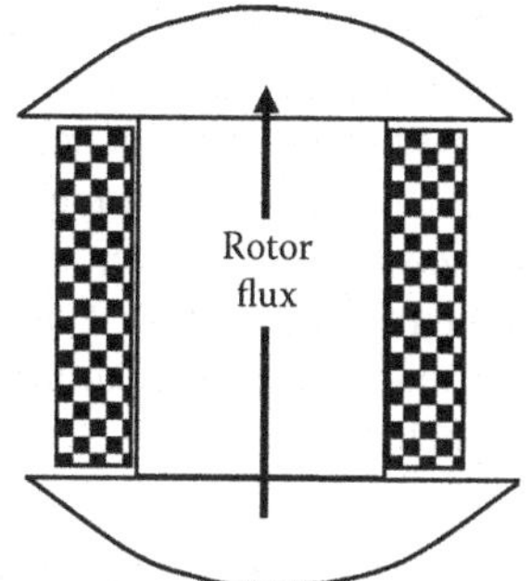

(b) Salient pole rotor with separately wound field coil around poles

FIGURE 3.6 Round rotor and salient pole constructions.

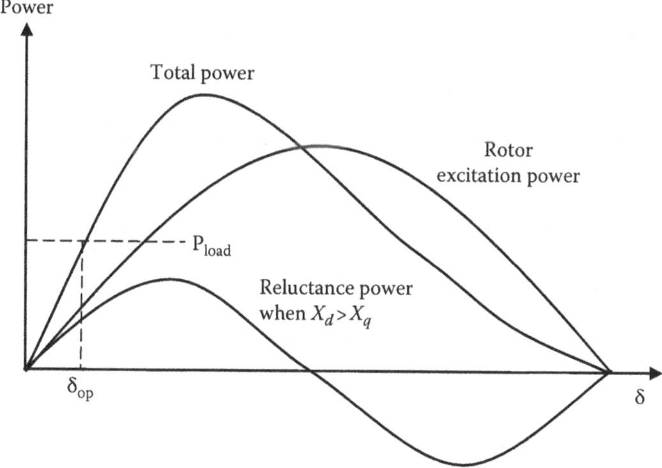

FIGURE 3.7 Salient pole power output versus power angle δ.

q-axis components, both rotating with the rotor, but each with its own steady magnetic path. The d- and q-axis analysis is beyond the scope here, but the final result, which supports what we reasoned above, is given below:

$$P_{\text{salient-pole}} = \frac{E_f V_T}{X_d} \sin\delta + V_T^2 \frac{\left(X_d - X_q\right)}{2X_d X_q} \sin 2\delta \text{ watts/phase} \tag{3.9}$$

where X_d and X_q are the d- and q-axis synchronous reactance of the machine. For the salient pole machines, the value of X_d is always greater than that of X_q. For a cylindrical rotor machine, the magnetic circuits in d- and q-axes are identical, giving

$X_d = X_q$. This makes the second term in Equation 3.9 vanish, leaving only the first term as in Equation 3.5.

Figure 3.7 shows that the salient pole machine increases the total power generated at a small power angle and has a higher P_{max} that occurs earlier in power angle δ. Therefore, for the same load power, the salient pole machine would naturally run at lower δ. Also, having higher P_{max}, it can take a slightly greater step load change than the round rotor machine of comparable design before losing the transient stability limit discussed below.

3.4 TRANSIENT STABILITY LIMIT

The maximum power limit P_{max} at δ = 90° in Figure 3.8a for the cylindrical rotor is called the steady-state stability limit. Any swing in δ beyond 90° may cause the rotor to lose synchronism and its power generation capability. Therefore, it is desirable to keep δ below 90° under all conditions, including any transient that may be

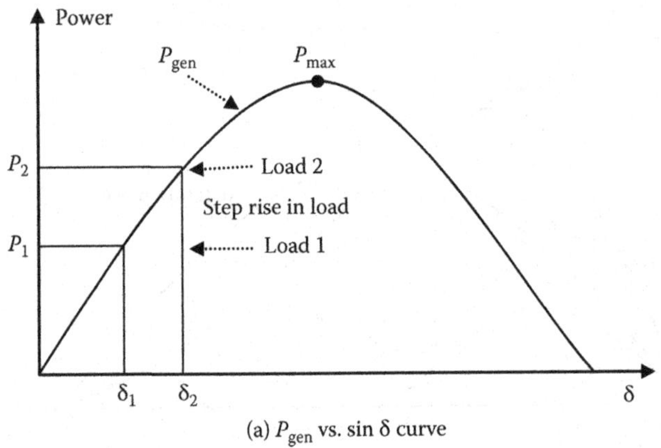

(a) P_{gen} vs. sin δ curve

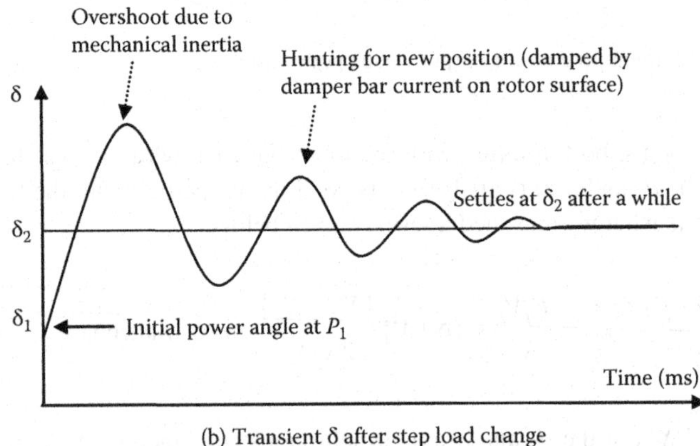

(b) Transient δ after step load change

FIGURE 3.8 Transient oscillations of rotor power angle following step load rise.

encountered during normal and abnormal operations. For example, if the generator load were suddenly changed from P_1 to P_2 in one step, the rotor power angle would increase from δ_1 at the old load P_1 to δ_2 at the new load P_2. This takes some time due to mechanical inertia of the rotor. No matter how short or long it takes, the rotor inertia and the electromagnetic restraining torque will set the rotor in a mass–spring–damper-type oscillation, swinging the rotor power angle beyond its new steady-state value of δ_2 as shown in Figure 3.8b. If δ exceeds 90° any time during this swing, the machine stability and the power generation capability may be lost.

For this reason, the machine can be loaded only to the extent that even under the worst-case load step, planned or accidental, or during all possible faults, the power angle swing remains below 90° with sufficient margin. This limit on loading the machine is called the transient or dynamic stability limit, which is generally the power output for which δ is about 25° to 35° under normal steady-state operation.

For damping the transient oscillations of the rotor following a step load change, each pole face is provided with copper bars running along the length of the machine (Figure 3.9). All bars on each pole face are shorted at both ends, forming a partial squirrel cage of copper conductors on each pole surface. When the rotor oscillates

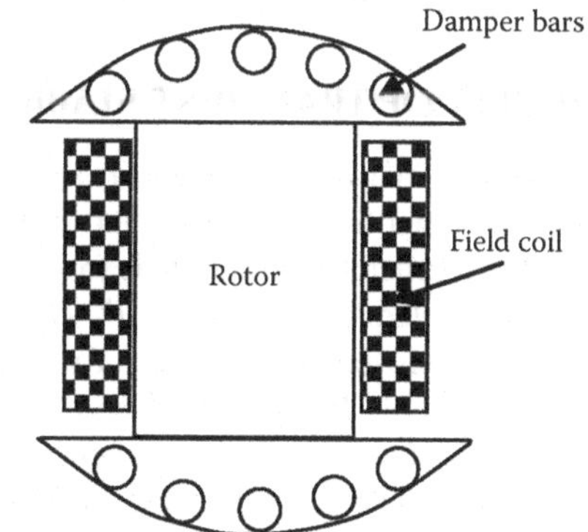

(a) Cross section at mid-length of machine

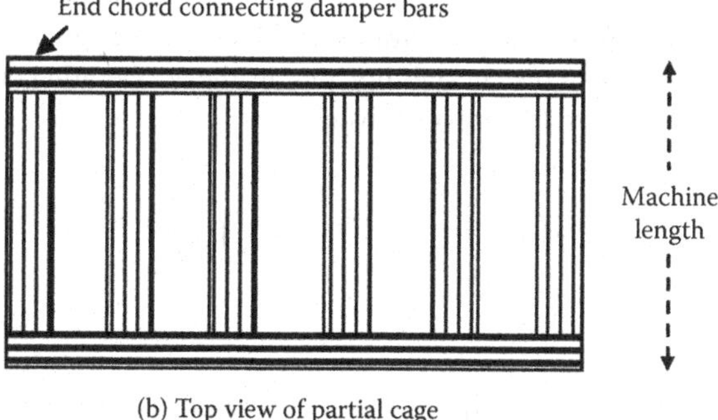

(b) Top view of partial cage

FIGURE 3.9 Damper bars near top surface of salient poles (used in round rotors also).

around the synchronous speed, these bars see a relative slip with respect to the stator flux that runs exactly at the synchronous speed. This slip induces currents in the bars, as in the squirrel cage induction motor. The resulting I^2R power loss in the bars depletes the oscillation energy, cycle by cycle, until the oscillations are completely damped out and the induced currents subside.

Thus, there is a small induction motor superimposed on the synchronous generator. It contributes damping only when the rotor oscillates around the constant synchronous speed. It is also used to start the machine as an induction motor to bring the generator near full speed before applying the dc excitation to the rotor and making it a synchronous machine then onward.

Equation 3.5 shows that the stability limit at a given voltage can be increased by designing the machine with low synchronous reactance X_s, which largely comes from the stator armature reaction component under steady-state (synchronous) operation.

Many loading events may constitute step load on the generator, such as turning on a large motor, high-power weapon on combat ships, and tripping a large load circuit breaker accidentally or for fault protection purpose. To maintain the dynamic stability under such transients, sudden loading on the generator should be limited to no more than 25% to 30% of the rated power in one step, but the exact limit can be determined by the following equal area criteria.

3.5 EQUAL AREA CRITERIA OF TRANSIENT STABILITY

If the generator load is suddenly increased in one step from P_1 to P_2 as shown in Figure 3.10a, and the mechanical input from the prime mover is also changed from P_1 to P_2 to match with the load, the generator power angle (rotor lead angle) will start changing from δ_1 to δ_2 under the accelerating power $P_a = P_2 - P_1$. The rotor with large mechanical inertia will take a relatively long time to reach from δ_1 to δ_2, overshoot δ_2, and reach to δ_3, which can be beyond the P_{max} point at $\delta = 90°$. Above δ_2, the generator supplies more power than the prime mover input P_2; hence, the rotor decelerates back to δ_1 and undershoots below δ_1. The generator now has more mechanical input than electrical output, accelerating the rotor again. Such swings around δ_2 continue until damped out by the power loss in the damper bars. For simplicity, if we ignore the damper bar effect during the first swing, the machine will be stable if the accelerating energy equals the decelerating energy. In a torsional system, since the energy is equal to (torque × angle δ) and the torque equals (power ÷ angular speed), we can write the condition for the transient stability, that is, the accelerating energy = decelerating energy. In the integral form, it is given by

$$\int_{\delta 1}^{\delta 2} \frac{(P_2 - P)}{\omega} \cdot d\delta = \int_{\delta 2}^{\delta 3} \frac{(P - P_2)}{\omega} \cdot d\delta \tag{3.10}$$

where $P = P_{max} \sin \delta$ = actual electrical power generated at power angle δ.

The two integrals are areas under the generator's $P - \delta$ curve. The first one represents the accelerating area A_a, and the second one represents the decelerating area A_d

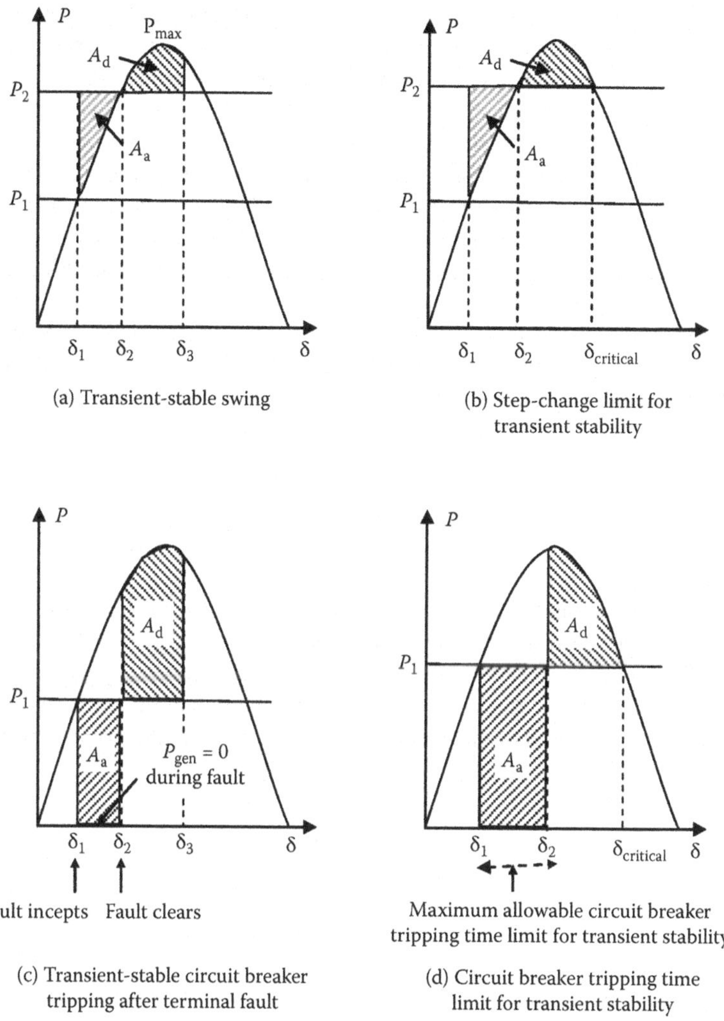

FIGURE 3.10 Equal area criteria of transient stability limit.

in Figure 3.10a. The generator will be stable during the transient oscillations if the two areas are equal, hence the name *equal area criteria of transient stability*.

Figure 3.10b depicts limiting step load change ($P_2 - P_1$) that will maintain the transient stability. It shows that the maximum value of $\delta_3 = \delta_{critical}$ can be the intersection of the generator P–δ curve and new load P_2. It can be determined by equating the two areas A_a and A_d. Knowing the initial load P_1, we can determine P_2 and the maximum permissible step load change ($P_2 - P_1$).

The generator can also lose the stability if the circuit breaker trips under a short circuit, delivering no electrical power to the load as shown in Figure 3.10c. The rotor will accelerate under the prime mover input power, accelerating to increase the rotor angle to δ_2. If the fault is cleared in time and the circuit breaker is closed under the automatic reclosure scheme (common on most systems), the generator starts delivering electrical power in excess of the prime mover input and starts decelerating to its original power angle δ_1 after the inertial overshoot to δ_3. The transient swing will be such that the accelerating area A_a is equal to the decelerating angle A_d. The maximum allowable fault clearing time for stability can be determined by applying the same equal area criteria by equating A_a and A_d in the limiting case shown in Figure 3.10d and determining $\delta_{critical}$. The duration of time it takes for the machine to move from δ_1 to $\delta_{critical}$ comes from the equation of mechanical motion, which equates the difference between the mechanical power input and the electrical power output. The difference goes in accelerating the rotor. Ignoring damping, we can write the following in terms of the torques:

$$T_{mech} - T_{elec} = J\frac{d\delta^2}{dt^2} \tag{3.11}$$

where J = polar moments of inertia of the rotor, $T_{elec} = T_{max}\sin\delta$ = back torque on the generator shaft corresponding to the electric load, and T_{mech} = mechanical torque input from the prime mover.

Equation 3.11 is called the swing equation. The solution of this differential equation, modified by power = torque × ω = torque × 2πf, leads to the critical fault clearing time:

$$t_{critical} = \sqrt{\frac{2H(\delta_{cr} - \delta_o)}{\pi f P_{mech}}} \text{ seconds} \tag{3.12}$$

where all δ's are in mechanical radians per second, and H = rotor inertia in terms of kilojoule (kilowatt seconds) kinetic energy stored at rated speed per kilowatt rating of the machine. The rotor inertial constant H is defined as the number of seconds the machine can deliver the rated power using its kinetic energy stored in the rotor inertia alone with no mechanical input from the prime mover. It includes the turbine rotor inertia as well.

Example 3.7

A 100-MVA, four-pole, 60-Hz synchronous generator that operates at 0.90 power factor load has a cylindrical rotor 5 m long and 1.5 m in diameter. The average mass density of the rotor's copper and magnetic steel combined is 8 g/cm^3. Determine the kinetic energy of the generator rotor running at rated speed. Assuming that the turbine rotor has inertia 1.5 × generator rotor inertia, determine the inertia constant H of the turbine–generator system.

Solution:

The rated speed of the four-pole, 60-Hz generator = 120 × 60 ÷ 4 = 1800 rpm

$$\text{Therefore, } \omega_{mech} = 2\pi \times 1800 \div 60 = 188.5 \text{ rad/sec.}$$

The polar moment of inertia of the cylindrical rotor, J = 1/2 mass × radius^2, where mass = 8 × π × (75 cm)^2 × 500 cm = 70.69 × 10^6 g or 70,690 kg. Therefore, J = 1/2 70,690 × 0.75^2 = 19,880 kg m^2

Kinetic energy = 1/2 $J\,\omega^2$ = 1/2 × 19,880 × 188.5^2 = 353.2 × 10^6 J = 353.2 MJ.

Turbine rotor kinetic energy = 1.5 × 353.2 = 529.8 MJ, and the total kinetic energy of the turbine–generator rotor = 529.8 + 353.2 = 883 MJ.

Generator power output P = 100 MVA × 0.90 power factor = 90 MW, which can be supplied from the rotor kinetic energy for 883 ÷ 90 = 9.81 s.

Therefore, by definition, H = 9.81 s.

The swing equation also leads to the undamped mechanical natural frequency of rotor oscillations, which is given by

$$f_{mech} = \frac{1}{2\pi}\sqrt{\pi f_{elec} \frac{\left(\frac{dP_{elec}}{d\delta}\right)_o}{H}} \text{ hertz.} \tag{3.13}$$

This equation is similar to that for the mechanical mass–spring oscillations, namely,

$$f = \frac{1}{2\pi}\sqrt{\frac{spring\ cons\tan tK}{Mass\ M}} \text{ hertz.} \tag{3.14}$$

Comparing Equations 3.13 and 3.14, we see that $dP_{elec}/d\delta$ is like a spring constant, which is nonlinear, stiffest near $\delta = 0$, soft at operating δ's, and totally resilient at $\delta = 90°$. It provides no restraining stiffness at $\delta = 90°$, beyond which it is unable to operate stably, and hence becomes unstable under steady-state loading. In determining $P_{elec} = P_{max} \sin\delta$ for swing equation, P_{max} must use the transient reactance X_d' of the generator (discussed later in Chapter 8).

Example 3.8

A 60-Hz round rotor generator has E_f = 1.5 pu, $V_{T(bus)}$ = 1.0 pu, X_d = 1.3 pu, and H = 5 s. Determine the mechanical natural frequency of the turbine–generator oscillations under transient disturbance when operating at 50% load.

Solution:

Again, it is recommended to read this example after covering the pu system in Chapter 5. Until then, we just say here that 1.0 pu = 100% of rated value and work in the pu system. The initial power angle δ_o is derived from Equation 3.5, that is,

$$P = 0.50 = \frac{1.5 \times 1.0}{1.3} \sin \delta_o = 1.154 \sin \delta_o, \text{ which gives } \delta_o = 25.68°.$$

The synchronizing power coefficient (generator spring constant) that provides restraining force for transient swings in δ is given by

$$K = dP/d\delta = P_{max} \cos \delta = 1.154 \cos \delta_o = 1.04.$$

The mechanical frequency of rotor oscillations following a step load change or any other transient disturbance is given by Equation 3.13, that is,

$$f_{mech} = \frac{1}{2\pi}\sqrt{\frac{\pi \times 60 \times 1.04}{5}} = 1 \text{ Hz}.$$

The mechanical transients, which are much slower than electrical transients, occur in a synchronous generator supplied by pulsating-power prime movers (e.g., diesel engine) or synchronous motor driving pulsating loads (e.g., compressor loads). If the transient oscillations are small around small δ, $\sin \delta \approx \delta$ in radians, and the torque becomes linear with δ, that is, $T = T_{max} \times \delta$. The machine appears as a torsional spring with spring constant $K = T/\delta = T_{max}$ N·m/rad. For large oscillations, the linearity no longer holds, and the solution must be obtained by a step-by-step numerical method on the computer. Moreover, as δ approaches 90°, the spring becomes softer, and the machine may lose synchronism.

For a large turbine generator, typical $t_{critical}$ given by Equation 3.12 is less than 1 s for automatic reclosure of the circuit breaker after a fault for maintaining transient stability, and the mechanical oscillation frequency given by Equation 3.13 is generally less than 1 Hz.

3.6 SPEED AND FREQUENCY REGULATIONS

When the load torque rises, the automatic speed-regulating governor increases the fuel input to maintain the speed. However, it does not fully compensate for the load increase, and the prime mover speed drops slightly in an approximately linear manner. This can also be explained in terms of the mechanical reaction torque (back torque) on the rotor, which is proportional to the stator current. As the load is increased, the rotor speed drops under increased back torque on the rotor. The generator speed regulator (governor) will allow more fuel to the prime mover to maintain the rotor speed constant. The practical speed governor must have a certain dead band (tolerance band), or else it would go through hunting oscillations in response to any load change. The dead band allows some decrease in speed without a response. Thus, as the load increases on the generator, the steady-state speed of the rotor will decrease slightly due to the dead band in the prime mover speed governor. Thus, even with an automatic speed control, practical prime mover governors cannot maintain perfectly constant speed. As a result, the prime mover speed drops slightly with increasing load.

The speed regulation of a mechanical power source—steam or gas turbine, diesel engine, or motor—is defined in a manner similar to the voltage regulation of an electrical power source:

$$\text{speed regulation} = \frac{n_{\text{no load}} - n_{\text{rated load}}}{n_{\text{rated load}}}. \tag{3.15}$$

Since the ac generator frequency is directly related with the prime mover speed, power engineers working with ac generators usually define the generator frequency regulation (GFR) in terms of the governor speed regulation (GSR):

$$\text{GFR} = \text{GSR} = \frac{f_{\text{no load}} - f_{\text{rated load}}}{f_{\text{rated load}}}. \tag{3.16}$$

The frequency droop rate (FDR) or the governor droop rate (GDR) is defined as the rate of frequency drop per kilowatt or megawatt of the output load, that is,

$$\text{FDR} = \text{GDR} = \frac{\Delta f}{\Delta P} \text{ hertz per kilowatt or hertz per megawatt} \tag{3.17}$$

Example 3.9

A 60-Hz synchronous generator rated 100 MVA at 0.90 power factor has the no-load frequency of 61.5 Hz. Determine the prime mover speed regulation, the GFR, and the generator FDR.

Solution:

We first note that under rated operation, the generator output power would be 100 × 0.90 = 90 MW and frequency is 60 Hz (implied since the rated frequency of the generator is 60 Hz). Then,

GFR or GSR = (61.5 – 60.0) ÷ 60 =.025 pu or 2.5%.

Since the frequency drops from 61.5 to 60 Hz from no load to rated load of 90 MW,

FDR = $\Delta f/\Delta P$ = (61.5 – 60.0) ÷ (90 – 0) = 1.5 Hz ÷ 90 MW = 0.01667 Hz/MW.

3.7 LOAD SHARING AMONG AC GENERATORS

The ac generator voltage is controlled by the field current and the speed, but the frequency is controlled only by the speed. Since the generator power ultimately comes from the prime mover, the prime mover speed and the generator frequency both droop at the same rate with increasing load on the generator. With two ac generators sharing load in parallel, their terminal voltages and frequencies must both be equal, and the frequency droop lines of two ac generators determine their share of the total load. The frequencies of generators 1 and 2 at their own load powers P_1 and P_2 can be written in terms of their respective FDRs as follows:

$$f_1 = f_{1\text{no-load}} - \text{FDR}_1 P_1 \text{ and } f_2 = f_{2\text{no-load}} - \text{FDR}_2 P_2 \tag{3.18}$$

where f = operating frequency, $f_{\text{no-load}}$ = no-load frequency, and FDR = frequency droop rate. By parallel connection, we impose on the bus that $f_1 = f_2 = f_{\text{bus}}$, that is,

$$f_{1\text{no-load}} - \text{FDR}_1 P_1 = f_{2\text{no-load}} - \text{FDR}_2 P_2 = f_{\text{bus}} \tag{3.19}$$

$$\text{Total load } P_T = P_1 + P_2. \tag{3.20}$$

Solving Equations 3.19 and 3.20 simultaneously for P_1 and P_2 (with all other parameters known), we get the load shared by generators 1 and 2. Figure 3.11 depicts the one-step method of determining the ac generator load sharing in a manner similar to that we discussed for the static power sources in Section 2.8.1.

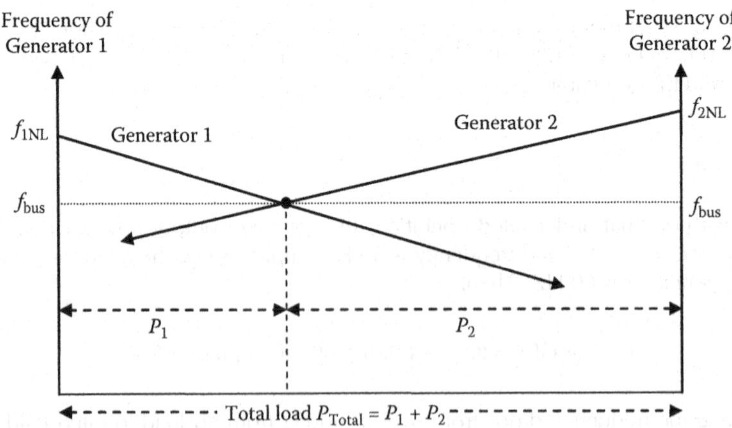

FIGURE 3.11 One-step method of determining two ac generators load sharing.

Example 3.10

The no-load frequency of two ac generators is the same, 60.5 Hz. The FDR of generator-1 is 0.0006 Hz/kW, and that of generator-2 is 0.0008 Hz/kW. If the two generators are in parallel supplying a total load of 1500 kW, determine the load shared by each generator and the operating frequency of the bus.

Solution:

Denoting kilowatt power P_1 and P_2 shared by generator-1 and generator-2, respectively, we write the two generator frequency droop line equations as

$$f_1 = 60.5 - 0.0006\ P_1 \text{ and } f_2 = 60.5 - 0.0008\ P_2.$$

In parallel operation, $f_1 = f_2 = f_{bus}$ and $P_1 + P_2 = 1500$ kW. Therefore, we write $f_{bus} = 60.5 - 0.0006\ P_1 = 60.5 - 0.0008\ (1500 - P_1)$, which gives $P_1 = 857$ kW and $P_2 = 1500 - 857 = 643$ kW.

The bus frequency is then $f_{bus} = 60.5 - 0.0006 \times 857 = 59.986$ Hz or $f_{bus} = 60.5 - 0.0008 \times 643 = 59.986$ Hz, the same as above.

We note that generator-1 is stiffer (droops less) than generator-2; hence, it shares greater load, as expected.

It is noteworthy that a stiffer generator with low FDR (flatter line) shares a heavier load, and a weaker generator with high FDR (more drooping) shares a lighter load. This is analogous to load sharing between two mechanical springs in parallel. A stiffer spring, which droops less with load, shares greater load. Also, the machine with higher no-load speed (droop line shifted upward) would share greater load. When we change the governor setting, we essentially shift the fuel-input rate, hence the speed line and the frequency line up and down, whereas the GDR remains the same. In practice, the load sharing is controlled manually or automatically by adjusting the prime mover's governor setting, which controls the input valve of the fuel (steam or diesel). The governor's automatic control system varies the fuel input rate and the speed directly proportional to the load. When the load increases, the fuel is increased, and vice versa.

With two generators working in parallel, if one generator momentarily slows down for any reason, it delivers less power and subsequently speeds up. The other generator takes greater load and slows down. Such adjustment takes place until both generators run at the exact same speed to generate the exact same frequency and the exact same terminal voltage. We can say that parallel generators have a great team spirit in helping each other run at the same speed as determined collectively by the total load demand on the team.

Caution:

In parallel operation, it is important that each machine shares the load (both in kilowatts and in kVARs) within its own rated limit. The kilowatt load can be balanced by adjusting the GSR, and the kVAR load can be balanced by the field excitation of each machine by adjusting the field rheostat in the voltage regulator. After such adjustments in both machines, the armature currents should be about the same percentage of their individual rated values and at about equal power factors.

3.8 ISOSYNCHRONOUS GENERATOR

If the frequency of generator-1 were maintained constant regardless of its load as shown in Figure 3.12, that is, if generator-1 were made infinitely stiff, its frequency droop under load is eliminated, and the bus frequency remains constant (flat) regardless of the total load. The constant frequency generator is known as the *isosynchronous generator*. Most governors have a set point adjustment that allows for varying the no-load speed of the prime mover. By adjusting this set point, one can adjust the frequency or the load shared by one generator with the other operating in parallel. In an isosynchronous generator, this adjustment is done by an automatic control system with precise compensation feedback. We note two points here when operating two generators in parallel.

1. Only one generator can be isosynchronous. Two isosynchronous generators cannot work in parallel since they would have no intersection that makes the common point of stable operation. In parallel, they would conflict with each other, may overload, and would self-destruct in a continuous search of an intersection.
2. Beyond the initial load sharing, all additional loads are taken by the isosynchronous generator, whereas the drooping generator load remains constant. This is analogous to two mechanical springs sharing a load, with one soft spring drooping with load and the other spring infinitely stiff like a solid metal block (we may call it an isoshape spring) as shown in Figure 3.13.

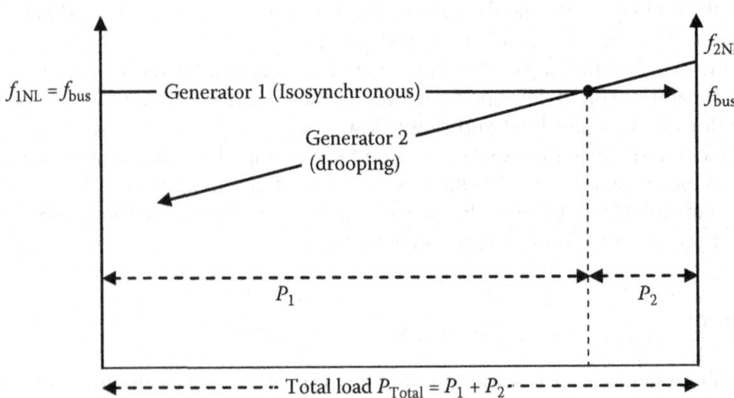

FIGURE 3.12 Load sharing between one isosynchronous generator and one drooping generator.

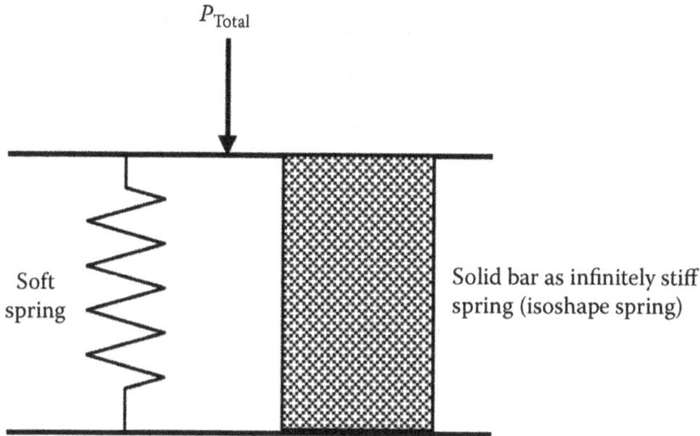

FIGURE 3.13 Load sharing between two springs in parallel, one soft and the other infinitely stiff.

> Regardless of the initial load sharing, any additional load will be taken by the solid metal block (isoshape spring), without additional drooping of either spring.

For this reason, the load shared by the isosynchronous generator is much greater than that shared by the drooping generator. Therefore, the prime mover governor for an isosynchronous generator is specifically designed for isochronous operation at any load from zero to 100%. Such a governor assures that the prime mover shares the load proportional to the generator kilowatt rating by using direct measurement of kilowatts. This governor provides rapid response to load changes, stable system operation, ability for paralleling dissimilar-sized turbine-generator sets, and fine speed regulation under 0.25%.

Droop is inherent in all prime mover speed controls, but in an isochronous generator, it is recovered in a short time. It is a temporary droop of transient nature—more commonly called the compensation. This gives a bus frequency of 60 Hz from no load to full load, as seen in Figure 3.14.

Example 3.11

Two 60-Hz generators, one isosynchronous and one drooping, share 15 MW each when supplying a total load of 30 MW with the bus frequency exactly at 60 Hz. If the total load is increased to 40 MW, determine the load on the isosynchronous generator and the bus frequency.

Solution:

Since the isosynchronous generator is infinitely stiff with no droop rate, it will take the entire additional load without a drop in frequency. Therefore, its new load will be 15 + (40 − 30) = 25 MW, and the new bus frequency will still be exactly 60 Hz.

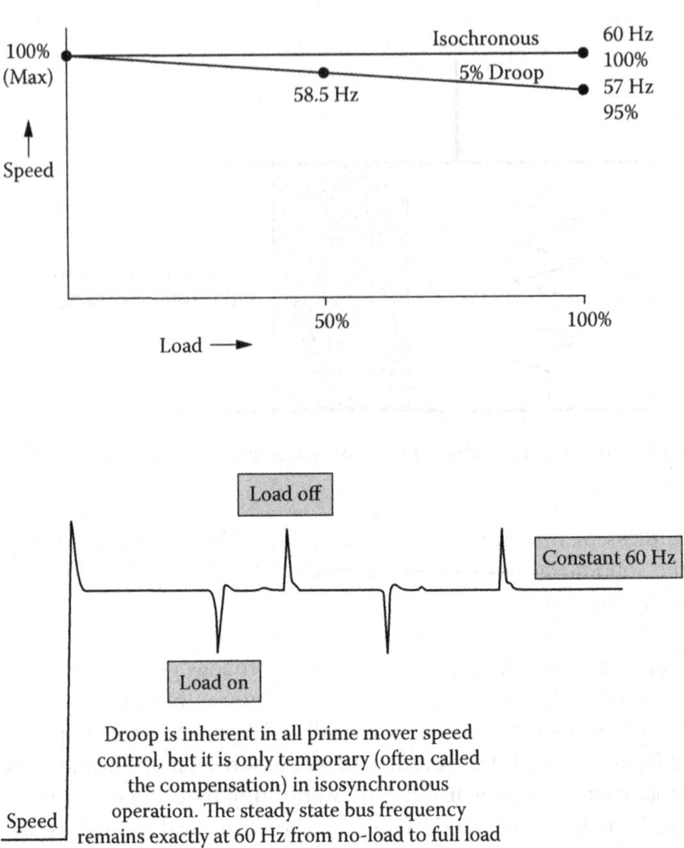

FIGURE 3.14 (Top) Load versus frequency lines of isosynchronous and drooping generators. (Bottom) Momentary frequency drop in isosynchronous generator. (Courtesy of Dana Walker, U.S. Merchant Marine Academy.)

3.9 EXCITATION METHODS

The synchronous generator excitation system is designed to produce the rotor magnetic field that can be varied to control the voltage and reactive power of the generator. In modern high-power machines, the synchronous reactance X_s is around 1.5 × base impedance of the machine. With such a high reactance, the phasor diagram in Figure 3.15 shows that E_f or the rotor field current required at rated load at 0.9 lagging power factor can be more than twice that at no load with the same terminal voltage. A typical excitation system has the corresponding current and voltage ratings, with a capability of varying the voltage E_f over a wide range of 1 to 3, or even more, without undue saturation in the magnetic circuit. Most excitation systems operate at 200 to 1000 V_{dc}. The excitation power to overcome the rotor winding I^2R loss ranges from 0.5% to 1% of the generator rating. For a large utility generator,

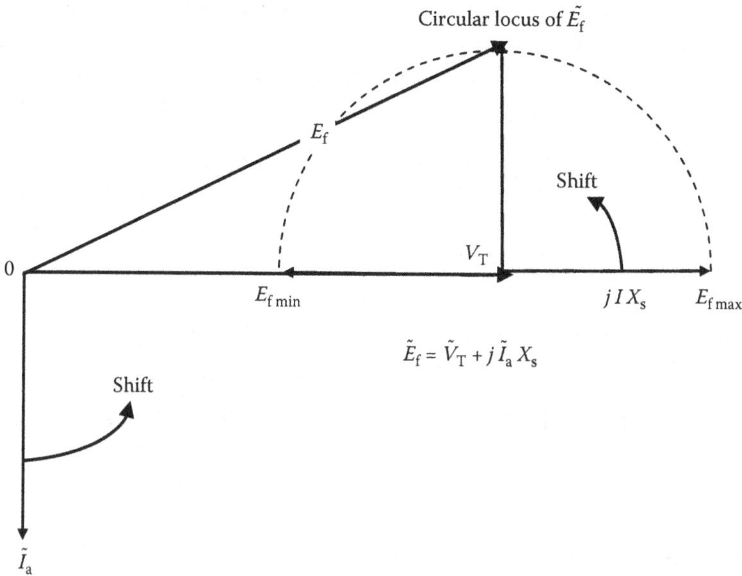

FIGURE 3.15 Variation in field excitation voltage E_f with load power factor varying from zero lagging to unity.

four types of excitation system—dc, ac, static, and brushless—as described below are possible.

DC exciter: A suitably designed dc generator supplies the main field winding excitation through conventional slip rings and brushes. Due to a low reliability and high maintenance requirement, the conventional dc exciter is seldom used in modern ac generators of large ratings.

AC exciter: It consists of a permanent magnet pilot exciter that excites the main exciter. The ac output of the pilot exciter is converted into dc by a floor-standing rectifier and supplied to the main exciter through slip rings. The main exciter's ac output is converted into dc by means of a phase-controlled rectifier, whose firing angle is changed in response to the terminal voltage variations. After filtering the ripples, the dc is fed to the main generator field winding.

Static exciter: It has no moving parts, as opposed to the rotating exciters described above. In the static exciter scheme, the controlled dc voltage is obtained from a suitable stationary ac source rectified and filtered. The dc voltage is then fed to the main field winding through slip rings. This excitation scheme has a fast dynamic response and is more reliable because it has no rotating exciter with mechanical inertia.

Brushless exciter: Most modern synchronous generators of large ratings use a brushless scheme of excitation to eliminate the need for slip rings and brushes. The brushless exciter is placed on the same shaft as the main generator. The ac voltage induced in the exciter is rectified by rotating diodes on the rotor and filtered into pure dc. The dc is then fed directly into the rotor field coil.

The excitation control system modeling for analytical studies must be carefully done as it forms multiple feedback control loops that can become unstable. The IEEE has developed an industry standard for modeling the excitation systems. The model must account for any nonlinearity due to magnetic saturation that may be present in practical designs. The control system stability can be improved by supplementing the main control signal by auxiliary signals, such as speed and power, as required by the feedback control system stability.

3.10 SHORT CIRCUIT RATIO

The short circuit ratio (SCR) of a synchronous generator is defined as the ratio of field current required to generate rated voltage at open circuit to the field current required to circulate rated armature current with short circuit. Thus, SCR is merely the reciprocal of the synchronous reactance X_s expressed in the pu system (covered in Chapter 5), that is,

$$\text{SCR} = \frac{I_f \text{ for rated voltage with open terminals}}{I_f \text{ for rated armature curent with shorted terminals}} = \frac{1}{X_{s(pu)}}. \quad (3.21)$$

Thus, the SCR is a measure of the steady-state short circuit current in case of a terminal fault; it is the ratio of the fault current to the rated current of the generator, that is,

$$\text{steady-state terminal short circuit current} = \text{SCR} \times \text{generator rated current}. \quad (3.22)$$

However, for the generator circuit breaker selection, the transient fault current using the generator's transient reactance—not the synchronous reactance—must be used, as covered in Chapter 8.

The SCR also indicates the machine's sensitivity to the change in the rated load. A machine with a higher SCR is larger in physical size and weight and hence costs more. However, it has smaller X_s, resulting in smaller internal voltage drop and smaller voltage regulation. The machine design engineer balances the SCR value for an optimum design.

3.11 AUTOMATIC VOLTAGE REGULATOR

Excitation control is required to (1) maintain the normal operating voltage, (2) vary kVAR generation to match with the load, and (3) increase the steady state and dynamic stability. The manual control with field rheostat can be adequate for small generators, but the automatic voltage regulator (AVR) is common for large machines.

As discussed in Chapter 1, the complex power $\tilde{S}$ delivered by the generator must match the complex power drawn by the load. In $\tilde{S} = P + jQ$, the real power P is balanced by the prime mover fuel flow rate that is controlled by the prime mover governor. The reactive power Q, on the other hand, is balanced by the field current controlled by the AVR. The governor and the AVR are two independent controllers in an ac generator. The AVR is a part of the excitation system, which works as follows in the brushless machines.

The AVR senses the voltage in the main generator winding and controls the excitation to maintain the generator output voltage within the specified limits, compensating for load, speed, temperature, and power factor of the generator. Three-phase rms sensing is employed for finer voltage regulation. The excitation current is derived from a dedicated three-phase permanent magnet generator to isolate the AVR control circuits from the effects of nonlinear loads and to reduce radio frequency interference on the generator terminals. Protection of the exciter against sustained generator short circuit current is another feature of the permanent magnet rotor used in the AVR. Additional features found in some AVRs are as follows:

- A frequency measuring circuit continually monitors the shaft speed of the generator and provides underspeed protection of the excitation system by reducing the generator output voltage proportionally with speed below a presettable threshold.
- The maximum excitation is limited to a safe value by internal shutdown of the AVR output device. This condition remains latched until the generator has stopped.
- Provision is made for the connection of a remote voltage trimmer, allowing the user to finely control the generator output. The AVR has the facility to allow parallel running with other similarly equipped generators.
- Typical transient response times are the AVR itself in 10 ms, field current to 90% in 80 ms, and machine voltage to 97% in 300 ms. The AVR also includes a stability or damping circuit to provide good steady-state and transient performance of the generator.
- The AVR includes a soft start or voltage ramp-up circuit to control the rate of voltage buildup when the generator runs up to speed. This is normally preset and sealed to give a voltage ramp-up time of approximately 3 s. If required, this can be adjusted between the limits defined in the AVR specifications.

PROBLEMS

1. A three-phase, Y-connected, 480-V generator is powering 1000-kW balanced three-phase Δ-connected load at 0.90 pf lagging. Determine (1) the line-to-line voltage and line current, (2) the generator phase voltage and phase current, and (3) the load phase voltage and phase current.
2. A balanced Y-connected load with $3 + j5$ Ω/ph (L-N) is connected in parallel with a balanced Δ-connected load with $12 + j15$ Ω/ph (L-L). Determine the combined total equivalent Y-connected impedance per phase. If these loads were powered by a Y-connected 2400 V_{LL} generator, determine the current drawn from the generator lines.

3. A three-phase, 11-kV, 60-Hz, 10-MVA, Y-connected synchronous generator has $X_s = 1\ \Omega$/ph and negligible R_a. Determine its power angle δ when delivering rated power at unity power factor. (Hint: First calculate $\tilde{E}_f$ and P_{max}.)
4. A three-phase, Y-connected synchronous generator is rated 15 MVA, 13.8 kV. Its resistance is negligible, and the synchronous reactance is 2 Ω per phase. Determine the field excitation voltage E_f when delivering rated megavolt-amperes at 0.90 power factor lagging.
5. A 3-MVA, 60-Hz, 6.6-kV, four-pole, 96% efficient round rotor synchronous generator has the synchronous reactance of 4.0 Ω/phase and negligible armature resistance. When operating at unity power factor, determine (1) the maximum power it can deliver under steady state with no step loading, (2) the rotor lead angle (power angle δ), and (3) the diesel engine horsepower output.
6. A large cruise ship generator was repaired and brought back online by synchronizing with five other generators, which collectively make an infinite bus for the incoming generator. The incoming generator has the synchronous reactance $X_s = 1.2$ pu, $V_T = 1.0$ pu, and the field current $I_f = 800$ A at the time of synchronization. After synchronization, while keeping the field current unchanged, the steam valves are adjusted such that the generator delivers 0.3 pu of its rated power. (1) Determine the armature current in pu value when delivering 30% power with rated field current $I_f = 800$ A. (2) If the field current is now increased by 50% while keeping the power output the same, determine the new armature current in pu value.
7. An 80-MVA, four-pole, 60-Hz synchronous generator operates at 0.90 power factor lagging. Its cylindrical rotor of 5-m length and 1.5-m diameter has the average mass density of 8.3 g/cm^3. Assuming that the turbine rotor has inertia 1.5 × generator rotor inertia, determine the inertia constant H of this turbine-generator set.
8. A 50-Hz round rotor generator has $E_f = 1.4$ pu, $V_T = 1.0$ pu, $X_d = 1.5$ pu, and $H = 7$ s. Determine the mechanical natural frequency of the turbine–generator oscillations under a transient disturbance when operating at 80% load.
9. A 60-Hz synchronous generator rated 30 MVA at 0.90 power factor has the no-load frequency of 61 Hz. Determine (1) the prime mover speed regulation, (2) the GFR, and (3) the generator FDR.
10. The no-load frequency of two ac generators is the same, 61 Hz. The FDR of generator-1 is 0.001 Hz/kW, and that of generator-2 is 0.0005 Hz/kW. If the two generators are in parallel supplying a total load of 1000 kW, determine the load shared by each generator and the operating frequency of the bus.
11. Two 40-MW, 60-Hz generators, one isosynchronous and one drooping, share an equal load of 25 MW each when supplying a total load of 50 MW with the bus frequency exactly at 60 Hz. If the total load is increased to 70 MW, determine the load on the isosynchronous generator and the bus frequency.
12. A 4160-V, 60-Hz, three-phase, Y-connected generator has a synchronous reactance of 0.9 pu and negligible armature resistance. At rated armature

current, determine the field excitation current range for a power factor changing from zero lag to unity to zero lead. Express the range in terms of the field current required when the generator is delivering the rated current at the unity power factor.

13. A three-phase, 500-kVA, 480-V, 60-Hz, *Y*-connected synchronous generator gave these test results at rated speed: (1) open circuit voltage at rated field current = 560 V_{LL} and (2) short circuit current at rated field current = 305 A. When cold at 20°C, the average dc resistance of three armature phase coils measured by ohmmeter was 0.20 Ω. Determine the armature ac resistance and synchronous reactance per phase at an operating temperature of 90°C.

QUESTIONS

1. Explain the difference between electrical degree and the mechanical degree in rotating electrical machines.
2. List three reasons that may cause the generator to fail in building up the terminal voltage.
3. Explain the counter torque developed in the generator when the load is connected to the stator terminals.
4. Explain why the field excitation current of the generator alters the power factor as seen by the machine.
5. In light of Equation 3.6, discuss the generator maximum power capability at lower frequency all the way up to zero frequency (dc). At very low frequency, what would limit the maximum power capability?
6. Explain the construction and performance differences between the round rotor and salient pole synchronous machines.
7. Identify two problems you may encounter in applying a large load in one step on a synchronous generator.
8. What loading events in a steel mill may cause large step load on the standalone generator of the mill?
9. What causes damping in the rotor oscillations following a sudden step change in the generator load?
10. In the synchronous machine, where the damper bars are located, what is their purpose, and how do they work?
11. A 1000-kVA, 460-V, 60-Hz, three-phase, four-pole generator has squirrel cage damper bars of 1-cm diameter on each of the rotor pole surface. What would be the damper bar current under the steady synchronous speed operation?
12. Why can two generators at different voltage and frequency not work in parallel?
13. List the conditions that must be met before a synchronous generator can be paralleled with other generators or with the power grid.
14. While transferring load from one generator to another, what changes really take place during this process in terms of the prime mover's GSR settings

and the droop lines? Which way the droop lines get adjusted and how do these changes transfer the load?

15. Clearly identify the benefits of using an isosynchronous generator. If two such generators are placed in parallel, what would you expect to happen?
16. Two generators are operating in parallel, one with a droop and the other isosynchronous. Explain their load sharing behavior when the total bus load increases and decreases.
17. If the synchronous generator suddenly gets shorted at its terminals, how would the speed change before the turbine fuel supply is reduced or cut off?

FURTHER READING

Chapman, S.J. 1999. *Electric Machinery Fundamentals*. Boston: McGraw Hill.
Say, M.G. 1983. *Alternating Current Machines*. New York: John Wiley & Sons.

4 AC and DC Motors

Of the total electrical energy generated worldwide, about 58% is used by all motors combined, about 7% for lighting, and the remaining 35% for heating and other uses. Major types of motor are the synchronous motor, induction motor (also known as asynchronous motor), and dc motor. All have two sets of coils with different currents, say, I_1 and I_2. The electromagnetic interaction between two currents produces motor torque $T_m = KI_1I_2$. If the motor shaft has a load torque $T_{Load} < T_{motor}$, the motor would accelerate to a speed at which $T_{Load} = T_{motor}$, where it would stop accelerating and run at steady speed. The motor armature produces *back voltage*, or its equivalent, and draws current from the source that is given by

$$\text{armature current} = \frac{\text{applied voltage} - \text{back voltage}}{\text{effective impedance of armature}}. \tag{4.1}$$

It is important to understand that the armature draws just enough current that is required to develop torque to meet the load torque at the steady running speed. A 100-hp rated motor does not always deliver full 100 hp regardless of the shaft load. The motor delivers what is needed to drive the load and draws power from the source equal to what it delivers to the load plus the internal losses. Thus, the power drawn from the source may be less or more than the rated load depending on the mechanical load on the shaft. However, if continuously overloaded without added cooling, the motor would heat up and burn.

The shaft horsepower, torque, speed, and kilowatt power delivered by the motor are related as follows:

$$\text{HP} = \frac{T_{\text{lb/ft}} \cdot n_{\text{rpm}}}{5252} \quad \text{or} \quad \text{kW} = \frac{T_{\text{N}\cdot\text{m}} \cdot 2\pi n_{\text{rpm}}}{60 \times 1000}. \tag{4.2}$$

Table 4.1 gives a breakdown of motor types and their energy usage in various horsepower ratings. It shows that about 98% of all motors are induction motors, which use about 93% of the electrical energy used by all motors rated 5 hp and higher. Smaller motors do not use much energy because their use is intermittent, often less than an hour in a day. We now discuss three most widely used motors in the industry.

TABLE 4.1
Motor Types by Approximate Number and Electrical Energy Usage

Motor Category	Number in Category as Percentage of All Motors	Energy Usage as Percentage of Total Used by All Motors	Typical Usage Pattern and Applications
<5 hp induction motors	88%	5%	Intermittent usage in small appliances
5–125 hp induction motors*	7%	50%	Heaviest-energy-usage motors
>125 hp induction motors	3%	43%	Continuous usage in industry
Synchronous motors	1%	1%	Mostly in large sizes (>5000 hp)
DC motors	1%	1%	Where easy speed control is needed

* Since induction motors in the 5–125 hp range consume about 50% of all the electrical energy used by all motors combined, the U.S. Department of Energy has focused on improving the design and efficiency of 5–125 hp induction motors.

4.1 INDUCTION MOTOR

The induction motor has been a reliable workhorse of the industry ever since it was invented by Nicola Tesla in 1888. It is the most widely used motor because of its simple, brushless, low-cost, and rugged construction. Most induction motors in use are three-phase motors in large ratings or single-phase motors in small ratings. The three-phase induction motor has three stator coils wound with wires and the rotor in a squirrel cage configuration as shown in Figure 4.1. The cage rotor is generally made of cast aluminum bars running along the machine length and two end rings shorting all the bars. The rotor cage in a high-efficiency motor is often made of copper, which has much better conductivity than aluminum. There is no electrical connection between the stator and the rotor. The three-phase current in three stator coils wound in P-poles configuration and powered at frequency f creates a magnetic flux rotating at constant speed n_s (called the synchronous speed), which is given by

$$\text{synchronous rpm} = \frac{120 \times \text{frequency}}{\text{Number of poles}}. \tag{4.3}$$

The number of magnetic poles created by the stator winding depends on the coil span, that is,

$$\text{number of poles } P = \frac{\pi \times \text{stator diameter}}{\text{stator coil span}}. \tag{4.4}$$

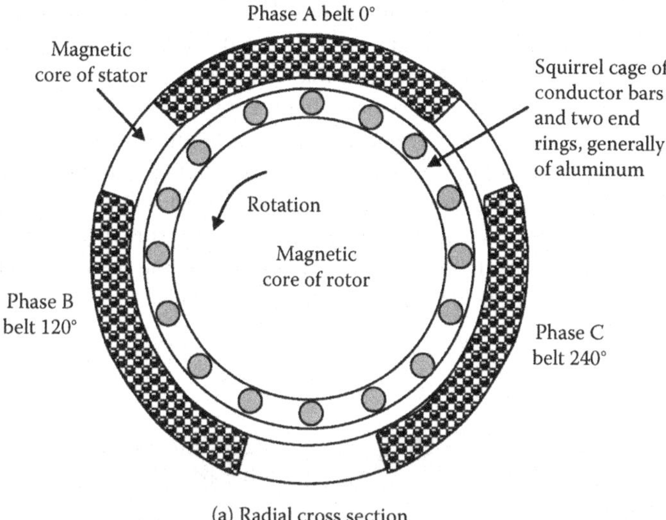

(a) Radial cross section

(b) Axial cross section of high efficiency motor with copper cage rotor

FIGURE 4.1 Three-phase squirrel cage induction motor construction.

Injecting the rotor current from an outside source, as in the synchronous or dc motor, is not required in the induction motor. The sweeping (cutting) flux of the stator *induces* current in the rotor bars, which, in turn, produces the mechanical torque, hence the name *induction* motor. This simplifies the construction a great deal, as no brushes or slip rings are required. The power transfer from the stator to the rotor is brushless; it is done by the magnetic flux as in the transformer. The cage rotor conductors with end rings constitute numerous shorted coils for the induced currents to circulate. From an analytical point of view, the induction motor is essentially a transformer with a shorted secondary coil that can rotate.

Most induction motors have a horizontal shaft, but vertical-shaft induction motors (Figure 4.2) are also available for places where the footprint has a premium, such as in city centers. Most induction motors have a squirrel cage rotor, but some special three-phase induction motors have rotor wound with wires—just like the stator—with all three phase terminals brought out through slip rings. The rotor coil terminals are shorted outside the slip rings in normal operation or with an external resistance to increase the starting torque or for speed control.

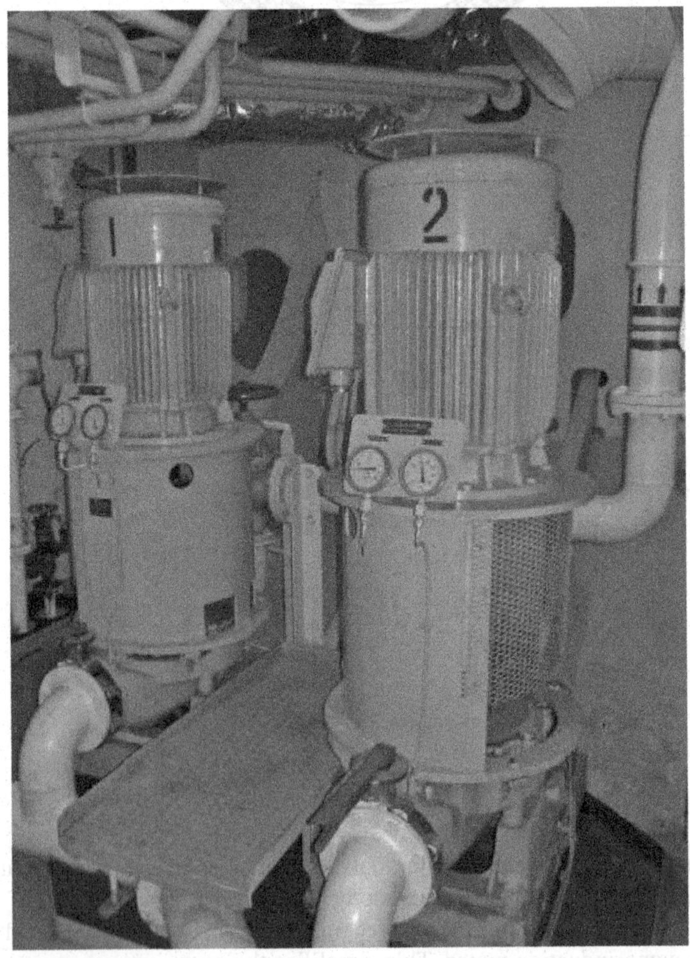

FIGURE 4.2 Vertical shaft motor saves footprints in ships and in city centers. (Courtesy of Raul Osigian, U.S. Merchant Marine Academy.)

TABLE 4.2
Synchronous Speed of 50-Hz and 60-Hz AC Motors

No. of Poles	2	4	6	8	12
50-Hz motor	3000	1500	1000	750	500
60-Hz motor	3600	1800	1200	900	600

For 50- and 60-Hz motors, the synchronous speed n_s of the stator flux is given in Table 4.2. The rotor speed n_r is always less than n_s, that is, the induction motor always runs at subsynchronous speed ($n_r < n_s$). The motor performance primarily depends on the rotor slip defined as

$$\text{slip speed} = n_s - n_r = \text{rotor slippage rpm relative to stator flux synchronous speed} \qquad (4.5)$$

$$\text{slip } s = \frac{n_s - n_r}{n_s} = \text{slip speed per unit of stator flux synchronous speed.} \qquad (4.6)$$

Example 4.1

A four-pole, 60-Hz, three-phase, 1740-rpm induction motor runs at 1790 rpm at no load. Determine at rated load (1) the slip speed in revolutions per minute, (2) the slip in perunit and percentage, and (3) the speed regulation.

Solution:

$$\text{Synchronous speed} = 120 \times 60 \div 4 = 1800 \text{ rpm.}$$

The 1740 rpm stated on the nameplate is the speed at rated load, and no-load speed is given as 1790 rpm. Using Equation 4.6, slip speed at rated load = 1800 – 1740 = 60 rpm.

$$\text{Slip at rated load} = (1800 - 1740) \div 1800 = 0.0333 \text{ pu or } 3.33\%.$$

Using Equation 3.15,

$$\text{speed regulation} = (1790 - 1740) \div 1740 = 0.0287 \text{ pu or } 2.87\%.$$

Under normal running operation, the slip is typically a few percent of n_s. If all loads were removed, the motor speed would rise approximately by the slip %, although the exact speed regulation is determined by Equation 3.15, which applies to any source of mechanical power. Therefore, in the first approximation, *motor speed regulation = slip at rated load.* Furthermore, since the rotor sees the stator flux rotating at slip speed, the rotor current frequency is

$$f_r = s \times f, \quad \text{where } f = \text{frequency of the stator supply power (main lines).} \tag{4.7}$$

The motor with load on the shaft, no matter how small, cannot run at synchronous speed. At synchronous speed, $s = 0$, and there would be no rotor current induced and no torque produced on the motor shaft. At a supersynchronous speed ($n_r > n_s$), the rotor sees the flux sweeping in the other direction, reversing the current direction, making the machine work as a generator, converting the shaft mechanical power into electrical power delivered out of the stator terminals to the power lines. Most wind power installations use an induction machine as the generator, driven at supersynchronous speed by the low-speed wind turbine with a high gear ratio.

4.1.1 Performance Characteristics

The induction motor equivalent electrical model has a resistance R and leakage inductance L in both the stator and rotor circuits. There is a substantial amount of stator leakage flux that cannot be ignored. It links only the stator conductors and does not cross the air gap. R and L are constant for the stator but vary in the rotor with the rotor frequency due to a skin effect in large rotor conductor bars. These variations are small and hence ignored here for simplifying the performance analysis without losing much accuracy. Denoting the constant rotor resistance R_r and leakage inductance L_r at stator frequency f, the rotor reactance $X_r = 2\,\pi f_r L_r$ varies directly with the slip frequency $f_r = s \times f$. The three-phase stator voltages produce rotating magnetic flux of amplitude determined by the basic voltage (Equation 2.2). The stator flux, less the leakage flux, flows in the magnetic circuit through the air gap and produces voltage in the rotor conductors equal to (slip × V_{BR}) where V_{BR} = induced voltage in the rotor at blocked rotor (i.e., speed = 0 or slip = 1.0). Since V_{BR} is linearly proportional to the stator voltage V by a constant factor K (equivalent to the rotor-to-stator turn ratio), the rotor current can be expressed as the rotor voltage divided by the rotor impedance at rotor frequency $s \cdot f$, that is,

$$I_r = \frac{sV_{BR}}{\sqrt{R_r^2 + (2\pi s f L_r)^2}} = K_1 \frac{sV}{\sqrt{R_r^2 + (2\pi s f L_r)^2}}. \tag{4.8}$$

Since the torque is given by power divided by speed, the rotor torque = $I_r^2 R_r / \omega_r$. Using Equation 4.8 in this torque expression, we get the motor torque:

$$T_m = K_2 \frac{s^2 V^2}{R_r^2 + (2\pi s f L_r)^2} \frac{R_r}{(1-s)}. \tag{4.9}$$

Even with R_r and L_r constant (i.e., ignoring the skin effect), the motor torque varies with slip in a complex nonlinear manner. The complete torque-speed characteristic has a maximum torque point T_{max} in the middle (called the pull-out or break-down torque) as shown in Figure 4.3. Note that the speed scale has zero on the left-hand side, and the slip scale has zero on the right-hand side. The torque increases linearly with slip on the right-hand side of the hump but inversely with slip on the left-hand side of the hump, which leads to unstable operation. Therefore, the induction motor always runs on the right-hand side of the hump near the synchronous speed where the motor torque equals the load torque at the steady operating speed n_{op}. Figure 4.3 also depicts the line current versus speed. The motor draws high current at the start, which falls as the speed builds up to the steady-state operating speed.

At the start (or at standstill, or locked rotor, or blocked-rotor condition), speed = 0 and $s = 1$, and R_r is negligible compared to high rotor reactance due to high rotor frequency. This simplifies Equation 4.9 to give the approximate starting torque

$$T_{start} = K_m \cdot \frac{V^2}{f^2}. \tag{4.10}$$

Under normal running, with typical low value of $s < 0.05$, the effective rotor reactance is negligible due to low rotor frequency. The running torque, therefore, is approximately given by

$$T_{run} = K_m \cdot s \cdot \frac{V^2}{f}. \tag{4.11}$$

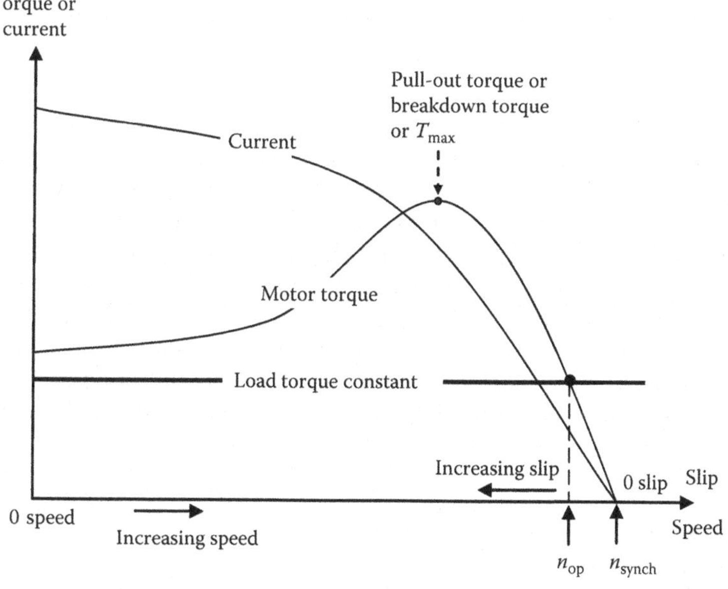

FIGURE 4.3 Torque and current versus speed and slip characteristics of induction motor.

In Equations 4.10 and 4.11, K_m = machine constant, s = rotor slip, V = supply line voltage, and f = supply line frequency. We note in Equation 4.11 that the motor running from constant-voltage and constant-frequency sources produces *torque linearly proportional to the **slip*** (emphasized *slip*, not the speed). There is no simple direct relation between the torque and the speed. Therefore, we must always derive the actual rotor speed from the slip found from the performance analysis.

Example 4.2

A 60-Hz, three-phase, four-pole induction motor delivers full load torque at 3% slip. If the load torque rises by 20% (overload condition), determine the new speed.

Solution:

The 60-Hz, four-pole motor has the synchronous speed of 1800 rpm. Its full load speed at 3% slip would be (1 − 0.03) × 1800 = 1746 rpm.

Figure 4.3 and Equation 4.11 show that the motor torque is linearly proportional to the rotor *slip* in the normal running range of an induction motor. The torque has no direct relation with speed; it must be derived only via the slip relation. Therefore, with a 20% rise in torque, the *slip* will also rise by 20% to new slip = 1.20 × 3 = 3.6%.

The new speed will then be (1− 0.036) × 1800 = 1735.2 rpm. This is a drop of 1746 − 1735.2 = 10.8 rpm, which is a 0.6% drop.

We note here that 20% overload causes the motor speed to drop by merely 0.6% (not by 20%). Obviously, the motor will deliver about 20% higher horsepower at 20% higher load torque at about the same speed.

Thus, the induction motor is not a constant horsepower motor; its horsepower output depends linearly on the torque loading.

Example 4.2 indicates that the induction motor speed drops by only a few percent from no load to full load, that is, it runs essentially at constant speed near the synchronous speed. Thus, the induction motor is not a constant horsepower motor; it delivers horsepower proportional to the torque load. Its speed can be changed only by changing the synchronous speed, which depends on the number of poles and the supply frequency. The motor speed in modern medium- and high-power installations is typically changed by the variable frequency drive (VFD) using power electronics converters. The power is supplied at a constant V/f ratio in order to create the rated flux to have rated torque under the rated armature current at all speeds starting from zero to full running speed. The need for maintaining the constant V/f ratio to avoid magnetic saturation is evident from Equation 2.2.

Example 4.3

A 100-hp, three-phase, 60-Hz, 460-V, four-pole, 1750-rpm induction motor is used in an application, which would load the motor to only 65% of the rated torque (this is obviously a higher-rated motor than we need). Determine (1) whether the motor at 65% load torque runs faster or slower than the rated speed of 1750 rpm, (2) the actual speed, and (3) the actual horsepower delivered to the load.

Solution:

(1) Since the induction motor slip increases with torque load, the speed decreases—although slightly—with increasing load, and vice versa. Thus, the motor would run faster at 65% load than at full-load rated speed of 1750 rpm.

(2) From Equation 4.2, we have

rated torque = rated hp × 5252 ÷ rpm = 100 × 5252 ÷ 1750 = 300 lb/ft.

For a four-pole motor, synchronous speed = 120 × 60 ÷ 4 = 1800 rpm, so the rated slip = (1800 − 1750) ÷ 1800 = 0.0278 pu. Since the induction motor slip linearly varies with torque, at 65% load torque, the actual slip will be 0.65 × 0.0278 = 0.01806 pu, and the speed = 1800(1 − 0.01806) = 1767.5 rpm.

(3) Actual horsepower delivered = (300 × 0.65) × 1767.5 ÷ 5252 = 65.55 hp, which is 65.55% of the rated 100 hp. Thus, we note that the induction motor horsepower output varies linearly with the load torque.

The induction motor efficiency depends on horsepower rating because large motors consuming high power are designed for better efficiency than small ones. As seen in Figure 4.4, large motors over the 500-hp range have efficiency around 93% in the standard design and around 96% in the high-efficiency design. The difference of

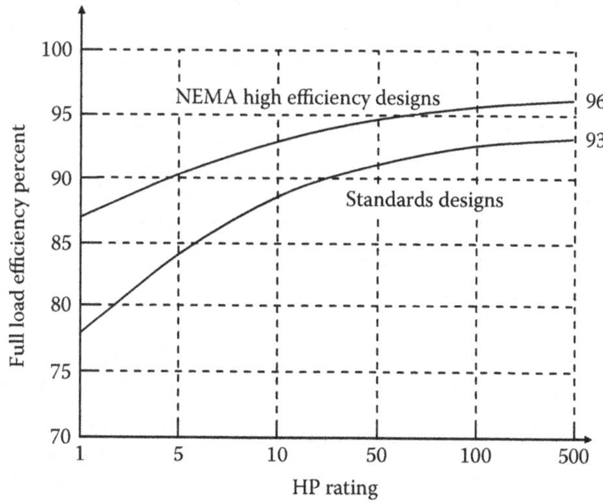

FIGURE 4.4 Induction motor efficiency at full load versus horsepower rating.

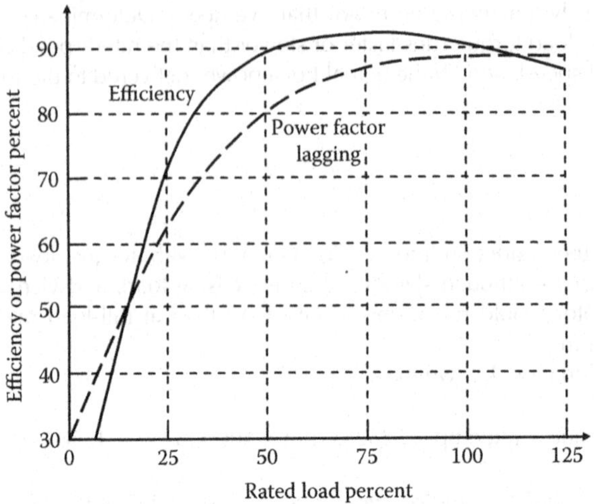

FIGURE 4.5 Standard 100-hp induction motor efficiency and power factor versus load.

3% in efficiency improvement saves significant electrical energy in a motor running for 2000 to 3000 h or longer during the year.

The efficiency and the power factor of a given induction motor depend on the loading level, as shown in Figure 4.5, which is for a standard 100-hp, three-phase, four-pole motor. The efficiency typically peaks at 75% to 80% of the rated power output.

4.1.2 Starting Inrush Kilovolt-Ampere Code

Like all motors, the induction motor starting directly at full line voltage draws high inrush current several times greater than the normal rated current, as seen in Figure 4.3. Such high current leads to the following ill effects:

- Unnecessary motor fuse melting or circuit breaker tripping (a nuisance since motor starting is not an abnormal fault condition).
- Momentary voltage drop in the cable that may disturb other sensitive components around the motor and may cause light flickers.
- Excessive kilovolt-ampere drawn during starting causes heavy stress on the power lines and the source.

The source, cable, and circuit protection, all must be sized for such inrush current, which is usually estimated from the code letter included on the motor nameplate. The code letter is expressed as the starting kilovolt-ampere per horsepower of the motor rating and is known as the locked-rotor or blocked-rotor kilovolt-ampere of the motor. It indicates the severity of the starting stress in terms of the kilovolt-ampere per horsepower it draws from the source on direct line starting. The range of codes found on motor nameplates is given in Table 4.3, which varies from a low

TABLE 4.3
Motor Starting Kilovolt-Ampere Codes

Code Letter	Starting Kilovolt-Ampere per Horsepower	Code Letter	Starting Kilovolt-Ampere per Horsepower
A	0–3.14	L	9.0–9.99
B	3.15–3.54	M	10.0–11.19
C	3.55–3.99	N	11.2–12.49
D	4.0–4.49	P	12.5–22.99
E	4.5–4.99	R	14.0–15.99
F	5.0–5.59	S	16.0–18.99
G	5.6–6.29	T	18.0–19.99
H	6.3–7.19	U	20.0–22.39
J	7.2–7.99	V	>22.4
K	8.0–8.99	–	–

Note: Adapted from NEMA.

of 3.14 to a high of 22.4 kVA/hp. A motor with a higher code letter produces higher stress on the power system design.

Example 4.4

A 100-hp, three-phase, 60-Hz, 460-V, 1710-rpm, four-pole, letter code J, squirrel cage induction motor with Δ-connected stator has efficiency of 90% and power factor 0.85 lagging when operating at rated load. Determine the input kilowatt, kilovolt-ampere, and line current during normal operation and the line current on direct start in Δ.

Solution:

$$\text{Output power} = 100\text{ hp} \times 746 \div 1000 = 74.6\text{ kW}$$

$$\text{Input power} = 74.6 \div 0.90 = 82.89\text{ kW}$$

$$\text{Apparent power} = 82.89 \div 0.85 = 97.52\text{ kVA}$$

$$\text{Line current} = 97.52 \times 1000 \div (\sqrt{3} \times 460) = 122.4\text{ A.}$$

The starting current at rated voltage (i.e., direct online start in Δ, *without* using a *Y*–Δ or reduced voltage or reduced frequency starter) is given by the letter code found on the motor nameplate. For this letter code J motor, Table 4.3 gives 7.2–7.99 kVA/hp at starting, so we use the conservative number 7.99. The starting inrush kVA = 7.99 × 100 = 799 kVA.

$$I_{\text{direct}\,\Delta\,\text{start}} = 799 \times 1000 \div (\sqrt{3} \times 460) = 1004 \text{ A}.$$

This is 8.2× the normal rated current of 122.4 A. However, 1004 A will gradually decay to 122.4 A as the motor gains speed.

4.1.3 Torque–Speed Characteristic Matching

Matching the torque versus speed characteristic of an induction motor with that of the load is an important consideration for the system design. The motor connected to a high torque load during the start must have a high starting torque, or else it would not start. Some motors have high starting torque but lower efficiency compared to others. The high starting torque requires a rotor with a high resistance, resulting in poor efficiency under running conditions. The National Electrical Manufacturers Association (NEMA) classifies motor designs according to the starting torque and the maximum breakdown (pull-out) torque as depicted in Figure 4.6 and listed in Table 4.4. The motor design features depicted in Figure 4.5 are as follows.

- Design B is a general-purpose motor used in centrifugal pumps, fans, blowers, and machine tools requiring low starting torque.
- Design A is also for general purpose, but it has a higher pull-out torque than the Design B motor.
- Design C motor offers a good compromise between high starting torque and high running efficiency. It is suitable for plunger pumps and compressors.

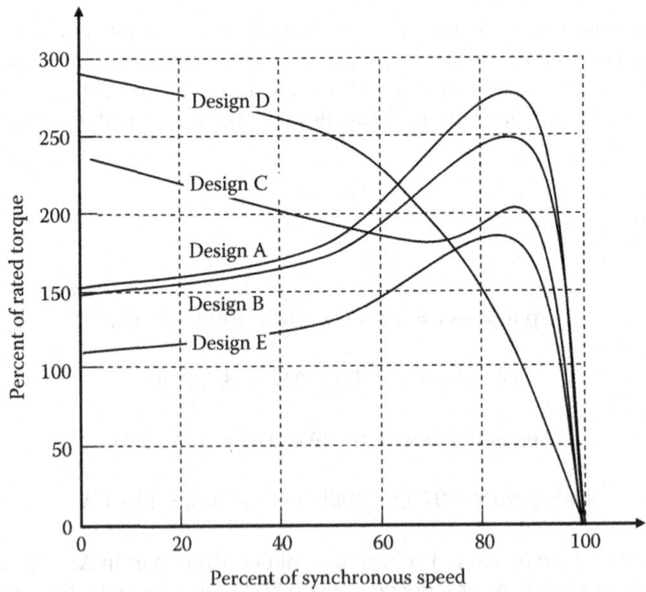

FIGURE 4.6 Induction motor torque versus speed characteristics by design classification.

TABLE 4.4
Starting Torque and Maximum Pull-Out Torque of Induction Motors in Selected HP Ratings

	Starting Torque (% of Rated Torque)			Pull-Out Torque (% of Rated Torque)		
HP Rating	**2-pole**	**4-pole**	**8-pole**	**2-pole**	**4-pole**	**8-pole**
			Design A, B			
1	-	275	135	250	300	215
5	150	185	130	215	225	205
10–200	135–100	165–100	125–120	200	200	200
250–500	70	80	100	175	175	175
			Design C			
5	-	255	225	-	200	200
25–200	-	200	200	-	190	190
			Design E			
1–20	180–140	190–150	150–120	200	299	180
50–100	120–100	130–110	120–100	180	180	170
300–500	80–75	90–75	90–75	160	160	160

- Design D motor, having high rotor resistance, has the highest starting torque but has the lowest efficiency under full-load running condition. It is primarily used with high-inertia loads, such as flywheel punch presses, elevators, cranes, hoists, barge positioning, rail car pulling, ship and barge mooring, etc. Capstans and winches in ships also use Design D motors for smooth starting under load and 300% starting torque.
- Design E motor, having low rotor resistance, gives the lowest starting torque but highest running efficiency. It costs more but pays back in energy savings where the load runs for extended periods of time over the year. We see in Table 4.4 that Design E motors in 300–500 hp ratings have starting torque less than the rated torque. These motors, therefore, must be brought to speed under light load and then fully loaded after acquiring full speed. For this reason, they are primarily used for centrifugal pumps, fans, blowers, and machine tools, all requiring low starting torque.

Since the induction motor consumes 93% of all the electrical energy used by all motors combined, its efficiency reporting on the nameplate has become the industry standard. However, various international standards for testing and reporting the induction motor efficiency on the nameplate can be significantly different. For example, the table below compares the efficiency and power factor of a 70-hp induction motor derived from three international standards, namely, the American Standards for testing induction motor efficiency (IEEE-112), the International Electrotechnical

Commission Standards (IEC34-2), and the Japanese Electrotechnical Commission Standards (JEC-37):

	IEEE-112	IEC34-2	JEC-37
Efficiency	90.0	92.7	93.1
Power factor	86.2	86.2	86.3

Although the pf test results are consistent, the efficiency tests leave much to be desired. The IEEE-112 standard results in the lowest efficiency. The difference comes from the different treatments of the stray-load loss (miscellaneous losses due to stray leakage flux not easy to quantify), which vary approximately with the load squared. The IEEE method derives it indirectly from tests, the IEC34-2 assumes it to be fixed at 0.5% of the rated power, whereas the JEC-37 ignores it altogether, resulting in the highest efficiency. The point here is that a high-efficiency motor made in the United States cannot be compared with a high-efficiency motor made in Japan for the energy-saving considerations.

4.1.4 Motor Control Center

The motor control center design requires specifying the following:

(1) The continuous and sustained short-circuit amperes that the bus must withstand without thermal and mechanical damage. Typical continuous current ratings of the motor control centers are 600, 800, and 1000 A, with typical short-circuit ampere rating of 40, 65, and 100 kA.
(2) The motor starter size (usually Y–Δ starter) as specified by the NEMA frame size below:

Motor hp rating:	<2	5	10	25	50	75	150	300
NEMA frame size:	00	0	1	2	3	4	5	6

As an example, we read from the table above that a 30-hp motor would need a NEMA frame size 3 starter (rounded up to the nearest standard size).
(3) Enclosure type, that is, ventilated dip proof or sealed watertight.
(4) Protection type, that is, fuse or molded case circuit breaker. The molded case breaker trips all three phases together to eliminate single phasing and is usually of a single element type without the thermal element. For this reason, the fuse protection, having a quick fusible link that melts quickly at higher current, has generally higher short-circuit current ratings and limits the damage.

4.1.5 Performance at Different Frequency and Voltage

The induction motor is often used with VFDs. At lower-than-rated frequencies, the drive also lowers the voltage to maintain the V/f ratio constant. This gives constant air-gap flux and constant torque at the rated armature current, as shown in

Figure 4.7a. At higher-than-rated frequencies, the motor is operated at rated voltage to lower the flux and torque as shown in Figure 4.7b. This maintains constant horsepower to avoid high armature current and overheating.

At constant frequency and reduced voltage, the maximum torque point shrinks in voltage-squared proportions as shown in Figure 4.7c. Such variable voltage operation is often used for speed control of small motors in fans and pumps, where the load torque is proportional to speed squared. At constant frequency and constant voltage, the motor torque varies with rotor resistance as shown in Figure 4.7d. High rotor resistance gives high starting torque, although with poor running efficiency. In a wound rotor induction motor, an external resistance is inserted in the rotor via slip

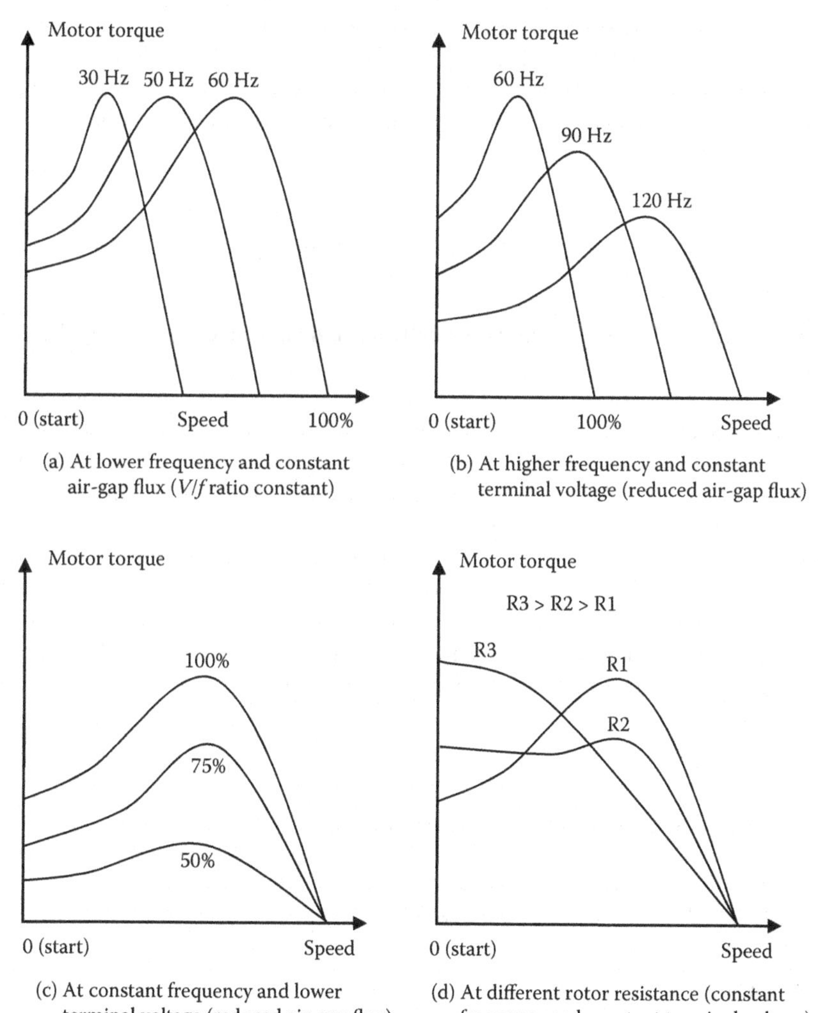

FIGURE 4.7 Induction motor torque versus speed curves at different frequencies and voltages.

rings and brushes to get high starting torque for heavy loads. Once the motor comes to full speed, the external resistance in the rotor circuit is reduced to zero by shorting the slip rings to gain better running efficiency.

4.2 SYNCHRONOUS MOTOR

The synchronous motor has the same construction as the synchronous generator, except that the current and energy conversion directions are reversed. The three-phase stator currents produce the rotating magnetic field as in the induction motor, and dc current is injected in the rotor coil from outside through brushes and slip rings. Therefore, it costs more and requires more maintenance compared with the induction motor. The synchronous motor works on the magnetic attraction between the stator and the rotor magnetic fields, as opposed to the current induced in the rotor by the sweeping stator flux in the induction motor. With a stator with P number of poles and supply frequency f, the synchronous motor runs exactly at synchronous speed at all loads, which is given by

$$\text{synchronous speed in rpm} = \frac{120f}{P}. \tag{4.12}$$

Since the synchronous motor is typically used in large size, it is usually driven by a VFD for speed control.

The electrical model is the same as that for the synchronous generator we covered in Chapter 3, except that the armature current now reverses, and Kirchhoff's voltage law (KVL) gives $\tilde{V}_T = \tilde{E}_f + j\tilde{I}X_s$. The phasor diagram, therefore, changes accordingly, as shown in Figure 4.8. The excitation current that determines E_f controls the motor power factor. Overexcitation results in a leading power factor, and underexcitation results in a lagging power factor. Some synchronous motors are designed to continuously operate in the overexcitation mode, providing leading kVARs to improve the system power factor, like capacitors. For this reason, an overexcited synchronous motor with no mechanical load—running only to supply leading kVARs to the system—is called the synchronous condenser (capacitor).

Example 4.5

A 10,000-hp, 60-Hz, 1200-rpm, 4.1-kV, three-phase, Y-connected synchronous motor draws the armature current of 1500 A at 0.80 leading power factor under rated load. It has synchronous reactance of 1.2 Ω/ph and negligible resistance. Determine the field excitation voltage (counter emf) $\tilde{E}_f$ and the torque angle δ.

Solution:

We make per-phase calculations. For Y-connection, $V_T = 4100 \div \sqrt{3} = 2367$ V, which we take as the reference phasor. The armature current at 0.8 power factor leading is $\tilde{I}_a = 1500\angle\cos^{-1} 0.80 = 1500\angle 36.87°$ A.

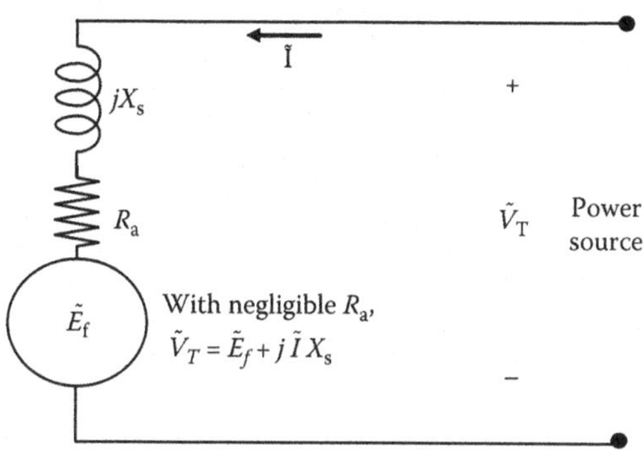

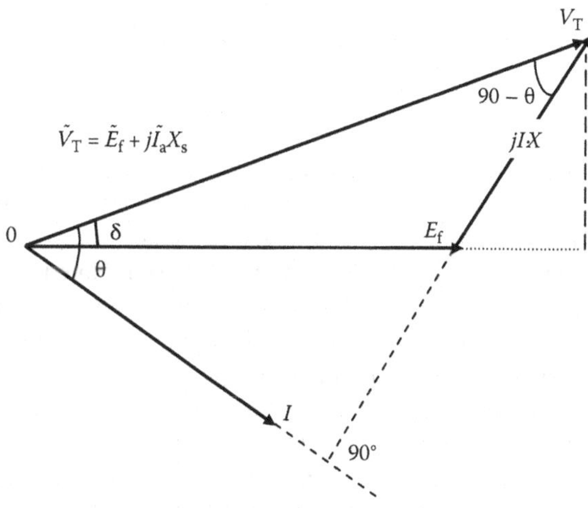

FIGURE 4.8 Synchronous motor electrical model and phasor diagram.

Based on the phasor diagram in Figure 4.8, modified for leading pf, the counter emf in the motor is given by

$$\tilde{E}_f = \tilde{V}_T - \tilde{I}_a \times jX_s = 2367\angle 0° - 1500\angle 36.87° \times 1.2\angle 90°$$
$$= 2367\angle 0° - 1800\angle 126.87° = 3447 - j1440 = 3735.7\angle -22.67°.$$

Therefore, the torque angle $\delta = -22.67°$ at rated load. The negative sign indicates that the counter emf lags the voltage applied at the terminals, that is, the rotor magnetic field lags (follows) the stator magnetic field. In the synchronous generator, the opposite is true.

The synchronous motor performance analysis under mechanical load is similar to that of the synchronous generator, resulting in the torque output at the motor shaft given by

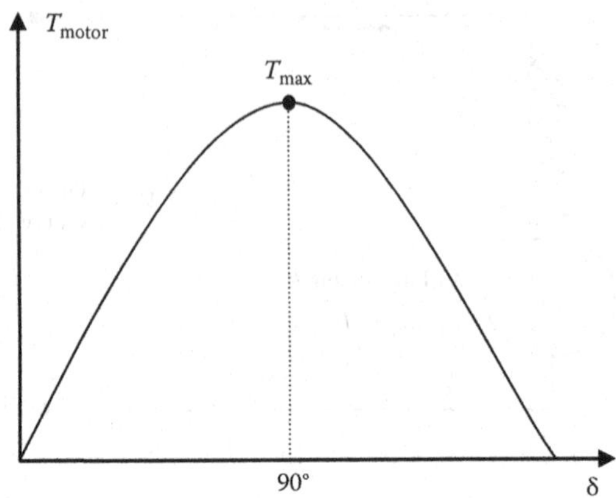

FIGURE 4.9 Synchronous motor torque versus torque angle δ.

$$T_{motor} = T_{max} \sin \delta \tag{4.13}$$

where T_{max} = maximum (pull-out) torque the motor can develop, and δ = torque angle (electrical degrees). The T_{motor} versus δ plot shown in Figure 4.9 is similar to that of P_{gen} versus δ for the synchronous generator, and the same steady-state and transient stability issues with step load changes apply here as well. Note that δ is called the torque angle in motor, as opposed to the power angle in generator, but it is still the angle between the center lines of the stator and the rotor magnetic fields.

Equation 4.13 gives the maximum motor torque $T_{motor} = T_{max}$ at δ = 90°. Beyond this point, the motor becomes unstable, falls out of synchronous operation, and stalls. The motor is normally designed to operate at δ equal to 25°–35° at rated load, and step load changes are kept below 30%–35% of the rated torque to assure transient stability. The damper bars here serve the same purpose as in the synchronous generator. The current induced in them during transient oscillations around the synchronous speed damps the electromechanical oscillations following a step load change. The partial cage formed by the damper bars and end chords is used for starting the synchronous motor as the induction motor under a reduced voltage and/or frequency before exciting the rotor field that eventually pulls the rotor in the synchronous mode. Then onward, the rotor runs at the synchronous speed in lock step with the stator rotating magnetic field. The torque is produced by the magnetic attraction between the two fields, with the rotor field trailing the stator field by angle δ. This is the reverse of the synchronous generator, where the rotor field leads the stator field by angle δ. One can think this way: the magnetic field of the primary source of power leads the other field by angle δ; the one that has the energy to supply leads the other, as found in most other places around us, such as in machines or in people.

Example 4.6

A synchronous motor is running at 70% rated load torque when the torque angle is 20°. Determine its pull-out torque T_{max} (steady-state stability limit torque) in percentage of the rated torque. If the load torque is increased to 100% rated torque in one step, determine the worst-case transient peak value of the torque angle. Ignore the damping bar contribution for a conservative estimate and assume the torque versus δ relation approximately linear in the operating range.

Solution:

The synchronous motor torque is related with the torque angle δ by $T = T_{max} \sin\delta$. At 70% load operation, $0.70\ T_{rated} = T_{max} \sin 20° = 0.342\ T_{max}$, which gives

$$T_{max} = (0.70 \div 0.342)\ T_{rated} = 2.05\ T_{rated}, \text{ or } 205\% \text{ of rated torque.}$$

Under the step change in the load torque from 70% to 100%, the new δ after the transient swing subsides will be $T = T_{max} \sin \delta_{new}$, that is, $1.0 \times T_{rated} = 2.05 \times T_{rated} \sin\delta_{new}$.

$$\text{Therefore, } \sin\delta_{new} = 1.0 \div 2.05 = 0.488, \text{ which gives } \delta_{new} = 29.2°.$$

If we ignore damping in the first cycle, and assume sin δ approximately linear, then the worst overshoot due to the rotor's mechanical inertia will be 100% of the change in δ, that is, $\delta_{overshoot} = 29.2° - 20° = 9.2°$, and $\delta_{peak} = 29.2 + 9.2 = 38.4°$.

The rotor angle will swing from 20° to 38.4° and will come back to 20° and then swing again upward. Such oscillations with 9.2° amplitude around the new. steady-state value of δ = 29.2° will continue until the swing energy is gradually dissipated by I^2R power loss in the rotor damper bars, cycle by cycle.

Since the worst-case swing peak of 38.4° is well below the 90° that is required for the transient stability, the machine operation would be stable during and after the rotor oscillations following the step load change.

The rotor of the synchronous motor can have any one of the following configurations:

- Conventional rotor with dc excitation current injected from outside via slip rings and brushes
- Brushless rotor with exciter and diodes on the rotating shaft of the main motor
- Permanent magnet rotor (brushless, as no dc current is needed on the rotor)
- Reluctance rotor (brushless, as no dc current is needed; it works on self-induced magnetism in the ferromagnetic rotor)

The rotor with slip rings and brushes is the least desirable design because of its low reliability and high maintenance. Most large synchronous motors are of brushless design. The permanent magnet synchronous motor has been widely used in small and medium power applications but has recently drawn a good deal of interest in large high-power density motors for electric propulsion in navy and cruise ships. The neodymium–iron–boron permanent magnet is common in such designs since it offers high magnetic strength.

4.3 MOTOR HORSEPOWER AND LINE CURRENT

The motor delivers horsepower to drive load torque T at speed n. In an ac or dc motor, the three are related as follows:

$$\text{hp} = \frac{T_{\text{lb/ft}} n_{\text{rpm}}}{5252} = \frac{T_{\text{N·m}} \omega_{\text{mech rad/s}}}{746}. \tag{4.14}$$

The nameplate fixed on the induction and synchronous motors includes the following key motor ratings: horsepower or kilowatt output, single-phase or three-phase, frequency, line-to-line voltage, speed (rpm), temperature rise (°C), service factor (SF), power factor (PF), starting kilovolt-ampere code, efficiency, and design type, where rpm = shaft speed at rated voltage and frequency when delivering rated horsepower; temperature rise = average temperature rise of the conductor above standard 40°C ambient air under continuous rated load; and SF = service factor = ratio of the maximum permissible continuous overload to the rated load.

The full-load line current I_L drawn by the three-phase induction or synchronous motor connected to line-to-line voltage V_{LL} can be determined from the following relation:

$$\sqrt{3} \cdot V_{\text{LL}} \cdot I_{\text{L}} \cdot \text{PF} = \frac{746 \cdot \text{hp}}{\text{efficiency}} = \frac{1000 \cdot \text{kW}}{\text{efficiency}}. \tag{4.15}$$

The output rating of a motor pumping fluid at a flow rate of Q m³/h against the total pressure head H meters of fluid with specific gravity SG and pump efficiency η_{pump} is given by

$$\text{motor kW} = \frac{Q \cdot H \cdot \text{SG}}{367 \cdot \eta_{\text{pump}}}$$

and

$$\text{motor HP} = \frac{\text{motor kW}}{0.746}, \tag{4.16}$$

where the total pressure head H = static pressure + friction loss in pipes + loss in pump.

The specific gravity of water is 1.0, but that of the crude oil varies from 0.76 to 0.85 depending on the type and temperature as listed in Table 4.5.

Example 4.7

A 100-hp, three-phase, 460-V, 60-Hz, ac motor has a power factor of 0.90 lagging and efficiency of 94% on its nameplate. Determine the line current it will draw while delivering rated horsepower.

TABLE 4.5
Specific Gravity of Crude Oil at Various Temperatures

	Temperature (°F)	Temperature (°C)	Specific Gravity*
Crude oil 48° API	60 130	15.6 54.4	0.79 0.76
Crude oil 40° API	60 130	15.6 54.4	0.825 0.805
Crude oil 35.6° API	60 130	15.6 54.4	0.847 0.824
Crude oil 32.6° API	60 130	15.6 54.4	0.832 0.84

* Specific gravity = 1.00 for water at 60°F.

Solution:

Power drawn from the source = output ÷ efficiency = 100 hp × 746 W ÷ 0.94 efficiency = 79,362 W. Using the general three-phase power relation $P_{3\text{-ph}} = \sqrt{3}\ V_{LL} I_L$ pf,

$$\text{motor line current } I_L = 79{,}362 \div (\sqrt{3} \times 460 \times 0.90) = 110.68 \text{ A.}$$

Example 4.8

A water pumping system in a ship is designed to pump 90 m³/h of fresh water against 80 m of total pressure head. Determine the motor horsepower rating and the motor control center kilovolt-ampere rating, assuming 94% pump efficiency, 96% motor efficiency, and 90% pf lagging.

Solution:

The total head of 80 m may be partly due to elevation difference between the water entrance and discharge and partly due to pipe friction; it does not matter to us how it is divided between the sources of the pumping head.

Using Equation 4.16,

$$\text{motor kW} = \frac{90 \times 80 \times 1.0}{367 \times 0.94} = 20.87 \text{ and motor HP} = \frac{20.87}{0.746} = 28.$$

Motor input power in kilowatts = 20.87 ÷ 0.96 = 21.74; motor control center kVA rating = 21.74 ÷ 0.90 = 24.16.

4.4 DUAL-USE MOTORS

Dual-frequency motors, often used on ships, are designed to operate either from 50- or 60-Hz supply around the globe. The motor must be operated at the design frequency, or else its performance will severely degrade. For example, if a 60-Hz motor is connected to 50-Hz lines, it will run at 50/60 = 0.8333 or 83.33% of the 60-Hz rated speed, that is, 16.67% slower. If the same voltage is applied, its magnetic core would have 16.67% higher flux and would saturate, drawing significantly higher magnetizing current, leading to severe overheating. To avoid such overheating, 60-Hz motor voltage and horsepower ratings must also be reduced by 16.67% when operating at 50 Hz. Under such reduced load, however, the starting torque and the pull-out torque remain essentially the same as in 60-Hz operation.

In the reverse, a 50-Hz motor can be operated at 60 Hz with appropriate derating. However, its speed will be 60/50 = 1.2 × rated speed, or 20% over the design speed; hence, the torque load must be reduced by 20% to keep the horsepower load the same. Dual-frequency motors are designed to operate at either 50 or 60 Hz safely, but with appropriate derating.

Example 4.9

A three-phase, 460-V, 60-Hz induction motor with a Δ-connected stator is designed to have the magnetic flux density in the core at the saturation limit of 1.6 T. If this machine is connected to 50-Hz lines, determine the voltage for limiting the flux density to the same 1.6-T saturation limit.

Solution:

Each phase of the Δ-connected stator sees the line voltage. Using Equation 2.2 at 60 Hz, we have $V_{60\ Hz} = 460 = 4.444 \times 60 \times N \times \phi_m$. At 50 Hz, we have $V_{50Hz} = 4.444 \times 50 \times N \times \phi_m$.

Taking the ratio of the two with the same number of turns N and to keep the same flux ϕ_m, we have $V_{50\ Hz}/460 = 50/60 = 0.8333$, or $V_{50\ Hz} = 0.8333 \times 460 = 383$ V. Thus, the voltage should be reduced in the same proportion as the frequency reduction.

The motor at 50 Hz will run at 50/60 = 0.8333 × 60-Hz rated speed. To avoid overheating, the current must be limited to the same 60-Hz rated value, which would give the same 60-Hz rated torque under constant air-gap flux.

Since horsepower = torque × speed, the horsepower loading at 50 Hz must be reduced to 50/60 = 0.833 or 83.33% of the 60-Hz horsepower rating.

Dual-speed motors are designed with multiple groups of stator coils that can be reconnected in two configurations in which the number of poles can be changed by a factor of 2. Therefore, the dual-speed motors are also known as the pole-changing motors. They are used for house air-blowers, fluid transfer pumps, ballast pumps, and saltwater pumps on ships.

Dual-voltage motors are designed with multiple groups of stator coils that can be connected in one of the two alternative—series or parallel—configurations, such that the stator terminal voltage can be changed by a factor of 2. The motor is normally connected to high voltage to result in low current and low I^2R power loss. For example, a 230–460 V dual-voltage motor is normally connected for 460-V operations. In small sizes, they are used in single-phase portable fan motors.

4.5 UNBALANCED VOLTAGE EFFECT

The single-phasing (losing one line voltage completely) is an extreme example of voltage unbalance. Small unbalance in three-phase voltages occurs when single-phase loads are not uniformly distributed on three phases. The unbalance in three-phase voltage is defined as the maximum deviation in three line-to-line voltages from the average of three line voltages, that is,

$$\text{unbalance voltage UBV} = \frac{\text{max voltage deviation from average}}{\text{average line voltage}} = \frac{V_{\text{max deviation from average}}}{\left(V_{ab} + V_{bc} + V_{ca}\right)/3}. \tag{4.17}$$

Experience indicates that the additional temperature rise under unbalanced voltage is given by

$$\%\Delta T = 2\ (\%\text{UBV}) \text{ and } \Delta T = (\%\Delta T/100) \times T_{\text{rated}}. \tag{4.18}$$

Even a few percent of unbalance in three-phase voltages can degrade the motor efficiency and torque and increase the heating. The resulting overheating can be avoided by reducing the motor horsepower load, that is, the motor must be derated, as shown in Figure 4.10. Otherwise, the motor life will get shorter.

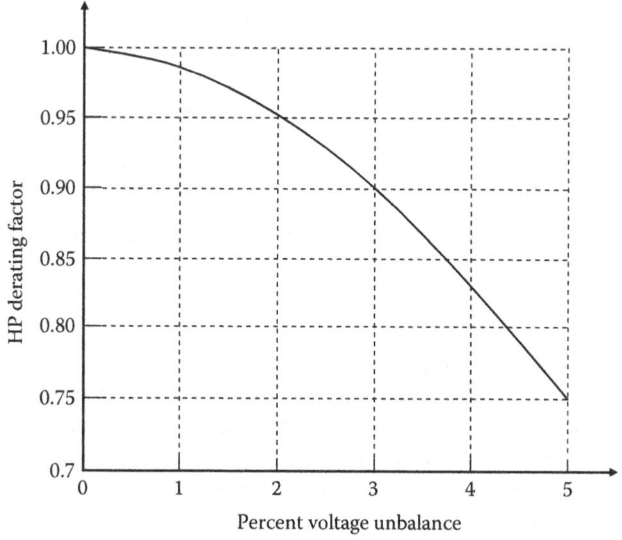

FIGURE 4.10 Three-phase motor derating factor versus voltage unbalance.

Example 4.10

A 150-hp, three-phase, 460-V induction motor has the design life of 20 years at 110°C rise in conductor temperature above the ambient air. If it is continuously operated under unbalanced line voltages of 460, 440, and 420 V, determine (1) the percent voltage unbalance, (2) the additional temperature rise when operating at rated load, (3) the expected life at rated load, and (4) the required horsepower derating for the motor to last the design life of 20 years.

Solution:

(1) V_{avg} = 1/3 (460 + 440 + 420) = 440 V. The voltage deviations are |460 – 440| = 20, |440 – 440 | = 0, and |460 – 440| = 20. Therefore, using Equation 4.17, unbalance in voltage, UBV = 20 ÷ 440 = 0.0455 pu or 4.55%. (2) Additional temperature rise is $\%\Delta T = 2(\%UBV) = 2(4.55\%) = 9.1\%$. Therefore, ΔT_{UBV} = (0.091) 110°C = 10°C, that is, the motor would run 10°C hotter than the design temperature rise of 110°C. (3) Using Equation 2.24

$$\text{expected life} = \frac{20}{2^{\left(10/10\right)}} = 10 \text{ years.}$$

The horsepower derating required for maintaining its design life of 20 years is derived from Figure 4.10, which shows that at 4.55% UBV, the motor horsepower should be derated to 0.78 of the normal rating, that is, 150 × 0.78 = 117 hp.

The adverse effects of single-phasing or unbalanced voltages can be analyzed in terms of symmetrical components presented in Appendix A. The analysis briefly runs as follows. The unbalanced voltages are resolved into symmetrical component voltages in the positive sequence (rotating counterclockwise), negative sequence (rotating clockwise), and zero sequence (not rotating at all). These sequence components are derived by the following equation set, where the operator a = 120° phase shift (similar to operator j = 90° phase shift we routinely use to analyze ac power circuits):

$$\begin{aligned}
&\text{positive sequence voltage of phase A, } \tilde{V}_{a1} = 1/3\left(\tilde{V}_a + a\cdot\tilde{V}_b + a^2\cdot\tilde{V}_c\right)\\
&\text{negative sequence voltage of phase A, } \tilde{V}_{a2} = 1/3\left(\tilde{V}_a + a^2\cdot\tilde{V}_b + a\cdot\tilde{V}_c\right)\\
&\text{zero sequence voltage of phase A, } \tilde{V}_{a0} = 1/3\left(\tilde{V}_a + \tilde{V}_b + \tilde{V}_c\right)
\end{aligned} \tag{4.19}$$

Note that operator a = 120° phase shift and a^2 = 240° phase shift. Once we calculate the sequence components in phase A, the positive sequence component voltages in phase B will be shifted by 120°, and that in phase C will be shifted by 240°, that is, $\tilde{V}_{b1} = a\cdot\tilde{V}_{a1}$ and $\tilde{V}_{c1} = a^2\cdot\tilde{V}_{a1}$. The negative sequence components are $\tilde{V}_{b2} = a^2\cdot\tilde{V}_{a1}$ and $\tilde{V}_{c2} = a\cdot\tilde{V}_{a2}$. The zero sequence components are $V_{b0} = V_{a0}$ and $V_{c0} = V_{a0}$. All these voltages are per phase (not line-to-line).

We see here that (1, a, a^2) represents a balance symmetrical three-phase phasors of unit magnitudes; V_{a1} is a phase A component of the unbalance voltages (V_a, V_b, V_c)

on the balance positive sequence frame (1, a, a^2); V_{a2} is a phase A component of the unbalance (V_a, V_b, V_c) on the balance negative sequence frame (1, a^2, a); and V_{a0} is a phase A component of the three unbalance voltages (V_a, V_b, V_c) on the zero sequence frame (1, 1, 1). For balanced three-phase voltages, the phasor sum $\left(\tilde{V}_a + \tilde{V}_b + \tilde{V}_c\right)$ adds up to zero, giving $V_{a0} = 0$, which we know to be true. For balanced three-phase voltages, V_{a2} will also reduce to zero, leaving only V_{a1} to have a nonzero value, which will be the actual value of $\tilde{V}_a$.

The sequence voltages applied to the motor will produce sequence currents and corresponding sequence fluxes. However, since most motors are connected in Δ, or ungrounded Y on the stator side, the three line currents must add up to zero in the absence of any return path. Therefore, no zero sequence current can flow into the motor, and this need not be considered any further. The performance of the motor with both positive and negative sequence voltages can be determined by using the superposition theorem, that is, by calculating the motor performance under the positive and negative sequence voltages separately and then superimposing the results. It can also be realized by the following reasoning.

The positive sequence voltage produces the magnetic flux rotating at synchronous speed in the positive direction and torque in that direction. The induction motor runs in the positive direction of rotating flux with a small slip. If a 60-Hz motor is running at a 0.05 perunit slip, the rotor current frequency is 0.05 × 60 = 3 Hz, which will produce rated rotor current and I^2R loss. However, the negative sequence voltage produces the flux rotating backward at synchronous speed. For the rotor running in the forward direction at 95% of the synch speed, the slip from the negative sequence flux is 1.95. For such a high slip, Equation 4.8 gives much higher rotor current, producing correspondingly high I^2R loss and negative torque. Both of these effects result in wasted energy and overheating, costing more in the energy bill and reducing the motor life.

Example 4.11

A three-phase induction motor in a 690-V distribution system under an abnormal condition is working on highly unbalanced three-phase voltages $\tilde{V}_a = 450\angle 0°$, $\tilde{V}_b = 450\angle{-90}°$, and $\tilde{V}_c = 900\angle{-225}°$. Determine its symmetrical component voltages V_a^+, V_a^-, and V_a^0, and comment on approximate motor current and power losses in the rotor and stator.

Solution:

The symmetrical components of the unsymmetrical voltages are derived from an equation similar to Equation A.3 in Appendix A, with all angles in degrees:

$$\begin{pmatrix} V_a^0 \\ V_a^+ \\ V_a^- \end{pmatrix} = \frac{1}{3}\begin{pmatrix} 1 & 1 & 1 \\ 1 & a & a^2 \\ 1 & a^2 & a \end{pmatrix}\begin{pmatrix} V_a \\ V_b \\ V_c \end{pmatrix} = \frac{1}{3}\begin{pmatrix} 1 & 1 & 1 \\ 1 & 1\angle 120 & 1\angle 240 \\ 1 & 1\angle 240 & 1\angle 120 \end{pmatrix}\begin{pmatrix} 450\angle 0 \\ 450\angle{-90} \\ 900\angle{-225} \end{pmatrix}.$$

The above matrix equation leads to

$$V_a^0 = 1/3\{450\angle 0 + 450\angle{-90} + 900\angle{-225}\} = 87.75\angle{-225°}$$

$$V_a^+ = 1/3\{450\angle 0 + 450\angle 30 + 900\angle 15\} = 590\angle 15°$$

$$V_a^- = 1/3\{450\angle 0 + 450\angle 150 + 900\angle{-105}\} = 222.3\angle{-105°}.$$

The positive sequence voltage is 590 ÷ 690 = 0.855 pu of the rated voltage, which will produce 85.5% flux rotating in the positive direction. The negative sequence voltage of 222.3 ÷ 690 = 0.322 pu will produce 32.2% of rated flux rotating in the reverse direction. The zero sequence voltage of 87.75 ÷ 690 = 0.127 will produce 12.7% of rated flux not rotating at all (alternating but stationary in space).

If the rotor were running under such unbalanced voltage in the positive (forward) direction at 0.05 pu slip, then the rotor slip would be 2 – 0.05 = 1.95 pu under the negative sequence flux and 0.95 pu under the zero sequence flux. Such high slips of 1.95 and 0.95 pu will produce excessive currents and I^2R power losses in both the rotor and the stator, soon burning the motor if the situation is not corrected.

4.6 DC MOTOR

The placement of the armature and the field coils in a dc machine is reversed from that in the ac machine. The dc motor has the field coil on the stator and the armature coil on the rotor. The dc is injected in the rotor (armature) via the commutator and brushes. Such construction features result in high capital and maintenance costs.

The electrical model of the most widely used dc motor—the shunt motor—is shown in Figure 4.11. The rotating armature in the dc flux produces voltage E_a that works as the back voltage. The applied terminal voltage V_T must overcome E_a and the armature voltage drop I_aR_a to push the current in. The motor speed would settle when the following Kirchhoff's voltage law (KVL) equation is satisfied:

$$V_T = E_a + I_aR_a \text{ where } E_a = K_m\,\phi_p\,n \tag{4.20}$$

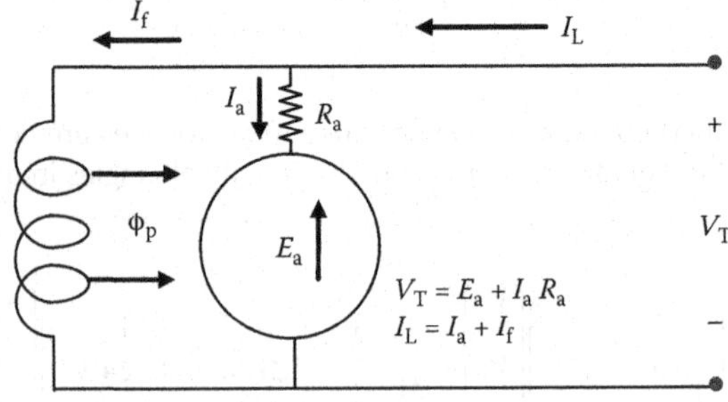

FIGURE 4.11 DC shunt motor electrical model.

and I_a = armature current, R_a = armature resistance, ϕ_p = flux per pole that depends on the field current I_f, n = rotor speed, and K_m = machine constant. Equation 4.20 can be rearranged in terms of the armature current or the speed as follows:

$$I_a = \frac{V_T - K_m \phi_p n}{R_a}$$

or

$$\text{speed } n = \frac{V_T - I_a \cdot R_a}{K_m \phi_p}. \tag{4.21}$$

Thus, the speed can be controlled by varying the applied voltage or by changing the flux per pole ϕ_p, which depends on the field current I_f in a linear nonsaturated magnetic core. Since the dc motor has relatively simple speed–torque–current relations, its speed control and the associated power electronics are simpler and lighter than those for the induction and synchronous motors. This is the main reason for the dc motor's wide use in the past where the speed control was required.

The starting current in the armature derived from Equation 4.21 with $n = 0$ is given by $I_{start} = V_T/R_a$, which is high since R_a is small. This high inrush current decays to the rated value (Figure 4.12) as the motor builds up the speed and the back voltage. To avoid overheating due to high starting current, large motors with high inertia need reduced-voltage starting until the speed builds up and the current approaches the normal rated value.

The steady-state full-load line current I_L drawn by a dc motor from line voltage V_T is given by the following equation:

$$V_T \cdot I_L = \frac{746 \cdot \text{hp}}{\text{motor efficiency}}. \tag{4.22}$$

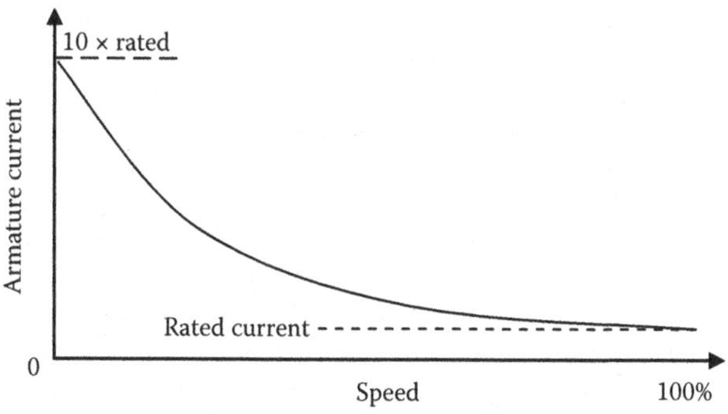

FIGURE 4.12 DC shunt motor current from start to full speed.

Example 4.12

A 50-hp, 240-V dc motor runs at 2350 rpm at rated conditions. If its field current is decreased by 20% while keeping the load current the same, determine its approximate and exact speed.

Solution:

The motor is assumed to be shunt type, as it is the most widely used dc motor. Its speed changes inversely with the field current under the same loading. Therefore, a 20% decrease in the field current will increase the motor speed approximately by 20% to 2350 × 1.2 = 2820 rpm.

For exact speed, new field current = 0.80 × rated field current:

$$\text{Therefore, exact new speed} = \frac{\text{rated speed}}{0.80} = 2937.5 \text{ rpm.}$$

The difference between the approximate and exact answers is 2937.5 − 2820 = 117.5 rpm, which is significant. The difference would be small if the change were small, that is, a few percent, but can be significant for a large change in the field current, as in this case. In general, using an exact method—which is simple enough—is recommended.

Typical efficiencies of dc machines are as follows:

HP rating	5	10	50	100	500	1000	5000
% Efficiency	75	82	88	91	93	95	97

4.7 UNIVERSAL (SERIES) MOTOR AC OR DC

If the stator and rotor are connected in series (Figure 4.13), both carry the same current, positive or negative, and the motor torque is always positive, as was seen in Equation 2.6, regardless of ac or dc supply. The series connection between the stator

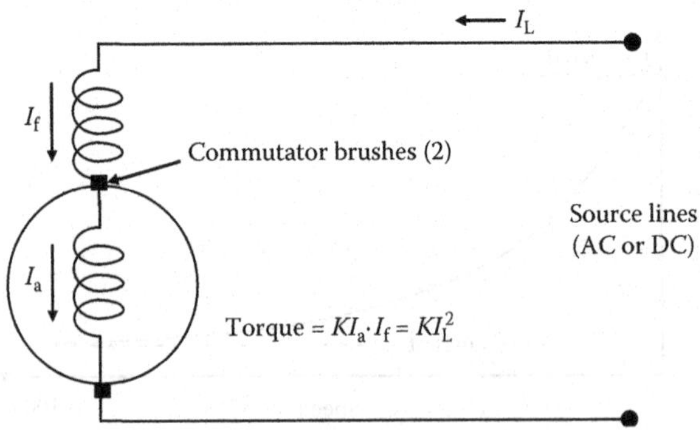

FIGURE 4.13 Universal series motor for ac or dc operation.

and the rotor coils must necessarily use commutator and brushes, limiting the practical horsepower rating of this motor to small in order to avoid sparking at the brushes. Small-appliance-size series motors in fractional hp ratings are called universal motors, whereas large motors in several hp ratings are called series motors. Since the torque is proportional to the current squared, and since the starting current is high (as in all motors), the series motor gives very high starting torque. For this reason, it is often used for moving large inertia load from rest, such as in trains, cranes, hoists, elevators, etc. However, the series motor has low flux at light load current, which speeds up the rotor as per Equation 4.21 and maintains constant horsepower output. The high speed at light load can self-destruct the rotor under centrifugal force, which is proportional to speed squared. For this reason, the series motor must not be used at very light load and must be protected from accidental overspeed by centrifugal switch in the supply lines. The dc shunt motor, on the other hand, works at essentially constant flux and constant speed and produces horsepower that is proportional to the load torque.

4.8 TORQUE VERSUS SPEED COMPARISON

Figure 4.14 displays the speed versus torque load characteristics of four major types of motor. It shows that the synchronous motor maintains its speed constant regardless of the load, whereas the series (dc or ac) motor speed drops inversely with the load. The induction motor and dc shunt motor speeds, on the other hand, droop slightly down with increasing load.

PROBLEMS

1. A two-pole, 60-Hz, three-phase, 3460-rpm induction motor runs at 3580 rpm at no load. Determine at rated load (1) the slip speed in revolutions per minute, (2) the slip in perunit and percent, and (3) the speed regulation.

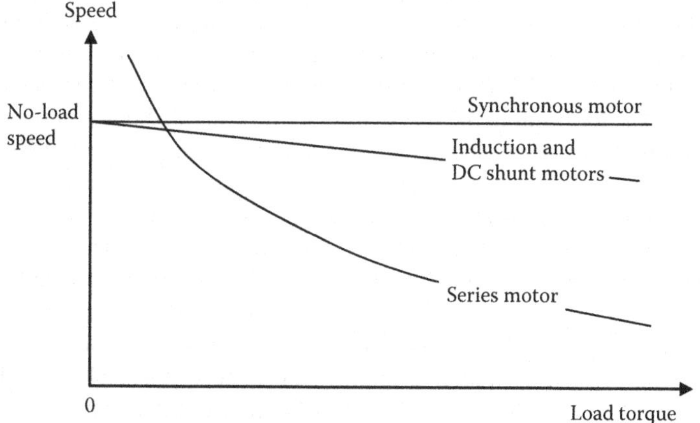

FIGURE 4.14 Speed droop versus torque load in ac and dc motors.

2. A 75-hp, three-phase, six-pole, 60-Hz induction motor delivers full load torque at 5% slip. If the load torque is increased by 15% as allowed by the service factor on the nameplate, determine the new speed and horsepower delivered.
3. A 150-hp, three-phase, 60-Hz, 460-V, four-pole, 1750-rpm induction motor is used in an application that loads the motor to only 70% of the rated torque. Determine (1) the actual speed and (2) the actual horsepower delivered to the load.
4. A 125-hp, three-phase, 60-Hz, 460-V, four-pole, 1720-rpm, letter code M, squirrel cage induction motor with a Δ-connected stator has efficiency 92% and power factor 0.90 lagging when operating at rated load. Determine the input kilowatt, kilovolt-ampere, and line current during normal operation and the line current on direct start in Δ.
5. A 5000-hp, 60-Hz, 1200-rpm, 4.1-kV, three-phase, *Y*-connected overexcited synchronous motor draws the armature current of 800 A at 0.85 leading power factor under rated load. It has a synchronous reactance of 1.5 Ω/ph and negligible resistance. Determine (1) the counter emf $\tilde{E}_f$, (2) the torque angle δ, and (3) the mechanical angle between the center lines of the rotor and stator magnetic fields.
6. A 1000-hp synchronous motor running at 80% rated load has a torque angle of 25°. Determine its pull-out torque T_{max} in percentage of the rated torque. If the motor load is increased to 100% in one step, determine the worst-case transient peak value of torque angle δ. Assume the torque versus δ relation approximately linear in the operating range and ignore the damping bar contribution.
7. A 250-hp, three-phase, 460-V, 60-Hz, ac motor has power factor of 0.90 lagging and efficiency of 92%. Determine the line current it will draw under 10% overload and the percent change in stator conductor temperature rise above ambient air.
8. A water pumping system in a ship is designed to pump 250 m^3/h fresh water against 70 m of total pressure head. Determine the motor horsepower rating and the motor control center kilovolt-ampere rating, assuming 95% pump efficiency, 92% motor efficiency, and 85% pf lagging.
9. A three-phase, 125-hp, 460-V, 50-Hz induction motor with a Δ-connected stator is designed to have a magnetic flux density in the core at the saturation limit of 1.65 T. If this machine is connected to 60-Hz lines, determine the voltage for limiting the flux density to the same 1.65-T saturation limit.
10. A 200-hp, three-phase, 460-V induction motor has a design life of 25 years at 110°C rise in conductor temperature above the ambient air. If it is continuously operated under unbalanced line voltages of 430, 450, and 480 V, determine (1) the percentage of voltage unbalance, (2) the additional temperature rise when operating at rated load, (3) the expected life at rated load, and (4) the required horsepower derating for the motor to last the design life of 25 years.
11. A three-phase induction motor in a 440-V distribution system under an abnormal condition receives highly unbalanced three-phase voltages $\tilde{V}_a = 360\angle 0°$, $\tilde{V}_b = 360\angle{-90°}$, and $\tilde{V}_c = 720\angle{-225°}$. Determine the symmetrical component voltages V_a^+, V_a^-, and V_a^0, and comment on approximate motor current and power losses in the rotor and stator.

12. A 100-hp, 240-V dc shunt motor runs at 3450 rpm at rated conditions. If its field current is decreased by 15% while keeping the same load current, determine its new speed.
13. A three-phase induction motor running from 460-V supply lines is drawing 60-A rated current at 0.90 power factor lagging. The power losses estimated from the manufacturer's data sheet at this load are 3000 W in the conductor, 1024 W in the magnetic core, and 1000 W in friction and aerodynamic windage. Determine (1) the output power and efficiency at rated load and (2) the maximum efficiency and the corresponding percentage of load.
14. A three-phase, 50-hp, 460-V, letter code-G induction motor is connected to 480 V_{LL} lines via a transformer with $0.034 + j0.097$ Ω/ph impedance and cable with $0.006 + j0.003$ Ω/ph in series. Assuming the locked rotor power factor 0.20, determine the percentage of voltage drop on start with and without the Y–Δ starter.
15. Determine the kilovolt-ampere load on a generator for powering a crude oil pump if (1) the pumping rate at discharge manifold is 5000 m^3/h against 100-m pressure head, (2) the pump pressure head to manifold pipeline is 20 m, and (3) the pump's internal pressure drop is equivalent to the 10-m head of the fluid. Assume the pump efficiency to be 88%, motor efficiency 95%, motor power factor 90%, and oil specific gravity 0.85.
16. Determine the running speed of a 100-hp, 440-V, four-pole, 60-Hz synchronous motor at full load. If the voltage and frequency both were reduced to 80% and the load is reduced to 64%, determine its new speed.
17. For a three-phase, 60-Hz induction motor with phase values of stator voltage $V = 120$ V, $R_{rotor} = 1$ Ω, and $L_{rotor} = 1$ mH, compute and plot using EXCEL the rotor current and motor torque for slip varying over a wide range of $s = -1$ to $+2$ pu, that is, from backward rotating to supersynchronous speed. Then, in reference to the chart, discuss in each speed range the torque, current, and power variations, and what they mean in terms of using the induction machine as a motor, generator, or brake.
18. For the motor of Problem 17, compute and plot using EXCEL the rotor current and motor torque from slip $s = 1$ (starting) to 0.05 (full load) with rotor resistance varying from 0.5 to 2 Ω/phase. Then, in reference to the chart, discuss the starting torque, full load speed, and full load current as the rotor resistance varies over the range. Also discuss what those variations mean in terms of the induction motor design for various starting torque and running efficiency requirements.

QUESTIONS

1. Discus the principal difference in construction and operation of the induction motor and the synchronous motor.
2. Which two factors jointly produce a rotating magnetic field in three-phase induction and synchronous motors?
3. Why is the slip necessary for the operation of the induction motor?

4. At what speed would an ideal induction motor on frictionless bearings run in vacuum?
5. Identify the advantage and disadvantage of using a Design D motor where high starting torque is required.
6. Identify the adverse effects of motor current during starting.
7. Explain the difference between the electrical degrees and mechanical degrees in four-pole ac machines.
8. What would happen if a 460-V, 60-Hz motor is connected to 460-V dc supply lines?
9. What would happen if a 240-V dc shunt motor is connected to 240-V, 50-Hz supply lines?
10. Explain the difference between dual-frequency and dual-voltage motors in construction and operation.
11. Identify the effect of unbalanced voltage on motor operation and the remedy to maintain its normal service life.
12. Identify the major advantage of a dc motor.
13. Why have dc motors fallen out of favor now?
14. What is required to change the direction of rotation of the induction motor and the dc shunt motor?
15. List a few candidate applications where high power series motors can be beneficial and the reason thereof.

FURTHER READING

Alger, P.L. 1981. *The Nature of Polyphase Induction Motor.* New York: John Wiley & Sons.

Boldea, S.A. 2010. *The Induction Machine Design Handbook.* Boca Raton: Taylor & Francis/CRC Press.

Krishnan, R. 2010. *Permanent Magnet Synchronous and Brushless DC Motor Drives.* Boca Raton: Taylor & Francis/CRC Press.

U.S. Department of Energy, 1980. *Classification and Evaluation of Electric Motors and Pumps.* Washington, DC.

5 Transformer

The transformer changes the system voltage from one level to another using two mutually coupled coils on a magnetic core as shown in Figure 5.1. The two coils are electrically isolated—not connected by wires. The power transfer from one coil to the other takes place via the alternating magnetic flux in the core that links both coils. The core is made of magnetic steel that has high permeability (low reluctance) to the magnetic flux. The magnetic steel is also known as electrical steel for its use in electrical machines. It is basically mild steel with a small percentage made of alloys that greatly improves the magnetic permeability. The magnetic core is made of thin laminations to keep eddy current loss low. It takes the flux created by the source side coil (usually high voltage, called the primary side, denoted by suffix 1) to the other coil (usually low voltage, called the secondary side, denoted by suffix 2). Almost all flux created by the primary coil reaches the secondary coil via the magnetic core, except a small percentage of leakage in air between the coils.

The polarities of the voltage and current on two sides are opposite as per Lenz's law mentioned in Section 2.1. This means that the current is going into the positive polarity in the primary coil (bringing the power inward from the source), whereas it is coming out of the positive polarity in the secondary coil (delivering power out to the load).

The fundamental voltage Equation 2.2 applied to two coils with N_1 and N_2 turns and negligible resistance and leakage flux gives the following rms voltages V_1 and V_2 on two sides of the transformer (ignoring the negative sign):

$$\frac{V_1}{V_2} = \frac{N_1}{N_2}. \tag{5.1}$$

Ignoring small power loss in core and coils, the conservation of energy require the volt-ampere (power) balance on two sides, that is, $V_1 \cdot I_1 = V_2 \cdot I_2$, and also the ampere-turns balance, that is, $I_1 \cdot N_1 = I_2 \cdot N_2$, leading to

$$\frac{I_1}{I_2} = \frac{V_2}{V_1} = \frac{N_2}{N_1}. \tag{5.2}$$

Although these relations are for the ideal transformer, they give fairly accurate results in practical transformers as well. The above transformer relations in rms values are emphasized in words below:

- Voltage ratio = turn ratio of the two coils.
- Current ratio = reciprocal (flip) of the turn ratio of the two coils.
- Volt-ampere product (VA or kVA or MVA) and power on both sides are equal.
- Ampere-turn products (magnetomotive force) on both sides are equal.

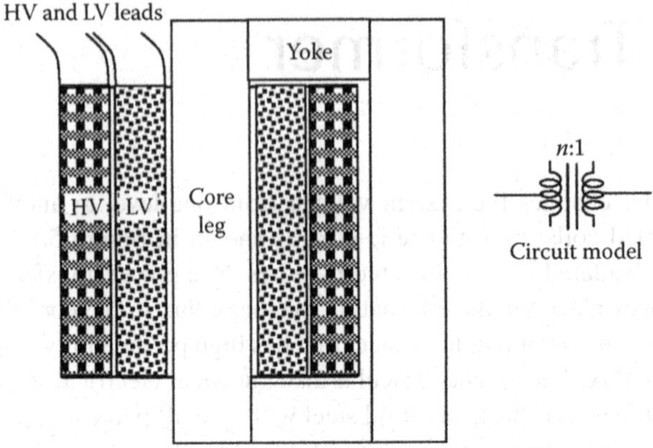

FIGURE 5.1 Single-phase transformer cross section with concentric HV and LV coils on one core limb.

The circuit symbol for a transformer with turn ratio $N_1/N_2 = n$ is shown in Figure 5.1 by two coils with n:1 placed in between, meaning that for every n turns in the primary coil, there is 1 turn in the secondary coil. Sometimes, vertical lines are placed between two coils to denote the laminated magnetic core. When the turn ratio is not given, it is assumed to be the voltage ratio as per Equation 5.1 without much loss of accuracy.

5.1 TRANSFORMER CATEGORIES

All transformers work on the same principle—Faraday's law of electromagnetic induction—and obey the same voltage, current, power, and ampere-turn relations on two sides as shown in Equations 5.1 and 5.2. However, transformers are categorized based on their primary purpose in a given application as follows:

(1) A *power transformer* has the sole purpose of transferring power from one voltage level on the primary side to another voltage level on the secondary side. We deal only with the power transformer in this chapter and call it just the transformer. More than 95% of the power transformers have turn ratio $n > 1$, making them step-down transformers. In a power grid, a step-up transformer is used to raise the generator voltage to HV (hundreds of kilovolts) to gain high efficiency for long-distance power transmission.

(2) A *voltage transformer (VT)* has the sole purpose of stepping down voltage from a very high level to a safe low level suitable for measurement by voltmeter. It is also called the *potential transformer (PT)* and has $n \gg 1$. Each VT and its fuse are mounted in a separate steel compartment. The high-voltage side connection to VT is appropriately insulated where it enters the compartment through porcelain bushings.

(3) A *current transformer (CT)* has the sole purpose of stepping down current from a very high level to a safe low level suitable for measurement

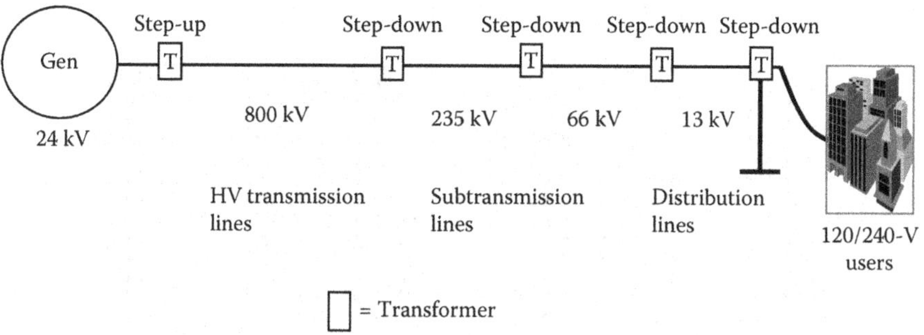

FIGURE 5.2 Typical voltage levels in land-based power grid from power plant to end users.

by ammeter. It has $n \ll 1$. The low-current side of CT has high voltage and must have sufficient insulation. Since CT can see high current during short-circuit faults, it is designed to withstand the thermal and mechanical stresses for peak current rating of the circuit breaker.

(4) An *isolation transformer* has the sole purpose of electrically isolating the secondary side (user side) from the primary side (source side) without changing the voltage level. This is done for safety or for other reasons. It has $n = 1$.

The VT and CT are jointly known as the *instrument transformers*. In addition to providing safety in measurements, they provide precise measurements since their output has a highly precise linear relation with input under all loading—unlike the power transformer.

Our main focus in this chapter is the power transformer. We need numerous power transformers because bulk power transmission is economical at HV with low current, but we must use power at various voltage levels—medium voltage for medium power and low voltage for small power for personal safety. A typical land-based utility power system is shown by the one-line diagram in Figure 5.2. Every kilovolt-ampere generated at the central power plant gets transformed five to six times before it is consumed at the user end. For this reason, over 95% of the transformers are step-down transformers.

Both HV and LV coils in the power transformer have tapping points around the rated number of turns. Typical tapping points are +5%, +2.5%, rated, –2.5%, and –5%, permitting some adjustment for low or high line voltage where the transformer is located. The transformer at low end of the line voltage, say, 5% less than the nominal voltage, can boost the voltage on the load side by connecting to +5% taps on the secondary side or to –5% taps on the primary side of the transformer. Either way, the turn ratio gets boosted by 5%.

5.2 TYPES OF TRANSFORMER

The power transformers' construction falls in various types. The single-phase unit is made with two coils (HV and LV, often concentric) on the core, as shown in Figure 5.1. The three-phase transformer is shown in Figure 5.3, where two coils

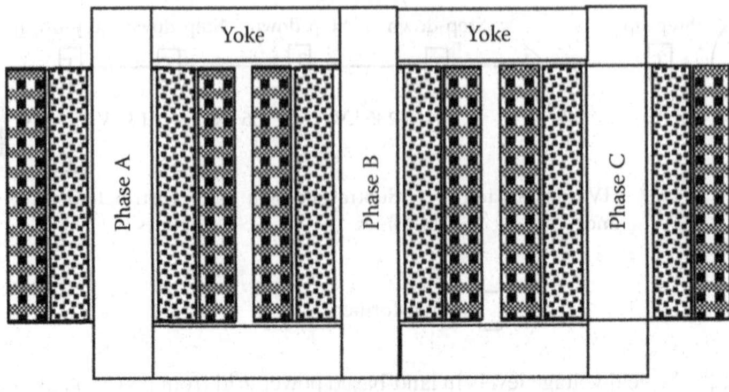

(a) Cross section

(b) Exterior view

FIGURE 5.3 Three-phase transformer with three pairs of concentric HV and LV coils on three core limbs.

(HV and LV) in each phase are always concentrically wound and placed on one core limb. All three limbs of the core meet at the top and bottom yokes, where the phasor sum of three-phase flux is zero due to 120° phase difference from each other, thus requiring no return path for the three-phase flux. The concentric HV and LV coils for each phase minimize the flux leakage. All three phases wound on a three-limb core are placed in one enclosure. They need less core material and hence cost less. Almost all transformers used in a land-based power grid are three-phase units except those providing the power to the residential and small commercial customers.

Power transformers (and circuit breakers) are categorized by the insulation medium they use, which also works as the cooling medium that takes the internal power loss to the enclosure surface for dissipation to the ambient air. They can be of the following types:

Dry-type transformers use air as the insulation and cooling medium. Since the air is a relatively poor insulator and coolant, these units are bulky and costly. However, they are safer than other types for indoor use. The enclosure can be ventilated or sealed where necessary for added safety. They are made up to 35-kV primary side voltage.

Oil-filled transformers have electrical-grade mineral oil as the insulation and cooling medium inside a sealed or breathing enclosure. They are extensively used in power grid and outdoor industrial and commercial installations. They are not used indoors because of possible oil spills and fire hazard following an accidental damage.

Gas-filled transformers work in pressurized gas (SF6 or nitrogen), which is much better insulation than air, requiring less separation between HV and LV parts. Moreover, it provides a better cooling medium. As a result, these transformers are compact and often used where space is at a premium, such as indoors in a factory or outdoors in a center city substation.

Nonflammable oil-filled units are available for indoor or other places where fire risk needs to be minimized. They offer compact low-cost design compared to dry-type units.

The key features of these transformers are listed in Table 5.1.

TABLE 5.1
Construction and Performance Features of Various Transformer Types

Performance Feature	Dry-Type (Air Ventilated)	Oil-Filled	Pressurized Gas-Filled
Typical location	Indoor, close to the loads	Outdoors in a yard	Substation in city with space at a premium
Construction	Coils impregnated with varnish or cast in epoxy	Coils and core immersed in insulating mineral oil	Coils and core in pressurized SF6 or nitrogen
Fire risk	Low	High due to oil	Low
Cooling mechanism	Open loop air flow through coils by convection and radiation, or forced air by fans	Circulating oil by convection inside and by air convection and radiation outside, or forced air by fans	Circulating gas by convection inside and by air convection and radiation outside
Average conductor temperature above 40°C ambient	80°C (costs more) to 150°C (costs less)	55°C	80°C (costs more) to 150°C (costs less)
Volume and weight	High	Low	Moderate
Cost	High	Low	High

5.3 SELECTION OF KILOVOLT-AMPERE RATING

The transformer size (kilovolt-ampere rating) for a given load profile is determined based on the following factors:

- The peak load it would supply, taking into account the National Electrical Code (NEC) load factors if numerous small loads with intermittent use are connected to it. Large loads are accounted for their individual on and off timing.
- For transformers supplying harmonic-rich power electronics loads, the additional harmonics heating is accounted for by selecting correct *K*-rating (discussed in Section 16.3.3).
- About 30% margin is then added for the future growth, which will make the transformer operate at about 1/1.3 = 0.77 or 77% of the rated capacity until the load grows to full capacity. This indirectly makes the transformer operate initially at peak efficiency since most power equipments have maximum efficiency at about 70%–80% of the rated load.

The transformer kilovolt-ampere, voltage, and frequency ratings must meet the IEEE Standard C57.12.01, and the mechanical and thermal integrity under short-circuit faults must meet the IEEE Standard 45 Section 12.01 and ANSI Standard C57.

Example 5.1

Determine the voltage and kilovolt-ampere ratings of Y–Δ connected transformer to power a three-phase Y-connected load drawing 800-A line current at 480 V from a three-phase, 4160-V source.

Solution:

For a three-phase transformer, kilovolt-ampere rating = $\sqrt{3} \times 480 \times 800 \div 1000 =$ 665 kVA.

Three-phase voltage rating = line voltages = 4160/480 V Y–Δ connection.

Alternatively, this load can also be supplied by three single-phase transformers, each rated 1/3 × 665 = 221.67 kVA. The single-phase transformer voltage rating depends on the transformer connections.

For the three single-phase units to be connected in Y on the HV side and Δ on the LV side:

$$\text{HV side voltage rating} = 4160 \div \sqrt{3} = 2402 \text{ V}$$

$$\text{LV side voltage rating} = \text{line voltage} = 480 \text{ V}$$

$$\text{Therefore, transformer voltage rating} = 2402/480 \text{ V}.$$

Since transformers in such small sizes are rarely custom made, and are more expensive, we round the above single-phase kilovolt-ampere size to the nearest standard rating readily available off the shelf, such as 225 kVA or more if margin is needed. The voltage ratings, however, cannot be rounded, but we can use units with HV ratings higher than 2402 V and HV/LV voltage ratios the same as 2402/480 V.

Example 5.2

A 5000-kVA, three-phase transformer bank steps down line voltage from 12.5 kV to 480 V. The HV side is connected in Δ and the LV side in grounded *Y*. Determine (1) the HV side line and phase currents and (2) the LV side line and phase currents.

Solution:

First we work with kilovolt-ampere per phase, which is ⅓ 5000 = 1666.7 kVA/phase for both HV and LV sides.

(1) On the HV side in Δ, phase voltage = line voltage = 12,500 V, which gives phase current = 1000 × 1666.7 ÷ 12,500 = 133.33 A and line current = √3 × 133.33 = 230.94 A.
(2) On the LV side in *Y*, phase voltage = 480 ÷ √3 = 277.1 V, phase current = 1000 × 1666.7 ÷ 277.1 = 6015 A, and line current = phase current = 6014 A.

Alternatively, we can work with three-phase kilovolt-ampere relations as follows:

(1) On the HV side in Δ, 5000 × 1000 = √3 × 12,500 × I_L, which gives line current I_L = 230.94 A and phase current = 230.94 ÷ √3 = 133.33 A.
(2) On the LV side in *Y*, 5000 × 1000 = √3 × 480 × I_L, which gives I_L = 6014 A and phase current = line current = 6014 A.

5.4 TRANSFORMER COOLING CLASSES

The power transformers are classified by the method used for cooling core and coils, that is, how the internal power loss is dissipated in ambient air to keep the temperature rise within the allowable limit set by the insulation material used in the construction. The temperature rise can be as high as 180° in some dry-type transformers used indoors and as low as 55°C in oil-immersed transformers widely used on land. In the increasing order of better cooling and compact size for the same kilovolt-ampere rating, the oil-filled transformer cooling classes are as follows:

- OA = oil–air self-cooled by natural convection and radiation
- FA = forced air cooled by fans
- FOA = forced oil–air cooled by oil pumps
- OW = oil–water cooled by water tubes or hollow conductors

Many transformers have multiple kilovolt-ampere ratings with alternative cooling methods, such as OA/FA, OA/FA/FOA, OW, etc.

5.5 THREE-PHASE TRANSFORMER CONNECTIONS

The three-phase transformer can be in a one three-phase unit or a bank of three single-phase units. Typical off-the-shelf three-phase transformer ratings readily available are from 9 to 5000 kVA. Typical single-phase transformer ratings range from 3 to 2500 kVA.

The primary coils in a three-phase transformer are typically connected in Δ, whereas the secondary coils are connected in either Δ or *Y*-Gr (*Y* with neutral grounded). The performance features of these two connection methods are as follows:

(1) Primary Δ–secondary *Y* with neutral grounded (common on land)
- (i) No neutral available on the primary side Δ.
- (ii) Each secondary line and neutral can be used for single-phase loads.
- (iii) Introduces 30° phase shift between the primary and secondary line voltages.
- (iv) Zero-sequence and triplen harmonic currents are blocked in the primary lines but can flow in the primary phase coils and the secondary lines, causing extra harmonic heating that requires *K*-rated transformers.
- (v) One ground fault on the secondary side causes the short-circuit current to flow and trip the circuit breaker.

(2) Δ primary–Δ secondary (common on ships)
- (i) No neutral on either primary or secondary side.
- (ii) First ground fault on either side causes no short circuit.
- (iii) First ground fault detection is required for reliable operation.
- (iv) No phase shift between the primary and secondary line voltages.
- (v) Zero-sequence and triplen harmonic currents are blocked on both sides.

Both the HV and LV side coils in a three-phase transformer can be connected in *Y* or Δ, but at least one side is generally connected in Δ for the following reason. Since the commercial transformer core is designed near the magnetic saturation point, the magnetizing current wave gets a hump near the peak value. Such a non-sinusoidal wave gives rise to the third harmonic current of the 3 × fundamental frequency, superimposed on the fundamental frequency magnetizing current in each phase. The third harmonic currents in each phase are not three-phase currents with the usual 120° phase difference. They have 3 × 120° = 360° = 0° phase difference from each other, that is, they are in phase and require a return or circulating path. The neutral wire in the *Y* connection carries the sum of the third harmonic currents in three phase lines back to the source, and the secondary is not impacted. The line voltages on both sides remain sinusoidal. In the absence of neutral wire, the line voltages on the LV side get distorted with the third harmonic voltages. One way of suppressing the third harmonic voltages is to have the phase coils on either the HV or LV side connected in Δ, which provides an internal path for circulating the third harmonic currents without impacting the outside line voltages. In a typical three-phase *Y*-connected source with or without neutral, the HV side of the transformer is usually connected in *Y*, and the LV side in Δ. In some applications, both sides are connected in Δ. In addition to suppressing the third harmonic currents, Δ–Δ connection improves the availability of power as explained below.

5.6 FULL-Δ AND OPEN-Δ CONNECTIONS

The three-phase power systems on land generally use three-phase transformers for economy, with three phases connected in Y or Δ as needed for the application. Where the maximum reliability is desired with minimum spares, such as on ships, three separate single-phase transformers are connected in Δ. With one of the three phases damaged, the entire three-phase unit with all phases in one enclosure becomes unavailable until it is repaired or replaced. With three single-phase units in separate enclosures, the damaged unit can be removed from the bank, leaving two healthy units operating in open-Δ (also called V–V) connection without impacting the line voltages. The kilovolt-ampere capacity of the open-Δ bank, however, gets reduced to 57.7% of the full-Δ capacity until the damaged unit is replaced.

Consider three single-phase transformers connected in Δ as shown in Figure 5.4, supplying three line currents of equal magnitudes I_L in all three lines at line-to-line voltages V_{LL}. If phase C transformer gets damaged and removed, the two remaining transformers in open Δ still maintain the same balanced three-phase voltages both on the load side as well as on the source side. However, the currents get adjusted for the new connection pattern. The phase A and B transformers now carry 1/2 I_L in open Δ instead of $I_L/\sqrt{3}$ before in full Δ. In order not to overload the feeder line cable, I_L must have the same value as before. This degrades the kilovolt-ampere capability of the full-Δ bank when operated in open-Δ as follows:

$\text{kVA}_{\text{full-}\Delta} = \sqrt{3} \times V_{LL} \times I_L$ and $\text{kVA}_{\text{open-}\Delta} = V_{LL} \times 1/2\ I_L$ in phase A + $V_{LL} \times 1/2\ I_L$ in phase B = $V_{LL} \times I_L$.

The ratio of the two kilovolt-amperes is therefore

$$\frac{\text{kVA}_{\text{open-}\Delta}}{\text{kVA}_{\text{full-}\Delta}} = \frac{V_{LL} I_L}{\sqrt{3} V_{LL} I_L} = 0.577\text{pu} = 57.7\%. \tag{5.3}$$

The open-Δ kilovolt-ampere loading, therefore, must be reduced to 57.7% of the full-Δ kilovolt-ampere capability in order to maintain the same ampere loading on the transformer feeder line cable. However, the line voltages on the load sides are still balanced three-phase voltages of full value as before. Just the load current is reduced to limit the primary side line current to its rated value to avoid overheating.

Example 5.3

Three single-phase transformers, each rated 100 kVA, are connected in Δ–Δ to power a three-phase 300-kVA load. If one of them is removed for repairs, determine (1) the line voltage on the load side and (2) the maximum kilovolt-ampere load the remaining two transformers can provide.

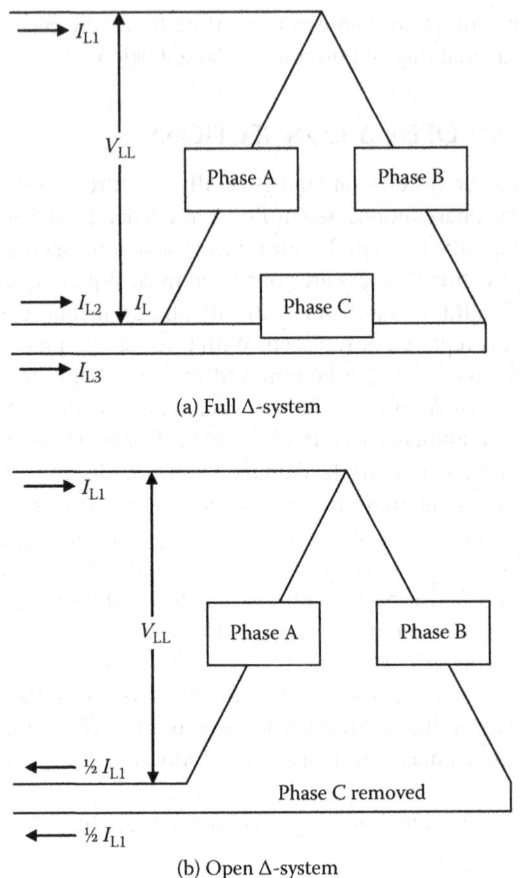

FIGURE 5.4 Full-Δ and open-Δ connections of three single-phase transformers.

Solution:

(1) By removing one transformer, the Δ–Δ connected bank of three single-phase transformers becomes V–V connected bank, but the line voltage remains the same as in Δ–Δ connection.
(2) The total kilovolt-ampere load, however, must be reduced to 57.7% in order not to overload the cables.

Therefore, maximum permissible three-phase load = 0.577 × 300 = 173.1 kVA.

5.7 MAGNETIZING INRUSH CURRENT

At the instant of closing the breaker on the transformer primary side and connecting to the supply lines, the primary side draws a heavy inrush current for initial magnetization of the core even with no load connected to the secondary side (Figure 5.5a). This can be simply explained by Faraday's law, which is (ignoring the – sign) $v(t) = N$

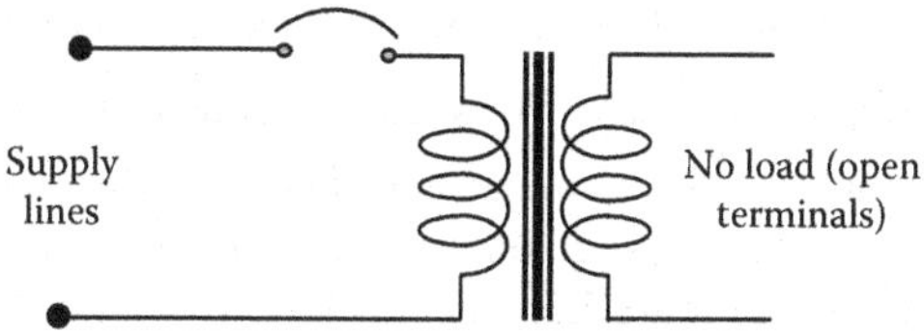

(a) Transformer with line breaker open and no load

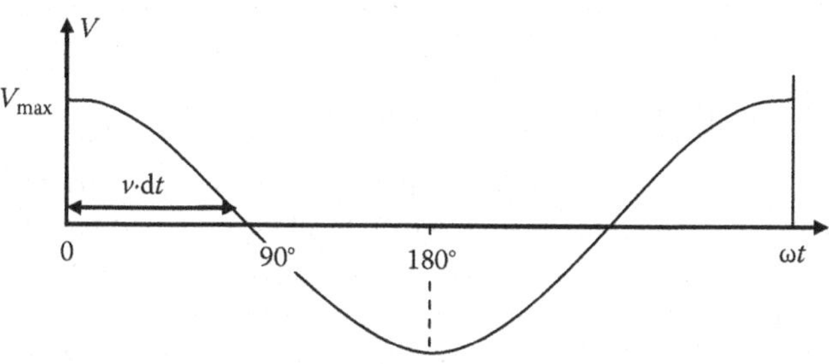

(b) Breaker closed at V_{max} instant

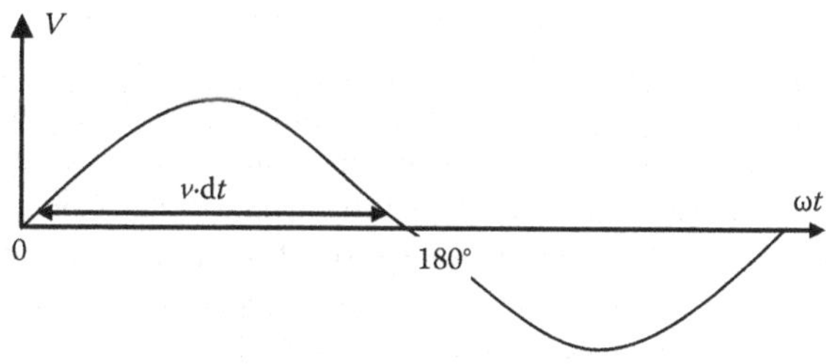

(c) Breaker closed at $V = 0$ instant

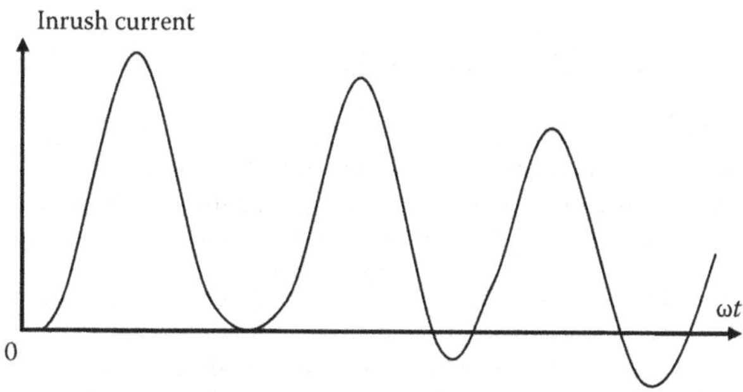

(d) Inrush current on connecting primary coil (three cycles)

FIGURE 5.5 Magnetizing inrush current on connecting transformer to supply lines with no load on the secondary side.

$d\varphi/dt$, where $v(t)$ is the sinusoidal supply voltage. The flux build-up in the core with an N-turn coil is then $\varphi = \frac{1}{N}\int v \cdot dt$, where the integral is the area under the $(v \times t)$ curve. The flux adds under positive voltage and subtracts when the voltage becomes negative. If the circuit breaker is closed on the instant of voltage passing to its natural maximum value as shown in Figure 5.5b, the flux rises from time 0 to 90° to the peak value $\varphi_{peak1} = \frac{1}{\omega N}\int_0^{90} v \cdot d(\omega t)$ and then starts falling back from φ_{peak1}. On the other hand, if the circuit breaker is closed at the instant of line voltage passing through its natural zero as shown Figure 5.5c, the flux rises from time 0 to 180° to the peak value $\varphi_{peak2} = \frac{1}{\omega N}\int_0^{180} v \cdot d(\omega t)$ and then starts falling back from φ_{peak2}. Clearly, φ_{peak2} covers twice the area compared to φ_{peak1}, giving $\varphi_{peak2} = 2\ \varphi_{peak1}$. Normally the core is designed to carry flux equal to φ_{peak1}, at which level the core is at the saturation knee point on its magnetization (B-H) curve. With φ_{peak2}, the flux density reaches two times the saturation limit, where the magnetic permeability of the core is extremely low, matching with that of air, which requires very high magnetizing current. For example, if φ_{peak1} requires the magnetizing current equal to 2% of the rated load current, then φ_{peak2} would require over 1000 times more, that is, 2000% or 20 times rated load current for a very short time in the first cycle. It then gradually decays to the normal 2% value in several cycles with oscillations that are damped out by the coil resistance and the power losses in the core and coil, as shown in Figure 5.5d.

Thus, the inrush magnetizing current can peak to 10 to 20 × rated load current and may trip the breaker or melt the fuse on the primary side. Connecting the transformer to the supply lines is a normal operation. Therefore, such tripping would be a nuisance that can be avoided by using the time-delay fuse or setting the circuit breaker trip time with intentional delay. Although the magnetizing inrush current does not impact the transformer kilovolt-ampere size determination, it does impact the fuse selection and circuit breaker setting to avoid nuisance tripping.

5.8 SINGLE-LINE DIAGRAM MODEL

In the single-line diagram of a three-phase power system model, an ideal transformer can be represented by an ideal wire with no power loss or voltage drop due to leakage flux. Practical transformers are designed with remarkably high efficiency, often approaching 99% in large units. However, small power losses occur in the conductor resistance and magnetic core. Also, some flux leakage in the gap between two coils causes a small voltage loss (drop), usually represented by the leakage inductance L. In a continuous one-line diagram, the transformer is then represented by its total coil resistance R and leakage reactance $X = \omega L$ in series, combined in impedance $Z = R + jX$, as shown in Figure 5.6. The magnetizing current and power loss in the core are drawn from the source side and do not enter in the series impedance between the primary and secondary coils for determining the line voltage drop due to Z. However, they are separately accounted for in the efficiency and temperature rise calculations.

Transformer nameplate typically includes kilovolt-ampere, number of phases (i.e., single-phase or three-phase), frequency, HV and LV side voltages, whether

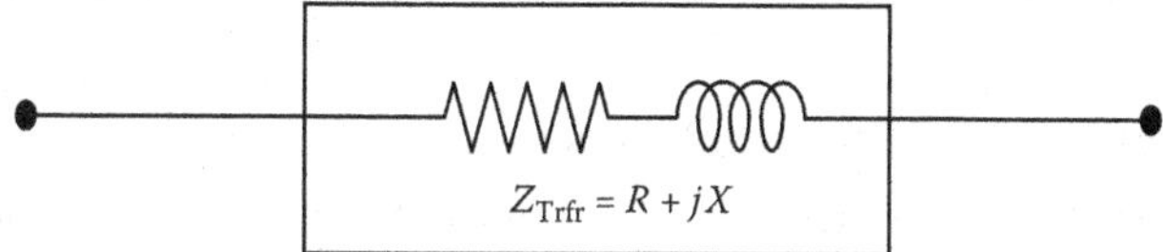

FIGURE 5.6 Equivalent electrical model of transformer in one-line diagram.

Y- or Δ-connected, percentage of taps, cooling class (air, oil, natural, or forced), average conductor temperature rise (in degrees Celsius), basic insulation level (BIL) against lightening voltages, percentage impedance *Z*, and sometimes percentage efficiency.

Table 5.2 gives representative values of % *Z* for different kilovolt-ampere and voltage class of single-phase distribution transformers. We note that smaller transformers have lower % *Z* than those of the larger units. The transformer is very efficient in power transfer, and its design efficiency improves with size, as seen in Table 5.3.

Since the transformer remains connected to the supply voltage regardless of the load current, the magnetic flux and the associated core loss are continuously present. For this reason, the transformer is designed to have power loss in core = 1/3 total power loss, and the loss in conductor = 2 × 1/3 total power loss. With such distribution of the total loss, the maximum efficiency occurs at √1/2 = 0.707 or 71% of the rated load when *conductor loss = core loss.* From the nameplate value of impedance *Z*, the equivalent series resistance *R* and *X* can then be estimated from efficiency η (known or estimated) and impedance *Z* as follows (in per unit values):

$$R = 2 \times \frac{1}{3}(1 - \eta)$$

TABLE 5.2
Typical Impedance Values of Single-Phase Transformers

	HV 460–4100 V		HV 4200 V–11 kV	
Kilovolt-Ampere Rating	**% *R***	**% *X***	**% *R***	**% *X***
10	1.5	1.8	1.5	2.0[a]
50	1.3	2.3	1.3	2.5
100	1.2	2.8	1.2	3.5
250	1	4.7	1	5.2
500	1	4.8	1	5.4
1000	0.9	5.0	0.9	6.0
10,000	0.8	6.0	0.8	7.0[b]

[a] Users of small transformers often specify impedance not less than 4% in order to keep the fault current below the available protection device rating.

[b] Large high-voltage transformers have higher impedance since the insulation gap between the HV and LV coils needs to be wider, resulting in greater leakage flux.

TABLE 5.3
Typical Values of Transformer Efficiency

Kilovolt-Ampere Rating	100 kVA	1000 kVA	10 MVA or Higher
% Efficiency	96%–97%	97%–98%	98%–99%

and

$$X = \sqrt{(Z^2 - R^2)} \tag{5.4}$$

where all values here are in perunit (pu). In % values, the number 1 in R above becomes 100.

5.9 THREE-WINDING TRANSFORMER

The transformer with one primary coil and two secondary coils, all three wound on a common core, is sometimes used for two purposes: (1) to supply two loads at different voltages—main load at high voltage and auxiliary load at low voltage; or (2) the third (called tertiary) winding is connected in delta, but no terminals are brought out for load connection (it is buried inside the enclosure). The buried Δ-connected tertiary coils carry the internally circulating third harmonic currents, which are all in phase. This prevents the third harmonic currents from entering the external lines and improves the quality of power in *Y*-connected, four-wire distribution systems.

Dual-voltage output from one transformer is obtained by using two secondary coils as shown in Figure 5.7a. In such a transformer, one output voltage is influenced by load change in the other via magnetic coupling. A short circuit in one secondary would also impact the output voltage of the other secondary. Analytically predicting such mutual influence can save the prototyping cost and time. The three-winding transformer equivalent circuit shown in Figure 5.7b can facilitate such analysis. It is based on the premises that between any two coils, there is a leakage flux and the corresponding leakage reactance *X*. The equivalent circuit shows the total impedance—made of the winding resistance and the leakage reactance—denoted by Z_{12} between coils 1 and 2, Z_{23} between coils 2 and 3, and Z_{31} between coils 3 and 1. The value of Z_{12} is determined by a short-circuit test between coils 1 and 2, and so on. The analysis is significantly simplified by representing the three coil-to-coil impedances in an equivalent *Y*-circuit shown in Figure 5.7c. Through a circuit analysis, the impedance values in an equivalent *Y*-circuit can be derived as follows:

$$\begin{aligned} Z_1 &= \frac{1}{2}\left(Z_{12} + Z_{13} - Z_{23}\right) \\ Z_2 &= \frac{1}{2}\left(Z_{12} + Z_{23} - Z_{13}\right). \\ Z_3 &= \frac{1}{2}\left(Z_{13} + Z_{23} - Z_{12}\right) \end{aligned} \tag{5.5}$$

These calculations become easier by expressing the impedances in percentage of the rated base impedance of the transformer, as discussed in the next section. It

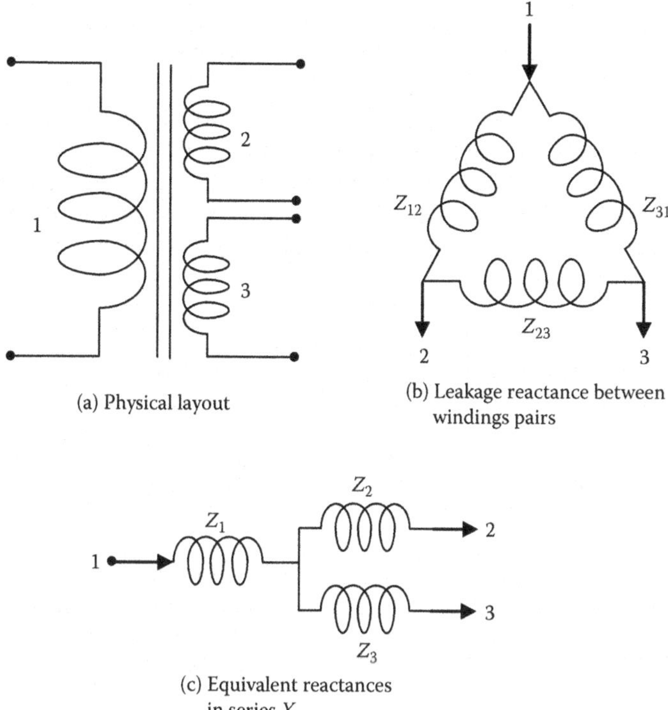

(a) Physical layout

(b) Leakage reactance between windings pairs

(c) Equivalent reactances in series Y

FIGURE 5.7 Equivalent circuit models of three-winding transformer powering two secondary loads at different voltages.

should be emphasized that these equivalent impedances have no real meaning. Any one impedance can be zero or negative, as long as the sum of impedances of any two branches is equal to the short-circuit impedance between those two windings. By using the Y-circuit, it is easy to see that the terminal voltage V_2 depends not only on the secondary load but also on the tertiary load because of the voltage drop in Z_1. It is also easy to see that when load I_2 is shed, the voltage of the other load will rise by I_2Z_1, and that Z_{31} can influence the short-circuit current in coil-2 terminals. The transformer with more than three windings can be similarly analyzed using the same general principle, although with added complexity.

5.10 PERCENT AND PERUNIT SYSTEMS

When we express any quantity in percent, we must have a base that is considered 100%. In a pu system, we expressed any quantity as a fraction of the base value that is considered 1.0 unit. Since 100% = 1.0 pu, 100 × pu value = % value, or pu value = % value/100:

$$\%Z \text{ of transformer} = \frac{\text{series } Z \text{ of transformer in ohms}}{\text{base } Z \text{ of transformer in ohms}} \times 100 \tag{5.6}$$

$$\text{perunit } Z \text{ of transformer} = \frac{\text{series } Z \text{ of transformer in ohms}}{\text{base } Z \text{ of transformer in ohms}} = \frac{\%\, Z}{100}. \tag{5.7}$$

In electrical power system studies, we deal with kilovolt-amperes, volts, amperes, and ohms. Generally, the rated values of equipment are taken as the base values in expressing percent or pu values. However, we cannot select all four base values since they are not independent. If the kilovolt-ampere and voltage bases are defined (rated values or otherwise), then the ampere and ohm bases are indirectly defined from the following relations:

$$\text{In a single-phase system, } I_{\text{base}} = \frac{1000\ \text{kVA}_{\text{base}}}{V_{\text{base}}}\ \text{A/phase}$$

$$Z_{\text{base}} = \frac{V_{\text{base}}}{I_{\text{base}}} = \frac{V_{\text{base}}^2}{1000\ \text{kVA}_{\text{base}}} = \frac{\text{kV}_{\text{base}}^2}{\text{MVA}_{\text{base}}}\ \Omega/\text{phase}. \tag{5.8}$$

In large, high-voltage, three-phase systems, it is customary to state three-phase kilovolt-ampere and line-to-line voltage in kV_{LL}, but the power system analysis is still done on a per-phase (line-to-neutral) basis, where in a three-phase system

$$I_{\text{base}} = \frac{\text{kVA}_{\text{3ph.base}}}{\sqrt{3}\text{kV}_{\text{LLbase}}}\ \text{A/phase}$$

$$Z_{\text{base}} = \frac{1000\ \text{kV}_{\text{LLbase}}^2}{\text{kVA}_{\text{3ph.base}}} = \frac{\text{kV}_{\text{LL.base}}^2}{\text{MVA}_{\text{3ph.base}}}\ \Omega/\text{phase}. \tag{5.9}$$

The values of base volts, base amperes, and base ohms are different on the LV and HV sides, but the kilovolt-amperes and % Z remain the same on both sides.

Example 5.4

Determine the base current and base impedance of a single-phase system based on 500 kVA and 277 V.

Solution:

V_{base} = 277 V (generally the rated value at the point of interest in the system)

Therefore, $I_{\text{base}} = 500 \times 1000 \div 277 = 1805$ A and $Z_{\text{base}} = V_{\text{base}} \div I_{\text{base}} = 277 \div 1805 = 0.1535\ \Omega$.

Alternatively, using Equation 5.8 for a single-phase system

$$Z_{base} = V^2_{base} \div (kVA_{base} \times 1000) = 277^2 \div (500 \times 1000) = 0.1535\ \Omega.$$

Example 5.5

A single-phase, 24-kVA, 480/120-V transformer has $Z = 4\%$ stated on its nameplate. Determine Z in ohms looking from (1) the HV side source when connected as a step-down transformer (2) the LV side source when connected as a step-up transformer.

Solution:

Looking from the HV side source, $V_{base} = 480$ V; $I_{base} = 24{,}000 \div 480 = 50$ A.

$$Z_{base} = 480 \div 50 = 9.6\ \Omega \text{ therefore } Z_{trfr} = 0.04 \text{ pu} \times 9.6 = 0.384\ \Omega.$$

Looking from the LV side source, $V_{base} = 120$ V; $I_{base} = 24{,}000 \div 120 = 200$ A.

$$Z_{base} = 120 \div 200 = 0.6\ \Omega \text{ therefore } Z_{trfr} = 0.04 \text{ pu} \times 0.6 = 0.024\ \Omega.$$

We note that $0.384 \div 0.024 = 16$. This is the same as $(480 \div 120)^2$, which is the voltage ratio squared, as is always the case.

We notice in Example 5.5 that the transformer series impedance on the LV side is 0.024 Ω, whereas that on the HV side is 0.384 Ω. We also notice that the ratio (HV side ohms/LV side ohms) = (HV side volts/LV side volts)^2. Thus, the transformer has the same % Z series impedance but different ohm values on the HV and LV sides.

Example 5.6

A single-phase, 150-kVA, 460/120-V transformer has 5% impedance and 97% efficiency. Find its series R and X in percent and in ohms looking from the HV side source.

Solution

Since efficiency $\eta = 97\%$, the total power loss is approximately 3% [actually $(1/0.97 - 1) \times 100 = 3.09\%$ to be exact]. About 2/3 of it is typically in resistance R of the coils and the remaining 1/3 in the core. Therefore, we take $R = 2\%$ in the series model of the transformer. Then, from the transformer impedance $Z = 5\%$, we derive the transformer reactance $X = \sqrt{(Z^2 - R^2)} = \sqrt{(0.05^2 - 0.02^2)} = 0.0458$ pu. For this calculation, we can also work in percent to get $\% X = \sqrt{(Z^2 - R^2)} = \sqrt{(5^2 - 2^2)} = 4.58\%$, as the percent signs get carried forward with the results.

For base values on the HV side, $V_{HVbase} = 460$ V, $I_{HVbase} = 150 \times 1000 \div 460 = 326$ A, and $Z_{HVbase} = 460 \div 326 = 1.41\ \Omega$. Therefore, the ohm values *looking from the HV side* are

$$Z = Z_{pu} \times Z_{HVbase} = 0.05 \times 1.41 = 0.076\ \Omega$$
$$R = R_{pu} \times Z_{HVbase} = 0.02 \times 1.41 = 0.282\ \Omega$$
$$X = X_{pu} \times Z_{HVbase} = 0.0458 \times 1.41 = 0.0656\ \Omega.$$

Another advantage of stating the transformer series impedance Z in percent—as opposed to in ohms—is that it quickly quantifies the magnitude of the short-circuit current in case of a short-circuit fault at its secondary terminals. This, in turn, leads to selecting the short-circuit current interruption rating of the circuit breaker or fuse, which is

$$\text{short-circuit current} = \left(\frac{100}{\%Z} \text{ or } \frac{1}{Z_{pu}} \right) \times \text{rated current.} \tag{5.10}$$

Although the rated currents on the HV and LV sides are different, the transformer impedance Z in percent or pu remains the same on both sides.

Example 5.7

Determine the rms fault currents on both sides of a 50-kVA, single-phase, 480/120-V transformer with 5% impedance as shown in the figure below. Assume that all generators behind the bus are large enough to make it an infinite bus and all cable impedances are negligible. Assume 100% voltage before the fault and ignore the normal load current.

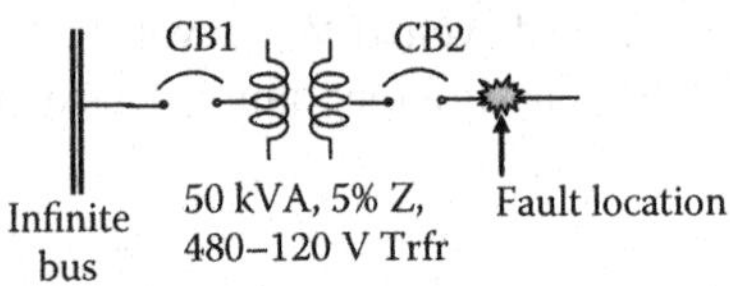

Solution:

For this transformer,

HV side base (rated) current = 50 × 1000 ÷ 480 = 104.2 A

LV side base (rated) current = 50 × 1000 ÷ 120 = 416.8 A.

The transformer will see the maximum fault current if it gets a dead short at its output terminals, where the only impedance from the infinite bus to the fault is that of the transformer itself (5% in this case). A fault on the transformer primary side, close to the bus, would produce greater current (theoretically infinite), but that current would not go through the transformer. In fact, such current would not be truly infinite but limited by the generator and cable impedance behind the bus, which we have ignored in this example.

The rms fault current through the transformer = 100% voltage ÷ 5% impedance = 20 per unit. Note that 20 is pu, as the percent sign gets cancelled out from the numerator and denominator, leaving the result in pu. It is safer to work in pu throughout, such as 1 pu voltage ÷ 0.05 pu impedance = 20 pu current. Thus, the fault current magnitudes on two sides are

$$\text{HV side fault current} = 20 \times 104.2 = 2084 \text{ A}$$

$$\text{LV side fault current} = 20 \times 416.8 = 8336 \text{ A.}$$

5.11 EQUIVALENT IMPEDANCE AT DIFFERENT VOLTAGE

Consider a 12-Ω heater connected to a 120-V room outlet as shown in Figure 5.8a. It will draw 10-A current and 120 × 10 = 1200 W. We know that this power eventually comes—via a step-down transformer—from a large regional power plant that generates power at high voltage, say, at single-phase 24 kV. Now, assume for a moment that no one else is drawing power from this plant except the 12-Ω heater. Using the balance of power, the current drawn from the 24-kV generator for the sole 1200-W heater would be 1200 W ÷ 24,000 V = 0.05 A. Therefore, the equivalent heater resistance as seen by the generator is 2400 V ÷ 0.05 A = 480,000 Ω as shown in Figure 5.8b. Thus, the actual 12-Ω resistance connected at a 120-V outlet looks like 480,000 Ω at the 24-kV generator terminals, that is, it gets multiplied by 480,000 ÷ 12 = 40,000. The multiplier is also $(24{,}000 \div 120)^2$, which is

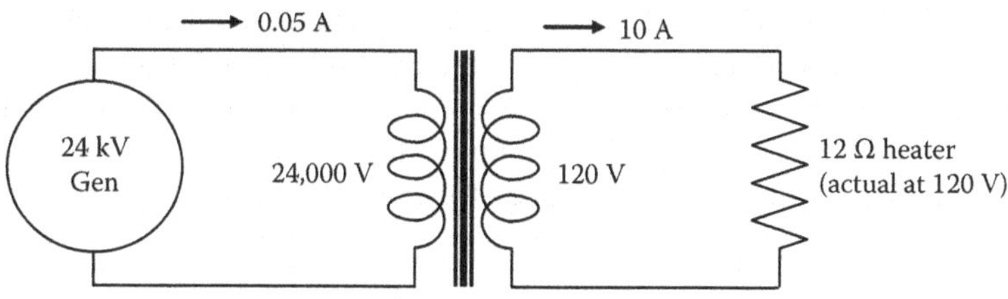

(a) Actual circuit with 12 Ω resistance at 120 V

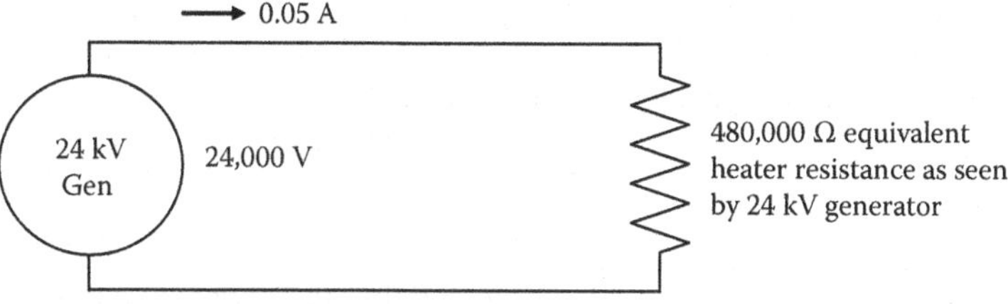

(b) Equivalent continuous circuit for same power from generator

FIGURE 5.8 Equivalent impedance in ohms at different voltage levels.

the voltage ratio squared. This concept is known by different names, such as the *impedance transformation*, *reflected impedance*, or *impedance referred to different voltage*. It gives *the equivalent impedance value at different voltage* that will absorb the same power (equivalency in the power system studies always means *for equal power*). The concept is needed in the fault current analysis later in Chapter 8 and can be generalized in the transformer or anywhere else in the power system as follows.

The equivalent value of given impedance in ohms looking from HV or LV side of a transformer changes by the voltage ratio squared; it looks high from the HV side and low from the LV side. The general expression for such *impedance transformation* of impedance Z_{V1} at voltage level V_1 to Z_{V2} at voltage level V_2 is

$$\frac{Z_{V2}}{Z_{V1}} = \left(\frac{V_2}{V_1}\right)^2. \tag{5.11}$$

This can also be reasoned in another way. The ac power absorbed in impedance Z ohms at voltage level V is $S = V^2 \div Z$. If we wish to convert the ohm value of Z_{V1} at voltage V_1 to the equivalent ohm value Z_{V2} at voltage V_2, which would absorb the same ac power S (condition for the equivalency), then we must have $S = V_1^2 \div Z_{V1} = V_2^2 \div Z_{V2}$, which leads to the same results as in Equation 5.11.

If the 12-Ω heater load resistance in the above example was expressed in the pu of Z_{base} on 1200 W and 120 V base, then $I_{\text{baseLV}} = 1200\text{ W} \div 120\text{ V} = 10\text{ A}$, $Z_{\text{baseLV}} = 120\text{ V} \div 10\text{ A} = 12\ \Omega$, and $Z_{\text{Load.pu}} = 12 \div 12 = 1.0$ pu. The pu value then remains the same on the HV side, which has $I_{\text{baseHV}} = 1200\text{ W} \div 24{,}000\text{ V} = 0.05\text{ A}$, and $Z_{\text{baseHV}} = 24{,}000\text{ V} \div 0.05\text{ A} = 480{,}000\ \Omega$. The equivalent ohm value of the heater resistance looking from the generator side is 480,000 Ω, which is 480,000 Ω equivalent ÷ 480,000 Ω base = 1.0 pu on the HV side, the same pu value as on the LV side. The percent and pu values remaining the same on both HV and LV sides is the main advantage of using percent or pu values in power system studies.

Example 5.8

A single-phase, 100-kVA, 480/120-V step-down transformer powers a load impedance of 12 Ω on the LV side. Determine (1) the current on the HV side from conventional transformer calculations and (2) the equivalent load impedance looking from the HV side in a continuous equivalent circuit.

Solution:

(1) Current on LV side = 120 ÷ 12 = 10 A, and the current drawn from the HV side is then 10 × flip of the voltage ratio, that is, 10 × (120 ÷ 480) = 2.5 A.
(2) The equivalent load impedance looking from the HV side is $Z_{\text{Load HV}} = 12\,(480 \div 120)^2 = 192\ \Omega$. Then, the current drawn from the HV side = 480 ÷ 192 = 2.5 A, which is the same as above.

5.12 CONTINUOUS EQUIVALENT CIRCUIT THROUGH TRANSFORMER

Consider a 480-120 V, single-phase transformer connected to load impedance Z_{Load} ohms via secondary side cable impedance Z_{LVcable} ohms and primary side cable impedance Z_{HVcable} ohms, all connecting to the source as shown in Figure 5.9a. We can reduce this circuit into a continuous circuit with one total impedance powered directly from the HV side source as follows.

The equivalent ohm values of Z_{Load} and Z_{LVcable} looking from the source side get transformed by n^2, that is, the voltage ratio squared. The ohm value of Z_{HVcable} remains the same as given; it does not get transformed since it is already on the source side in a physical location. Therefore, the total equivalent impedance looking from the source on the HV side is

$$Z_{\text{TotalHVside}} = Z_{\text{HVcable}} + n^2\,(Z_{\text{LVcable}} + Z_{\text{Load}})\text{ ohms} \tag{5.12}$$

where $n = (V_{\text{HV}} \div V_{\text{LV}}) = (480 \div 120) = 4$ in this case.

The current drawn from the source is then $I_{\text{HV}} = 480\text{ V} \div Z_{\text{TotalHVside}}$ amperes.

If we work in a pu system—expressing the values in pu of their respective base values—the voltage ratio multiplier n^2 does not apply, and the total impedance looking from the HV or LV side is merely the sum of all pu impedances, that is,

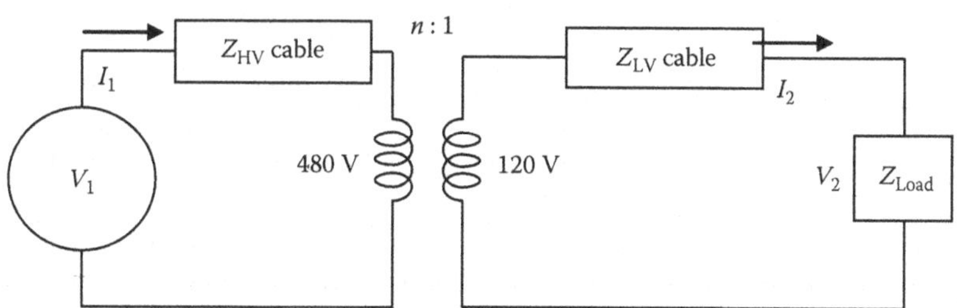

(a) Actual circuit with load powered from transformer secondary side (all impedances in ohms)

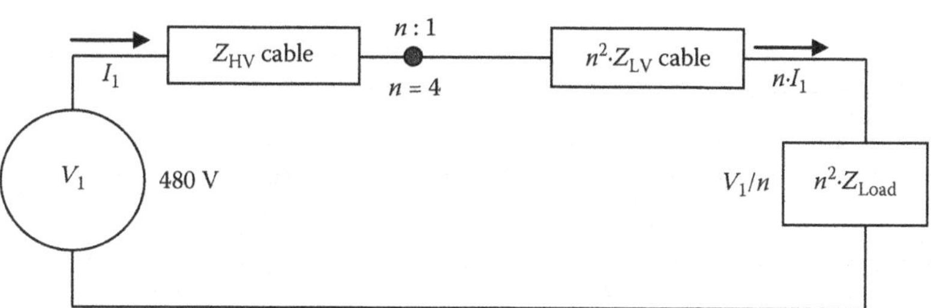

(b) Equivalent continuous load circuit powered from transformer primary side (all impedances in ohms)

FIGURE 5.9 Equivalent continuous circuit model with electrical continuity through transformer.

$$Z_{Total.pu} = Z_{HVcable.pu} + Z_{LVcable.pu} + Z_{Load.pu}$$

and

$$I_{pu} = V_{pu} \div Z_{Total.pu}. \tag{5.13}$$

At full 100% primary voltage, in a pu system

$$I_{pu} = 1.0 \div Z_{Total.pu}. \tag{5.14}$$

In amperes

$$I_{HVamp} = I_{pu} \times I_{HVbase} \text{ and } I_{LVamp} = I_{pu} \times I_{LVbase}. \tag{5.15}$$

Example 5.9

Consider a transformer rated 15 kVA, single-phase, 480/120 V supplying 9.6-Ω load as shown in the following figure (top). In calculations specific to a given transformer, it is customary to take the rated voltage on two sides as their respective base voltage and the rated kilovolt-ampere as the base kilovolt-ampere (which remains the same on both sides). Then, on the secondary (suffix 2) and primary (suffix 1) sides, we have

I_{2base} = 15 kVA × 1000 ÷ 120 V = 125 A and Z_{2base} = 120 V ÷ 125 A = 0.96 Ω

I_{1base} = 15 kVA × 1000 ÷ 480 V = 31.25 A and Z_{1base} = 480 V ÷ 31.25 A = 15.36 Ω.

By the way, we note that $(480 \div 120)^2 \times 0.96 = 15.36$ Ω, which means that the base ohms get transformed by the voltage ratio squared, as we have learned earlier.

Secondary current I_2 = 120 V ÷ 9.6 Ω = 12.5 A, which is 12.5 ÷ 125 = 0.10 pu or 10% of the base current on the LV side.

Primary current I_1 = 12.5 × (1 ÷ 4) = 3.125 A, which is also 3.125 ÷ 31.25 = 0.10 pu or 10% of the base current on the HV side.

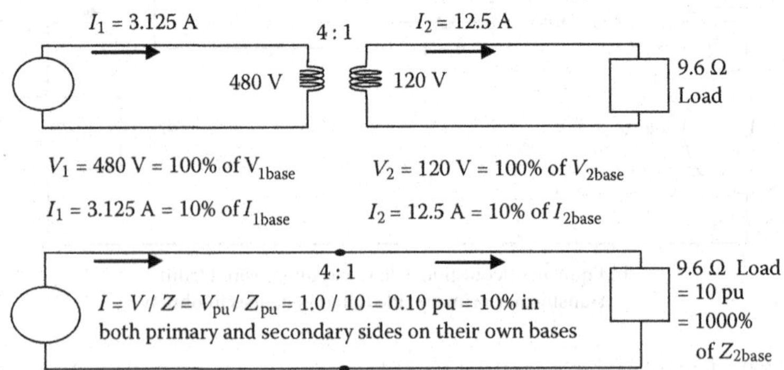

Alternatively, we express the load impedance 9.6 Ω on the LV side as 9.6 ÷ 0.96 = 10 pu on LV base. Then, I = 1.0 pu V ÷ 10 pu Z = 0.10 pu current, which will be the same in pu on both the LV and HV sides.

Although volts and amperes are different, the % voltage and % current are the same on both the LV and HV sides, as shown in the bottom figure. It means that the current is continuous in percentage from the source side to the load side, although the base of expressing percentage changes on two sides. Therefore, the entire circuit when expressed in percentage looks as if the two sides were connected by wire at the heavy dots shown in the bottom circuit, with the continuity of % voltage and % current on two sides. This simplifies the system analysis by a great deal when there are multiple transformers in series in the system, which can be reduced to one continuous circuit starting from the generator to the load. Such continuous circuit calculations are possible only if we work in percent or pu system.

5.13 INFLUENCE OF TRANSFORMER IMPEDANCE

With fixed primary voltage equal to the supply line voltage, the secondary voltage in percentage equals the primary voltage minus the internal voltage drop in the transformer series impedance Z. As the voltage drop varies with load current and power factor, so does the secondary side voltage. The voltage regulation is a measure of how close the secondary voltage remains to the rated value with varying load and power factor. It is desirable to have a low voltage regulation, which requires using a transformer with low impedance. It can be calculated at any load, but the full-load (worst-case) regulation is of most interest to power engineers, which is defined as

$$\%\text{ voltage regulation} = \frac{V_{2\text{ no load}} - V_{2\text{ full load}}}{V_{2\text{ full load}}} \times 100. \tag{5.16}$$

If the secondary voltage needs to be maintained constant regardless of the load, it must be regulated by using an automatic tap changer over the range equal to the % voltage regulation. Alternatively, the secondary loads must allow that much % variation in their input voltage without adversely impacting the performance.

The exact value of the voltage regulation at any load and pf can be determined from Equations 2.14 and 2.15. However, for typical loads at all practical power factors, Equation 2.18 gives fairly approximate voltage drop due to the transformer series impedance:

$$V_{\text{drop}} = I\left(R\cdot \text{pf} + X\sqrt{1-\text{pf}^2}\right) \text{ and } \%\text{V.R.} = \%\text{V}_{\text{drop}} \tag{5.17}$$

where R and X are the transformer series impedance derived from the nameplate impedance and efficiency values using Equation 5.4. One can use volt, ampere, and ohm units, all in percent or pu. In Equation 5.17, the + sign with X is for lagging pf, which becomes – sign for leading pf.

A low impedance gives a desirable low regulation, but it adversely impacts the short-circuit current in case of a dead short on the secondary terminals. The actual short-circuit current depends on all the equipment connected in series from the generator to the fault location, including the transformer. However, a conservative estimate of the fault current can be made by ignoring the generator and cable impedances, with only the transformer % Z in the circuit with 100% voltage at the transformer primary and solid three-phase short at the secondary terminals. Full (100% or 1.0 pu) source voltage is now available for the short through the transformer impedance, giving $I_{sc.pu} = 1.0 \div Z_{Trfr.pu} = 100\% \div \% Z_{Trfr}$. Note that percent divided by percent gives pu, as two percent signs cancel out, leaving the quotient a dimensionless multiplier. The actual short circuit (sc) amperes on the two sides of the transformer are then $I_{HVsc.amp} = I_{sc.pu} \times I_{HVbase}$ amperes and $I_{LVsc.amp} = I_{scpu} \times I_{LVbase}$ amperes, for which the circuit breakers must be rated. The probability of mechanical or thermal damage on the transformer during a dead fault increases directly with the short-circuit current or inversely with the transformer impedance.

Thus, a lower % Z gives lower voltage regulation on one hand but gives higher short-circuit current on the other hand. These two effects are carefully traded by the system engineer to meet the voltage regulation requirement while keeping the circuit breaker rating and the short-circuit damage risk reasonably low.

Example 5.10

A 5000-kVA transformer has series impedance of $R = 3\%$ and $X = 7\%$. Determine the percentage of rise in the output voltage on unloading the transformer from full load at 0.8 pf lagging to zero load, which is also known as the full-load voltage regulation (VR) of the transformer.

Solution:

Full load means $I = 1.0$ pu, and 0.85 power factor lagging means $\theta = \cos^{-1} 0.85 = 31.8°$, so $\tilde{I} = 1.0\angle{-31.8°}$ pu. The full-load voltage is taken as 100% or 1.0 pu. We will work out this example in three ways.

(1) Transformer series impedance $\tilde{Z} = 0.03 + j0.07 = 0.076\angle 66.8°$ pu. Taking $V_{2\text{ full load}} = 1.0\angle 0°$ pu as the reference phasor,

$$\tilde{V}_{2\text{ no load}} = \tilde{V}_{2\text{ full load}} + \tilde{I} \times \tilde{Z} = 1.0\angle 0° + 1.0\angle{-31.8°} \times 0.076\angle 66.8 = 1.0\angle 0° + 0.076\angle 35° = 1.063\angle 2.35° \text{ pu}.$$

Using the voltage magnitudes, we get VR = (1.063 − 1) ÷ 1 = 0.063 pu or 6.3%.

(2) We can also use exact Equation 2.14 in the pu system to obtain the following:

$$V_{2\text{no load}} = V_s^2 = \left(V_L + I \cdot R\cos\theta + I \cdot X\sin\theta\right)^2 + (I \cdot X\cos\theta - I \cdot R\sin\theta)^2$$
$$= (1.0 + 1\times 0.03\times 0.8 + 1\times 0.07\times 0.527)^2 + (1\times 0.07\times 0.8 - 1\times 0.03\times 0.527)^2$$
$$= 1.063 \text{ pu}.$$

Then, VR = (1.063 − 1.0) ÷ 1.0 = 0.063 pu or 6.3%, same as above.

(3) Alternatively, using approximate Equation 2.18 with pu value, we obtain voltage drop $= 1.0\left(0.03\times 0.85 + 0.07\sqrt{1-0.85^2}\right) = 0.0624$ pu or 6.24%, which compares well with the exact value of 1.003 − 1.0 = 0.063 pu.

In this percent range, the secondary voltage must be regulated by using an automatic tap charger if a constant load voltage is desired. Otherwise, the secondary loads must allow this much percent change in their input voltage and still deliver specified performance.

PROBLEMS

1. Determine the kilovolt-ampere and voltage ratings of a Δ–*Y* connected transformer to power a three-phase *Y*-connected load drawing 600-A line current at 480 V from a three-phase, 4160-V source using (i) one 3-phase transformer, or (ii) three single-phase transformers.
2. A 3000-kVA, three-phase transformer bank steps down line voltage from 12.5 kV to 480 V. The HV side is connected in Δ, and the LV side in grounded *Y*. Determine (1) the HV side line and phase currents and (2) the LV side line and phase currents.
3. Three single-phase transformers, each rated 75 kVA, are connected in Δ–Δ to power a three-phase 225-kVA load. If one of them is removed for repairs, determine (1) the line voltage and line current on the load side and (2) the maximum kilovolt-ampere load that the remaining two transformers can provide.
4. Determine the base current and base impedance of a single-phase system based on 50 kVA and 120 V.
5. A single-phase, 36-kVA, 480/120-V transformer has series impedance Z = 5% stated on its nameplate. Determine Z in ohms looking (1) from the HV side source when connected as a step-down transformer and (2) from the LV side source when connected as a step-up transformer.
6. A single-phase, 225-kVA, 460/120-V transformer has 6% impedance and 96% efficiency. Find its series R and X in percent and in ohms looking from the HV side source.
7. Determine the rms fault currents on both sides of a 100-kVA, single-phase, 460/120-V transformer with 6% impedance as shown in the following figure. Assume the generator bus infinite and all cable impedances negligible.

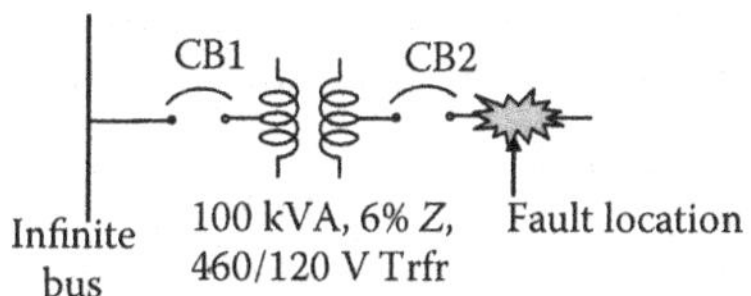

8. A single-phase, 100-kVA, 480/120-V step-down transformer powers a load impedance of 8 Ω on the LV side. Determine (1) the current on the HV side from conventional transformer calculations and (2) the equivalent load impedance looking from the HV side in the continuous equivalent circuit.
9. A 25-kVA, single-phase, 460/120-V transformer powers 6-Ω load as shown in the figure below. Determine the HV and LV side currents in amperes and also in percent based on the rated voltage on two sides as their respective base.

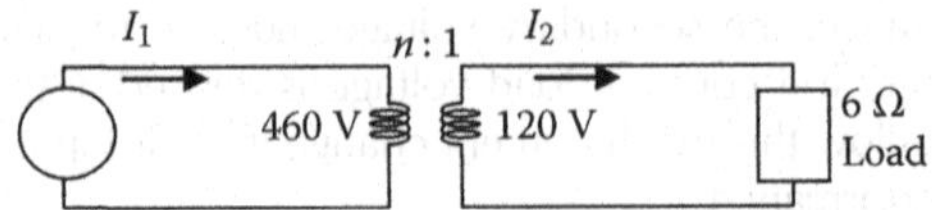

10. A 2-MVA, 4160/480-V transformer has series impedance of $R = 2\%$ and $X = 6\%$. Determine the (1) full-load voltage regulation of the transformer at 0.85 pf lagging and (2) secondary output voltage on unloading the transformer completely.
11. A three-phase, 500-kVA, 4160/480-V transformer bank has $R = 1.0\%$ and $X = 5\%$. Using Equation 2.15 or 2.18 with pu values, determine the % voltage rise in the secondary voltage on unloading from the following load conditions: (1) 80% load at the unity power factor, (2) full load at 0.85 power factor lagging, and (3) full load at 0.90 pf leading. Assume that the secondary voltage was equal to the rated voltage in all load conditions.
12. A single-phase, 100-kVA, 460/120-V transformer with 3% R and 6% X is powering a 3-KW heater. Determine the total continuous circuit impedance in ohms looking from the source side and the current drawn from the source.
13. Determine a three-phase 480V/208Y–120V transformer kilovolt-ampere rating if the daily variations in the secondary load current are 150 A for 5 h, 170 A for 1 h, and 130 A for 18 h, taking into account the service factor of 1.15 for 2 h. Allow 30% margin for future growth.

QUESTIONS

1. Explain the difference between power transformer, voltage (potential) transformer, and current transformer.
2. Why does the three-phase transformer not need the fourth core leg for the return flux?
3. Which type of transformers are used in indoor power systems and why? How do they compare in cost, volume, and weight with other types commonly used outdoors?
4. What are the instrument transformers, and how do they differ from the power transformer?
5. Why is electrical equipment using pressurized gas (or vacuum) as the insulating mediums more compact than the dry-type equipment?
6. If you place a dry-type transformer in an oil tank, it can deliver greater kilovolt-ampere load and withstand higher lightening voltages. Discuss why.
7. The three-phase transformer costs less than three single-phase transformers in a three-phase system. Even then, three single-phase transformers are used in Δ–Δ connected to form the three-phase bank in some applications. Explain why.
8. Which part of the system would get overloaded if the kilovolt-ampere loading is not reduced after one of the three single-phase transformers connected

in Δ–Δ is removed for service? What will be the heat generation rate in that part?

9. Why is the magnetizing current at the instant of connecting the transformer primary to the supply lines very high even when there is no short or no load connected on the secondary side?
10. Explain in your own words the concept of reflected impedance or the impedance retransformation.

FURTHER READING

Chapman, S.J. 1999. *Electric Machinery Fundamentals*. Boston: McGraw Hill.
Smith, S. 1985. *Magnetic Components*. New York: Van Nostrand Reinhold.

6 Power Cable

The term *wire* generally means one or more insulated conductors in small size (solid or stranded for flexibility), whereas *cable* means one or more insulated conductors of large size grouped in a common insulation jacket, often with a ground shield. Figure 6.1 shows a single-conductor, 2-kV class cable, whereas Figure 6.2 is a three-conductor, three-phase cable suitable for 5- to 15-kV applications.

6.1 CONDUCTOR GAGE

Each conductor size is measured in American wire gage (AWG), British wire gage (BWG), or in metric gage designated by the conductor cross-section area in square millimeters. The AWG and BWG numbers are log-inverse measures of the conductor diameter. For example, the AWG is set on log scale as

$$\text{AWG} = 20 \log\left(\frac{0.325}{\text{dia}_{\text{inch}}}\right) \quad \text{or} \quad \text{dia}_{\text{inch}} = \frac{0.325}{10^{\frac{\text{AWG}}{20}}} \tag{6.1}$$

where dia = conductor diameter in inches, bare solid, or one equivalent diameter of the conducting area of all strands combined. A decrease in one gage number increases the diameter by a factor of 1.1225 and area by a factor of 1.26. The diameter doubles every six gages down, whereas the area doubles every three gages down. One can visualize the AWG number as the approximate number of wires that can be placed side by side in 1-in. width. For example, 14 bare conductors of AWG 14 can be placed side by side in 1-in. width, or the AWG 14 conductor has approximately 1/14-in. diameter.

The conductor is typically made of numerous thin strands for flexibility in handling and bending in installation. The strand diameter varies from 10 to 22 mils (1 mil = 1/1000th in. = 25.4 μm) depending on the wire gage. The strands are generally tin-plated to avoid oxidation that normally occurs on pure uncoated copper. The thin film of lubricant used on strands in the cable-making machine acts as the inter-strand insulation that keeps the skin effect (discussed later in this chapter) confined to the individual strand.

The cable sizes heavier than AWG 4/0 (also written as 0000) are designated by the net conductor cross-section area in kilocircular mils (kcmil), which is also known as MCM (the first M for 1000 in Roman numerals), where

$$\text{MCM} = \text{kcmil} = \frac{(\text{wire diameter in mils})^2}{1000} = 1000 \times (\text{wire diameter in inch})^2. \tag{6.2}$$

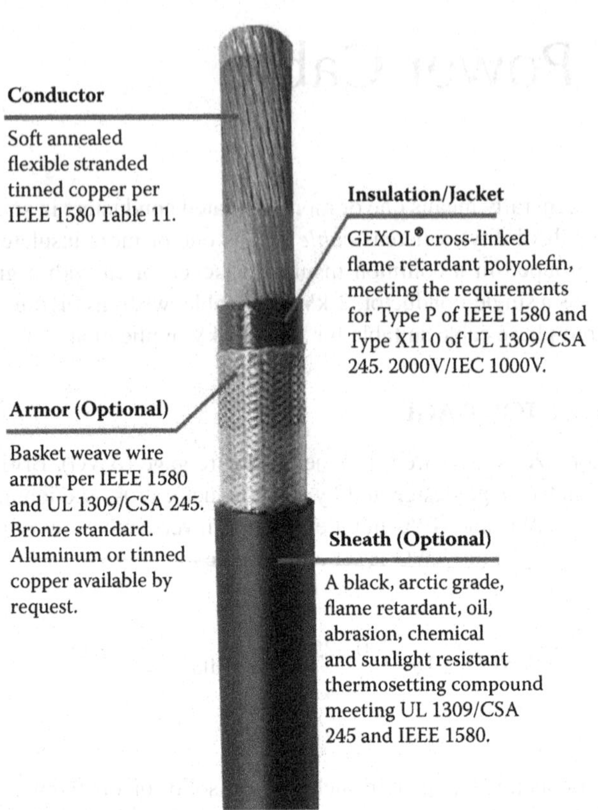
Conductor
Soft annealed flexible stranded tinned copper per IEEE 1580 Table 11.
Insulation/Jacket
GEXOL® cross-linked flame retardant polyolefin, meeting the requirements for Type P of IEEE 1580 and Type X110 of UL 1309/CSA 245. 2000V/IEC 1000V.
Armor (Optional)
Basket weave wire armor per IEEE 1580 and UL 1309/CSA 245. Bronze standard. Aluminum or tinned copper available by request.
Sheath (Optional)
A black, arctic grade, flame retardant, oil, abrasion, chemical and sunlight resistant thermosetting compound meeting UL 1309/CSA 245 and IEEE 1580.

FIGURE 6.1 Single-conductor, 2-kV class power cable showing strands, insulation jacket, and outer sheath/shield. (From Gexol-insulated cable, a product of AmerCable, Inc. With permission.)

The IEEE standard uses kcmil to specify all conductor sizes, which increase linearly with the conductor cross-section area. The European and international standards measure wire in metric gage equal to the square millimeter of the net conductor area. The AWG, IEEE, and metric gages are all listed in Table 6.1, along with the number of strands, the strand diameter, and the total uninsulated conductor diameter in the last column.

The most common conductor material is copper, although aluminum finds special applications for its light weight and low cost. The physical properties of copper and aluminum are compared in Table 6.2. For the same electrical resistance per meter length, the aluminum conductor will weigh in the (resistivity × mass density) ratio, which is (2.83 × 2.7) ÷ (1.724 × 8.89) = 0.50 or 50% compared to copper.

37-105

Type MMV Medium Power Cable

Single Conductor: 5kV – 15kV, 100% & 133% Insulation Levels. Rated 90°C

Multi-Conductor: 5kV – 15kV, 100% & 133% Insulation Levels. Rated 90°C

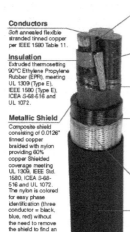

Conductors
Soft annealed flexible stranded tinned copper per IEEE 1580 Table 11.

Insulation
Extruded thermosetting 90°C Ethylene Propylene Rubber (EPR), meeting UL 1309 (Type E), IEEE 1580 (Type E), ICEA S-68-516 and UL 1072.

Metallic Shield
Composite shield consisting of 0.0126" tinned copper braided with nylon providing 60% copper Shielded coverage meeting UL 1309, IEEE Std. 1580, ICEA S-68-516 and UL 1072. The nylon is colored for easy phase identification (three conductor = black, blue, red) without the need to remove the shield to find an underlying colored tape.

Conductor Shield
A combination of semi-conducting tape and extruded thermosetting semi-conducting material meeting UL 1309, IEEE 1580, ICEA S-68-516 and UL1072.

Insulation Shield
Semi-conducting tape, with overlap, for fast and easy termination meeting UL 1309, IEEE 1580, ICEA S-68-516 and UL 1072.

Grounding Conductor (optional)
One uninsulated soft annealed flexible stranded tinned copper conductor per ASTM B 33 and sized according to Table 21.1 of UL 1072.

Jacket
A black, arctic grade, flame retardant, oil, abrasion, chemical and sunlight resistant thermosetting compound meeting UL 1309, IEEE 1580, ICEA S-68-516 and UL 1072. Colored jackets for signifying different voltage levels are also available on special request (ie. yellow = 5kV, orange = 8kV and red = 15kV).

Armor (optional)
(Optional) 0.0126" bronze braid providing 88% minimum coverage meeting UL 1309 and IEEE Std. 45-1998.

Sheath (optional)
A black, arctic grade, flame retardant, oil, abrasion, chemical, and sunlight resistant thermosetting compound meeting UL 1309, IEEE 1580, ICEA S-68-516 and UL 1072. Colored jackets for signifying different voltage levels is also available on special request (ie. yellow = 5kV, orange = 8kv and red = 15kV).

Ratings & Approvals

- UL Listed as Marine Shipboard Cable (E111461)
- American Bureau of Shipping (ABS)
- Det Norske Veritas (DNV) Pending
- Lloyd's Register of Shipping (LRS) Pending
- 90°C Temperature Rating
- Voltage Rating – 5kV to 15kV

Applications

AmerCable's Type MMV marine medium voltage cables are for use aboard commercial ships, mobile offshore drilling units (MODUs), and fixed or floating offshore facilities.

Features

- These cables utilize flexible stranded conductors, braided shields and a braided armor (when armored) which make them very suitable for applications involving repeated flexing and high vibration.
- These cables have a small minimum bending radius (6xOD for unarmored cables and 8xOD for armored cables) for easy installation.
- Optional uninsulated grounding conductors sized per UL 1072.
- The increased flexibility of this cable allows for termination of one end and coiling on multiple module offshore platforms. Then coiling and terminating other end when modules are mated at sea thereby reducing installation time.
- Passes IEC 332-3 Category A and IEEE 1202 flame tests.

FIGURE 6.2 Three-phase medium voltage power cable with ground conductor, shield, armor, and outer sheath for 5 to 15 kV applications. (From Gexol-insulated cable, a product of AmerCable, Inc. With permission.)

6.2 CABLE INSULATION

A variety of insulations are used in manufacturing wires and cables depending on the temperature rating desired in various applications. Major insulation materials and their operating temperature ranges are given in Table 6.3. Obviously, higher temperature rating of the insulation allows higher current carrying capacity of the conductor. In addition to an appropriate operating temperature rating of the insulation, the marine cable also needs high moisture resistance to withstand the damp and even wet conditions normal on ships and around oil drilling platforms. The industry has devised letter designations to indicate the temperature and moisture resistance

TABLE 6.1
American Wire Gauge (AWG), IEEE, and Metric Conductor Sizes with Stranding Profile

Size AWG/ kcmil	Number of Strands	Individual Strand Diameter (in.)	Closest IEEE 45 Standard Size	Equivalent Metric Size (mm^2)	Uninsulated Conductor Diameter (in.)
18	19	0.0100	2	0.96	0.049
16	19	0.0117	3	1.32	0.059
14	19	0.0147	4	2.08	0.074
12	19	0.0185	6	3.29	0.093
10	37	0.0167	10	5.23	0.113
8	37	0.0201	16	7.57	0.136
6	61	0.0201	26	12.49	0.175
4	133	0.0177	41	21.11	0.258
2	133	0.0223	66	33.51	0.324
1	209	0.0201	83	42.79	0.361
1/0	266	0.0201	106	54.45	0.407
2/0	342	0.0201	133	70.01	0.461
3/0	418	0.0201	168	85.57	0.510
4/0	532	0.0201	212	108.91	0.575
262	646	0.0201	262	132.25	0.654
313	777	0.0201	313	159.06	0.720
373	925	0.0201	373	189.36	0.785
444	1110	0.0201	444	227.23	0.860
535	1332	0.0201	535	272.68	0.941
646	1591	0.0201	646	325.70	1.029
777	1924	0.0201	777	393.87	1.132
1111	2745	0.0201	1111	561.94	1.354

TABLE 6.2
Copper and Aluminum Conductor Properties

Characteristic	Copper	Aluminum
Resistivity, Ω m at 20°C	1.724×10^{-8}	2.830×10^{-8}
Mass density, g/cm^3	8.89	2.70
Specific heat, J/kg °C at 20°C	377	900
Thermal conductivity, W/m °C	395	211
Temperature coefficient of resistance, α per °C	3.93×10^{-3}	3.90×10^{-3}
Melting point, °C	1083	660
Flex life (relative)	1	0.5
Thermal coefficient of expansion (relative)	1	1.4
Creep rate at 65°C (relative)	1	1000

TABLE 6.3
Cable and Wire Insulations and Operating Temperatures

Cable and Wire Insulation (Material Identifying Letter)	Operating Temperature Range	
	°C	°F
Thermoplastic (T)	–40°C to 60°C	–40°F to 140°F
Rubber (R)	–40°C to 75°C	–40°F to 167°F
Vinyl	–20°C to 80°C	–4°F to 176°F
Cross-linked polyethylene (X)	–60°C to 80°C	–76°F to 176°F
Neoprene	–30°C to 90°C	–22°F to 194°F
Polypropylene	–20°C to 105°C	–4°F to 221°F
Fluorinated ethylene propylene (PF)	–40°C to 150°C	–40°F to 302°F
Teflon	–70°C to 200°C	–94°F to 392°F
Silicon rubber (S)	–70°C to 200°C	–94°F to 392°F

TABLE 6.4
Letter Designation for Operating Conditions

Operating Conditions	Letter Designation
High temperature 70°C (158°F)	H
Very high temperature 90°C (194°F)	HH
Wet or damp	W
Oil resistance	M
Flexible (stranded)	F

of the cable insulation as listed in Tables 6.3 and 6.4. Some examples of using the letter designations follow:

- RH cable is made from rubber that can withstand a high temperature up to 70°C (158°F).
- RHH cable is made from rubber that can withstand a very high temperature up to 90°C (194°F).
- TW cable is made from thermoplastic that is moisture resistant and can operate under wet conditions up to 60°C (140°F).
- XHHW cable is made from cross-linked polyethylene that can withstand a very high temperature in wet conditions up to 90°C (194°F).

6.3 CONDUCTOR AMPACITY

Each conductor in the cable is sized to meet the required current carrying capacity (ampacity) and the voltage drop limitation under normal and inrush currents. The cable ampacity is limited by the continuous operating temperature limit, which, in turn, is limited by the insulation type. The cable surface temperature is typically limited to 60°C to 75°C for safety that depends on the application. The ampacity depends also on the ambient air temperature, which can vary from 40°C to 65°C depending on the cable routing (e.g., indoor, outdoor, in sun, in boiler room, etc.) and on grouping in the raceway because of the mutual heating of the neighboring conductors. A cable with three or four conductors in one jacket can carry less current than a single conductor of the same size, and several cables side by side in a raceway can carry even less current. Both the higher ambient temperature and the raceway grouping require derating of the wire ampacity from the normally rated values given in the vendor technical data sheets.

Ampacities of three-phase cables with various insulation temperature ratings in 30°C (86°F) ambient cooling air are listed in Table 6.5 for low-voltage cables. At higher ambient air temperature, the ampere rating of a given cable must be reduced, and vice versa. Such derating factors are listed in Table 6.6.

TABLE 6.5
Permissible Ampacity of Selected Low-Voltage Three-Phase Insulated Copper Cables in Dry Location in Ambient Air at 30°C (86°F)

	Temperature Rating of Insulation			
AWG or kcmil[a]	60°C (140°F) Type T, TW, RUW	90°C (194°F) Type TA, FEP, THHN, XHHW[a]	125°C (257°F) Insulation	200°C (392°F) Special Use
14	15	25	30	30
10	30	40	50	55
6	55	70	85	95
4	70	90	115	120
2	95	120	145	165
1/0	125	155	200	225
2/0	145	185	230	250
3/0	165	210	265	285
4/0	195	235	310	340
250	215	270	335	–
350	260	325	420	–
500	320	405	500	–
750	400	500	620	–
1000	455	585	730	–
1500	520	700	–	–
2000	560	775	–	–

[a] kcmil = 1000 circular mils = MCM in old usage (first M for 1000 in Roman numerals).

TABLE 6.6
Cable Derating Factors in Table 6.5 Ampacity for Ambient Air Temperature Higher than 30°C (86°F)

Ambient Temperature		Temperature Rating of Insulation			
°C	°F	60°C (140°F)	90°C (194°F)	125°C (257°F)	200°C (392°F)
40	104	0.82	0.90	0.95	–
50	122	0.58	0.80	0.89	–
60	140	–	0.67	0.83	0.91
70	158	–	0.52	0.76	0.87
80	176	–	0.30	0.69	0.84
90	194	–	–	0.61	0.80
100	212	–	–	0.51	0.77
140	284	–	–	–	0.59

Source: Adapted from NEC.

Example 6.1

A cable made with polypropylene insulation has nominal ampacity of 200 A in standard 40°C ambient air. Determine its ampacity for use in the ship's engine room uptake where the ambient air is 70°C.

Solution:

As per Table 6.3, polypropylene insulation is good up to 105°C, and that leaves 105 – 40 = 65°C temperature rise with nominal ampacity in 40°C ambient air. If we use this cable in 70°C ambient air, we must limit the conductor temperature rise to 105 – 70 = 35°C instead of 65°C. We write Equation 2.23 in a ratio form to cancel out the cable heat dissipation geometry constant that does not change with the ambient temperature:

$$\frac{\Delta T_{\text{new}}}{\Delta T_{\text{old}}} = \frac{35}{65} = \frac{K(I_{\text{new}}^2 R)^{0.8}}{K(I_{\text{old}}^2 R)^{0.8}} = \left\{\frac{I_{\text{new}}}{I_{\text{old}}}\right\}^{1.6},$$

which gives

$$I_{\text{new}} = 200 \times \left(\frac{35}{65}\right)^{\frac{1}{1.6}} = 200 \times 0.68 = 136 \text{ A}.$$

This cable can be used only for 136-A maximum in the engine room uptake, that is, a derating factor of 136 ÷ 200 = 0.68 factor or 68% of its nominal ampacity of 200 A.

Alternatively, Table 6.6 can be used with some interpolation. For example, for 105°C class polypropylene insulation, we read between 90°C and 125°C columns, which for 70°C ambient air gives a derating factor between 0.52 and 0.76. The 105°C being approximately the midpoint in the 90°C–125°C range, the derating factor would be 1/2 (0.52 + 0.76) = 0.64 approximately, which is a close match with 0.68 derived above from the heat transfer fundamentals.

6.4 CABLE ELECTRICAL MODEL

In the electrical model of power systems, each phase conductor of the cable is represented by its resistance R and leakage reactance X placed in series in the single-line diagram (Figure 6.3). The reactance $X = \omega L$ where inductance L is due to the magnetic flux leakage in air between the current carrying conductors. The X effectively causes a proportional voltage drop, which is 90° out of phase with that in R. Therefore, the total cable impedance $Z = R + jX$. The cable also has a capacitance to ground, which is small and is generally ignored.

The cable R and X values are often provided by the cable manufacture or can be found from tables in reference books or from the general formulas given below.

The conductor resistance varies directly with length and inversely with the cross-section area, that is,

$$R = \frac{\text{resistivity} \times \text{length}}{\text{conductor cross-section area}}$$

or

$$R = K\frac{\text{length in feet}}{\text{area in circular mils}}\ \text{ohms.} \tag{6.3}$$

For dc or low-frequency currents at 20°C, $K = 10.372$ for copper and 18.046 for aluminum, both of electrical grade. The conductor resistance changes with temperature. Its value R_2 at higher temperature T_2 increases from R_1 at lower temperature T_1 as follows:

$$R_2 = R_1 \times \{1 + \alpha\,(T_2 - T_1)\} \tag{6.4}$$

where α = temperature coefficient of resistance = 0.0039 per degree Celsius for both copper and aluminum.

The leakage reactance of the three-phase cable shown in the cross section in Figure 6.4 is given by

$$X = 52.9\frac{f}{60}\log_{10}\left(\frac{\text{GMD}_\phi}{\text{GMR}_\phi}\right)\ \text{micro-ohms per phase per foot of cable} \tag{6.5}$$

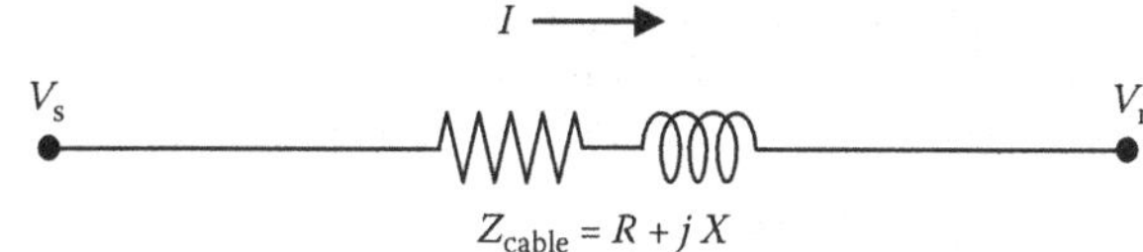

(a) Cable impedance between source and load

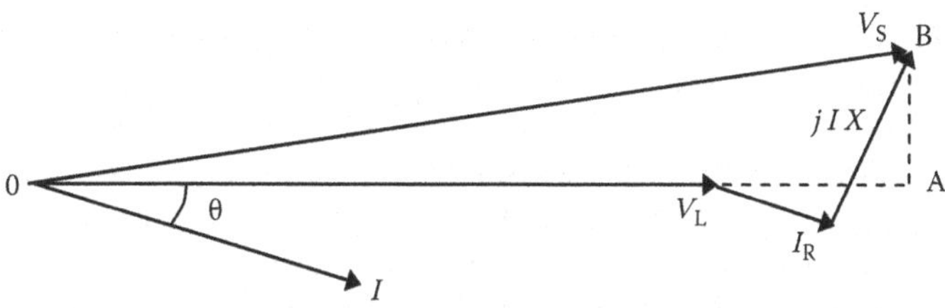

(b) Phasor diagram for voltage drop

FIGURE 6.3 Cable electrical model and phasor diagram.

where f = supply line frequency; $GMD_\varphi = (D_{ab}D_{bc}D_{ca})^{1/3}$ = geometrical mean distance between phase conductors separated by center-to-center distance of D_{ab}, D_{bc}, and D_{ca} between phase conductors a–b, b–c, and c–a, respectively; and $GMR_\varphi = (r_1 r_2 \ldots r_n)^{1/n}$ = geometrical mean radius of one phase conductor made of n strands placed at radial locations $r_1, r_2, r_3, \ldots, r_n$ (not shown) from the conductor center.

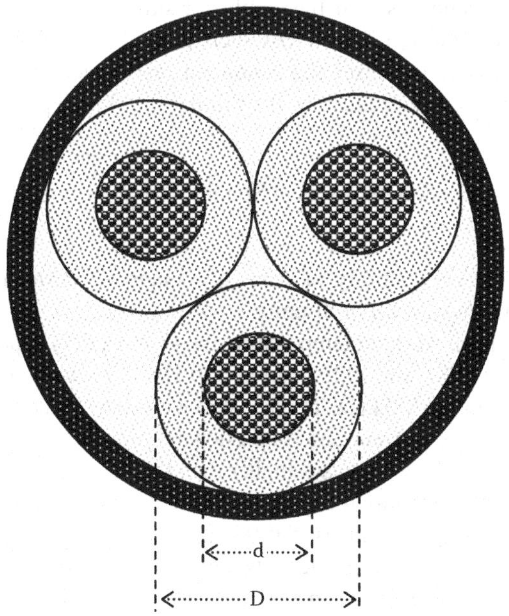

FIGURE 6.4 Three-phase cable cross section and GMD and GMR definitions.

The GMD and GMR can be in any unit (centimeters or inches), as they are in a ratio. For cable sizes used in medium power ranges, GMD_φ and GMR_φ can be approximated as follows:

$GMD_\varphi = D$, where D = outer diameter of each insulated phase conductor in single-phase (two-wire) or three-phase (three- or four-wire) cable
$GMR_\varphi = 0.375 \times d$, where d = outer diameter of the bare phase conductor.

The values in Equations 6.3 and 6.5 are per phase that are used in balanced three-phase *Y*-connected system analysis, where only one wire per phase carries the load current. There is no return current, so the neutral wire (even if provided) does not contribute in the voltage drop. In a single-phase system, the values in Equations 6.3 and 6.5 are multiplied by 2 to account for the lead and return conductors carrying the same current.

For twisted pairs of small wires touching each other, $D = (d + 2 \times$ wire insulation thickness). Since the insulation thickness is proportional to the wire radius for mechanical reasons, the GMD/GMR ratio remains approximately constant, making the cable inductance somewhat insensitive to the wire gage. For AWG 4/0 to AWG 30 wires, for example, the leakage inductance remains in the range of 0.5–0.7 μH per meter length of round-trip twisted pair up to hundreds of kilohertz frequency. Twisting does not reduce the inductance per meter but adds 10%–15% in the length.

6.5 SKIN AND PROXIMITY EFFECTS

Equation 6.3 gives dc resistance of a conductor in which the current is uniformly distributed. High-frequency current distribution is not uniform over the conductor cross section. It concentrates near the conductor skin as shown in Figure 6.5 for a round conductor, leaving the inner cross section not fully utilized in carrying the current. This effectively increases the conductor resistance and hence the I^2R loss and heating. The higher the frequency, the thinner the skin depth, which is given by

$$\text{skin thickness in SI units, } \delta = \sqrt{\frac{\rho}{\mu \pi f}} \tag{6.6}$$

where ρ = electrical conductivity of wire, μ = magnetic permeability of the flux medium, and f = frequency, all in SI units.

For copper at typical working temperature, the skin depth is approximately 10 mm (0.394 in.) at 60 Hz, 3.87 mm (0.152 in.) at 400 Hz, 0.5 mm (0.020 in.) at 20 kHz, and 0.25 mm (0.010 in.) at 100 kHz. Thinner skin with concentrated current means less cross-section area for the current and proportionally higher conductor resistance. For this reason, the ac resistance is always higher than the dc resistance depending on the frequency. For example, a 5 mm (0.20 in.) thick bus bar would have the ratio R_{ac}/R_{dc} = 1.1 at 60 Hz, 1.32 at 400 Hz, and 20 at 100 kHz.

The conductor adds power loss due to its own skin effect and also due to eddy currents induced by the leakage flux of neighboring current carrying conductors. In case of two conductors carrying single-phase current in opposite directions, the currents in both conductors concentrate near their adjacent faces as shown in Figure 6.6. The

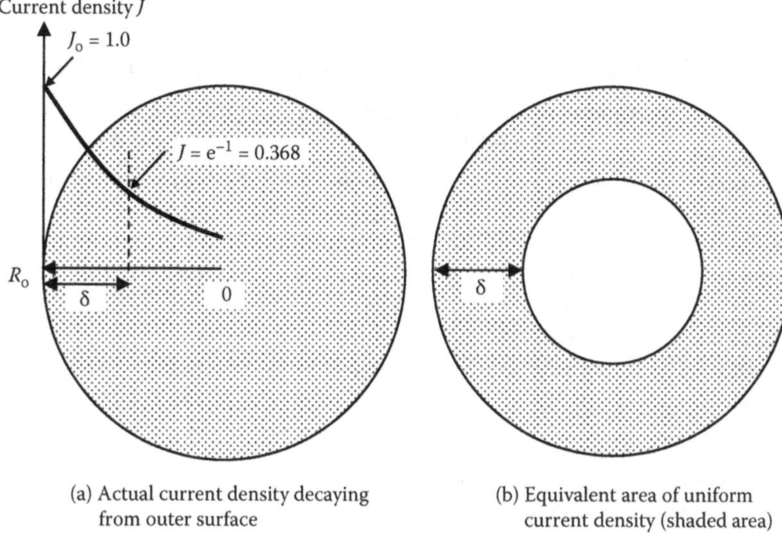

(a) Actual current density decaying from outer surface

(b) Equivalent area of uniform current density (shaded area)

FIGURE 6.5 Skin effect and resulting nonuniform current distribution in round conductor.

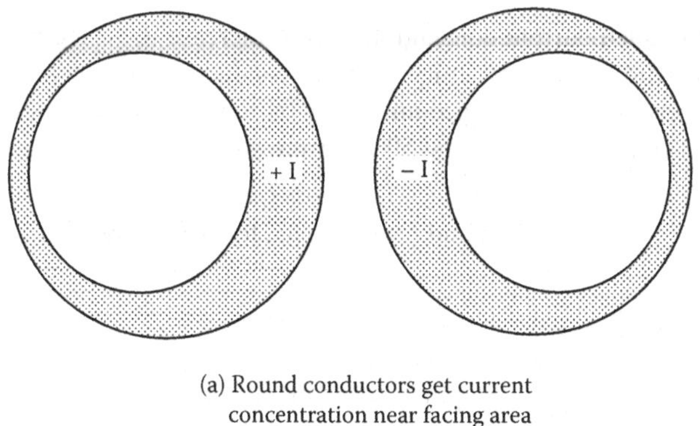

(a) Round conductors get current concentration near facing area

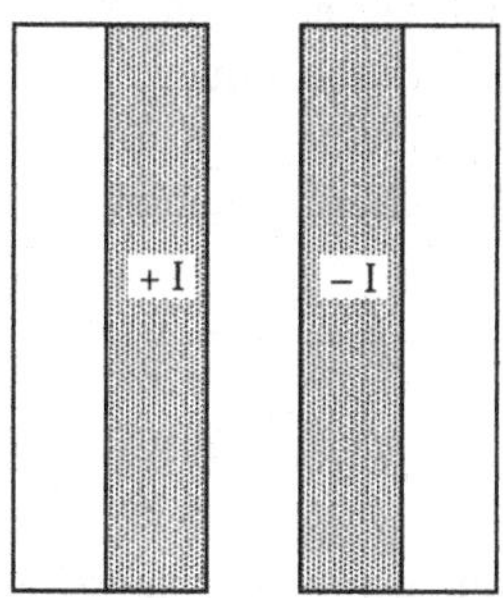

(b) Bus bars with facing flats get current concentration near facing strips

FIGURE 6.6 Proximity effect in two adjacent current carrying conductors.

TABLE 6.7
Combined Skin and Proximity Effect on AC/DC Resistance Ratio at 60 Hz

Conductor Size AWG or kcmil[a]	Cable with Nonmetallic Shield in Air or Nonmetallic Conduit		Cable with Metallic Shielded or Metallic Raceway	
	Copper	Aluminum	Copper	Aluminum
Up to AWG #1 (84 kcmil)	1.000	1.000	1.000	1.000
250 kcmil	1.005	1.002	1.06	1.02
500 kcmil	1.018	1.007	1.13	1.06
1000 kcmil	1.067	1.026	1.30	1.19
1500 kcmil	1.142	1.058	1.53	1.36
2000 kcmil	1.233	1.100	1.82	1.56

Source: Selected data from NEC.
Note: 60-Hz resistance = dc resistance × factor above = approximate 50-Hz resistance.
[a] kcmil = 1000 circular mils = MCM in old usage (first M for 1000 in Roman numerals).

virtual effect of such current concentration is to increase the effective resistance of the conductor working in proximity of other conductors. The combined skin and proximity effects are generally accounted for by multiplying factors listed in Table 6.7 for various conductors at 60 and 50 Hz, from which

$$R_{ac} = R_{dc} \times \text{skin and proximity factor from Table 6.7.} \tag{6.7}$$

6.6 CABLE DESIGN

The cable design primarily requires selecting the conductor size with required ampacity at the operating temperature that will also meet the voltage drop limitation over the feeder length under steady-state current and the motor starting inrush current. Both calculations require the values of the conductor current I and power factor, resistance R, and leakage reactance X, all values per phase. Then, the analysis is similar to that in Section 2.7, where Equations 2.15 and 2.17 gave

$$V_{drop} = I \times (R\cos\theta + X\sin\theta) = I \times Z_{eff} \tag{6.8}$$

where

$$Z_{eff} = (R\cos\theta + X\sin\theta). \tag{6.9}$$

All calculations can be done using volts, amperes, and ohms per phase, or in pu or percent values, and taking θ positive for lagging pf and negative for leading pf.

Equation 6.9 is a convenient definition of Z_{eff} because the product of I and Z_{eff} simply gives the voltage drop magnitude or the voltage drop per ampere in the cable:

$$V_{drop/amp} = Z_{eff} \text{ volts/ampere.} \tag{6.10}$$

Many cable manufacturers list Z_{eff} as the cable impedance at typical pf of 0.85 lagging. Therefore, cable Z values listed in a manufacturer's catalog are typically

$$Z_{cable} = R \times 0.85 + X \times \{1 - 0.85^2\}^{1/2} = 0.85\, R + 0.527\, X\ \Omega/\text{phase.} \tag{6.11}$$

Note: If the pf is much different than 0.85 lagging, Equation 6.10 is not valid. Equation 6.8 gives better results for a wide range of power factors from 0.2 lag to 0.9 lead if $R \ll X$.

The cable size is usually selected to meet the ampacity with a 20% to 30% margin that will also limit the steady-state voltage drop generally below 3%–5% from the controlled switchboard to the load that may include a transformer.

Many cable vendors provide ac resistance R_{ac} and leakage reactance X_L values for various cable sizes, as illustrated in Table 6.8, which includes the skin and proximity

TABLE 6.8
Impedance of 600-V Three-Phase Cables with Copper Conductors at 60 Hz and 75°C (167°F), Including Skin and Proximity Effects

	Ohms per Phase (Line to Neutral) per 1000-ft. Copper Wires at 60 Hz[a], 75°C, with Skin and Proximity Effects						
	Reactance X_L		R_{ac} for Copper Conductors		Z_{eff} at 0.85 pf Lagging[b]		
AWG or kcmil	PVC or Aluminum Conduit	Steel Conduit	PVC Conduit	Steel or Aluminum Conduit	PVC Conduit	Aluminum Conduit	Steel Conduit
10	0.050	0.063	1.2	1.2	1.1	1.1	1.1
4	0.048	0.060	0.31	0.31	0.29	0.29	0.30
1	0.046	0.057	0.15	0.16	0.16	0.16	0.16
00	0.043	0.054	0.10	0.10	0.11	0.11	0.11
0000	0.041	0.051	0.062	0.065	0.074	0.078	0.080
250	0.041	0.052	0.052	0.055	0.066	0.070	0.073
300	0.041	0.051	0.044	0.047	0.059	0.063	0.065
400	0.040	0.049	0.033	0.037	0.049	0.053	0.056
500	0.039	0.048	0.027	0.031	0.043	0.048	0.050
750	0.038	0.048	0.019	0.023	0.036	0.040	0.043
1000	0.037	0.046	0.015	0.019	0.032	0.036	0.040

Source: Modified from NEC data.

[a] Approximate 50-Hz values $R_{50\text{ Hz}} = R_{60\text{ Hz}}$ and $X_{L50\text{ Hz}} = (50/60)\, X_{L60\text{ HZ}}$.

[b] $Z_{eff} = R \cos\theta + X \sin\theta$, where $\cos\theta$ = pf of the cable load. The cable voltage drop per phase (line to neutral) is approximately equal to $I_{Line} \times Z_{eff}$. The table values above are for 0.85 power factor lagging.

effects and the conduit type as well. The skin and proximity effects reduce the cable leakage inductance slightly. The difference between dc and high-frequency inductance is small and is generally ignored in power system studies.

Example 6.2

A 1-kV, AWG 4 copper conductor has dc resistance of 0.30 Ω per 1000 ft. at 25°C. Using the tables provided in this chapter, determine its 60-Hz resistance at an operating temperature of 110°C.

Solution:

For a 1000-ft. length of AWG 4 conductor, Table 6.8 gives R_{ac} = 0.31 Ω at 60 Hz and 75°C. We correct it for 110°C using Equation 6.4:

$$R_{ac} \text{ at } 110°\text{C} = 0.31\ \{1 + 0.0039\ (110 - 75)\} = 0.352\ \Omega.$$

Alternatively, Table 6.7 gives an R_{ac}/R_{dc} ratio of 1.00. With a temperature correction factor of {1 + 0.0039 (110 – 25)} = 1.3315, we have R_{ac} at 110°C = 0.30 × 1.00 × 1.3315 = 0.399 Ω.

Seeing such a difference (0.352 versus 0.399 Ω per 1000 ft.) is possible, as data come from different sources with somewhat different considerations. In such a situation, the engineer must use a conservative value or one in accordance with a specific standard cited in the contract.

Example 6.3

A three-phase, *Y*-connected, 4.16 kV_{LL} feeder has *R* = 50 mΩ and *X* = 30 mΩ, both per phase per 1000-ft. length. Determine the voltage drop in the feeder at (1) 0.8 power factor lagging and (2) unity power factor. Establish the V_{drop} per 1000-ft. run for a 1-MVA load so that it can be easily reused for various megavolt-ampere loading and feeder lengths.

Solution:

$$V_{LN} = 4160 \div \sqrt{3} = 2400\ \angle 0°\ \text{V}.$$

For 1-MVA three-phase load, phase current *I* = 1000 kVA ÷ (√3 × 4.16 kV) = 138.8 A/ph.

For 1000-ft. run, *R* = 0.05 Ω/ph and *X* = 0.07 Ω/ph.

Using Equation 2.15 at 0.8 power factor lagging, that is, cos θ = 0.8 and sin θ = 0.6

V_{drop} = 138.8 (0.05 × 0.8 + 0.07 × 0.6) = 11.38 V/phase per MVA per 1000-ft. cable.

At unity pf, $\cos\theta = 1.0$, $\sin\theta = 0$, and

$$V_{drop} = 138.8\ (0.05 \times 1.0 + 0.07 \times 0) = 6.94 \text{ V/phase per MVA per 1000-ft. cable.}$$

At leading pf, it would be even less than 6.94 V/phase and can be zero at a certain leading pf. This illustrates the effects of pf on the voltage drop.

The cable must be selected to operate reliably in an applicable environment, which the cable manufacturers take into account. A few technical data sheets of flexible power cables are given in Tables 6.9 through 6.14, with permission from Gexol-insulated cable, a product of AmerCable, Inc.

The cable manufacturers also offer special designs to meet specific requirements in certain applications, for example, for variable frequency drives, mobile substation, magnetic crane for loading and unloading ship cargo, etc. Some such requirements are listed in Table 6.14.

Example 6.4

Select a three-phase 4.6-kV cable to carry 240 A/phase in single-banked tray and determine its outer sheath diameter and weight per 1000 ft. The cable must limit the voltage drop below 30 V/phase per 1000 ft. at an operating temperature of 90°C.

Solution:

As the first trial, we use Table 6.11 for a 5-kV class cable and choose AWG 4/0 that has 260-A ampacity in a single-banked tray and voltage drop of 0.136 V/A. This amounts to the actual voltage drop of 240 A (our actual current) × 0.136 = 32.64 V per 1000 ft. The ampacity of this cable meets the requirement, but the voltage drop does not.

We, therefore, select the next higher size, a 262 kcmil, 5-kV class cable that has the ampacity of 296 A in single-banked tray and voltage drop of 0.118 V/1000 ft., amounting to 240 × 0.118 = 28.32 V, now meeting the requirement.

For this 262-kcmil cable, Table 6.11 gives the nominal diameter as 2.608 in. with an outer sheath and weight as 5654 lb. per 1000 ft. The overhead trays should be designed to support this weight.

Example 6.5

Select a three-phase, 11-kV cable to power a 20,000-hp propulsion motor that has 97% efficiency and 95% pf lagging. Determine the % voltage drop if the distance from the switchboard to the motor is 200 ft. You may place more than one cable in parallel, if needed.

TABLE 6.9
Two-Conductor Flexible Power Cable with 1-kV, 110°C Class Insulation

			Unarmored		Armored (B)		Armored and Sheath (BS)						Ampacity		
Size AWG/ kcmil	mm^2	Part No. 37-102	Nominal Diameter (in.)	Weight (lb./ MFt.)	Nominal Diameter (in.)	Weight (lb./ MFt.)	Nominal Diameter (in.)	Weight (lb./ MFt.)	DC Resistance at 25°C (Ω/1000 ft.)	AC Resistance 110°C, 60 Hz (Ω/1000 ft.)	Inductive Reactance (Ω/1000 ft.)	Voltage Drop 110°C (V/A/1000 ft.)	110°C	100°C	95°C
16	1.3	–501	0.350	75	0.400	141	0.540	202	4.610	6.121	0.039	8.511	20	19	20
14	2.1	–507	0.380	84	0.430	165	0.561	230	2.907	3.859	0.036	5.379	33	31	27
12	3.3	–515	0.420	111	0.470	190	0.601	263	1.826	2.424	0.034	3.390	43	40	32
10	5.2	–553	0.460	146	0.510	230	0.641	307	1.153	1.530	0.032	2.151	53	49	43
8	7.6	–209	0.600	221	0.650	327	0.781	416	0.708	0.940	0.034	1.336	69	64	58
6	12.5	–210	0.690	308	0.730	424	0.903	559	0.445	0.590	0.032	0.850	91	85	77
4	21	–594	0.887	516	0.937	664	1.110	835	0.300	0.399	0.029	0.582	118	110	103
1/0	54	–216	1.243	1128	1.293	1334	1.466	1562	0.117	0.156	0.028	0.245	213	199	184
4/0	109	–219	1.593	2003	1.643	2271	1.878	2680	0.059	0.080	0.026	0.138	329	307	285

Source: From Gexol-insulated cable, a product of AmerCable, Inc. With permission.
Note: Cable diameters shown as nominal are subject to a ±5% manufacturing tolerance.

TABLE 6.10
Three-Conductor Flexible Power Cable with 1-kV, 100°C Class Insulation

Size AWG/ kcmil	mm²	Part No. 37-102	Unarmored Nominal Diameter (in.)	Unarmored Weight (lb./ MFt.)	Armored (B) Nominal Diameter (in.)	Armored (B) Weight (lb./ MFt.)	Armored and Sheath (BS) Nominal Diameter (in.)	Armored and Sheath (BS) Weight (lb./ MFt.)	DC Resistance at 25°C (Ω/1000 ft.)	AC Resistance 110°C, 60 Hz (Ω/1000 ft.)	Inductive Reactance (Ω/1000 ft.)	Voltage Drop 110°C (V/A/1000 ft.)	Opt. Uninsulated Grounding Cond. Size AWG	Ampacity 110°C	Ampacity 100°C	Ampacity 95°C
16	1.3	–502	0.369	65	0.419	127	0.519	181	4.610	6.121	0.039	8.511	–	17	16	16
14	2.1	–508	0.401	102	0.451	176	0.583	228	2.907	3.859	0.036	5.379	–	27	25	22
12	3.3	–516	0.445	133	0.495	212	0.626	276	1.826	2.424	0.034	3.390	–	33	31	27
10	5.2	–308	0.488	189	0.538	281	0.669	352	1.153	1.530	0.032	2.151	–	44	41	36
8	7.6	–309	0.637	274	0.687	385	0.818	477	0.708	0.940	0.034	1.336	–	56	52	48
6	12.5	–310	0.723	390	0.773	519	0.946	650	0.445	0.590	0.032	0.850	8	75	70	64
4	21	–312	0.942	678	0.992	843	1.165	1004	0.300	0.399	0.029	0.582	8	99	92	85
2	34	–314	1.084	987	1.134	1160	1.307	1374	0.184	0.244	0.028	0.366	6	131	122	113
1	43	–315	1.206	1234	1.256	1458	1.431	1675	0.147	0.195	0.028	0.299	6	153	143	131
1/0	54	–316	1.326	1448	1.376	1781	1.550	2015	0.117	0.156	0.028	0.245	6	176	164	152
2/0	70	–317	1.422	1945	1.472	2082	1.645	2424	0.093	0.125	0.027	0.200	6	201	188	175
3/0	86	–318	1.528	2379	1.578	2720	1.814	3106	0.074	0.100	0.027	0.166	4	234	218	202
4/0	109	–319	1.765	2864	1.815	3233	2.050	3652	0.058	0.080	0.026	0.138	4	270	252	235
262	132	–320	1.980	3452	2.030	3880	2.266	4434	0.048	0.067	0.026	0.119	3	315	294	267
313	159	–321	2.131	4023	2.181	4434	2.418	4919	0.040	0.056	0.026	0.105	3	344	321	299
373	189	–322	2.231	4772	2.281	5219	2.517	5718	0.034	0.047	0.025	0.092	3	387	361	334
444	227	–323	2.394	5670	2.444	6176	2.680	6864	0.028	0.041	0.025	0.083	2	440	411	372
535	273	–324	2.637	6784	2.687	7492	2.986	8250	0.024	0.035	0.026	0.075	2	498	443	418
646	326	–326	2.958	7961	3.008	8414	3.301	9258	0.020	0.030	0.026	0.068	1	553	516	470
777	394	–327	3.168	9573	3.218	10065	3.511	10945	0.016	0.026	0.026	0.063	1	602	562	529

Source: From Gexol-insulated cable, a product of AmerCable, Inc. With permission.

TABLE 6.11
Three-Conductor Medium Voltage Cable with 5-kV Class Insulation (100/133% Insulation Level)

			Unarmored		Armored and Sheathed (BS)		Ampacity						
Size AWG/ kcmil	mm²	Part No. 37-105	Nominal Diameter (in.)	Weight (lb./1000 ft.)	Nominal Diameter (in.)	Weight (lb./1000 ft.)	In Free Air (A)	Single Banked in Trays (A)	DC Resistance at 25°C (Ω/1000 ft.)	AC Resistance at 90°C, 60 Hz (Ω/1000 ft.)	Inductive Reactance (Ω/1000 ft.)	Voltage Drop (V/A/1000 ft.)	Optional Grounding Conductor
8	7.6	–301	1.137	781	1.369	1218	66	56	0.708	0.885	0.048	1.275	8
6	12.5	–302	1.226	955	1.457	1424	88	75	0.445	0.556	0.044	0.815	6
4	21	–303	1.402	1307	1.625	1824	116	99	0.300	0.376	0.039	0.560	6
2	34	–304	1.538	1690	1.824	2372	152	129	0.184	0.230	0.036	0.356	6
1	43	–305	1.626	1974	1.911	2692	175	149	0.147	0.184	0.035	0.291	4
1/0	54	–306	1.783	2423	2.081	3232	201	171	0.117	0.147	0.034	0.239	4
2/0	70	–307	1.913	2884	2.210	3749	232	197	0.093	0.117	0.033	0.196	4
3/0	86	–308	2.007	3315	2.305	4220	266	226	0.074	0.094	0.032	0.163	3
4/0	109	–309	2.140	3937	2.438	4899	306	260	0.058	0.075	0.031	0.136	3
262	132	–310	2.310	4619	2.608	5654	348	296	0.048	0.063	0.030	0.118	3
313	159	–311	2.453	5319	2.796	6549	386	328	0.040	0.053	0.029	0.104	2
373	189	–312	2.589	6107	3.000	7402	429	365	0.034	0.045	0.029	0.092	2
444	227	–313	2.818	7280	3.161	8684	455	387	0.028	0.039	0.028	0.083	1
535	273	–314	2.974	8463	3.317	9964	528	449	0.024	0.033	0.028	0.074	1
646	326	–315	3.164	9814	3.507	11,407	584	496	0.020	0.028	0.027	0.067	1
777	394	–316	3.385	11,526	3.729	13,226	647	550	0.016	0.025	0.027	0.062	1/0

Source: From Gexol-insulated cable, a product of AmerCable, Inc. With permission.

TABLE 6.12
Three-Conductor Medium Voltage Cable with 8-kV Class Insulation (100%, Insulation Level)

			Unarmored		Armored and Sheathed (BS)		Ampacity						
Size AWG/ kcmil	mm²	Part No. 37-105	Nominal Diameter (in.)	Weight (lb./1000 ft.)	Nominal Diameter (in.)	Weight (lb./1000 ft.)	In Free Air (A)	Single Banked in Trays (A)	DC Resistance at 25°C (Ω/1000 ft.)	AC Resistance at 90°C, 60 Hz (Ω/1000 ft.)	Inductive Reactance (Ω/1000 ft.)	Voltage Drop (V/A/1000 ft.)	Optional Grounding Conductor
6	12.5	–317	1.338	1094	1.561	1589	88	75	0.445	0.556	0.046	0.818	6
4	21	–318	1.514	1462	1.779	2134	116	99	0.300	0.376	0.041	0.562	6
2	34	–319	1.650	1970	1.998	2725	152	129	0.184	0.230	0.038	0.357	6
1	43	–320	1.800	2263	2.085	3054	175	149	0.147	0.184	0.037	0.293	4
1/0	54	–321	1.895	2617	2.181	3454	201	171	0.117	0.147	0.036	0.241	4
2/0	70	–322	2.025	3100	2.310	3989	232	197	0.093	0.117	0.034	0.198	4
3/0	86	–323	2.119	3531	2.404	4458	266	226	0.074	0.094	0.033	0.165	3
4/0	109	–324	2.252	4162	2.537	5140	306	260	0.058	0.075	0.032	0.138	3
262	132	–325	2.422	4864	2.707	5913	348	296	0.048	0.063	0.031	0.119	3
313	159	–326	2.565	5581	2.914	6884	386	328	0.040	0.053	0.030	0.105	2
373	189	–327	2.704	6392	3.054	7760	429	365	0.034	0.045	0.030	0.093	2
444	227	–328	2.930	7582	3.280	9059	455	387	0.028	0.039	0.029	0.084	1
535	273	–329	3.096	8806	3.439	10,366	528	449	0.024	0.033	0.029	0.075	1
646	326	–330	3.267	10,137	3.611	11,780	584	496	0.020	0.028	0.028	0.068	1
777	394	–331	3.512	11,959	3.855	13,708	647	550	0.016	0.025	0.028	0.063	1/0

Source: From Gexol-insulated cable, a product of AmerCable, Inc. With permission.

TABLE 6.13
Three-Conductor Medium Voltage Cable with 15-kV Class Insulation (100% Insulation Level)

			Unarmored		Armored and Sheathed (BS)		Ampacity						
Size AWG/ kcmil	mm²	Part No. 37-105	Nominal Diameter (in.)	Weight (lb./1000 ft.)	Nominal Diameter (in.)	Weight (lb./1000 ft.)	In Free Air (A)	Single Banked in Trays (A)	DC Resistance at 25°C (Ω/1000 ft.)	AC Resistance at 90°C, 60 Hz (Ω/1000 ft.)	Inductive Reactance (Ω/1000 ft.)	Voltage Drop (V/A/1000 ft.)	Optional Grounding Conductor
2	34	–346	2.157	2759	2.443	3697	156	133	0.184	0.230	0.042	0.361	6
1	43	–347	2.239	3073	2.524	4045	178	151	0.147	0.184	0.040	0.296	4
1/0	54	–348	2.335	3466	2.620	4477	205	174	0.117	0.147	0.039	0.244	4
2/0	70	–349	2.461	3991	2.810	5242	234	199	0.093	0.117	0.037	0.201	4
3/0	86	–350	2.559	4466	2.910	5764	269	229	0.074	0.094	0.036	0.168	3
4/0	109	–351	2.691	5150	3.041	6513	309	263	0.058	0.075	0.035	0.141	3
262	132	–352	2.749	5749	3.152	7348	352	299	0.048	0.063	0.034	0.122	3
313	159	–353	2.881	6483	3.287	8184	389	331	0.040	0.053	0.033	0.107	2
373	189	–354	3.021	7331	3.365	8856	432	367	0.034	0.045	0.032	0.095	2
444	227	–355	3.183	8380	3.527	9983	456	388	0.028	0.039	0.031	0.086	1
535	273	–356	3.357	9599	3.701	11,285	528	449	0.024	0.033	0.031	0.077	1

Source: From Gexol-insulated cable, a product of AmerCable, Inc. With permission.

TABLE 6.14
Cable Requirement in Selected Special Applications

Application	Special Cable Requirements
Variable frequency drives	Braided shield to limit EMI emission
	Multiple ground conductors to reduce voltage unbalance and common mode noise back to VFD
	Lower dielectric constant of the insulation to reduce reflected wave peak voltages
	Insulation with higher breakdown voltage strength to resist 2 × rated voltage spikes
Magnetic cranes	Thinner strands for greater flexibility
Mobil substation	Greater protection against abrasion, impact, heat, oil, alkali, and acid
Portable power	
Shore-to-ship power (cold ironing)	Thinner strands for greater flexibility Greater abrasion resistance and tear strength in wet conditions

Solution:

The line current in the motor lines is derived from

$$\text{Motor input kVA}_{3\text{ph}} = \frac{20{,}000 \times 0.746}{0.97 \times 0.95} = \sqrt{3} \times 11 \times I_L,$$

which gives $I_L = 850$ A.

For 11-kV operating voltage, the next standard insulation class is 15 kV. Therefore, we use Table 6.13. The heaviest cable in the table is 535 kcmil with ampacity of 449 A in trays. Therefore, we must use two cables in parallel, each carrying 1/2 × 850 = 425 A, which is less than the cable ampacity of 449 A.

For a 535-kcmil cable, Table 6.13 gives V_{drop} of 0.077 V/A per 1000 ft. For a 200-ft. cable carrying 425 A, $V_{drop} = 0.077 \times 425 \times (200 \div 1000) = 6.545$ V/ph. The motor line-to-neutral voltage is $11{,}000 \div \sqrt{3} = 6351$ V/ph, and $V_{drop} = 6.545 \div 6351 = 0.001$ pu or 0.1%.

Although the voltage drop in this feeder is negligibly small, we cannot go to thinner cable to save cost since this cable design is determined by the ampacity and not by the voltage drop.

6.7 CABLE ROUTING AND INSTALLATION

Cables are often assembled in raceways that provide support and mechanical protection for the cable conductor and insulation. The most common raceways are conduits (metallic or nonmetallic) and rectangular trays, open from the top or totally enclosed. Cables in a flat bundle can also be supported from the ceiling by hooks without any enclosure. They should be adequately supported to withstand the peak mechanical force under the worst-case short circuit current.

TABLE 6.15
Cable Bend Radius Limit in Installations as per Various Industry Standards

Industry Standard	Unshielded Cable	Shielded Cable
IEEE-45	6 × diameter for all cables	8 × diameter for all cables
IEC-92	4 × diameter for < 1 in. (25 mm) diameter cable 6 × diameter for > 1 in. (25 mm) diameter cables	8 × diameter for all cables
Transport Canada	4 × diameter for < 1 in. (25 mm) diameter cable 6 × diameter for > 1 in. (25 mm) diameter cables	6 × diameter for all cables

The physical routing of cables may require many bends. The cable bending radius in the installation is limited to a certain value that depends on the cable type and diameter. Table 6.15 lists the maximum allowable bend radius in order to limit the bending stress below the damaging level. In general, cables should not be bent in a radius less than six to eight times the cable diameter.

PROBLEMS

1. A cable with neoprene insulation has nominal ampacity of 250 A in 30°C ambient air. Determine its ampacity for use in a boiler room where the ambient air is 50°C.
2. A 1-kV, 750-kcmil copper cable in a metallic tray has dc resistance of 0.016 Ω per 1000 ft. at 25°C. Using the tables provided in this chapter, determine its 60-Hz resistance at operating temperature of 110°C. What change would you expect in R_{ac} at 400 Hz?
3. A three-phase, *Y*-connected, 4.16 kV_{LL} feeder has $R = 200$ mΩ and $X = 35$ mΩ, both per phase per 1000-ft. length. Determine the V_{drop} per 1000 ft. per MVA load so that it can be easily reused for various feeder lengths and megavolt-ampere loadings at (1) 0.85 power factor lagging and (2) unity power factor.
4. Select a three-phase 6.6-kV cable to carry 525 A/phase in a single-banked tray and determine its (1) voltage drop per 100 ft. at operating temperature of 90°C, (2) outer sheath diameter, and (3) weight per 100 ft.
5. Select a three-phase, 13.8-kV cable to power a 24,000-hp propulsion motor that has 96% efficiency and 90% pf lagging. Determine the % voltage drop if the switchboard is 250 ft. away from the motor. Place more than one cable in parallel, if needed.
6. Select a three-phase, 460-V power distribution cable for a 100-hp induction motor that has 95% efficiency and 0.85 pf lagging. The motor is located 20 m from the switchboard. Determine the steady-state voltage drop in the cable. Allow 30% margin in cable ampacity.
7. A special three-phase cable is assembled as shown in Figure 6.4 using three one-conductor cables, each with a conductor diameter of 1 in. and an insulated outer diameter of 1.5 in. The net copper area of 20-mil-thin strands is 75% on a 1-in.-diameter base. Assuming an R_{ac}/R_{dc} ratio of 1.1, determine the cable *R* and *X* values per 1000 ft. at 50 Hz and 20°C.

QUESTIONS

1. Between AWG 30 and AWG 10 conductors, which is a heavier conductor, and how many of each can be placed in a 1-in. width?
2. In specifying the conductor size, what do terms MCM and kcmil stand for?
3. List two major factors that limit the cable ampacity.
4. Give two reasons why heavy conductors are made of thin strands as opposed to solid copper.
5. Explain the difference between the skin effect and the proximity effect and how they increase the conductor ac resistance over the dc resistance.
6. The term *effective* is used for the rms value of ac voltage or current and also for the cable impedance. Explain the term *effective impedance* and its use.

FURTHER READING

Ghandakly, A.A., Curran, R.L., and Collins, G.B. 1990. Ampacity ratings of bundled cables for heavy current applications. *IEEE Industry Applications Society Conference Records*, 2, Oct., 1334–39.

Neher, J.H. and McGrath, M.H. 2000. Calculation of the temperature rise and load capability of cable systems. *IEEE Paper*, ieeexplore.ieee.org/iel2/497/3988/00152358.pdf.

7 Power Distribution

The power distribution scheme for a small township starts from the township substation switchyard, from where the power is distributed to all users via multiple feeders going to various parts of the township. A large cruise ship also has a very similar scheme, which starts from the ship generator switchboard.

7.1 TYPICAL DISTRIBUTION SCHEME

Figure 7.1 depicts a typical power distribution system. The township power comes from transmission lines via a step-down substation transformer shown in the dotted box. It shows four three-phase feeders running *radially* from the central distribution station switchboard to various loads via circuit breakers. Feeders 1, 2, and 4 are shown in a single-line diagram, whereas Feeder 3 is shown in full detail with four-wire configuration (three-phase lines A, B, and C with neutral N). The lateral feeders are tapped off from each phase (line to neutral) of the main feeder such that single-phase service loads are about equally distributed on all three phases of the system. For example, the first two topmost lateral feeders—one going to the left and one going to the right—take power from phase A, the next two lateral feeders take power from phase B, and the next two from phase C.

The three-phase loads are, of course, connected to all three phases as shown near the end of Figure 7.1. If the voltage drop at the end of the main feeder exceeds the allowable limit with a cable that meets the ampacity requirement but not the voltage drop requirement, three-phase capacitors may be connected to improve the pf and henceforth reduce the voltage drop as shown in Equation 2.18. The power loss in capacitors can be minimized by switching them off during light load when the voltage drop remains within the limit and switching them back on only during heavy load demand. The capacitors can be remotely or automatically switched on and off in response to the feeder voltage at the far end.

If there are more continuing loads in a large system, the pattern shown in the upper part of Figure 7.1 is repeated after the sectionalizing switch that disconnects the lower part of the loads from the upper part if needed for maintenance or to isolate a fault.

Multiple voltages, where desired, can be obtained from the distribution transformer in two ways, shown in Figure 7.2. In the three-phase system shown in Figure 7.2a, the *Y*-connected transformer with 208-V line-to-line secondary with neutral can provide three different voltages: 120 V from line to neutral for small single-phase loads, 208 V for high-power single-phase loads connected from line to line, and 208 V three-phase for large three-phase loads. Figure 7.2b is typical for servicing small single-phase service loads in homes, where the distribution transformer with 240-V center-tapped secondary winding provides 120 V for low power loads

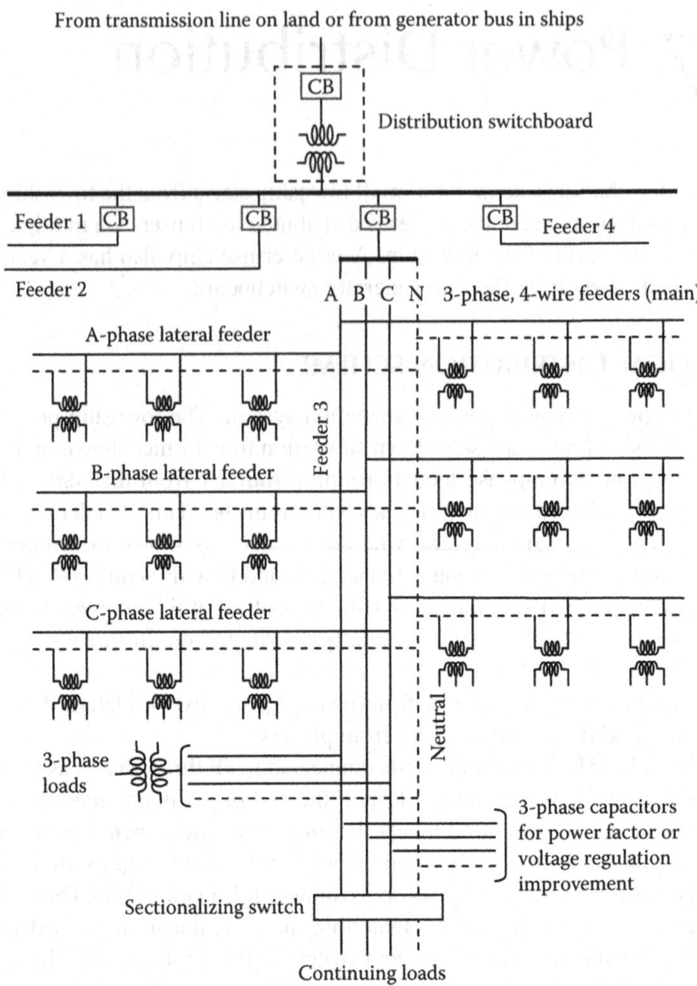

FIGURE 7.1 Typical distribution system from township substation switchyard.

(lights and small appliances) and 240 V for high power loads (space heater, cloth dryer, and kitchen stove).

Voltages below 120 V are required in some hands-on operations, such as arc welding and metal cutting. They are obtained by a special step-down transformer designed with high internal series impedance, which would cause the transformer output voltage to drop heavily in case of an accidental short circuit.

7.2 GROUNDED AND UNGROUNDED SYSTEMS

The three-phase *Y*-connected power distribution system can be three-wire without neutral or four-wire with neutral that can be either grounded or ungrounded. Land-based systems are generally grounded, whereas navy ships generally require an

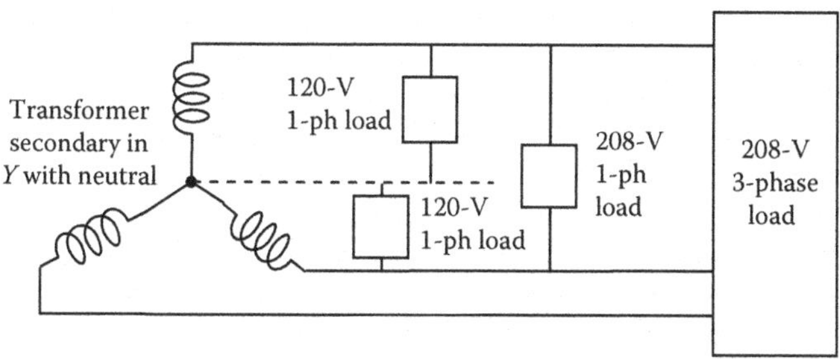

(a) 208-V 3-phase Y-connected transformer secondary for 120-V 1-phase and 208-V 3-phase loads (primary not shown)

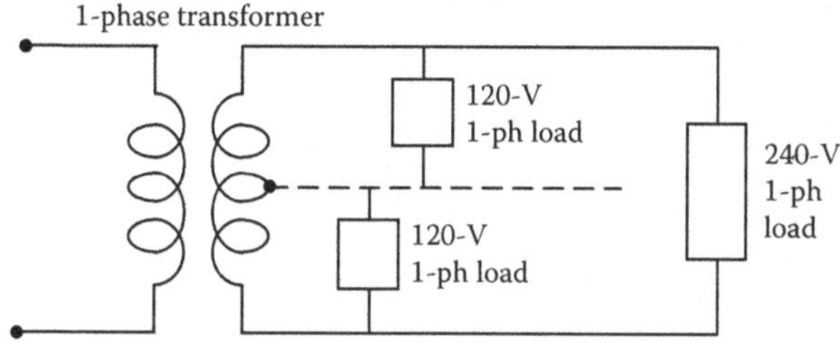

(b) 480-V/240-V 1-phase transformer with center-tapped secondary for 120-V and 240-V 1-phase loads

FIGURE 7.2 Multiple-voltage distribution systems for single-phase and three-phase loads.

ungrounded system coming out of the generator for reliability but a grounded distribution system for low-voltage service loads for personnel safety. As for the neutral, the following three alternatives may be considered:

1. Solidly grounded neutral via ground strap of negligible resistance and leakage reactance ($Z_{ground} \approx 0$)
2. High-impedance ground via a resistance and/or reactance inserted in series with the ground strap to control the ground fault current matching with the protection system
3. Ungrounded system where the neutral—although running full length along the phase wires—is not connected to the ground

In comparing the pros and cons of grounded versus ungrounded systems, we must recognize that about 90% of faults in three-phase systems are between one of the phases and the ground. In the grounded system shown in Figure 7.3a, a ground fault on one phase in the first instant results in high fault current that will trip the circuit breaker. However, in the ungrounded system shown in Figure 7.3b, the first ground fault does not complete the ground loop. Only the second ground fault completes the loop and results in high fault current that will trip the circuit breaker. Thus, the ungrounded system is a single-fault tolerant system. It offers an improved availability

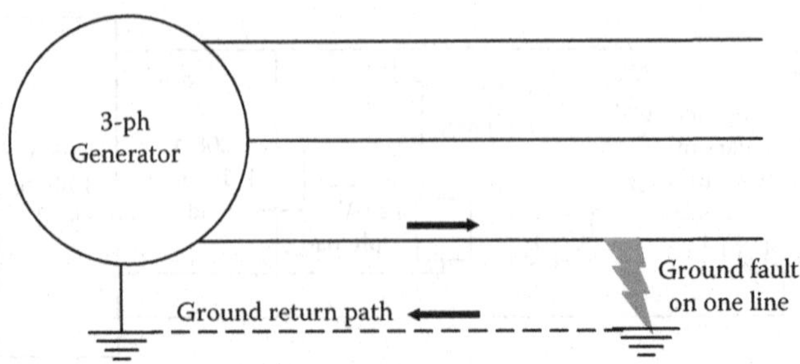

(a) Grounded 3-phase systems results in high current on first line to ground fault

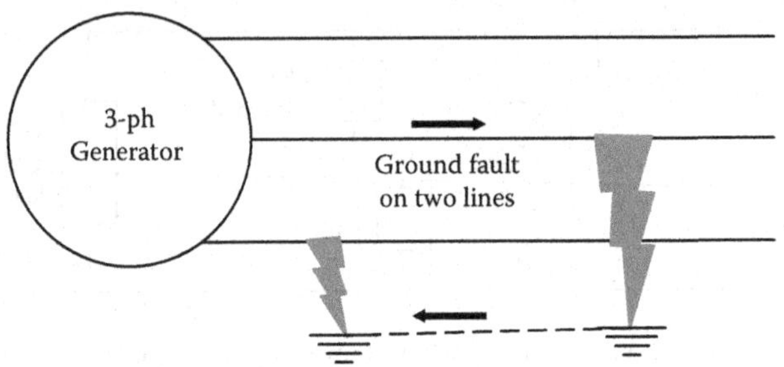

(b) Ungrounded 3-phase system is single-fault tolerant (second line to ground fault causes high current)

FIGURE 7.3 Ground faults in grounded and ungrounded (single-fault-tolerant) power systems.

of power, provided that we detect the first ground fault, locate it, and clear it before the second one occurs.

There is another reason why the ungrounded system is preferred in navy ships. With the generator neutral grounded, the triplen harmonic high-frequency currents due to power electronics loads and zero sequence currents due to unbalanced 3-phase loads—in phase in all three phases—must flow in the neutral to form the return path. In some systems, particularly with heavy-power electronics loads, the neutral current can exceed the fundamental frequency phase current. This may cause interference to sensitive low-frequency radio and sonar signals, which could compromise the navy operation.

In the grounded system, grounding the neutral of the *Y*-connected system makes the system grounded, whether or not the neutral is running along with the lines. It is also important to understand the difference between grounding the neutral and grounding the power equipment chassis, as they are not the same. Although both are often erroneously called ground wires, the former is correctly called the neutral ground, and the latter is called the chassis ground. The chassis ground is for human safety, and the power distribution neutral ground is for the equipment safety and desired performance. The system shown in Figure 7.4 becomes a grounded system

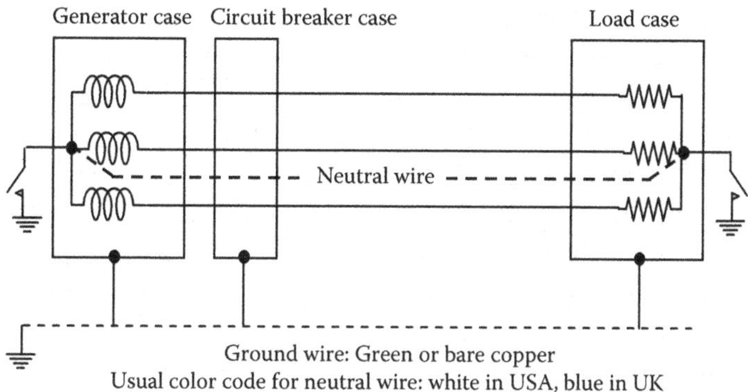

FIGURE 7.4 Neutral and chassis ground wires run separate and serve a different purpose.

when the switch in the neutral is closed such that the neutral wire is grounded, or else the system remains an ungrounded. Regardless of whether the system neutral is grounded or ungrounded, the equipment enclosures (chasses) are always grounded for personnel safety. The ground and neutral wires are of different colors and run separately as shown in Figure 7.4. The ground wire is always independent of the neutral wire. It basically keeps the enclosures at zero voltage under all operating conditions, even if an internal live conductor touches the enclosure wall. The general color code of the ground wire is green or bare copper, and the neutral wire color is white in the United States and blue in the United Kingdom and Europe.

The neutral current is zero with a balanced three-phase load with no harmonics. However, large power electronics loads typically found in modern power systems draw high triplen-harmonic currents that require the neutral wire to be as heavy as the phase conductor, or even heavier in some cases. The industry standards, therefore, require that a generator neutral wire must be sized to be equal to the largest phase conductor supplying the system with all generators connected in parallel.

7.3 GROUND FAULT DETECTION SCHEMES

In the ungrounded system, the first ground fault must be detected immediately and corrected before the second ground fault trips the generator. The generally used ground fault detection schemes are shown in Figure 7.5. Figure 7.5a is typically used in a dc or single-phase ac system, where a ground fault on line A causes lamp A connected to it to lose voltage and turn dark, while lamp B remains lit. Pressing the push button opens the normally closed contactor, breaks the ground loop, and brings lamp A back to light. The engineer can ascertain the ground fault by pushing the button on and off and seeing lamp A turn off and on in response. Figure 7.5b is for a low-voltage system to which the lamps can be directly connected. Lamp A will turn dark if line A gets grounded, which can again be ascertained by pressing the push button on and off. Figure 7.5c is for a high-voltage system, where a single lamp is connected inside the Δ of secondary coils of a suitable step-down transformer. When all phases are healthy, the phasor sum of all phase voltages in the Δ is zero, keeping the lamp

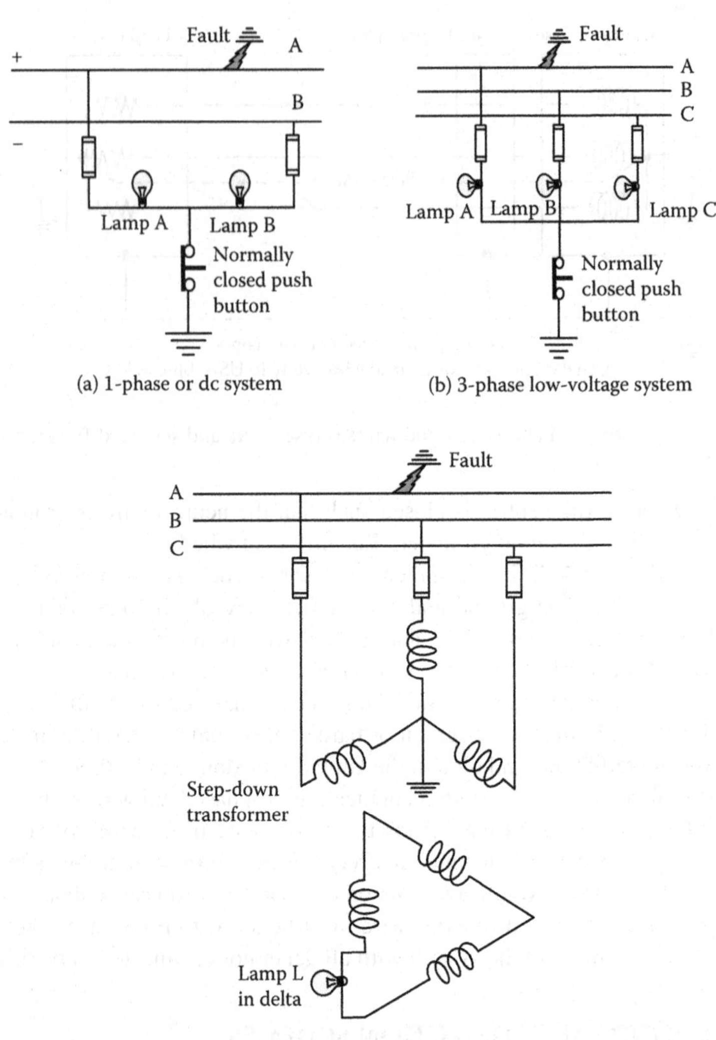

FIGURE 7.5 Ground fault detection schemes in LV and HV distribution systems.

dark. If phase C on the primary side gets grounded, phase C of the secondary side also loses the voltage, and the phasor sum of the remaining two voltages in the Δ is not zero. This results in lamp L being lit, indicating a ground fault in one of the lines, although it cannot indicate which line is faulted. Locating the fault in any of these schemes requires troubleshooting skills mixed with good logic.

7.4 DISTRIBUTION FEEDER VOLTAGE DROP

The distribution system has an automatic voltage regulation scheme at one or more control stations, such as the factory switchboard or township substation, to maintain constant voltage regardless of the load current. Various feeder cables deliver power from this controlled *sending end* to the user load at the *receiving end*, with the voltage gradually dropping from a maximum value near the sending end to the minimum value at the receiving end. Both the maximum and minimum values must be within the allowable limits for all users in the system. For this reason, the distribution system is sized not only for the required ampacity but also to stay below the allowable voltage drop limits under steady-state rated current and during transients such as inrush current during motor starting that generally puts the distribution system under stress. If the load current $\tilde{I}$ of lagging pf cos θ flows from the switchboard that is controlled to maintain constant bus voltage $\tilde{V}_s$ at the sending end to the load at the receiving end at voltage V_r, and the total series impedance of all transformers and cables between sending and receiving ends is $R + j\,X$ ohms/phase, then the magnitude of the voltage drop between the two ends is approximately given by Equation 2.15 or 2.18 for all practical power factors, that is,

$$V_{\text{drop}} = I\{\text{R} \cos\theta + \text{X} \sin\theta\} = I\left\{R\text{pf} + X\sqrt{1-\text{pf}^2}\right\}. \tag{7.1}$$

All calculations above are done using amperes, volts, and ohms per phase, and taking θ positive for lagging pf and negative for leading pf. One can also use R and X values in perunit based on the receiving-end kilovolt-ampere and voltage. Then, $V_r =$ 1.0 pu, and $I = 1.0$ pu when delivering rated load, or $I = 0.8$ pu when delivering 80% load, etc. Equation 7.1 gives perunit voltage drop if pu values of current, resistance, and reactance were used. Per unit calculations are preferred when there are multiple equipments in series.

7.4.1 Voltage Drop during Motor Starting

The power factor of motor starting current is very low, 0.2 to 0.3 lagging. If $R \ll X$, the phasor diagram of $\tilde{V}_r = \tilde{V}_s - \tilde{I} \times (R + jX)$ in Figure 7.6 with such poor pf shows that the voltage drop magnitude $(V_s - V_r) = I \times Z$, which is approximately in phase with

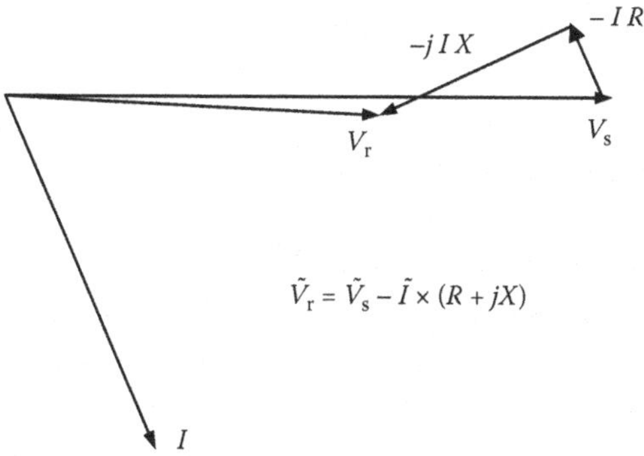

FIGURE 7.6 Feeder voltage phasor diagram under motor starting inrush current.

the source voltage. Therefore, V_{drop} under inrush current in a cable with impedance Z in ohms is simply given by

$$V_{drop} = I_{start} \times Z, \text{ during motor starting inrush current.} \quad (7.2)$$

Here, the cable impedance Z is not Z_{eff} often found in the vendor's data sheets but is the actual $Z = \sqrt{(R^2 + X^2)}$. The voltage drop can further be approximated as $I \times X$ in large cables where $X \gg R$ or as $I \times R$ in small cables where $R \gg X$.

7.4.2 Voltage Boost by Capacitors

We note that the system voltage drop depends also on the load power factor. A power factor of 0.85 lagging is common in many systems. Capacitors can improve the power factor close to 1.0 and, in turn, reduce the voltage drop by eliminating the $X \sin \theta$ term from Equation 7.1 and also by reducing the line current. The voltage drop at unity pf is then simply $I \times R$. A leading power factor has a $-\sin \theta$ term that can result in a negative voltage drop, meaning a higher voltage at the receiving end than at the sending end.

The voltage boost by the capacitors placed at the end of a feeder having the total system resistance R and leakage reactance X_L (inductive) is shown in Figure 7.7a, where all values are per phase. The new phase voltage $\tilde{V}_{LN}^{new} = \tilde{V}_{LN} - \tilde{I}_c \times (R + jX_L)$ or

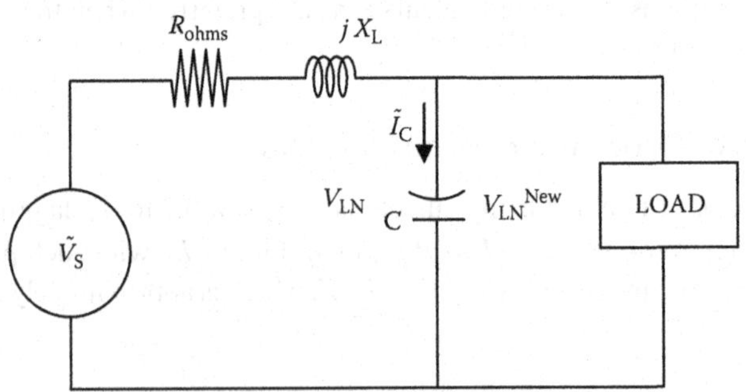

(a) Capacitor added at feeder end

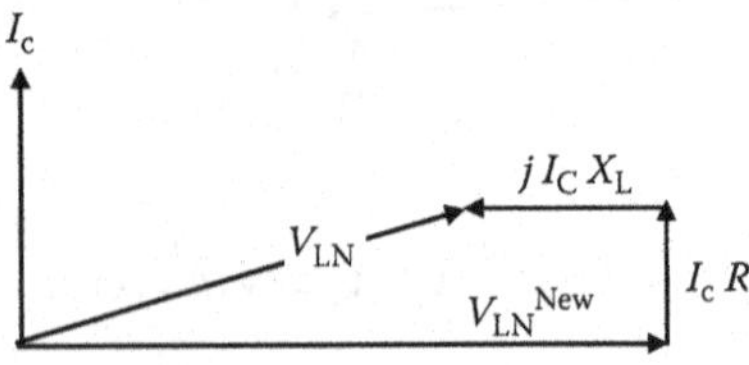

(b) Load voltage change with added I_c

FIGURE 7.7 Feeder voltage boost with pf improvement capacitors.

$\tilde{V}_{LN} = \tilde{V}_{LN}^{new} + \tilde{I}_c \times (R + jX_L)$, the phasor diagram of which is shown in Figure 7.7b. The voltage boost magnitude is then approximately equal to $V_{boost} = V_{LN}^{new} - V_{LN} = I_c \times X_L$. The capacitor bank's single-phase kVAR$_{cap}$ rating = $V_{LN} \times I_c$, giving I_c = kVAR$_{cap\cdot1ph}$ ÷ kV$_{LN}$, which leads to

$$V_{boost} = \frac{\text{kVAR}_{cap.1ph}}{\text{kV}_{LN}} \cdot X_L \text{ volts/phase} \quad \text{and} \quad \%V_{boost} = \frac{V_{boost}}{V_{LN}} \times 100. \tag{7.3}$$

Ignoring the favorable effect of line current reduction with pf improvement, the voltage drop as per Equation 7.1 can be zero for leading pf that gives ($R \cos \theta - X \sin \theta$) = 0 or $R/X = \tan \theta$ (when θ reverses the sign and enters in the leading region). Therefore, the receiving-end voltage will be constant when the leading pf angle $\theta = \tan^{-1}(R/X)$. The R/X ratio for a typical power distribution cable is around 0.2 to 0.3. Using its average value of 0.25, the pf angle $\theta = \tan^{-1}(0.25) = 14°$ leading, or pf = $\cos \theta = 0.97$ leading, gives a flat receiving-end voltage, regardless of the load current. Capacitors are typically used for improving poor lagging pf by bringing it close to unity, up to about 0.95 lagging, beyond which they give a diminishing rate of economic return on the capacitor capital cost. Therefore, making the pf 0.97 leading is generally uneconomical for pf improvement.

Example 7.1

A three-phase, 460-V, 60-Hz, 200-hp induction motor has design efficiency of 90% and pf of 0.85 lagging. The cable reactance from the control center to the motor is 0.015 Ω/ph. If the motor pf were corrected to unity by placing capacitors next to the motor, determine the % voltage boost at the motor terminals.

Solution:

$$\text{Motor full load current} = \frac{200 \times 746}{0.90 \times 0.85 \times \sqrt{3} \times 460} = 244.8 \text{ A/ph.}$$

phase voltage = 460 ÷ √3 = 265.6 V = 0.2656 kV$_{LN}$, and X_L = 0.015 Ω/ph.

The motor power triangle has three-phase power P = 200 × 0.746 = 149.2 kW, S = 149.2 ÷ 0.85 = 175.5 kVA, and $Q = \sqrt{\{S^2 - P^2\}}$ = 92.5 kVAR.

For unity pf, kVAR$_{cap/ph}$ = 1/3 × 92.5 = 30.83, and using Equation 7.3

$$V_{boost} = \frac{30.83}{0.2656} \times 0.015 = 1.74 \text{ volts/phase.}$$

$$\% V_{boost} = 1.74 \div 265.6 = 0.00656 \text{ pu or } 0.656\%.$$

7.4.3 System Voltage Drop Analysis

The voltage drop analysis for the entire system with multiple feeders is carried out using the method we just developed for one feeder. The drop in each feeder cable must be accordingly calculated to ascertain that it meets the limitation imposed by the system requirements. The system voltage drop data are usually compiled in a table. Table 7.1 is an example of such compilation for three feeder cables from three control centers 1, 2, and 3.

Since the voltage drop analysis is a major task in designing a power distribution system, we summarize below the important approximation that can be used for quick screening of the alternatives.

If the feeder delivers load current I at certain pf between two points A and B separated by total resistance R and inductive reactance X of all equipment in series, the approximate voltage drop between points A and B is given by

$V_{\text{drop per amp}} = R \cdot \text{pf} + X \cdot \sqrt{1-\text{pf}^2}$ for all practical power factors
$V_{\text{drop}} = I \times R$ at unity or very near unity pf (such as with pf improvement)
$V_{\text{drop}} = I \times Z$ at pf typical during motor starting inrush current $\{Z = \sqrt{(R^2 + X^2)}\}$
$V_{\text{drop}} = I \times Z_{\text{eff}}$ in cables with current at 0.85 pf lagging (Z_{eff} from cable vendor data sheet).

Remember that these are good approximations for quick designs. The exact voltage drop calculations that may be required to show the compliance with contractual standards or in a litigation case come from Equation 2.14 in Section 2.7.

TABLE 7.1
Proforma Table for Voltage Drop Analysis in Various Distribution Cables

Feeder Line >	From Motor Control Center #1	From HVAC Control Center #2	From Service Load Center #3
Horsepower or kilowatt rating			
Cable type			
Cable rated voltage			
Cable rated ampacity			
Full-load amperes			
Conductor area			
Cable length			
Resistance R ohms			
Reactance X ohms			
Starting current			
Amperes trip			
Voltage drop (volts)			
Voltage drop (%)			

Example 7.2

A three-phase, 100-hp, 460-V, letter code G induction motor is started directly from 480-V lines. The internal (Thevenin) source impedance and the cable impedance add up to a total of 0.01 + j0.02 Ω/phase. If the starting pf of the motor is 0.35 lagging, determine the exact and approximate voltage drops in percentage of the line voltage.

Solution:

A letter code G motor draws 5.6 to 6.3 kVA/hp on starting as per Table 4.3. Therefore, the worst-case starting current is 6.3 × 100 × 1000 ÷ (√3 × 460) = 791 A at $\theta = \cos^{-1} 0.35 = 69.5°$ lagging. The phase voltage at the source is 480 ÷ √3 = 277.1 V.

Using exact Equation 2.13, the voltage at motor load terminals is

$$277.1\angle 0° - 791\angle -69.5° \times (0.01 + j0.02) = 259.4\angle 0.2° \text{ V/ph}$$

$$\text{voltage drop magnitude} = (277.1 - 259.4) = 17.7 \text{ V/ph}$$

$$\%\, V_{\text{drop}} = (17.7 \div 277.1) \times 100 = 6.39\%.$$

By approximate Equation 2.15, V_{drop} = 791 (0.01 × 0.35 + 0.02 × 0.937) = 17.6 V/ph, a close match with the exact value of 17.7 V.

By approximate Equation 7.2, $V_{\text{drop}} = I \times Z_{\text{Total}} = 791 \times \sqrt{(0.01^2 + 0.02^2)} = 17.69$ V, which is even closer to 17.7 V derived from the exact formula.

This indicates that the approximate Equation 2.15, which is common for many voltage drop calculations, gives fairly accurate results for a wide range of power factors.

7.5 BUS BAR ELECTRICAL PARAMETERS

Heavy power in tens of megawatts requires a high current that is often distributed by bus bars of rectangular cross section, shown in Figure 7.8. For the two bus bars with facing flats as in Figure 7.8a, the R and L parameters per meter run of two bars combined (1 m for lead plus 1 m for return) are given below in SI units:

$$\text{for lead and return bars combined, } R = \frac{2\rho}{a \cdot b}\ \Omega/\text{m} \tag{7.4}$$

$$\text{leakage inductance between bars, } L = \frac{4\pi \times 10^{-7}}{b}\left(d + \frac{2a}{3}\right) \text{H/m} \tag{7.5}$$

where ρ = conductor resistivity = 0.01724 μΩ m for copper and 0.0282 μΩ m for aluminum both at 20°C, a = bar thickness, b = bar width, and d = separation between bar faces. These equations are valid for bars thinner than the skin depth at low operating

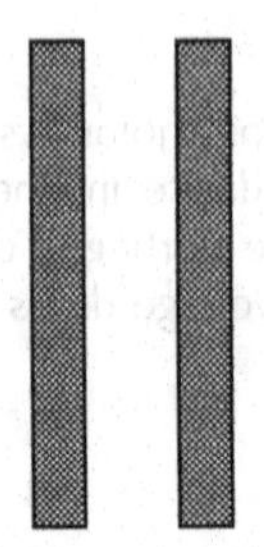

(a) Rectangular bus bars for 1-phase ac or dc with facing flats

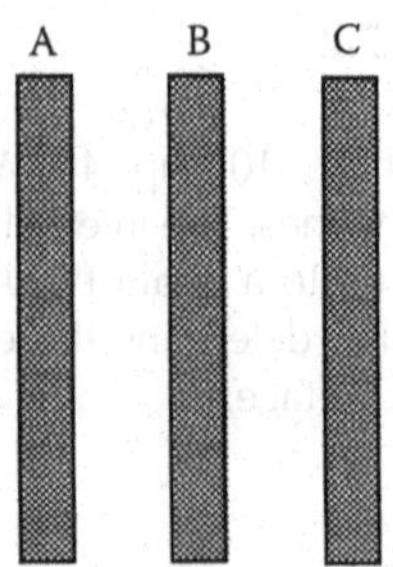

(b) Rectangular bus bars for 3-phase power

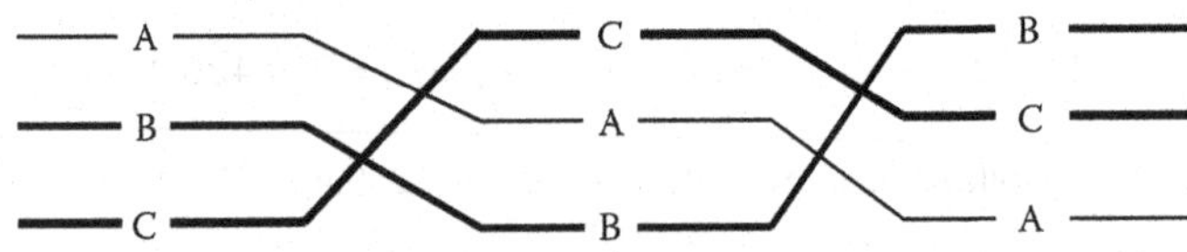

(c) Transposition of 3-phase bus bars for balanced reactance in each phase

FIGURE 7.8 Rectangular bus bars: single-phase and three-phase with transposition at 1/3 and 2/3 length.

frequency (50 or 60 Hz). For higher frequency, R will be higher and L will be slightly lower due to skin and proximity effects.

Copper bus bars with facing flats, 10 cm wide × 5 mm thick, separated by 1-cm insulation, would typically have $R = 69$ μΩ/m at 20°C and $L = 0.1675$ μH/m.

Long three-phase bus bars shown in Figure 7.8b may be used to carry high current from the source to a distant load. However, the unequal spacing between phase bars A–B, B–C, and C–A in this configuration results in unbalanced leakage reactance per phase, causing unbalanced voltage drops and unbalanced voltages at the load end. Such unbalance can be eliminated by transposing the bus bars as shown in Figure 7.8c such that each bar occupies all three possible positions in turn (one after another) relative to each other, with different positions in each one-third of the distribution length.

Sometimes, bus bars in HV equipment are encapsulated in epoxy and assembled with facing edges (Figure 7.9) to withstand high voltages. Such configuration results in high reactance per meter, which generally does not matter due to short distance within the enclosure.

7.6 HIGH FREQUENCY DISTRIBUTION

High frequency distribution systems, such as 400 Hz in aircraft and 20 kHz used in some U.S. Navy systems, are more difficult to design for low voltage drop. Since the

FIGURE 7.9 High-voltage bus bars in switchgear through porcelain bus ducts: cast epoxy insulated with joints covered by removable boots. (From Myers Power Products, Inc. With permission.)

leakage reactance $X = 2\pi f \times L$ increases with frequency, the cable has high reactance at high frequency. R also increases at high frequency due to skin and the proximity effects. The result is a high voltage drop per ampere. For a high-power bus, copper tubes with wall thickness less than the skin depth at operating frequency are often used to reduce ac resistance. Copper tube bus bars have added advantages of possible water cooling and structural integrity, compared to slender solid rectangular bars. However, hollow round conductors result in high leakage reactance, which increases the voltage drop per ampere.

In a design where a low reactance is desired, bars in Figure 7.8a can be made solid in a wider and thinner shape for the same cross section. For a very low reactance in a very high frequency distribution system, each phase bus bar can be made in a virtually concentric configuration, shown in Figure 7.10. The central bar carries the lead current to the load, and the two outer thinner bars carry the return current. The inductance in this case gets reduced to approximately one-half compared with that with two parallel bars. In SI units, it is

$$L = \frac{1}{2}\frac{4\pi \times 10^{-7}}{b}\left(d + \frac{2a}{3}\right) \text{ H/m} \tag{7.6}$$

where a = lead (center) bar thickness, b = bar width, and d = separation between the center bar and one of the outer bars.

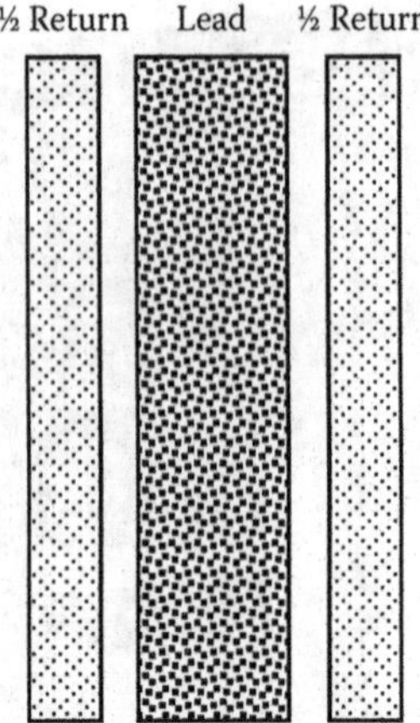

FIGURE 7.10 Virtually concentric low-reactance bus bars for high frequency power distribution.

Example 7.3

Two rectangular copper bus bars, each 1/4 in. thick × 4 in. wide with facing flats, are separated by a 1/2-in. space between the flats. Determine per foot run of bus bars (1) the resistance and leakage inductance at 75°C and (2) the mechanical force under a fault current of 50-kA peak, assuming the bar shape factor $K = 0.7$.

Solution:

1. Conductor cross-section area = 1/4 × 4 = 1 in.^{2}, which has the equivalent diameter = √ (4 × 1 ÷ π) = 1.128 in. or $1128^2 = 1.2732 \times 10^6$ circular mils = 1273.2 kcmil.

   Using Equation 6.3 in British units, $R = 10.372 \times 2 \div (1.2732 \times 10^6)$ = 16.29 μΩ/ft. at 20°C.

   In SI units, bus thickness a = 1/4 in. = 6.35 mm = 0.00635 m, width b = 4 in. = 101.6 mm = 0.1016 m, separation d = 1/2 in. = 12.7 mm = 0.0127 m, and length L = 1 ft. = 304.8 mm = 0.3048 m. Then, using Equation 7.4, we have

$$R = \frac{2 \times 0.01724 \times 0.3048}{0.00635 \times 0.1016} = 16.29\ \mu\Omega/\text{ft. at } 20°\text{C (same as above).}$$

   At 75°C, Equation 6.4 gives $R = 16.29\ [1 + 0.0039\ (75 - 20)] = 16.29 \times 1.2145 = 19.78$ μΩ/ft.

   Using Equation 7.5, the flat bus bar inductance—independent of temperature—is given by

$$L = \frac{4\pi \times 10^{-7}}{0.1016}\left(0.0127 + \frac{2 \times 0.00635}{3}\right) = 0.21 \times 10^{-6} = 0.21\,\mu\text{H/m}$$

   or $L = 0.21 \times 0.3048 = 0.0638$ μH/ft.

2. Using Equation 2.3, the mechanical force between the bars is given by

$$F = \frac{5.4 \times 0.7 \times 50{,}000 \times (-50{,}000)}{0.5} \times 10^{-7} = -1890\ \text{lbf/ft}.$$

The negative sign indicates repulsive force. The bus bar must be braced to support this force without exceeding the allowable bending stress or deflection between supports to avoid mechanical damage following a short-circuit fault.

Very high frequency power is often distributed by a *Litz cable*, which is made from numerous thin strands of magnet wire. A thin conductor with a hard polyimide-type film insulation often used for winding coils for electromagnets is known as the *magnet wire*. The strands in a Litz cable are continuously transposed so as to occupy every possible position in the cable over one transposition pitch. This way, the current distribution in the cable is forced to be uniform regardless of the cable diameter. An arbitrarily large-diameter Litz cable can have R_{ac}/R_{dc} ratio near 1.0 even at very high frequency if the individual strand diameter is thinner than the skin depth. The leakage reactance is another matter. A round cable, conventional or Litz, would have high leakage reactance due to high frequency since $X = 2\pi f L$. The leakage inductance can be reduced by making the Litz cable with rectangular straps with facing flats. It can further be reduced by using a three-strap design, as shown in Figure 7.10. A 20-kHz virtually concentric Litz cable (Figure 7.11) for a high-power distribution system for NASA's Space Station Freedom was designed, manufactured, and tested by Induction General, Inc.

FIGURE 7.11 Virtually concentric Litz cable for 20-kHz power distribution for space station. (Courtesy of M.R. Patel, NASA Report CR-175071, 1986.)

7.7 SWITCHBOARD AND SWITCHGEAR

The switchboard (SWBD) or switchgear (SWGR) is a general term that includes the assembly of circuit breakers, disconnect switches, fuses, relays, meters, instruments, controllers, and other equipment associated with the generation and distribution of electrical power. Examples of switchboard are the generator power panels, motor control center, or any other central power distribution center. One such switchboard's exterior view is shown in Figure 7.12. The switchboard in a large ship can be extensive enough to fill a large room, as shown in Figure 7.13. The following terms are often used in designing and discussing the switchboard:

FLA, full-load ampere: The nameplate rated load current of the motor or any other load

CAR, cable ampacity requirement: It determines the cable size in circular mils of a conductor that is required to carry the rated load current with some margin without overheating.

CM, circular mills: Square of the diameter of a conductor in mils (1/1000th of an inch). It represents the conductor area available to carry the current.

FIGURE 7.12 Generator switchgear (15 kV) in self-contained enclosure. (From Myers Power Products, Inc. With permission.)

FIGURE 7.13 Typical main switchboard on large cargo ship. (From Raul Osigian, U.S. Merchant Marine Academy.)

AT, ampere trip: The trip current setting of a circuit breaker that protects the cable by tripping out.

AF, ampere frame: The peak fault-current withstand capability of the circuit breaker without mechanical or thermal damage.

LVR, low-voltage release: It automatically reconnects the circuit when power is restored after a temporary loss. This is provided to auxiliary equipment critical to the system performance, where automatic restarting would not pose a hazard.

LVP, low-voltage protection: It will permanently shut down the circuit if power is lost and will not start again by itself when the power is restored. All large motors drawing very high starting current for a long time use LVP.

Red-V, reduced-voltage starting: It reduces the power drawn from a generator while starting a large load such as a heavy motor.

FLS, full-voltage starting: It connects small loads directly to the line at full voltage.

7.7.1 Automatic Bus Transfer

The automatic bus transfer (ABT) uses various types of circuit breakers and relays for automatic transfer of a power feeder from the main source to an alternate source when the main source detects a low or zero line voltage (Figure 7.14). All ABTs have a local manual transfer capability also, overriding the automatic operation. Some ABTs may have remote bus transfer (RBT) capability that can be operated from a remote location manually (but not automatically).

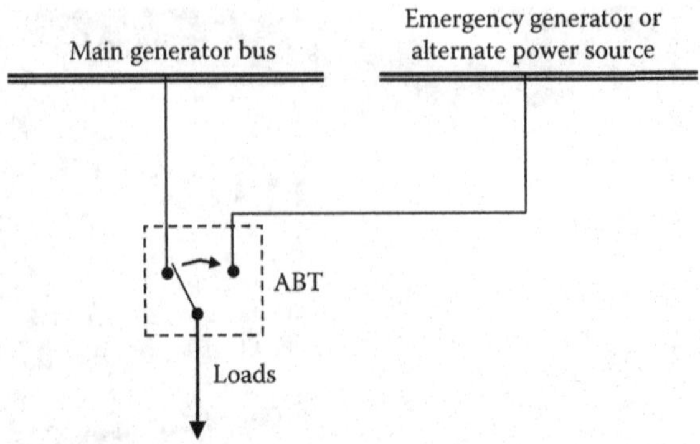

FIGURE 7.14 Automatic bus transfer switch.

7.7.2 Disconnect Switch

It is used to disconnect the circuit only under zero current after the circuit current is removed. It assures safety during maintenance work by manually disconnecting the power source from the line and keeping it locked open. It is not a circuit breaker; it cannot break the load current. If the disconnect switch is opened while carrying load current, it will surely spark heavily, may catch fire, or could even explode, posing a safety risk.

PROBLEMS

1. A three-phase, 460-V, 60-Hz, 500-hp induction motor has a design efficiency of 92% and pf of 0.80 lagging. The cable reactance from the control center to the motor is 0.015 Ω/ph. If the motor pf were corrected to unity by placing capacitors next to the motor, determine the % voltage rise at the motor terminals.
2. A three-phase, 150-hp, 460-V, letter code L induction motor is started directly from 480-V lines. The source impedance and the cable impedance add up to a total of $0.01 + j0.02$ Ω/phase. If the starting pf of the motor is 0.25 lagging, determine the exact and approximate voltage drops in percentage of the line voltage.
3. Two rectangular copper bus bars, each 8 mm thick × 100 mm wide with facing flats, are separated by 10-mm space between the flats. Determine per meter run of the bus bars (1) the resistance and leakage inductance at 75°C and (2) the mechanical force under a fault current of 35 kA peak, assuming a bar shape factor of 0.75.
4. Determine the % voltage regulation of a 100-kVA, 480/120-V distribution system having total impedance $Z_{Total} = (2 + j9)\%$ delivering 90% load at 0.8 pf lagging. Make your calculations using perunit values.
5. A 60-Hz single-phase cable having inductance of 0.3 μH/m and resistance of 2.5 mΩ/m is delivering 100 kW at 2400 V and 0.85 pf lagging. The load

is 40 m away from the generator. Determine the generator terminal voltage rise on removing this load.

6. A new 1000-kW, Δ-connected, 440-V, 60-Hz induction motor is proposed on your job. On a direct line start, it would draw 10 × rated current and cause transient voltage sag of 15%, which is unacceptable. Determine the line voltage sag with a *Y*–Δ starter.
7. A power distribution center serves single-phase load via cable and a 50-kVA 277/120-V step-down transformer. Based on the transformer kilovolt-ampere rating, the cable impedance is $2 + j1\%$, and the transformer impedance is $1 + j5\%$. The equipment is drawing 75% load at 86.6% power factor lagging. Using pu values, determine the % voltage regulation from the distribution center to the load point.
8. In a power distribution system from the generator to a pump, various equipment efficiencies are given as follows: generator 95%, transformer 97%, cable 99%, motor 94%, and pump 93%, all in series. If the pump output is 1000 hp, determine the contribution of this pump on the prime mover's output in kilowatts.
9. In a cruise ship that is 1000 ft. long and 300 ft. wide, a three-phase, 4160-V main feeder from the generator switchboard distributes 1200-kVA hotel load to 12 sectors fed by single-phase, 100-kVA, 120-V lateral feeders as shown in the following figure. Each junction of the main feeder and the lateral feeder has a single-phase, 100-kVA, step-down transformer for 120-V cabin loads. Determine (1) the size of the main and lateral feeders, (2) the voltage available to the last lateral feeder at point E, and (3) the voltage available to the last cabin at point EE. Assume (i) 0.85 pf lagging load and (ii) that both feeders have uniform size from the start to the end of the line (versus it having tapered cross section as the load current decreases from maximum at the start to minimum at the end).

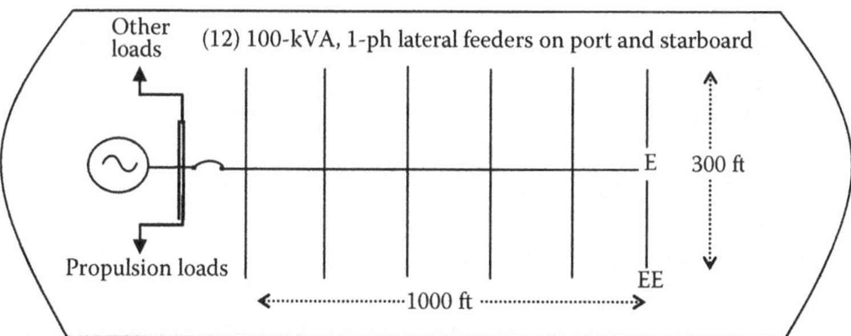

10. A 20-kHz high-power cable has a 10-mm conductor diameter. Determine its approximate R_{ac}/R_{dc} ratio if it is (1) made of one solid conductor, (2) made of numerous thin conductor strands that are not insulated from each other, and (3) made of Litz cable. Use the concept of skin depth in your calculations, and explain the reason why a Litz cable gives the best results at such high frequency.

QUESTIONS

1. How are two voltages, 120 and 240 V, obtained in residential power distribution from a single secondary coil of a single-phase step-down transformer?
2. How many voltages can be obtained from a three-phase *Y*-connected transformer?
3. Explain the advantage of an ungrounded system that is usually preferred in navy and commercial ships.
4. Explain the function of neutral ground and chassis ground.
5. On detecting a ground fault on an ungrounded system, how would you make sure that it is indeed a ground fault and not an anomaly?
6. Which bus bar configuration—facing flats or facing edges—results in lower voltage drop per ampere per meter length?
7. How do capacitors improve the voltage regulation (i.e., reduce the voltage drop)?
8. Sketch the magnetic flux pattern of two parallel bus bars and virtually concentric bus bars for a single-phase distribution system.
9. In dealing with high power at high frequency (e.g., 10 MW at 10 kHz), what type of the distribution line (round hollow tubes or rectangular bus bars) would you consider to keep the voltage drop low, and why?
10. Identify the difference between the circuit breaker and the disconnect switch.

FURTHER READING

Gonen, T. 2008. *Electric Power Distribution System Engineering*. Boca Raton: Taylor & Francis/CRC Press.

Momoh, J.A. 2008. *Electric Power Distribution, Automation, and Control*. Boca Raton: Taylor & Francis/CRC Press.

Patrick, D.R. and Fardo. S.W. 1999. *Electrical Distribution System*. Upper Saddle River: Prentice Hall.

Short, T. 2004. *Electric Power Distribution Handbook*. Boca Raton: Taylor & Francis/CRC Press.

8 Fault Current Analysis

The short circuit (called *fault* in this chapter) is defined as a live conductor touching another live conductor or ground in a grounded system. It can happen accidentally or when the insulation in between breaks down. The result is an abnormally high current that can cause mechanical and thermal damage to all equipment in series leading to the fault. The magnitude of prospective fault current at a given location in the system determines the current interrupting capability of the fuse or circuit breaker that must be placed between the power source and the fault location in order to protect the system from potential damage. For this reason, the power system engineer invests a great deal of effort to analyze the worst-case fault current at key locations in the entire system.

8.1 TYPES AND FREQUENCY OF FAULTS

The types of fault that can possibly occur in a three-phase, four-wire distribution system with neutral grounded are as follows, in the order of the frequency of occurrence:

L-G: One line shorting to the ground (not a fault in an ungrounded system)
L-L: Two lines shorting together but not to the ground
L-L-G: Two lines shorting together and to the ground
L-L-L: All three lines shorting together but not to ground (three-phase symmetrical fault)
L-L-L-G: All three lines shorting together and to ground (three-phase symmetrical fault to ground)

The L-L-L and L-L-L-G faults are called symmetrical faults because they involve all three lines symmetrically, resulting in fault currents that are balanced symmetrical three-phase currents. The other three types are called unsymmetrical faults, where the fault currents are not symmetrical in all three lines. The frequency of occurrence of electrical faults in practical power systems is shown in Figure 8.1. About 70% of faults in three-phase systems start as single-line-to-ground (L-G) fault. However, the subsequent heat generation breaks down the insulation between other lines in the cable as well, soon leading to three-phase symmetrical L-L-L-G fault. Similarly, faults that start as L-L (15%) and L-L-G (4%) also soon lead to symmetrical fault. For this reason, the fault current analysis presented below is focused on the three-phase symmetrical fault.

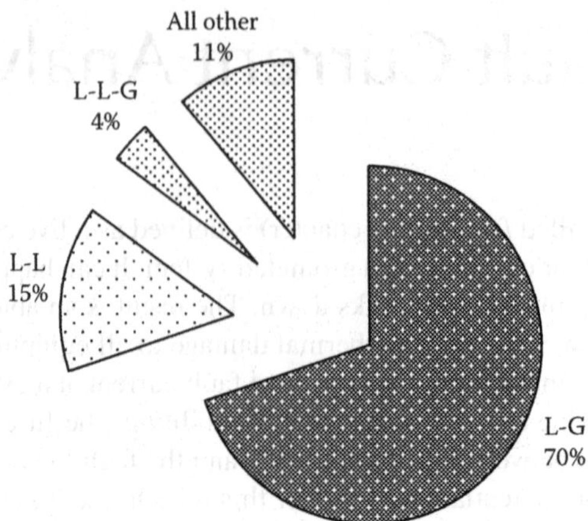

FIGURE 8.1 Types and frequency of occurrence of short-circuit faults in three-phase systems.

8.2 FAULT ANALYSIS MODEL

The system fault analysis model is developed by connecting the electrical models of all series components involved in the fault loop in a one-line diagram. Each component is represented by its equivalent series resistance R and leakage reactance X. Typically small R and X values of components like circuit breaker, fuse, relay contacts, etc., are ignored to simplify the analysis. Since the analysis is carried out on a per-phase basis in Y-connection (actual or equivalent), Δ-connected impedance values (if any) are divided by three to convert into equivalent Y-connected values. It is much simpler to carry out the fault current analysis in percent or pu system since the generator and transformer $\%\ Z$ are usually given on the equipment nameplate. The cable impedance, generally given in ohms per 1000 ft. on the manufacturer's data sheet, is converted to pu Z by dividing it by the base impedance per phase, which is

$$Z_{\text{base}} = \frac{V_{\text{base}}}{I_{\text{base}}} = \frac{V_{\text{base}}}{V_{\text{base}}} \frac{V_{\text{base}}}{I_{\text{base}}} = \frac{V_{\text{base}}^2}{VA_{\text{base}}}\ \Omega/\text{phase}. \tag{8.1}$$

Using kilovolt-ampere or megavolt-ampere per phase and line-to-neutral voltage V_{LN} or kV_{LN}, we have

$$Z_{\text{base}} = \frac{V_{\text{LN.base}}^2}{1000 \times \text{kVA}_{\text{1ph.base}}} = \frac{\text{kV}_{\text{LN.base}}^2}{\text{MVA}_{\text{1ph.base}}}\ \Omega/\text{phase}. \tag{8.2}$$

In a three-phase system, using line-to-line voltage kV_{LL}, we have

$$I_{\text{base}} = \frac{\text{kVA}_{\text{3ph.base}}}{\sqrt{3}\text{kV}_{\text{LL.base}}}\ \text{A/phase} \tag{8.3}$$

$$Z_{\text{base}} = \frac{1000\text{kV}^2_{\text{LL.base}}}{\text{kVA}_{\text{3ph.base}}} = \frac{\text{kV}^2_{\text{LL.base}}}{\text{MVA}_{\text{3ph.base}}}\ \Omega/\text{phase}. \tag{8.4}$$

For fault current analysis of the entire system involving many components in series, each having its own kilovolt-ampere and voltage ratings, one common base kilovolt-ampere and one base voltage are selected, which are generally the rated voltage and rated kilovolt-ampere of the first upstream transformer feeding the fault from the source side. All component impedance values either in ohms or in $\% Z$ based on their own voltage and kilovolt-ampere ratings are first converted into this common base selected for the system study. If the system impedances are all given in ohms, the conversion to a common voltage base (called the impedance transformation in Section 5.11) is done in the voltage square ratio, that is,

$$\text{for converting ohm values, we use } \frac{Z_{\text{base2}}}{Z_{\text{base1}}} = \left(\frac{V_{\text{base2}}}{V_{\text{base1}}}\right)^2 \Omega/\text{phase}. \tag{8.5}$$

With lower kilovolt-ampere base, the base current is lower, and the base impedance is higher; hence, $\% Z$ gets lower in the same proportion as the kilovolt-ampere base. Therefore, the $\% Z$ value changes linearly with the kilovolt-ampere base but is independent of the voltage base. If the $\% Z$ value on kilovolt-ampere base 1 is known, then its equivalent value on kilovolt-ampere base 2 is given by the following:

$$\text{for converting } \% Z \text{ value, we use } \frac{\% Z_{\text{base2}}}{\% Z_{\text{base1}}} = \left(\frac{\text{kVA}_{\text{base2}}}{\text{kVA}_{\text{base1}}}\right). \tag{8.6}$$

Equations 8.4 and 8.5 can be used with kVA_{1ph} or kVA_{3ph} and with V_{LN} or V_{LL} values since the conversion from one system to another is expressed in the ratios.

Example 8.1

Determine the base current and base impedance in a three-phase power system on the base of 10 $\text{MVA}_{\text{3-ph}}$ and 6.6 kV_{LL}, first using single-phase formulas and then using three-phase formulas, recognizing that both single-phase and three-phase formulas give the base current and base impedance values *per phase* (there is no such thing as three-phase amperes and ohms).

Solution:

From the fundamental definitions in Equation 8.1 on a per-phase basis,

$$\text{kVA}_{\text{base}} = 1/3\ 10{,}000 = 3333.3 \text{ kVA and kV}_{\text{base}} = 6.6 \div \sqrt{3} = 3.81 \text{ kV}.$$

Therefore, $I_{\text{base}} = 3333.3 \div 3.81 = 874.8$ A, and $Z_{\text{base}} = V_{\text{base}} \div I_{\text{base}} = 3810 \div 874.8 = 4.355\ \Omega/\text{ph}$.

Using single-phase Equation 8.2, $Z_{base} = kV_{LN}^2 \div MVA_{1ph} = 3.81^2 \div 3.3333 = 4.355\Omega/ph$.

Using three-phase Equation 8.3, $Z_{base} = kV_{LL}^2 \div MVA_{3ph} = 6.6^2 \div 10 = 4.355\ \Omega/ph$.

8.3 ASYMMETRICAL FAULT TRANSIENT

Consider a system model from the source to the fault location that has been reduced to a total equivalent *R-L* circuit, shown in Figure 8.2a. The initial load current, being small compared with the prospective fault current, can be ignored for simple but fairly accurate analysis. The fault on the system is applied by shorting the switch on the instant when the sinusoidal source voltage is $\sqrt{2}\ V_s \sin(\omega t + \theta)$, where t = time after the inception of fault and θ = voltage phase angle on the instant of closing the switch (fault inception angle on the sinusoidal cycle). The value of $\theta = 90°$ implies that the fault occurs precisely when the circuit voltage is passing through its natural peak as in Figure 8.2b, and $\theta = 0°$ implies that the fault occurs precisely when the sinusoidal circuit voltage is passing through its natural zero as in Figure 8.2c.

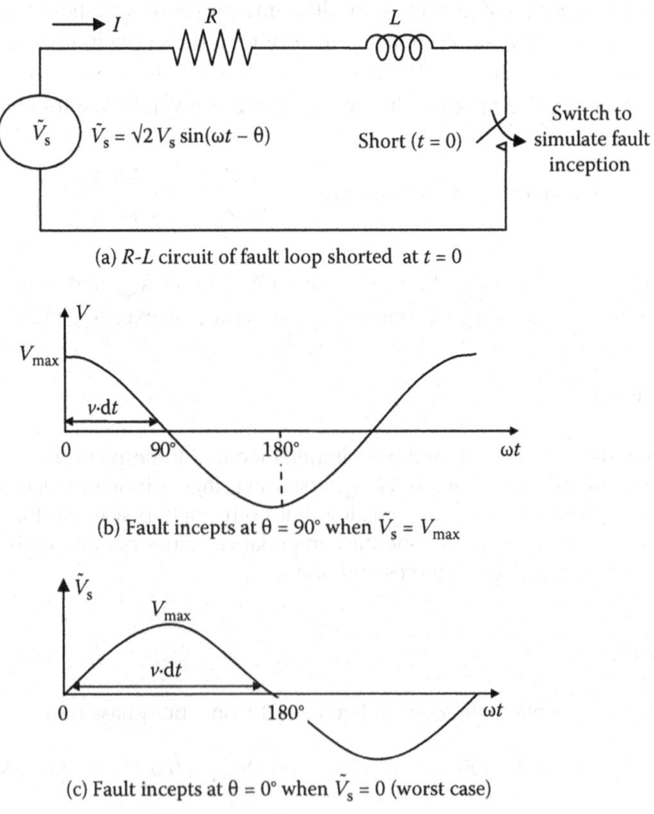

FIGURE 8.2 Transient current in *R-L* circuit after sinusoidal voltage is switched on.

8.3.1 Simple Physical Explanation

In practical power systems, the leakage inductance L dominates the resistance R. A simple physical insight to the fault current can be developed by ignoring R and writing the basic law of physics that gives the voltage drop in the inductor as

$$v(t) = L\,\mathrm{d}i/\mathrm{d}t \quad \text{or} \quad i = \frac{1}{L}\int v \cdot \mathrm{d}t. \tag{8.7}$$

Equation 8.7 expresses the circuit current as the integral of voltage with time, that is, the area under the v–t curve. If the fault incepts when the sinusoid voltage is passing through its peak (Figure 8.2b), then it stays positive for only one-fourth cycle. The current in this case keeps rising only during 0° to 90°, reaching a peak value of $I_{1\text{peak}} = \frac{1}{\omega L}\int_0^{90} v \cdot \mathrm{d}(\omega t)$, and then decreases to zero before moving to the equal peak on the negative side. On the other hand, if the fault incepts when the sinusoid voltage is passing through its natural zero (Figure 8.2c), then it stays positive for one-half cycle (twice as long as before) pumping twice the energy in the circuit. The current in this case keeps rising during 0° to 180°, reaching the peak value of $I_{2\text{peak}} = \frac{1}{\omega L}\int_0^{180} v \cdot \mathrm{d}(\omega t)$, and then decreasing to zero, but not moving to the negative side at all. Obviously, $I_{2\text{peak}} = 2 \times I_{1\text{peak}}$, that is, the fault current is pushed twice as high as shown in Figure 8.3b, which can be seen as the symmetrical ac superimposed on a

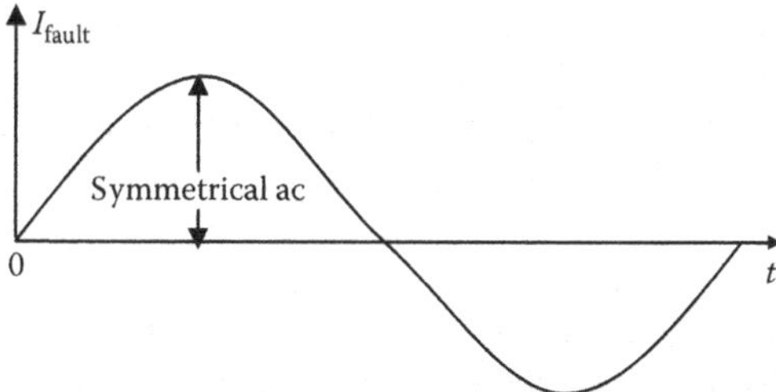

(a) Symmetrical fault current if fault incepts at θ = 90°

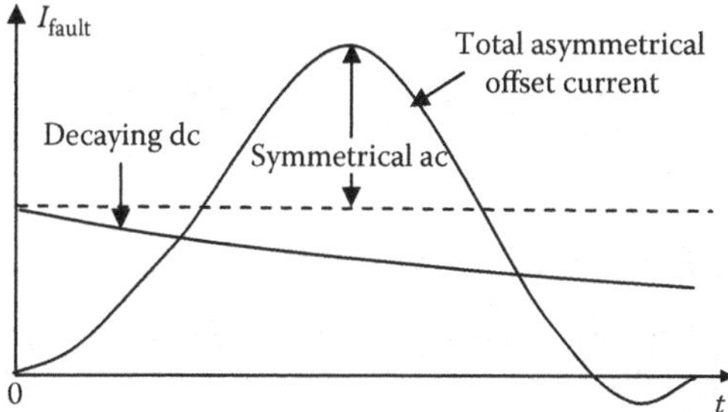

(b) Asymmetrical fault current if fault incepts at θ = 0° (worst case)

FIGURE 8.3 Transient current with fault inception angle θ = 90° and θ = 0° (worst case).

dc of initial value equal to the peak value of ac. We have seen similar analysis for the magnetizing inrush current in transformer in Section 5.7. It is analogous to the inertial energy stored in a large moving mass being dumped on a spring, or the *water hammer* building up the pressure in water pipe on the instant of suddenly shutting off the valve.

8.3.2 Rigorous Mathematical Analysis

For rigorous transient analysis of the *R-L* circuit in Figure 8.2a, we apply Kirchhoff's voltage law to write

$$\sqrt{2}\, V_s \sin(\omega t + \theta) = Ri + L\frac{di}{dt} \quad \text{for} \quad t > 0 \tag{8.8}$$

with the initial conditions $i(0) = 0$ and $v(0) = \sqrt{2}\, V_s \sin\theta$ at $t = 0$. (8.9)

The particular solution to Equation 8.8 with the initial condition 8.9 is the following transient fault current:

$$i(t) = \sqrt{2} I_{rms} \left(\sin(\omega t + \theta - \theta_z) - \sin(\theta - \theta_z) e^{-\frac{R}{X}\omega t} \right) \tag{8.10}$$

where R = total resistance in the fault circuit, $X = \omega L$ = total reactance, $\theta_z = \tan^{-1}(X/R)$ = impedance angle, and I_{rms} = rms value of the ac component superimposed on the exponentially decaying dc transient current. The fault current, therefore, is a complex function of the switch closing (fault inception) angle θ and the fault circuit time constant $\tau = L/R = X/(\omega R)$.

The fault current given by Equation 8.10 has the maximum peak value when $\theta = 0$, that is, if the fault incepts when the source voltage is passing through its natural zero during the sinusoidal cycle. On the other hand, the fault current peak is the lowest when $\theta = 90°$, that is, if the fault incepts when the source voltage is passing through its natural sinusoidal peak. Most faults incept when $\theta = 90°$, that is, when the system voltage is at its sinusoidal peak, because the insulation is more likely to break down at the peak voltage rather than at zero voltage. Fortunately, nature is on our side here.

The worst-case ($\theta = 0°$) transient current is closely (although not exactly) given by

$$i(t) = \sqrt{2} I_{rms} \left(e^{-\frac{R}{X}\omega t} - \cos\omega t \right). \tag{8.11}$$

8.4 FAULT CURRENT OFFSET FACTOR

The plots of $i(t)$ in Figure 8.3 with $\theta = 0$ and with $\theta = 90°$ indicate that the ratio of their respective first peaks

$$K_{peaks} = \frac{\text{first peak value if } \theta = 0°}{\text{first peak value if } \theta = 90°}$$

can approach 2.0 when $R = 0$. This ratio is known as the fault current *asymmetry factor, offset factor, or peak factor*, which is 2.0 when $\theta = 0$ (worst-case fault inception time and with $R = 0$). By plotting it for various X/R ratios, it can be seen that K_{peaks} factor is 1.0 for $X/R = 0$ (i.e., $X = 0$ or purely resistive circuit) and 2.0 for $X/R = \infty$ ratio (i.e., $R = 0$ or purely inductive circuit). For X/R ratio in between, the K_{peaks} value is between 1.0 and 2.0.

It is customary to calculate the rms value of short-circuit current for sizing the circuit breaker, which trips after the transient has somewhat subsided. However, the mechanical structure of the equipment and bus bars must be designed for the mechanical force at the first peak of the worst-case fault current. To relate these two currents of the design engineer's interest, the factor K is defined as

$$\text{Therefore, } K = \frac{\text{first asymmetrical peak of worst case fault current}}{\text{symmetrical rms value of fault current}} = \sqrt{2} \times K_{peaks}. \tag{8.12}$$

The values of K for various X/R ratios are listed in Table 8.1. It shows that a circuit with negligible reactance having $X/R = 0$ gives no asymmetry in the fault current, $K_{peaks} = 1.0$, and the first peak is just the usual $\sqrt{2}$ of the sine wave, which occurs at 1/4 of the 60-Hz period (i.e., 1/4 of 16. 667 ms). On the other hand, in a highly inductive circuit having $X/R > 100$, $K_{peaks} = 2.0$, and the first peak is $2\sqrt{2}$, which is twice the usual sine wave peak and occurs at 1/2 of the 60-Hz period (i.e., 1/2 of 16.667 ms).

TABLE 8.1
First Fully Offset Asymmetrical Peak Factor *K*

System *X/R* Ratio	Offset Factor *K*[a]	Time of First Peak in 60-Hz System (ms)
0	$1 \times \sqrt{2} = 1.414$ (zero asymmetry)	1/4 (1/60) s 4.167 ms
1	1.512	6.1
2	1.756	6.8
3	1.950	7.1
5	2.19	7.5
7	2.33	7.7
10	2.456	7.9
20	2.626	8.1
50	2.743	8.2
100	2.785	8.3
Infinity	$2 \times \sqrt{2} = 2.828$ (full asymmetry)	1/2 (1/60) 8.333 ms

[a] $K = \dfrac{\text{first fully offset asymmetrical peak value}}{\text{symmetrical ac rms value}}$

This means that with a large *X*/*R* ratio, the initial dc component is as great as the first symmetrical sinusoidal ac peak.

The asymmetrical fault current shown in Figure 8.3 can be characterized by (1) the sum of steady ac and decaying dc that starts from a high initial value, that is, I_{asym} ac = I_{sym} ac + exponentially decaying dc; (2) the decay rate of the dc component that increases with an increasing *R*/*X* ratio; and (3) asymmetry (offset) factor *K* that depends on the instant in the cycle at which the short circuit initiates.

8.5 FAULT CURRENT MAGNITUDE

Different faults identified in Section 8.1 produce different fault current magnitudes, with the worst-case values given by the symmetrical three-phase fault. The analytical method of calculating those magnitudes varies in complexity.

8.5.1 Symmetrical Fault Current

Figure 8.4 is a one-line representation of a three-phase symmetrical fault (L-L-L or L-L-L-G) in grounded or ungrounded *Y*-connected system with short at location F, where the voltage collapses to zero. The fault current from the source returns to the source via the path of least resistance through the short (zero resistance), thus completely bypassing the load. Therefore, the load is not considered at all in the fault current calculations. All of the source voltage is now consumed to circulate the fault current in the fault loop impedance. In rms values, Kirchhoff's voltage law (KVL) gives

$$I_{\text{fault.rms}} = \frac{\text{rms source voltage per phase}}{\text{sum of all impedances per phase in fault loop}}. \tag{8.13}$$

Since the generator, transformer, and cable are involved in series from the source to the fault location, $I_{fault} = V_{gen} \div (Z_{gen} + Z_{trfr} + Z_{cable})$,

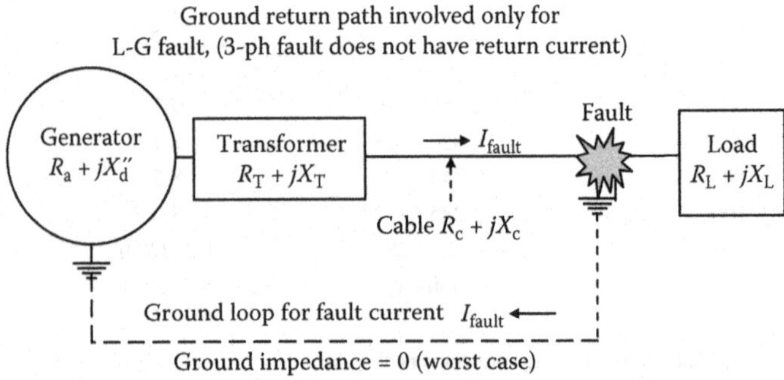

FIGURE 8.4 Short-circuit fault with generator, transformer, and cable in series.

$$\text{therefore, } I_{\text{fault.rms}} = \frac{V_{\text{LN.gen.rms}}}{(R_{\text{gen}} + R_{\text{trfr}} + R_{\text{cable}}) + j(X_{\text{gen}} + X_{\text{trfr}} + X_{\text{cable}})} \text{ A/ph.} \tag{8.14}$$

If the generator is very large compared to the transformer, it behaves very stiffly with effective $Z_{gen} = 0$, that is, R_{gen} and X_{gen} equal to zero in Equation 8.14. In other situations, where the system from the source up to the fault location can be represented by a Thevenin equivalent source model with V_{Th} and Z_{Th} known from calculations or by test, the fault current is simply

$$I_{\text{fault.rms}} = \frac{V_{\text{Th.rms}}}{Z_{\text{Th}}}. \tag{8.15}$$

8.5.2 Asymmetrical Fault Current

Equations 8.14 and 8.15 give the symmetrical rms current. The worst-case fault current occurs if the fault inception angle $\theta = 0°$, with its first peak magnitude given by the offset factor K from Table 8.1 times the above symmetrical value, that is,

$$I_{\text{asym.peak}} = I_{\text{fault.sym.rms}} \times \text{offset factor } K \text{ from Table 8.1.} \tag{8.16}$$

For a three-phase symmetrical fault, with or without the ground involved, the worst-case ($\theta = 0°$) fault current is given by Equation 8.16. However, only one of the three phases (say, phase A) can have $\theta = 0°$, which would see the fully offset asymmetrical peak. The other two phases, phases B and C, would have $\theta = 120°$ and $240°$, respectively. Therefore, the peaks of fault currents in phases B and C would have lower offsets with lower peak magnitudes than that in phase A. The average asymmetrical peak current in a three-phase fault is defined as

$$I_{\text{avg.assym.peak}} = \frac{1}{3}(\text{sum of first peaks in all three phases}). \tag{8.17}$$

This average value may be useful in estimating the total heat in the three-phase equipment before the fault current is interrupted by the circuit breaker. It is emphasized here that the mechanical force between conductors is proportional to the current squared, which is extremely high at the first fully offset asymmetrical peak of the fault current. The mechanical structure of the generator and transformer coils, bus bars, circuit breakers, etc., must be designed (braced) to withstand this force without exceeding the allowable bending stress and deflection between supports. Also, the equipment must be designed to limit the temperature rise under fault current within the permissible limit, assuming that the heat accumulates adiabatically in the conductor mass until the circuit breaker clears the fault in several cycles.

Example 8.2

A three-phase 13.8-kV distribution center has a source (Thevenin) impedance of $0.5 + j2.5\ \Omega$/phase. Determine the symmetrical rms and the worst-case asymmetrical first peak value of the fault current in a balanced three-phase fault at the distribution center bus.

Solution:

Since the source impedance at the bus is already given in ohms, we will work with ohm values. The bus impedance of the source behind the distribution center is as follows:

$$Z_{source} = \sqrt{(0.5^2 + 2.5^2)} = 2.55\ \Omega/\text{ph}.$$

The rms symmetrical fault current per phase $I_{sym} = (13{,}800 \div \sqrt{3}) \div 2.55 = 3125$ A.

The X/R ratio of the system is 2.5/0.5 = 5. From Table 8.2, the asymmetrical offset peak factor $K = 2.19$.

Therefore, the worst-case first asymmetrical peak current = 2.19 × 3125 = 6843 A.

All electrical equipment and bus bars in this distribution center must be designed to withstand 6843-A peak current without mechanical damage. The thermal withstand capability, however, depends on how fast the current is interrupted, which can be around 100 ms (several cycles).

Example 8.3

A three-phase, 480-V generator powers a load via single-phase feeders and 25-kVA, single-phase, 480/120-V step-down transformer with series impedances of $1 + j5\%$ as shown below. The single-phase feeder on the generator side has $0.002 + j0.006\ \Omega$ impedance. The generator is much larger than the 25-kVA transformer rating, making the generator bus infinitely stiff with zero internal impedance. Determine the symmetrical rms current rating of the transformer circuit breaker on the 120-V load side, assuming the fault location is close to the breaker LV terminals.

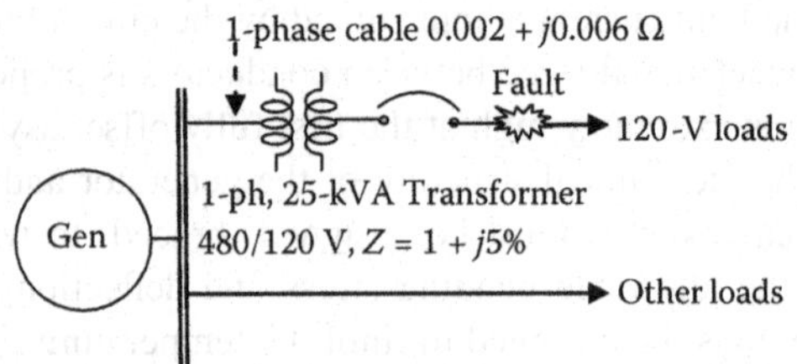

TABLE 8.2
Synchronous Generator Reactance (pu) and Time Constants (s)

Parameter Value	Large Solid Round Rotor Turbo-Generators	Medium-Size Salient Pole Generators with Dampers
Synchronous reactance	$X_d = X_q = 1.0\text{–}2.5$	$X_d = 0.8\text{–}1.5$ $X_q = 0.5\text{–}1.0$
Transient reactance	$X'_d = X'_q = 0.2\text{–}0.35$	$X'_d = 0.2\text{–}0.3$ $X'_q = 0.2\text{–}0.8$
Subtransient reactance	$X''_d = X''_q = 0.1\text{–}0.25$	$X''_d = 0.1\text{–}0.2$ $X''_q = 0.2\text{–}0.35$
Negative sequence reactance	$X_2 = 0.1\text{–}0.35$	0.15–0.50
Zero sequence reactance	$X_o = 0.01\text{–}0.05$	0.05–0.20
Time constant τ_{dc}	0.1–0.2 s	0.1–0.2 s
Transient time constant τ'	1.5–2.5 s	1.0–1.5 s
Subtransient time constant τ''	0.03–0.10 s	0.03–0.10 s

Solution:

We ignore the LV cable impedance due to its negligible length. Since the HV cable impedance is given in ohms and the transformer impedance is given in percent, we must convert both either in ohms or in percent. Here we arbitrarily choose to convert the transformer impedance in ohms based on the 120-V side. Therefore,

$$V_{base} = 120\ \text{V},\ I_{base} = \frac{25{,}000}{120} = 208.3\ \text{A},\ Z_{base} = \frac{120}{208.3} = 0.576\ \Omega.$$

Transformer impedance in ohms = $Z_{pu} \times Z_{base}$

$$= (0.01 + j0.05) \times 0.576 = 0.00576 + j0.0288\ \Omega.$$

Total impedance from the generator to the 120-V load side fault = $Z_{cable} + Z_{trfr}$

$$= (0.002 + j0.006) + (0.00576 + j0.0288) = 0.00776 + j0.0348\ \Omega$$
$$= 0.03565\ \Omega \text{ on the 120-V base.}$$

Symmetrical rms fault current on the transformer secondary side = 120 ÷ 0.03565 = 3365.6 A.

Circuit breaker rating must be a next standard step higher, say 3500 A_{rms} symmetrical.

Example 8.4

A three-phase, 10-MVA, 4.16-kV generator with subtransient reactance X''_d = 20% feeds a three-phase, 1-MVA, 4.16 kV/480 V transformer with X_{trfr} = 5% via a cable that has the leakage reactance X_{cable} = 2 mΩ/phase as shown below. Neglecting all resistances, determine the symmetrical rms and the worst-case first offset (asymmetrical) peak current in a three-phase fault at the transformer output terminals using (1) all reactance values in percent and (2) all reactance values in ohms.

X_{cable}= 2 mΩ/ph

10 MVA, 4.16 kV, $X_d'' = 20\%$

1 MVA, 4.16 kV/480 V, $X_{trfr} = 5\%$

(1) For percentage bases, we take the transformer secondary ratings of 1 MVA and 480 V as base. Using Equation 8.3, $I_{base} = 1000 \div (\sqrt{3} \times 0.480) = 1202.8$ A/ph, and $Z_{base} = (480 \div \sqrt{3}) \div 1202.8 = 0.2304$ Ω/phase.

Or, using Equation 8.4, $Z_{base} = 0.480^2 \div 1 = 0.2304$ Ω/phase.

X_{trfr} is given on its own base, so it does not need any adjustment. However, the generator X_d'' customarily given on its own megavolt-ampere base gets proportionately adjusted to 1-MVA base selected for the calculations here, that is, $X_d'' = (1 \div 10) \times 20\% = 2\%$, and the cable's 2 mΩ/phase is on the 4180-V side, which has $Z_{base} = 1000 \times 4.18^2 \div 1000 = 17.4724$ Ω, making $X_{cable} = (0.002 \div 17.47) \times 100 = 0.01145\%$ on 1 MVA, 480 V base.

$$\text{Therefore, } \% X_{Total} = X''_{d.gen} + X_{cable} + X_{trfr} = 2 + 0.01145 + 5 = 7.01145\%.$$

With full 100% voltage before the fault,

$$I_{fault} = 100\% \div 7.0145\% = 14.256 \text{ pu or } 14.256 \times 1202.8 = 17{,}147 \text{ A/ph symmetrical rms.}$$

(2) We now repeat the calculations using the ohm values looking from the 480-V base.

$$X_{trfr} = 0.05 \times 0.2304 = 0.01152\ \Omega \quad X_d'' = 0.02 \times 0.2304 = 0.004608\ \Omega.$$

X_{cable} = 2 mΩ gets adjusted by the voltage ratio square to become

$$X_{cable} = 0.002 \times (480 \div 4180)^2 = 0.000026\ \Omega/\text{ph}.$$

$$\text{Therefore, } X_{total} = 0.004608 + 0.000026 + 0.01152 = 0.0161544\ \Omega/\text{ph}.$$

$$I_{fault} = (480/\sqrt{3})\text{ V} \div 0.0161544\ \Omega = 17{,}155 \text{ A/ph}$$
(checks within rounding errors).

Furthermore, since the resistance in this fault loop is zero, the first peak will get fully offset by a factor of 2 × √2, that is,

$$\text{fault current at first peak } I_{asym\cdot offset.peak} = 2 \times \sqrt{2} \times 17{,}155 = 48{,}523\ A_{peak}.$$

The symmetrical fault current from the generator terminals can be simply calculated by the flip of the voltage ratio, that is, $I_{fault.gen} = 17{,}155 \times (480 \div 4180) = 1970$ A/ph symmetrical rms, and the first peak will be 2 × √2 × 1970 = 5572 A.

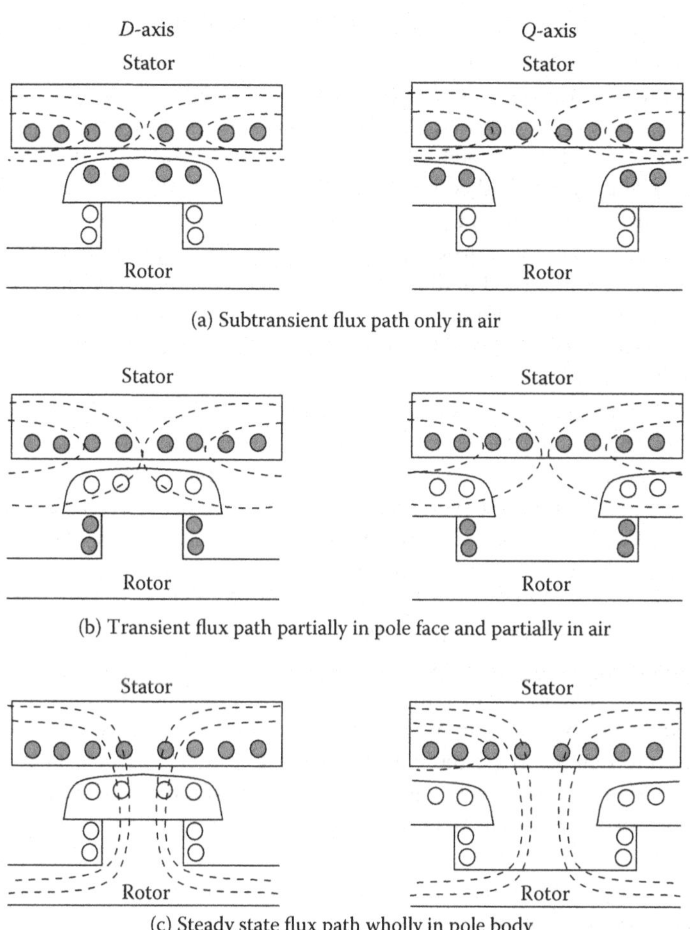

FIGURE 8.5 Generator subtransient and transient reactance and corresponding flux paths.

8.5.3 Transient and Subtransient Reactance

When the generator reactance has significant effect in determining the fault current, the initial value of the generator subtransient reactance should be used in the fault current analysis. During the fault, the generator reactance changes over a wide range, which can be explained by the constant flux theorem we covered in Section 2.1. On the instant of fault, the stator coils get shorted, and the stator current starts rising rapidly with corresponding sudden and fast increase in the flux. However, in reference to Figure 8.5, the damper bars shorted by end rings on the generator pole faces prevent the sudden increase in the flux with opposing current in the damper bars, forcing the stator flux in the air gap to take the path mostly in air, as shown in Figure 8.5a. In salient pole machines, the direct and quadrature axes (*d*- and *q*-axes) have different magnetic structure; hence, both *d*- and *q*-axes are shown. As the damper bar current decays exponentially in five subtransient time constants, the stator flux gradually penetrates deeper in the rotor pole surface as in Figure 8.5b but not through the rotor field coil. This decay rate is much quicker than what follows and is known as the subtransient phase of the short circuit. At the end of the subtransient phase starts the transient phase, in which the stator flux is maintained constant by the opposing current induced in the rotor field coil, which decays exponentially in five transient time constants. During the transient phase, the stator flux gradually penetrates the rotor field coil. At the end of the transient phase, the stator flux in the stator and rotor cores and the air gap is established in the normal pattern corresponding to the steady-state magnetic circuit shown in Figure 8.5c. During these three regions of time, the rotor conductors shown with shadow in Figure 8.5a through c carry the induced current to push back the rising stator flux.

Since the inductance of the coil is a measure of the flux it produces per ampere, and the flux, in turn, depends on the magnetic flux path, the effective stator inductance and reactance are the least in the subtransient phase shown in Figure 8.5a, small in the transient phase shown in Figure 8.5b, and the normally high synchronous reactance in steady-state phase shown in Figure 8.5c. These inductances multiplied by the angular frequency $\omega = 2\pi f$ give the subtransient, transient, and synchronous reactance denoted by X''_d, X'_d, and X_d, respectively. The X_d here is the normal steady-state synchronous reactance on the d-axis. The transient time constant T''_d in a typical synchronous generator is very short, between 5% and 10% of the transient time constant T'_d. The salient pole machine with different magnetic structure on q-axis has different values of X''_q, X'_q, and X_q. In large solid rotor turbo-generators, in absence of the discrete damper bars, the current induced in the solid pole faces virtually provides the damper bar effect. These reactances are not physically identifiable but are concepts formulated to deal analytically with the complex nature of the machine during sudden short circuit. Table 8.2 lists various d- and q-axis reactance and time constants for large solid rotor turbo-generators and medium-size salient pole generators with damper bars on the pole faces.

The change from subtransient to transient to steady-state time regions is gradual with exponential rise in reactance and corresponding decay in current. As shown in Figure 8.6a, the reactance drops from its steady-state synchronous value of X_d to X''_d on the instant of the fault and then gradually rises to X'_d and eventually returns to the normal value of X_d. Similar changes take place on the q-axis. The fault current magnitude, which is initially high with a low value of X''_d, decays with rising reactance and eventually reaches the steady-state value as shown in Figure 8.6b.

The actual asymmetrical current, however, has two components, a symmetrical sinusoidal superimposed on a decaying dc. It is shown in Figure 8.7 in three different regions of time during the fault, the subtransient, transient, and steady state. The peak value at the beginning of each region is equal to

$$I_\text{peak} = \frac{\sqrt{2}E_\text{f}}{d\text{-axis reactance at beginning of the time region}}, \tag{8.18}$$

where E_f = prefault field excitation voltage of the generator.

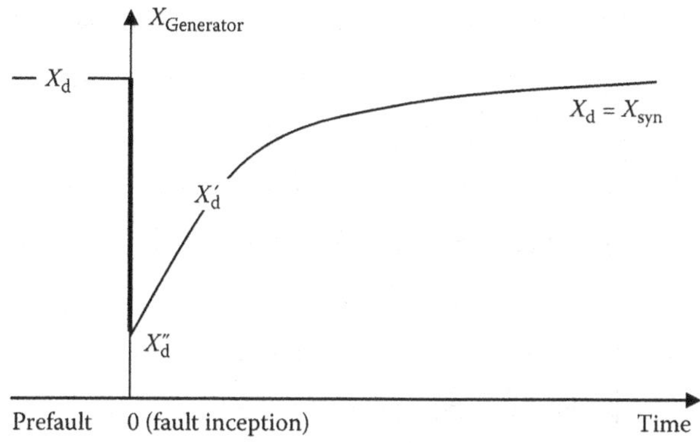

(a) Step fall and then rising leakage reactance of generator

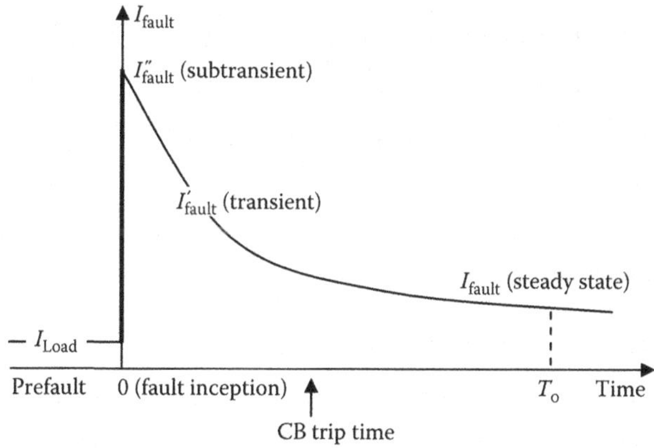

(b) Step rise and then decaying envelope of fault current peaks

FIGURE 8.6 Time-varying subtransient and transient reactance and fault current magnitudes.

In Figure 8.7, the circuit breaker tripping time is typically 0.1 to 0.2 s (6 to 12 cycles in 60-Hz systems and 5 to 10 cycles in 50-Hz systems). The time T_o for the fault current to reach a steady-state symmetrical value using X_d is less than 1 s for a 1-MVA machine and as high as 10 s in a 1000-MVA machine because of large rotor field coil inductance. The fault current we calculate for protection (i.e., for sizing the circuit breaker) is the initial subtransient fault current's ac rms value using X''_d, which we multiply by factor K to derive its worst-case value at the first fully offset peak.

Example 8.5

For a three-phase fault at the spark location shown in the figure below with all equipment three-phase and generator of finite size, determine the symmetrical

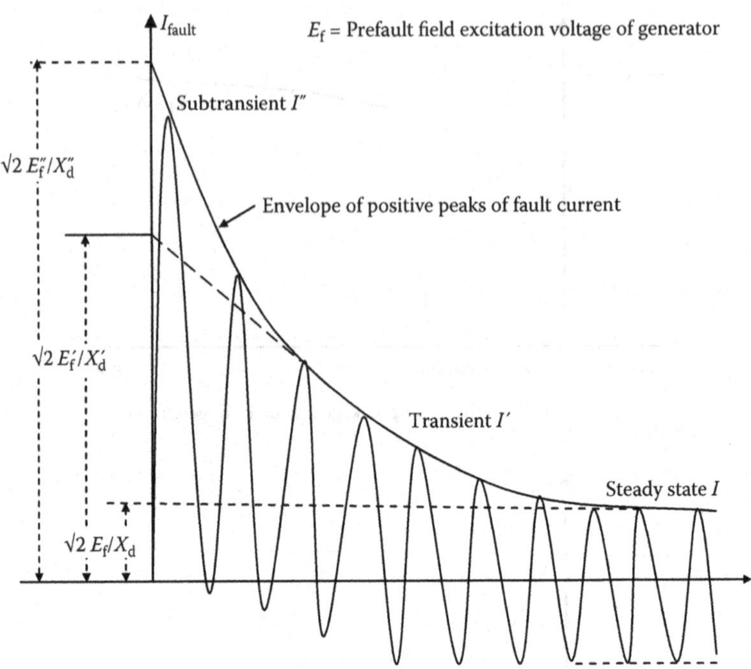

FIGURE 8.7 Exponentially decaying asymmetrical fault current and envelope of peak values.

rms fault current. Ignore the series resistance (but not the reactance) of the transformer and the generator. Assume all cables short with negligible R and X, and follow the following steps:

(1) As is commonly done by power engineers, convert the system to % values based on the kilovolt-ampere and voltage ratings of the transformer feeding the fault. Then, determine the fault current in amperes at the fault location and also at the generator terminals using the voltage ratio.
(2) Repeat the calculation by converting the system to % Z values based on the kilovolt-ampere and voltage ratings of the generator, and then determine the fault current in amperes at the generator and the transformer terminals and compare it with (1).

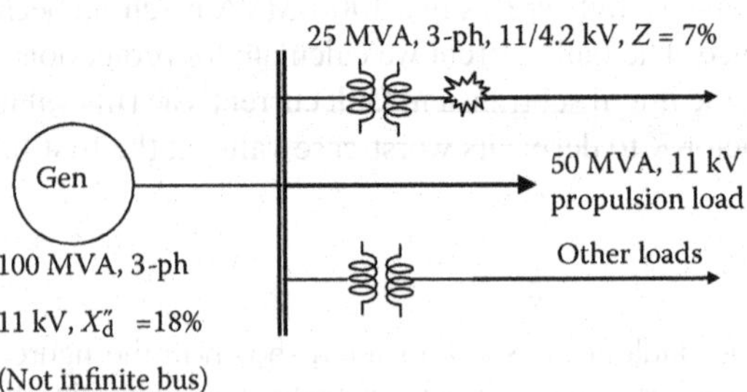

Solution:

The generator in this example is not large enough to be treated as an infinite bus; hence, it will have an effect on the fault current. The problem statement does not say how the transformer is connected. The generator is always *Y*-connected. We can assume that the transformer is also *Y*-connected and make the calculations per phase. If the transformer was actually Δ-connected, the line current would still be the same as in *Y*-connection. Only the phase current and voltage in the transformer coils would be different. The actual transformer connection does not matter in calculating the line current. Assuming it in *Y* gives the same line current (which is the phase current in *Y*).

We recall that the % impedance value prorates with the kilovolt-ampere base but does not change with the voltage base. Therefore,

(1) On a 25-kVA transformer secondary base: the generator X''_q of 18% is prorated on the transformer ratings as $Z_{gen} = 18\left(\frac{25}{100}\right) = 4.5\%$.

The total impedance per phase from the generator to the fault location = $Z_{gen} + Z_{trfr} = 4.5 + 7 = 11.5\%$.

$$\text{Therefore, } I_{fault} = \frac{100\% \text{ voltage}}{11.5\% \; Z_{Total}} = 8.696 \text{pu}.$$

$$\text{Transformer base current on the 4.2-kV side} = \frac{25{,}000/3}{4.2\text{kV}/\sqrt{3}} = 3437 \text{ A per phase.}$$

Therefore, the fault current coming out of transformer LV lines = 8.696 × 3437
= 29,900 A/ph.

Using the voltage ratio, the fault current coming out of generator = 29,900 × (4.2 ÷ 11) = 11,412 A/ph.

(2) On the generator base: the transformer *Z* of 7% prorated on the generator kilovolt-ampere rating = 7 × (100 ÷ 25) = 28%. Total impedance = $Z_{gen} + Z_{trfr}$ = 18 + 28 = 46%.

Therefore, the fault current on the generator base = 100% voltage ÷ 46% *Z* = 2.174 pu.

$$\text{Generator base current} = \frac{100{,}000/3}{11\text{kV}/\sqrt{3}} = 5249 \text{ A}.$$

Therefore, the Fault current coming out of the generator = 2.174 × 5249 = 11,412 A/ph.

Again, using the voltage ratio, the fault current coming out of the transformer LV line = 11.412 × (11 ÷ 4.2) = 29,900 A/ph. These values are the same as those derived in (1).

8.5.4 Generator Terminal Fault Current

In view of the complexities in transient fault current analysis, the IEEE standard on medium-scale electrical power systems allows using the following rms fault currents pending details of the equipment and system parameters for exact calculations.

The rms value of the average of first peaks in all three phases:

$$I_{\text{avg asymmetrical rms}} = 8.5 \times \text{generator rated rms current.} \tag{8.19}$$

The rms value of the first fully asymmetrical peak in the worst phase:

$$I_{\text{fully asymmetrical rms}} = 10.0 \times \text{generator rated rms current.} \tag{8.20}$$

The first peak of fully offset current in the worst phase = $\sqrt{2}$ × rms value from Equation 8.20 must be used in the mechanical force calculations for bracing the structure to avoid mechanical damage under the worst-case short-circuit fault.

8.5.5 Transformer Terminal Fault Current

When the generator and cable impedances are not known, one can make the most conservative estimate of the fault current using only the impedance of the transformer feeding the fault. This essentially treats the transformer primary side source as the *infinite bus*, defined as the bus of infinite kilovolt-ampere capacity with zero internal source impedance. This is fairly accurate when the transformer rating is small compared to the generator rating. It essentially ignores the generator and cable impedances up to the transformer primary side bus. Such simplification leads to the most conservative fault current estimate of the symmetrical rms fault current at the transformer's secondary terminals with full (1.0 pu or 100%) transformer voltage before the fault:

$$\text{worst-case fault current } I_{\text{sym.rms}} = \frac{100}{\% \, Z_{\text{trfr}}} \times \text{transformer secondary rated amps.} \tag{8.21}$$

$$\text{Or, } I_{\text{sym.rms}} = \frac{\text{MVA}_{3\text{ph}} 10^6}{\sqrt{3} V_{\text{LL}} Z_{\text{pu}}} \text{ A/phase} \tag{8.22a}$$

where $\text{MVA}_{3\text{ph}}$ = 3-phase megavolt-ampere rating of the transformer, V_{LL} = rated line voltage in volts, Z_{pu} = transformer series impedance in perunit (all from the nameplate).

The first peak of the worst-case fully asymmetrical fault current in a transformer connected to an infinitely large power source is K × value from Equation 8.22a, that is,

$$I_{\text{first.asym.peak}} = K \cdot \frac{\text{MVA}_{3\text{ph}} 10^6}{\sqrt{3} V_{\text{LL}} Z_{\text{pu}}} \text{ A/phase} \tag{8.22b}$$

where K = offset factor from Table 8.1 for transformer X/R ratio (K = 2.33 for X/R = 7 that is typical for large power transformers).

8.6 MOTOR CONTRIBUTION TO FAULT CURRENT

If the generator is powering a motor via cable as shown in Figure 8.8, and the cable gets shorted between the generator and the motor, the fault current comes not only from the generator but also from the motor. The kinetic and magnetic energy of the motor rotor feeds current back to the fault. It is complex to calculate, but the IEEE standard suggests an approximate estimate of the symmetrical rms fault current from the motor equal to

$$I_{\text{fault.motor.sym.rms}} = M \times \text{motor rated current}$$

where $M = 2$ for motor voltage < 240 V, $M = 3$ for motor voltage between 240 and 600 V, and $M = 4$ for motor voltage > 600 V. The total current going into the ground fault (not in the cables though) is then equal to that coming from the transformer, plus that coming from the motor. In terms of the symmetrical rms value, they are

$$I_{\text{fault.Total}} = I_{\text{fault.gen}} + I_{\text{fault.motor}} \tag{8.23}$$

where

$I_{\text{fault.Total}}$ = total fault current to L-L-L short;
$I_{\text{fault.gen}}$ = fault current from the generator side calculated with all known impedances; and
$I_{\text{fault.motor}}$ = fault current from motor side = 2 to 4 × rated motor current.

The motor contribution in the fault current can be explained this way. Before the fault, there was a back voltage in the rotor opposing the applied line voltage that was pushing the current into the motor. However, the motor line voltage collapses to zero after the fault, and the motor back voltage becomes the only voltage from the motor

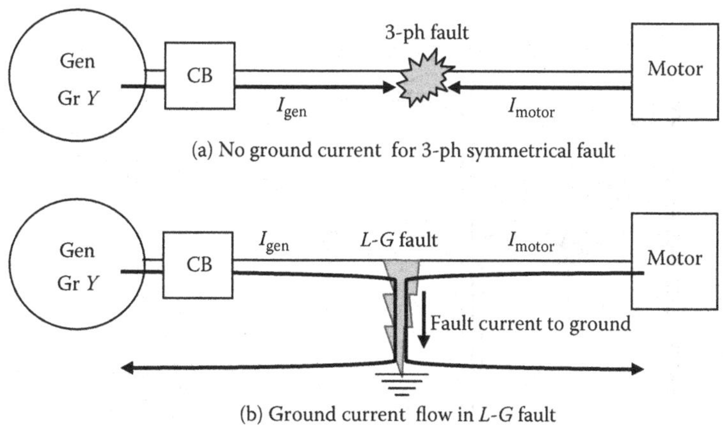

FIGURE 8.8 Motor contribution to the fault current.

side to the fault. This results in current flowing from the motor to the fault until the motor electromagnetic and kinetic energies get depleted.

The mechanical analogy of the above is a pump driving an air compressor. The pump has the pressure, and the compressor has the back pressure that is somewhat less than the pump pressure. If the air pipe between the pump and the compressor gets a fault (puncture) in the middle, the air rushes out from the hole to the atmosphere from both sides, from the pump as well as from the compressor.

Since the motor current comes from the load side of the circuit, the generator-side cable and circuit breaker do not see this current. As for the motor-side cable, the fault current from the motor side is generally less than the fault current from the generator-side and for a much shorter duration (only several cycles). For these reasons, the risk that the motor-side current poses to the cable or the circuit breaker appears irrelevant when compared to the generator-side fault current. One can then wonder why we include the motor contribution in the fault current estimate. It is for one reason: the fault current flowing from conductor to conductor through the cable insulation is the sum of fault currents coming from the generator and from the motor. It is this total current that produces heat in the insulation in very small local spots, posing risks of extreme overheating, explosion, and even fire. Therefore, it needs to be accounted for in the risk assessment.

Example 8.6

A three-phase, 6600-V main bus feeds a three-phase, 2-MVA, 6600/480-V transformer with 5% impedance. The transformer powers multiple loads, including a three-phase, 500-hp motor as shown in the following figure. Determine the symmetrical rms fault current through CB1, CB2, CB3, and CB4 under a three-phase fault at the spark location assuming (1) all cable impedances are negligible, (2) the bus has multiple generators of large capacity (i.e., infinite bus), and (3) the motor delivering rated load before the fault with 92% efficiency and 90% pf lagging.

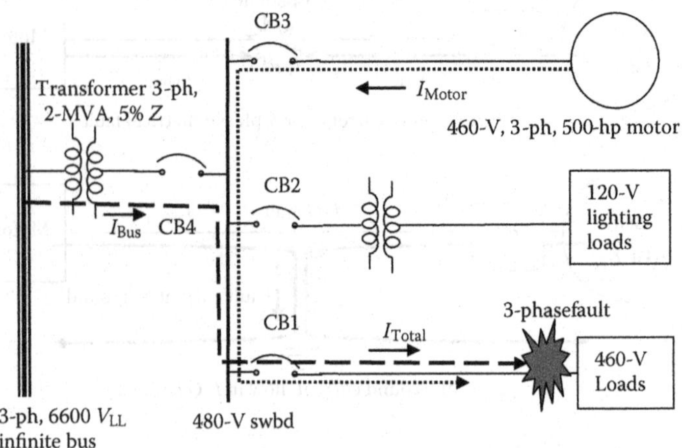

Solution:

From the infinite main bus to the fault, with all cable impedances negligible, the only impedance is that of the transformer, which is 5%. Therefore, the bus contribution to the fault current $I_{fault.bus}$ = 100% ÷ 5% = 20 pu (note that this is pu, not percent).

The rated current on the transformer secondary side = 2,000,000 ÷ $\sqrt{3}$ × 480 = 2406 A/ph. This and all currents below are symmetrical rms values.

Therefore, CB4 current = generator contributions through 2-MVA transformer secondary
= 20 × 2406 = 48,120 A.

$$\text{motor rated current} = \frac{500 \times 746\ \text{W}}{\sqrt{3} \times 460 \times 0.92 \times 0.90} = 565\,\text{A}.$$

Using multiple M = 3 for a 460-V motor as per Section 8.6, CB3 current = motor contribution to the fault = 3 × 565 = 1695 A.

CB1 current = CB4 current from bus + CB3 current from motor
= 48,120 + 1695 = 49,815 A.

CB2 current = 0, as it is not involved in this fault.

However, the CB2 rating must be determined by similar fault current calculations involving the CB2 branch. That is why the power engineer makes short-circuit current fault calculations at numerous possible fault locations (often using commercially available software) to size all CBs in the system.

8.7 CURRENT-LIMITING SERIES REACTOR

Protecting the system components from high fault current that can cause mechanical or thermal damages may require limiting the fault current by adding a series reactance in line. Series resistance is seldom used due to its high I^2R power loss, and an iron-core reactor would saturate and offer low reactance under high fault current. An air-core reactor with low resistance is, therefore, sometimes used for limiting the fault current. Such a reactor is carefully designed to minimize eddy currents due to leakage flux reaching the neighboring metal parts. The required current-limiting reactance value can be determined from the fault current analyses outlined in this chapter. Since a series reactance in a power line results in lower fault current, but higher steady-state voltage drop and some I^2R power loss, its design must be optimized with the overall system requirements. This is an example of another trade-off made by the system design engineer. A superconducting coil with zero resistance can be an ideal reactor for such application.

8.8 UNSYMMETRICAL FAULTS

For unsymmetrical faults (L-G, L-L, and L-L-G), the fault current analysis becomes even more complex. It involves the theory of *symmetrical components*, a brief summary of which is given in Appendix A. Industry standards give working formulas for such faults based on the analysis using symmetrical components for unsymmetrical faults. Such analysis includes resolving unsymmetrical (unbalanced) three-phase voltages into three sets of symmetrical (balanced) voltages, one with phase difference of +120° (positive sequence), one with phase difference of –120° (negative sequence), and one with no phase difference at all (zero sequence). The resulting three symmetrical component currents are then determined using the following:

$$\text{symmetrical component current} = \frac{\text{symmetrical component voltage}}{\text{symmetrical component impedance}} \tag{8.24}$$

$$\text{unsymmetrical phase currents} = \text{phasor sum of three symmetrical component currents.} \tag{8.25}$$

We give below the final results for the fault currents in the most frequent unsymmetrical fault (i.e., L-G fault) using the symmetrical component analysis presented in Appendix A. For line A to ground fault, we know that $I_b = I_c = 0$ and the neutral current $I_n = -I_a$. The line A current, however, will be

$$I_a = \frac{V_a}{\frac{1}{3}(Z_1 + Z_2 + Z_0) + Z_f + Z_g + Z_n} \tag{8.26}$$

where V_a = prefault voltage on line A; Z_1, Z_2, and Z_o = total positive, negative, and zero sequence impedance, respectively, in the fault loop; and Z_f, Z_g, Z_n = arcing fault impedance, actual earth impedance, and intentionally placed impedance in the neutral, respectively.

For special cases of dead fault with solidly grounded neutral, $Z_f = Z_g = Z_n = 0$ in Equation 8.26.

For L-L and L-L-G faults, see Appendix A.

8.9 CIRCUIT BREAKER SELECTION SIMPLIFIED

For determining the worst-case fault current to size the current interrupting capability of the circuit breaker, the fault current analysis can be simplified as follows, where all quantities are per phase and in per unit on a common base with the maximum generator capacity connected. First, consider a three-phase fault current, which is given by $I_{3ph\cdot sym.rms} = E_{gen}/X_1$. Then consider a single-phase L-G fault current with solid ground, which is given by $I_{LG.sym.rms} = 3E_{gen}/(2X_1 + X_0)$, where E_{gen} = prefault generator induced voltage per phase (also known as E_f), X_1 = total positive sequence reactance from the generator to the fault location, and X_0 = total zero sequence reactance from the generator to the fault location. In these totals, we must use the subtransient reactance for a synchronous generator. For selecting the circuit breakers interrupting capability, we use the greater of the two fault current values, that is,

$$\text{circuit breaker rating} = I_{3ph\cdot sym.rms} \text{ or } I_{LG.sym.rms}, \text{ whichever is greater.} \tag{8.27}$$

PROBLEMS

1. Determine the base current and base impedance in a three-phase power system on the base of 50 $MVA_{3\text{-ph}}$ and 11 kV_{LL}, first using single-phase formulas and then using three-phase formulas, recognizing that both single- and three-phase formulas give the same base current and base impedance values *per phase*.
2. A three-phase 6.6-kV distribution center has a source impedance of $0.1 + j0.5$ Ω/phase. Determine the symmetrical rms value and the worst-case asymmetrical first peak of the fault current in a balanced three-phase fault at the distribution center bus.
3. A three-phase, 600-V generator feeds a load via single-phase feeder and 50-kVA, single-phase 600/120-V step-down transformer with series impedances of $2 + j6\%$ as shown in the figure below. The single-phase feeder on the generator side has $0.003 + j0.025$ Ω impedance. The generator rating is much greater than 50 kVA, making it an infinitely stiff bus with zero internal impedance. Determine the symmetrical rms current rating of the transformer circuit breaker on the load side, assuming the fault location close to the breaker LV terminals.

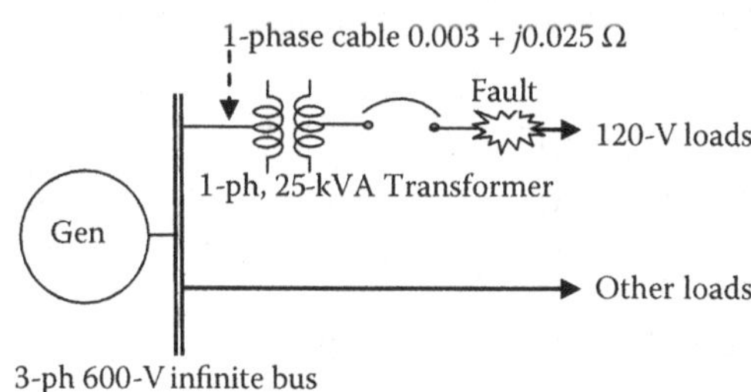

4. A 15-MVA, three-phase, 6.6-kV generator with subtransient reactance X''_d = 15% feeds a three-phase, 1-MVA, 6.6-kV/480 V transformer with X_{trfr} = 7% via cable that has a leakage reactance X_{cable} = 2.2 mΩ/phase as shown in the figure below. Neglecting all resistances, determine the symmetrical rms value and the worst-case first offset peak of the fault current for a three-phase fault at the transformer output terminals using (1) reactance values in percent and (2) all reactance values in ohms.

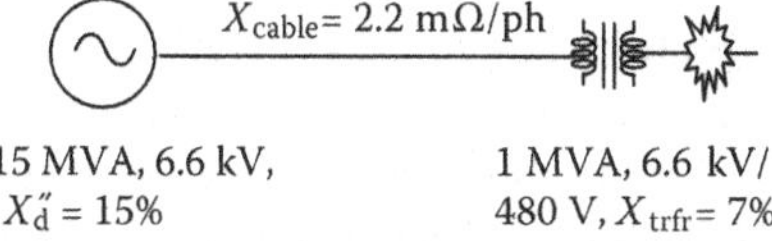

5. For a three-phase fault at the spark location in the figure below with all equipment three-phase and the generator of finite size, determine the symmetrical

rms fault current (1) in a 4.2-kV line coming from the transformer and (2) in an 11-kV line coming from the generator. Ignore the series resistances of the transformer and generator and assume all cables short with negligible R and X.

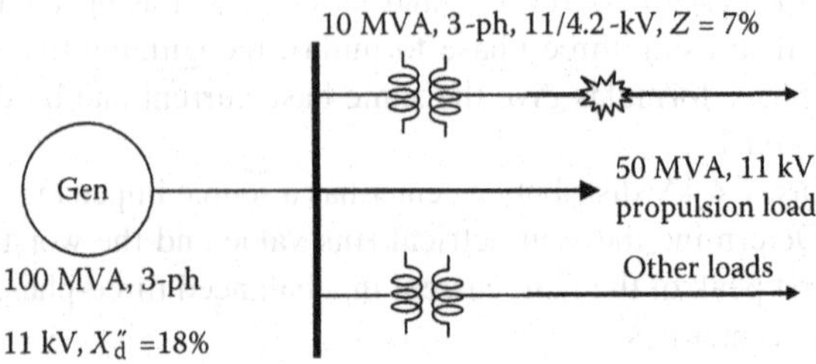

6. A three-phase, 4.16-kV main bus feeds a three-phase, 5-MVA, 4.16-kV/480-V transformer with 6% impedance. The transformer powers multiple loads, including a three-phase, 1000-hp motor as shown in the following figure. Determine the symmetrical rms fault current through CB1, CB2, CB3, and CB4 under a three-phase fault at the spark location assuming (1) all cable impedance are negligible, (2) the bus has multiple generators of large capacity to make it an infinite bus, and (3) the motor delivers rated load before the fault with 90% efficiency and 0.85 pf lagging.

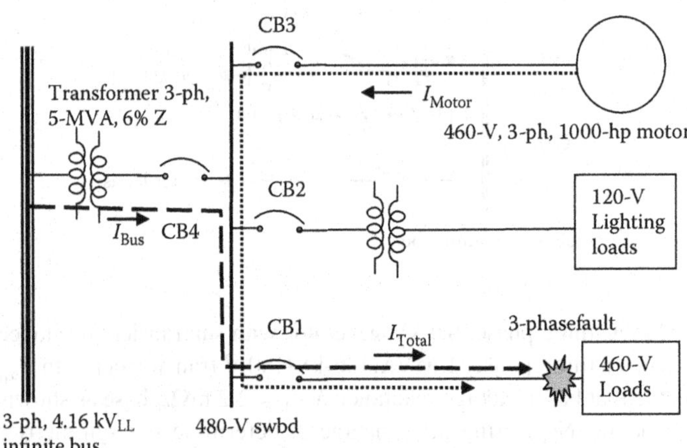

7. A three-phase, 1000-kVA, 4.16-kV Δ/480-V grounded-Y transformer with 5% impedance feeds a three-phase, 100-hp, 460-V motor. If a three-phase symmetrical fault occurs in the feeder, determine the fault current, including the motor contribution. Ignore the generator and cable impedances.
8. A 1500-kVA, three-phase, 460-V, Y-connected generator with neutral solidly grounded has R = 2% and subtransient $X''_d = 14\%$. It supplies two 750-kVA, three-phase, 208-V Y-connected feeders, one of which gets a three-phase short-circuit fault. The transformer in each feeder has R = 2% and X = 6% on the nameplate. Ignoring the short cable impedance, determine the symmetrical rms fault current with full generator voltage prior to the fault.

9. Estimate the peak asymmetrical and the average asymmetrical rms fault currents coming out of a 20-MVA, three-phase, 4160-V, GrY generator terminal if a symmetrical fault develops somewhere in the distribution system. You have no other information.
10. A three-phase distribution cable gets three-phase fault 150 ft. away from the 4160-V, *Y*-connected generator with neutral grounded. The subtransient impedance of the generator is 0.1 + *j*0.4 Ω/ph, and the cable impedance is 0.6 + *j*0.03 Ω/ph per 1000 ft. Determine the symmetrical rms fault current and the first offset peak, ignoring the motor contribution.
11. Determine the symmetrical rms fault current for a three-phase fault at the spark location shown in the following figure. Ignore all cable impedances.

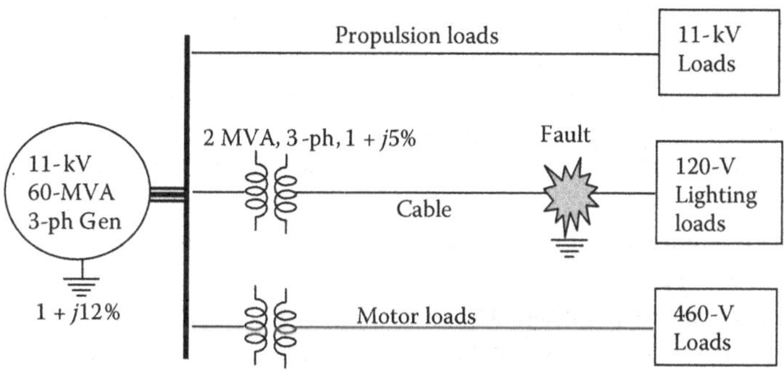

12. In the distribution systems shown in the figure below, the generator impedance is not known. Determine the symmetrical rms fault current for a three-phase fault at the spark location. The cable impedance 1 + *j*2% is based on the 200-kVA transformer ratings. Make conservative assumptions where needed.

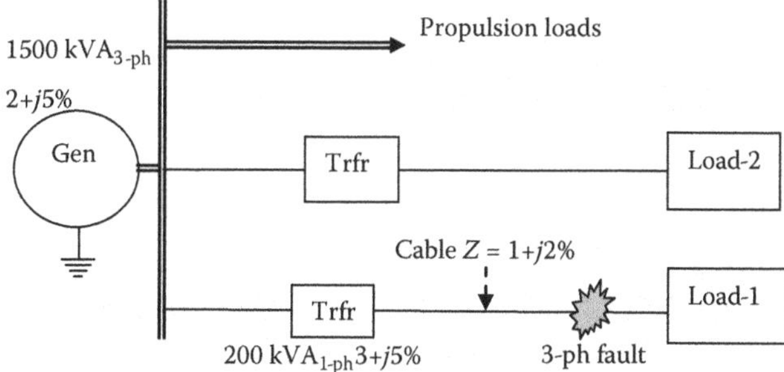

QUESTIONS

1. Explain how the most frequent unsymmetrical L-G fault soon leads to a three-phase symmetrical L-L-L-G fault.
2. What system parameters do you need in order to calculate the fault current magnitude?

3. By hydraulic analogy, explain how the first peak of the transient fault current gets much greater than the symmetrical peak.
4. Explain the R-L-C circuit physics leading to the asymmetrical fault current peak almost twice in the worst case than the symmetrical peak.
5. Explain the terms *d*-axis and *q*-axis in the synchronous generator and motor.
6. Explain the subtransient and transient reactance of the synchronous generator.
7. Between the two rough estimates of fault current as per the IEEE standard, which multiplier with the generator rated current, 8.5 or 10, should you use in the structural design of the bus bars and all electrical machines involved in series with the fault?
8. Transformer A has 5% impedance and transformer B has 7% impedance. Which one will see greater fault current at the LV terminals?
9. Describe in your own words how the motor load contributes to the fault current.
10. What trade-offs are involved in using and selecting the current-limiting series reactor?
11. Which mathematical method is used to analyze unsymmetrical (L-G, L-L, and L-L-G) faults?
12. Summarize in a paragraph the simple determination of the circuit breaker's symmetrical fault current interruption rating.

FURTHER READING

Bergen, A.R. and Vittal, V. 2000. *Power Systems Analysis.* Upper Saddle River: Prentice Hall.

Kersting, W.H. 2007. *Distribution Systems Modeling and Analysis.* Boca Raton: Taylor & Francis/CRC Press.

9 System Protection

The power system design includes protection from all credible fault currents and overvoltages that may occur in normal and abnormal operations. The fault (short circuit) occurs by accident or when the insulation breaks down due to aging and vibrations or due to overvoltages exceeding the strength of the insulation. Considering the working environment around electrical equipment (e.g., humidity, high temperature, and vibrations) it is important in selecting the right insulation type.

The heat generation rate and the mechanical force in conductors, both being square functions of the current, are the highest at the first subtransient peak of the asymmetrical fault current. The current then decays to the transient symmetrical value in several cycles and steady-state value in about 1 s, as the generator reactance rises from low subtransient to transient to steady-state synchronous reactance value. The mechanical damage is avoided by bracing the structural parts to withstand the first peak of the mechanical force without exceeding the allowable stress or deflection limits. If that is not possible or practical, the protection system must incorporate a suitable fault current-limiting fuse or circuit breaker, or insert a series reactor in the lines, to limit the current. The thermal damage is avoided by interrupting the fault, typically in several cycles, before it can overheat the equipment and burn the insulation.

Good protection starts with fast fault detection. The voltage, current, and temperature measurements generally detect faults in the system. For example, when the voltmeter or a sensor detects zero or a very low bus voltage, it may be a signal of ground fault on the bus.

The following tasks are part of the system protection design to minimize the extent of outage following a fault:

- Fault current analysis to determine the symmetrical rms and asymmetrical peak fault currents at various locations in the system where a circuit breaker or fuse is needed.
- Selecting each circuit breaker and fuse with required continuous and ampere interruption (AIC) ratings with proper margin.
- Strategically placing the fault detection sensors at key locations in the system.
- Protection coordination along the power flow line to assure selective tripping among all fuses and circuit breakers with recommended pick-up and delay settings.
- Arc flash analysis and assessment of the risk and hazard levels arising from fault current.

The fuse and circuit breaker are major devices that protect the power system from fault currents. The general design practice in using either one includes the following:

- All loads are individually protected to prevent a fault in any one load from damaging the feeder or the power source that is common (community property) for many loads.
- A protective device is also placed near the source to provide protection against faults in the cable from the source to the load.
- Large fault currents in the system must be interrupted before the generator can lose the transient stability. Relatively lower fault currents must be interrupted before the transient temperature of the conductor reaches the maximum allowable transient limit. For a copper conductor, this limit is typically 325°C—about one-half of its melting temperature—when copper starts softening.

9.1 FUSE

The fuse provides protection by melting away a thin metal link in the faulted circuit. The metal link may be of silver, copper, or nickel, with silver being more common for long-term performance stability. The fuse body is generally filled with sand-type filler (Figure 9.1) to suppress sparks when the fuse link melts and interrupts. The sparking occurs due to the inductive energy in the load and wire loops of the lead and return conductors. The fuse life is determined primarily by aging of the metal link under thermoelastic cycles of load on and off.

The ground insulation requirements on a fuse vary with the voltage rating. For example, the 120-V fuse may typically require the following:

- Dielectric withstand capability > 500 V measured by megger.
- Ground leakage current at rated circuit voltage < 1 mA.
- Withstand full recovery voltage equal to the open circuit voltage in steady state and up to 2× circuit voltage during transient that appears across the fuse terminals after clearing the fault.
- Resistance after fuse clearing >1 MΩ to 10 MΩ, depending on the rated current and the maximum fault current it can interrupt without exploding.
- After fuse clearing, the body and terminal surface temperatures must remain below 50°C for fuses up to 100 A. For larger fuses up to 500 A, the body temperature must remain below 50°C, but the terminal temperature may rise up to 75°C.

FIGURE 9.1 Typical fuse construction with metal link and sand filler.

9.1.1 Fuse Selection

Three key ratings to consider in the fuse selection are (1) the rated amperes it must carry continuously, (2) the rated circuit voltage it must support, and (3) the maximum current it must interrupt without exploding, which must be greater than the prospective fault current at the fuse location.

The fuse clears (melts or blows) at higher than rated current in time determined by its fault current versus clearing time characteristic (known as the *i–t* curve). Due to inherently large manufacturing variations, the *i–t* curve is typically given as a band shown in Figure 9.2. For example, at 5× rated current, the clearing time may be from 0.01 to 1.0 s. This must be accounted for in the system design to properly protect the circuit so that it clears when required and does not clear when not required.

The fuse ampere rating must be carefully selected because either a conservatively or liberally sized fuse is bad for the circuit protection. General requirements for selecting the current rating of a fuse are as follows. It must

- Carry 110% of rated current for at least 4 h without clearing
- Clear within 1 h at 135% of rated current
- Clear within 2 min at 200% of rated current
- Clear within 1 ms at 1000% (5×) rated current
- Have voltage dropped below 200 mV at rated current

The typical fuse voltage rating is the circuit voltage, and the current rating is 1.2 to 1.3 times the rated current in the load circuit it protects, rounded upward to the next standard available rating.

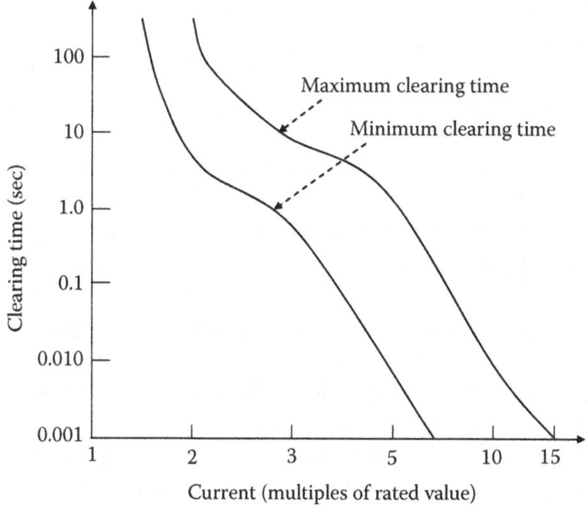

FIGURE 9.2 Fault current versus clearing time band of standard general-purpose fuse (normalized with rated current).

When the fuse clears, the break in the circuit current causes the full rated voltage to appear across the blown fuse. If the load is inductive, the transient voltage across the fuse may be substantially higher—up to twice the rated line voltage—due to the inductive energy kick (analogous to *water hammer* in water pipes). Under this over-voltage, a destructive arc may be formed across the fused element and may continue to grow. The resulting heat and pressure may cause the fuse to explode in the worst case. The rating of the fuse is therefore selected such that the fuse can interrupt a dead short without shattering or emitting a flame or expelling molten metal.

In locations around combustible materials, where explosive vapors may be present, the general-purpose nonsealed fuse poses a safety concern due to possible arcing when the fuse melts. The hermetically sealed fuse may be used in such applications. It has been developed for safe operation in an explosive mixture of chemical vapors classified by the National Electrical Code as Class 1 environment (hazardous). The hermetically sealed fuse eliminates the explosion possibility by containing the arcing inside. Moreover, it is fast and more predictable in clearing time since there is no arcing in open air and the associated plasma, which can linger on for a long time.

9.1.2 Types of Fuse

Three types of fuse are available in the industry for different applications.

1. *Standard (single-element) fuse*, which is a general-purpose fuse used in lighting and small power circuits. It has one element that melts when accumulated heat brings it to the melting temperature. Its *i–t* characteristic has a rather wide band, as shown in Figure 9.2.
2. *Time-delay (slow-blow or dual-element) fuse*, which is designed to ride through the starting inrush current drawn by certain load equipment—such as motor, transformer, capacitor, heater, etc.—for a short time immediately after turn-on. This fuse has two elements in series, one a heavy bead that takes time to heat up under moderate currents under overloads and the other a thin strip with a large dissipation area, which melts only when the current rises rapidly to a very high value under faults. The *i–t* characteristic of such a dual-element fuse also has a large manufacturing tolerance band.
3. *Current-limiting (fast-acting) fuse*, which clears before a large prospective fault current builds up to the first peak. Its fusible link has brief arcing and melting time such that it clears the fault in about one-fourth cycle, resulting in the let-through current being much less than the prospective peak fault current that a normal fuse would see. Such a fuse is used in delicate heat-sensitive power electronics circuits with diodes, thyristors, transistors, etc. Figure 9.3 shows a typical clearing time characteristic of a 200-A, 600-V current-limiting fuse in a branch circuit with a prospective symmetrical fault current of 10,000 A_{rms} (25,000 A first peak). It will interrupt the fault at a peak let-through value of 8000 A (instead of 25,000 A prospective first peak). Many current ratings for such a fuse are available under the trade name AmpTrap.

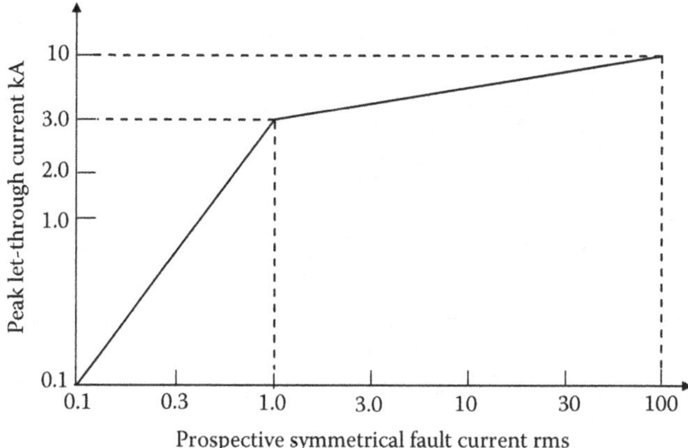

FIGURE 9.3 Prospective symmetrical fault current versus peak let-through current in current-limiting fuse.

Table 9.1 lists typical applications of the three fuse types, and Figure 9.4 compares the current versus average clearing time characteristics. For each type, significant manufacturing variations exist around the average clearing time, as seen in Figure 9.2 for the standard general-purpose fuse.

The fuse rating is based on factory tests conducted at room temperature. The ambient air temperature in actual operation has an influence on the actual rating of the fuse. The actual rating of the fuse must be lowered from the vendor's catalog rating (nominal rating) if the ambient air temperature is higher than standard room temperature, and vice versa. Tests suggest that the fuse must be derated or up-rated with the ambient temperature as shown in Figure 9.5. As a rule of thumb, the nominal current rating of fuse operating in high ambient air temperature must be derated by

0.5%/°C ambient above 25°C for a time-delay fuse
0.2%/°C ambient above 25°C for a standard general-purpose fuse.

TABLE 9.1
Fuse Types and Their Typical Applications

Fuse Type	Blow Time at 2 × Rated Current	Typical Applications
Fast-acting (current-limiting) fuse	< 1 s	Power electronics circuits and meter protection
Standard (general-purpose) single-element fuse	< 10 s	Most general-purpose standard circuits (lighting, small power circuits)
Time-delay dual-element (slow-blow) fuse	> 10 s	Circuits drawing high inrush current on switch-on to avoid nuisance blowing (motors, transformers, heaters, capacitor charging, etc.)

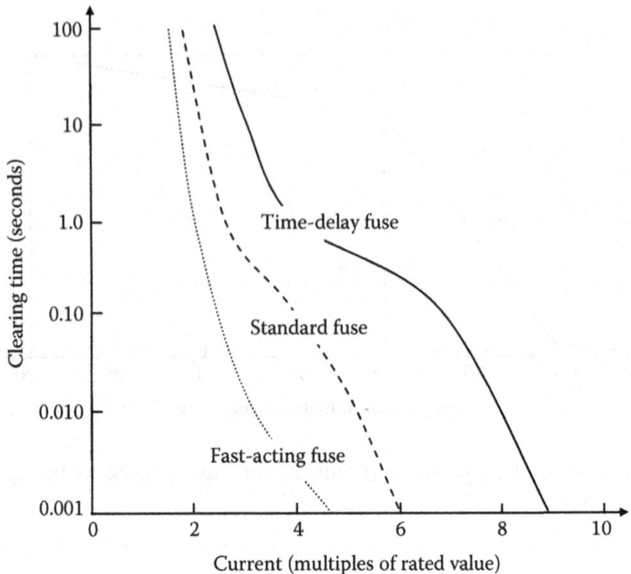

FIGURE 9.4 Fault current versus clearing time for three types of fuse (average values).

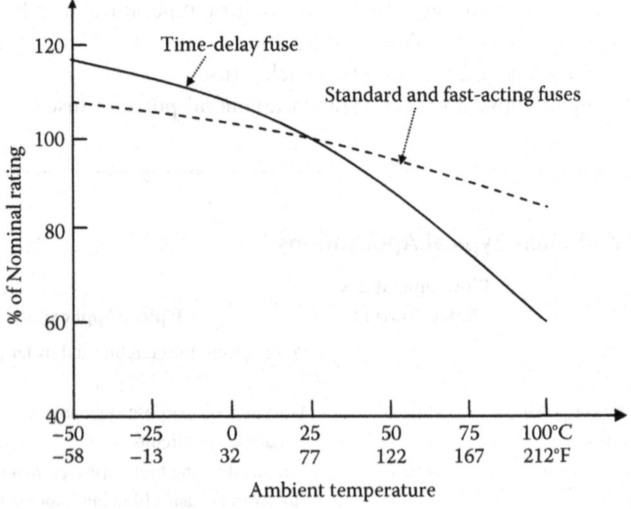

FIGURE 9.5 Derating and uprating factors for fuse versus operating ambient temperature.

Example 9.1

Determine the general-purpose fuse rating for a single-phase, 120-V, 2-kW load circuit in the boiler room uptake, which has an ambient air temperature of 65°C. Assume the load equipment efficiency of 90% and power factor 85%.

Solution:

Recalling that the equipment rating is the output, we have

$$\text{load current} = \frac{2 \times 1000}{120 \times 0.90 \times 0.85} = 21.8\text{ A}\,.$$

We design the load branch circuit for 30% overload.

Therefore, design current = 1.3 × 21.8 = 28.3 A.

Using Figure 9.5, the fuse rating for the boiler room uptake with 65°C ambient air should be derated to about 0.90 (hard-to-read scale).

Alternatively, using the rule of thumb given in Section 9.1.2,

derating of the general purpose fuse = 0.2%/°C above 25°C

= 0.002 × (65 – 25) = 0.08, or the derating factor = 0.92.

Therefore, fuse rating = 28.3 ÷ 0.92 = 30.8 A.

We choose a 30-A fuse, as we have a 30% margin in the current.

9.2 OVERLOAD PROTECTION

The fuse and circuit breaker clear fault only if the overcurrent is greater than 200% of its own rated current. That leaves a protection gap between 115% overload allowed in most equipment for an hour or two and 200% overload. This gap is bridged by the overload protection. In its simplest and often-used form shown in Figure 9.6, it consists of a pair of bimetallic links that remain closed in normal operation but deform differentially to open the contacts when overheated.

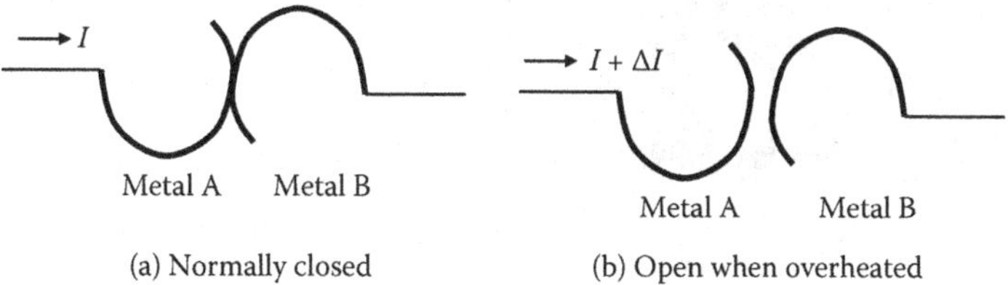

FIGURE 9.6 Overload protection by bimetallic link with normally closed contacts.

9.3 ELECTROMECHANICAL RELAY

Although not the protective device by itself, the electromechanical relay is an important part of the circuit breaker and various protective schemes in the electrical power system. The electromechanical relay detects the fault current and energizes the operating coil that opens the circuit breaker contacts. Relays are also used in many control circuits. They come in two types: (1) electromechanical relay for high-power circuits and (2) power electronics (solid-state transistor) relay for low-power circuits.

The electromechanical relay has metal contacts that open or close in response to a signal current in the operating coil as shown in Figure 9.7. The coil (called the solenoid) wound around a fixed iron core attracts a movable iron (called the armature) when current passes through the coil. The spring force reverses this action when the coil current is removed. The relay is designed to offer an inverse *i–t* characteristic just like the fuse, that is, shorter opening time at higher current. Compared to the power electronics relay, the electromechanical relay offers high power handling capability and negligible power loss. In this book, we will consider only the electromechanical relays generally used in high-power circuits.

Small relays used in control circuits come in holding or latching variety and are typically rated for 24 V, 28 V, 35 V, etc. at various current ratings up to 30 A_{rms}.

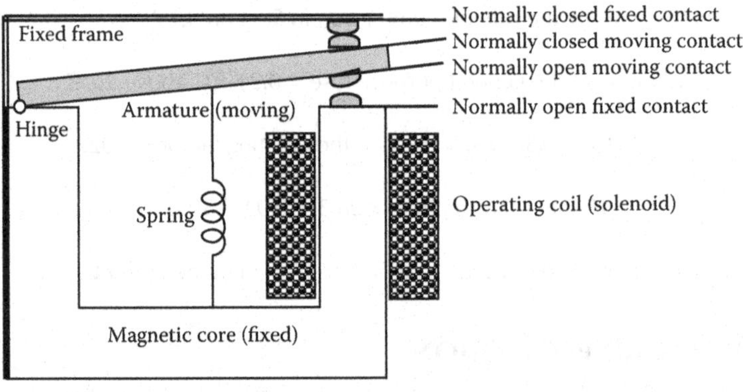

(a) Interior components

(b) Exterior view

FIGURE 9.7 Electromechanical relay construction and exterior view with connectors.

Contacts are designed such that the voltage drop at rated current is 0.1 to 0.2 V across good contacts at rated current. Large relays used in power circuits come in much higher voltage and current ratings.

When the relay contacts open, the inductive energy of the coil causes arcing until the energy is depleted. This results in much shorter contact life, electromagnetic interference (EMI), and fire hazard if a combustible vapor surrounds the relay. For safety, it is important to avoid arcing by absorbing and dissipating the inductive energy of the relay coil. Two schemes generally used for this purpose are described below.

Free-wheeling (arc-suppression) diode. In this scheme shown in Figure 9.8a, a diode is connected in the reverse voltage bias in parallel with the coil. When the relay contacts are closed, only the relay coil carries the current along the solid line. However, when the contacts are opened, dc current in the coil finds a continuing circulation path through the diode along the dotted ellipse. The current then decays exponentially in five time constants of the R-L circuit formed by the coil inductance and resistance, which is 5 × L/R. The diode is called a free-wheeling diode as it carries the current only to dissipate the stored energy in the relay coil with no other purpose except to prevent arcing when the relay contacts open.

Bifilar (fly-back) shorted coil. In this scheme shown in Figure 9.8b, the relay coil is wound with two identical wire filaments in parallel, one energizing the relay contacts and the other shorted on itself. It is like a transformer with two coils, one energized with dc and one coil shorted. Under normal dc current in the relay coil, the shorted coil carries no current, as the dc flux cannot induce any current in it. However, when the relay coil contacts open, the sudden change in its falling current induces the current in the shorted coil equal to the inversed turn ratio, which is 1.0 for the bifilar coils.

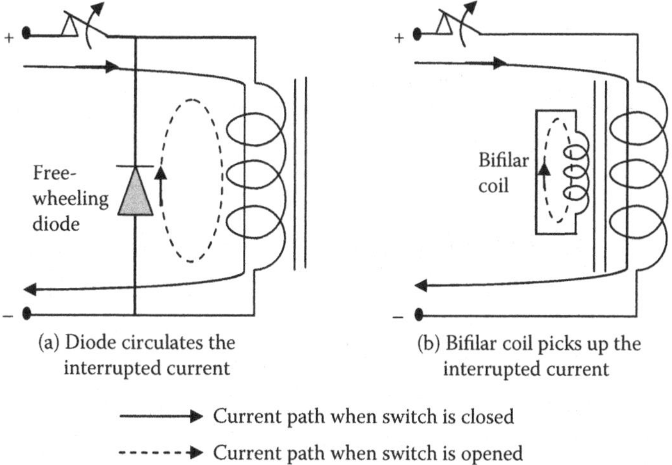

FIGURE 9.8 Relay contact arc-suppression schemes to absorb inductive energy after current interruption.

Therefore, the relay current essentially gets kicked back or fly-backed to the shorted coil, and the total magnetic energy of the system is preserved. The shorted coil current eventually decays in five time constants of its own R-L circuit. This serves the same purpose as the free-wheeling diode in the above scheme, that is, to prevent arcing when the relay contacts open.

9.4 CIRCUIT BREAKER

The circuit breaker opens and closes electrical contacts automatically in response to a fault (short circuit) or manually when needed. It is made of various relays with operating coils and heavy electromechanical contacts that open under fault current. The main power contacts of high-voltage (HV) circuit breaker, relay, and disconnect switch are usually silver plated for stable performance over the service life. The circuit breaker current versus opening time characteristic is similar to that of the fuse, that is, higher currents get interrupted in shorter times in the inverse *i*–*t* relation.

The fault is removed automatically by tripping the circuit breaker when the fault is detected by various current and voltage sensors (usually CTs and VTs) located at various locations strategically selected in the system. The abnormal current under a short-circuit fault causes the protective relay to operate, activating the tripping circuit in the circuit breaker and eventually opening the circuit breaker contacts and clearing the fault. Figure 9.9 depicts a typical overcurrent protection scheme using a current transformer (sensor), overcurrent relay, and a circuit breaker. In case of a fault, the resulting high line current *I* is sensed by the CT, the output of which flows through the operating coil of the overcurrent protection relay in its loop, which, in turn, closes the normally open contacts in the

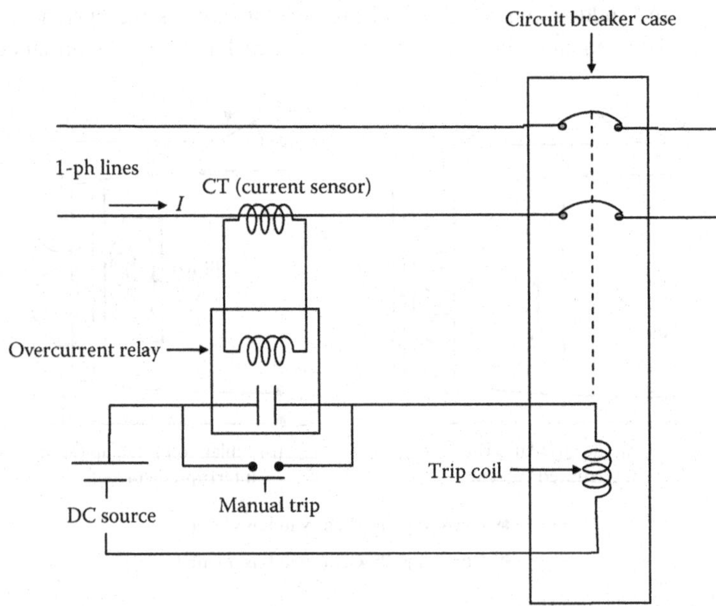

FIGURE 9.9 Circuit breaker scheme with overcurrent relay to open poles (contacts).

relay. If the relay is of a plunger type, it opens instantaneously; otherwise, after an intentional time delay, that can be adjusted by the system design engineer. Closing the relay contacts energizes the circuit breaker trip coil, which, in turn, opens the circuit breaker poles (contacts). The arc formed during the circuit breaker pole opening is blown away by blast air or transverse magnetic field between the poles. The arc extinguishes when the arc current comes to its natural zero on the sinusoidal cycle, and the insulating property of the interpole gap is resorted after the ionized medium is blown away. If the arc is not extinguished in time, it can create conductive ionized gases and molten or vaporized metal, potentially leading to explosion of the circuit breaker. Therefore, all circuit breakers must have some form of arc extinguishing feature incorporated in the design.

The manual trip push button in Figure 9.9 when pressed bypasses the overcurrent protection relay contacts and directly energizes the circuit breaker trip coil, opening the circuit breaker poles, even when there is no fault.

The desired time-delay characteristic of the overcurrent relay is obtained in two ways: (1) by adjusting the relay spring force and/or the contact positions and (2) by adjusting the relay coil taps. Various combinations of the two settings give a family of the time-delay characteristic shown in Figure 9.10. The trip settings of various circuit breakers in the system are coordinated such that the circuit breaker closest to the fault is tripped first by a relay circuit, keeping the rest of the system operating

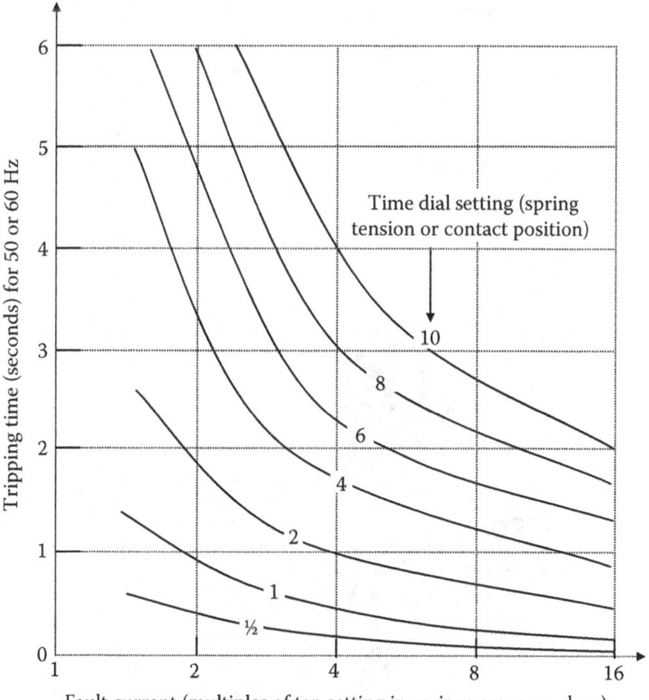

FIGURE 9.10 Two time-delay settings in circuit breaker gives a family of *i*–*t* curves to choose from.

normally. In a chain of circuit breakers, the generator breaker should have the longest trip time setting for a given fault current.

9.4.1 Types of Circuit Breaker

Molded case dual-element circuit breaker. This type is generally used for small LV circuits. It is made of two tripping elements and contacts enclosed in a common molded plastic case (Figure 9.11). Its trip is initiated by the thermal element in time inversely propositional to I^2 or by the electromagnetic force that trips the contact almost instantaneously. Thus, the overload current is tripped by the thermal element, and the fault current is tripped by the magnetic force. Therefore, it is also known as the dual-element thermal-magnetic circuit breaker. The thermal element trips when dissimilar bimetallic contacts are sufficiently heated to deform in a manner that opens the contacts. The magnetic element trips when the operating relay coil (solenoid) produces high electromechanical force on the actuator to open the contacts. A typical molded case circuit breaker has the current versus trip time characteristic (the inverse i–t curve) shown in Figure 9.12. As in the fuse, many manufacturing variations are accounted for by giving a band of clearing time for a given fault current. The engineer must assure that the minimum and maximum values of trip time

FIGURE 9.11 Dual-element molded case circuit breaker.

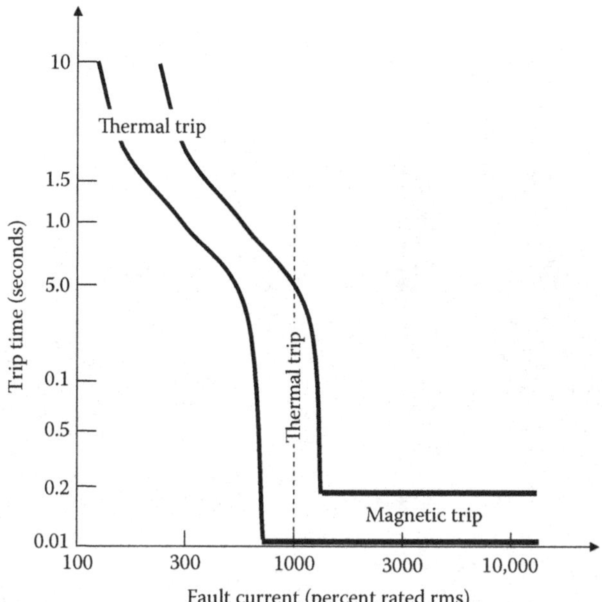

FIGURE 9.12 Current versus clearing time of dual-element molded case circuit breaker.

band meet the system requirements. Table 9.2 lists a few selected NEMA standard continuous current ratings of available molded case circuit breakers.

Air circuit breaker and air-blast circuit breaker. This type is suitable for medium voltages and is generally used indoors, where the fire risk associated with an oil-filled circuit breaker is not acceptable. The insulation and cooling medium between live conductors is air, which makes it economical but bulkier than other types. In the air-blast type of a circuit breaker, the arc at contacts during the current interruption is quickly blown away by a blast of air or magnetic flux.

Vacuum circuit breaker or pressurized gas-filled circuit breaker. The insulating medium between live conductors in this type of the circuit breaker is vacuum or

TABLE 9.2
NEMA Standard Frame Sizes for Continuous Ampere Ratings of Molded Case Circuit Breakers

Frame Size (Partial List)[a] (A)				
Frame100	**Frame200**	**Frame400**	**Frame800**	**Frame1200**
15	125	200	300	700
30	175	250	400	1000
40	200	300	500	1200

[a] Many other ratings are available but not listed here.

FIGURE 9.13 HV oil circuit breaker in outdoor substation yard.

pressurized gas (SF6 or nitrogen), both of which require less spacing between live conductors. It makes the circuit breaker design compact, suitable for indoors or outdoors in high-voltage (> 35 kV) and high-power-distribution systems where space is at a premium.

Oil-filled circuit breaker. It is widely used outdoors for high power distribution in the power grid and industrial plants in voltage ratings in hundreds of kilovolts (Figure 9.13). Oil is one of the best insulation and cooling mediums widely used in high-voltage equipment. It is generally not used indoors to avoid potential oil spill and fire hazard in case the oil tank ruptures following a fault.

All of the above circuit breakers use mechanical contacts that arc and ware when open to interrupt fault current. Power electronics circuit breakers with no mechanical contacts or moving parts are being developed for high-power applications and may become available in the near future.

The system with a high *X*/*R* ratio is hard on the circuit breaker operation since it results in high asymmetrical peak for the same symmetrical rms value of the fault current. When the current carrying contacts open, the magnetic energy stored in the system's leakage inductance keeps the current flowing until all the magnetic energy is diverted or dissipated in some form. The contacts continue arcing until the magnetic energy is depleted, or the current naturally comes to zero in its natural sinusoidal cycle in the ac system. The dc current is difficult to break because there is no natural sinusoidal zero in dc. The arcing in ac or dc must be minimized and diverted away from the contacts by blowing air at it, by magnetic force, or by absorbing the arc energy in an insulating fluid such as oil or pressurized gas. After the current is interrupted, the voltage between the contacts rises to the full system voltage after some time, but momentarily to an even higher value up to twice the system voltage rating (water-hammer effect), as shown in Figure 9.14. This is called the recovery voltage. At the peak of the recovery voltage, if the intercontact space is still ionized due to the just-extinguished arc, the contact may strike back and reestablish the fault. Adequate space between open contacts in the circuit breaker design is provided to avoid the arc striking back and reinitiating the fault.

The standard low-voltage (<2 kV) circuit breaker specifications in various frames are listed in Table 9.3. For example, the second column gives various specifications of the 4000-A circuit breaker that can be placed in line where the continuous operating current is 4000 A_{rms}, which can be set for momentary overcurrent from 4000 to 16,000 A for protection coordination. It can be used for rated system voltage from 800 to 1600 V line-to-line rms, with corresponding decrease in the short-circuit current ratings (both asymmetrical peak and sustained rms kiloampere ratings). That is, the short-circuit current interruption capability is higher at lower line voltage, and vice versa, as indicated in by slashes in rows 3 through 6.

The standard ratings of an indoor circuit breaker in the range of 4.76 to 38 kV are listed in Table 9.4. These circuit breakers have some limited flexibility in use at somewhat

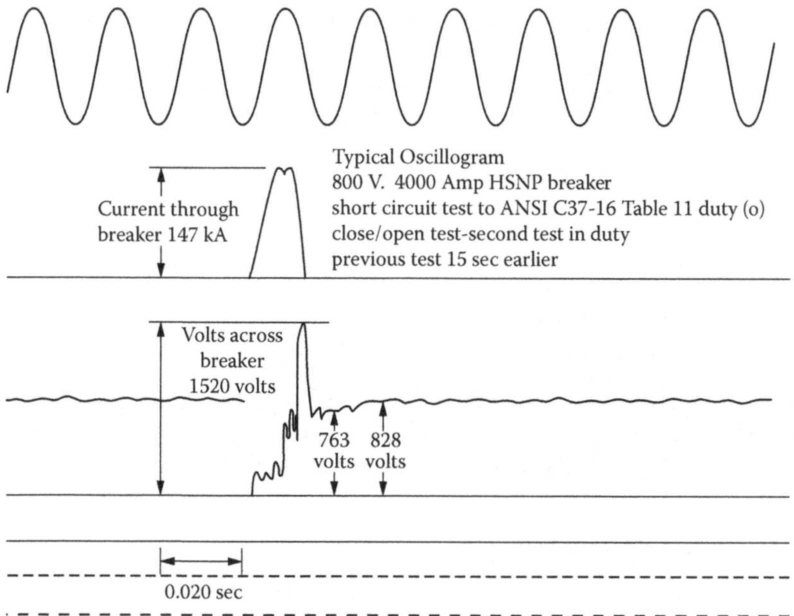

FIGURE 9.14 Voltage recovery and restriking voltage in circuit breaker following current interruption.

TABLE 9.3
Typically Standard Circuit Breaker Specification for <2 kV Lines

Continuous Current Rating Maximum (A_{rms})	4000	6000	8000	10,000	12,000
Overload current setting range (kA)	4–16	6–24	8–32	10–40	12.48
Voltage rating (V_{rms})	800/1600	800/1600	800/1600	800/–	800/–
Short-circuit rating peak (kA prospective)	200/100	200/100	200/100	160/–	135/–
Sustained (kA)	120/60	120/60	120/60	100/60	120/60
Short time withstand current ratings—250 ms	132/132	132/132	159/159	159/–	159/–
Operating coil (close and trip coils) control voltage (V_{dc})	125	125	125	125	125
Close coil current average at 125 V_{dc} (A)	38	38	38	38	38
Trip coil current average at 125 V_{dc} (A)	2	2	2	2	2

TABLE 9.4
Standard Ratings of Indoor Circuit Breaker in 4.76–38 kV Range

Rated System Voltage kV_{LLrms}	Rated Continuous Current kA_{rms}	Rated Symmetrical Fault Current Interrupt kA_{rms}	Rated Symmetrical Fault Current Interrupt MVA_{rms} (within *K*-factor)[a]
4.76	1.2	8.8	72.5
4.76	1.2, 2.0	29	240
4.76	1.2, 2.0, 3.0	41	340
8.25	1.2, 2.0	33	470
15	1.2, 2.0	18	470
15	1.2, 2.0	28	730
15	1.2, 2.0, 3.0	37	960
38	1.2, 2.0, 3.0	21	1380

[a] This column equals $\sqrt{3}\, kV_{LLrms} \times kA_{sym.fault.rms}$ using columns 1 and 3 data. The *K*-factor of the circuit breaker, typically 1.3, gives the range of applying lower voltage up to factor $1/K$ and increases the fault interrupt capacity by factor *K*, such that the fault megavolt-ampere capacity remains the same as listed in this column.

higher voltage at a proportionally reduced current within its own *K*-factor (not to be confused with the transformer *K*-rating). For example, if the line voltage is higher by a factor of $K \leq 1.3$, then these two kiloampere ratings must be lowered by a factor of $1/K$.

9.4.2 Circuit Breaker Selection

The circuit breaker has several ratings to select, such as

- Rated continuous current it carries within allowable temperature rise
- Rated line voltage it supports with adequate insulation to ground

- Symmetrical short-circuit megavolt-ampere or fault current it can interrupt without thermal damage
- Interrupting (tripping) time, which is typically between five and 15 cycles, and can be adjusted for coordination with other downstream and upstream circuit breakers
- Switching overvoltage withstand capability without restriking the contacts after tripping
- Lightning impulse voltage withstand capability
- First asymmetrical peak current withstand capability without mechanical damage to itself

The engineer must consider all ratings and factors listed above to select a proper circuit breaker for the application at hand. However, the key considerations in selecting the circuit breaker are as follows:

- Continuous current rating must be greater than the maximum load current it must carry.
- Interrupt capacity must be greater than the maximum fault current it can see (as calculated in Chapter 8).
- Voltage rating must be greater than the operating line voltage.

The required current interrupting capability of the circuit breakers can be determined as follows, where all quantities are per phase and in per unit on a common megavolt-ampere and kilovolt base with the maximum generator capacity connected. First, consider a three-phase fault current, which is given by $I_{3ph.sym.rms} = E_{gen}/X_1$. Then, consider a single-phase L-G fault current with solid ground, which is given by $I_{LG.sym.rms} = 3\,E_{gen}/(2X_1 + X_0)$, where E_{gen} = prefault generator induced voltage per phase (also known as E_f), X_1 = total positive sequence reactance from the generator to the fault location, and X_0 = total zero sequence reactance from the generator to the fault location. In these totals, use the subtransient reactance for a synchronous generator. For selecting the circuit breakers interrupting capability, use the greater of the two fault current values, $I_{3ph.sym.rms}$ or $I_{LG.sym.rms}$.

There are different settings on the circuit breaker that the engineer can effectively use for proper protection. The following is an example of choosing and setting various circuit breaker ratings for protecting a generator with rated current of 1600 A:

- Continuous rated current 1600 A.
- Interruption capacity 60 kA symmetrical rms.
- Long time delay: 10–30 s at 225% of rated current (3600 A). This is adjusted for protection from overload that is not high enough to trip the circuit breaker.
- Short time delay: 2–20 cycles in inverse relation with current at 4–10 kA. This is higher than the overload range but not high enough to cause an instantaneous trip.
- Instantaneous trip: at 48–60 kA. This is to protect from dead short at the generator terminals or close to it in the systems.

Example 9.2

The 8.25-kV circuit breaker in Table 9.4 can be used for rated fault current interruption of 33 kA at a rated 8.25-kV_{LL} system voltage. Determine the range of system voltage and interruption current it can be used for.

Solution:

The circuit breaker selection is usually facilitated by rating the circuit breaker by its megavolt-ampere capability, which is $\sqrt{3} \times 33$ kA $\times$ 8.25 kV = 470 MVA for this circuit breaker. It is called the *short-circuit megavolt-ampere capability of the circuit breaker,* as listed in the last column of Table 9.4. The megavolt-ampere capability can be split between the current and voltage within the factor of 1.3 (not any higher). Therefore, this circuit breaker can be used for a maximum of 33 × 1.3 = 43 kA at system voltage of 8.25 ÷ 1.3 = 6.35 kV, giving the fault current interruption range between 33 and 43 kA and the corresponding system voltage range between 8.25 and 6.35 kV, respectively, such that the product of kilovolts and kiloamperes remains the same at 470 MVA.

Therefore, this circuit breaker can be used for 33-kA interruption at 8.25 kV or 43-kA interruption at 6.35 kV and other current and voltage ratings in between.

9.4.3 Standard Ratings of LV Breaker

International Standards IEC 60898-1 and European Standards EN 60898-1 define the rated current I_n of a circuit breaker for low-voltage applications as the maximum current it is designed to carry continuously at 30°C. Commonly available rated currents for low-voltage circuit breakers vary from 6 to 125 A. These circuit breakers are labeled with rated current in amperes but without the unit symbol A. Instead, the ampere number is preceded by a letter B, C, or D that indicates the instantaneous tripping current, which is the minimum value of current that will cause the breaker to trip in less than 100 ms without intentional time delay, expressed in multiples of I_n. The letter B stands for I_n between 3 and 5, C for I_n between 5 and 10, and D for I_n between 10 and 20. Letter K stands for time-delay protection for I_n between 8 and 12 for protecting loads that have inrush current lasting about 0.5 to 2 s. Letter Z stands for current-limiting fast-acting protection for I_n between 2 and 3 for protecting loads with power electronics devices or instrument transformers.

In the United States, the Underwriters Laboratory (UL) uses a two-tier rating in low-voltage circuit breakers. For example, a 22/10 rating means that it has a 22-kAIC (kiloampere Interruption Capability) tenant breaker feeding 10-kAIC branches. Common ratings for such breakers are 22/10, 42/10, and 100/10.

9.5 DIFFERENTIAL PROTECTION OF GENERATOR

The generator is protected from external faults by a circuit breaker at its output terminals. A fault internal to the generator may not change the external line current to trip the circuit breaker but can damage the machine due to internal heating. A

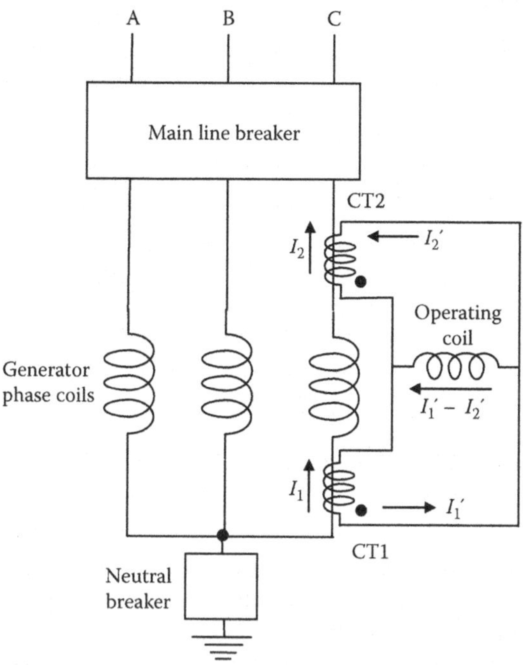

FIGURE 9.15 Differential protection scheme for internal fault in generator stator winding.

widely used scheme for protecting the generator from internal faults—applicable to transformer bank and bus bars also—is depicted in Figure 9.15. It shows the protection scheme for phase C only, but the same scheme is applied on phases A and B as well. With no internal fault, the incoming and outgoing currents I_1 and I_2 are equal in magnitude and in phase, and the differential current $I_1' - I_2'$ in the relay operating coil is zero, as is the trip coil current. However, when there is an internal fault diverting some incoming current to the ground fault in phase C, $I_1 \neq I_2$ and the differential current $I_1' - I_2'$ flows in the operating coil. This energizes the trip coil, leading to opening of the circuit breaker poles. The CTs must be identical for this scheme to work satisfactorily, or else the minimum pick-up current that will close the relay contacts must be set as a percentage of the generator rated current so that a small imbalance in the CTs will not cause nuisance tripping.

9.6 DIFFERENTIAL PROTECTION OF BUS AND FEEDERS

The differential protection scheme for protecting the bus and feeders is depicted in the one-line diagram of Figure 9.16. In three-phase implementation, three differential relays are required, one in each phase. All CTs used in such a scheme must be identical. The CT2 and CT3 are in parallel, adding their output currents. If there is no fault on the bus, the feeder current balance is maintained, that is, $I_2 + I_3 = I_1$, and so are the outputs of the CTs, that is, $I_1' = I_2' + I_3'$. The polarities of the CT windings are selected such that the relay operating coil draws the differential current $I_1' - (I_2' + I_3')$,

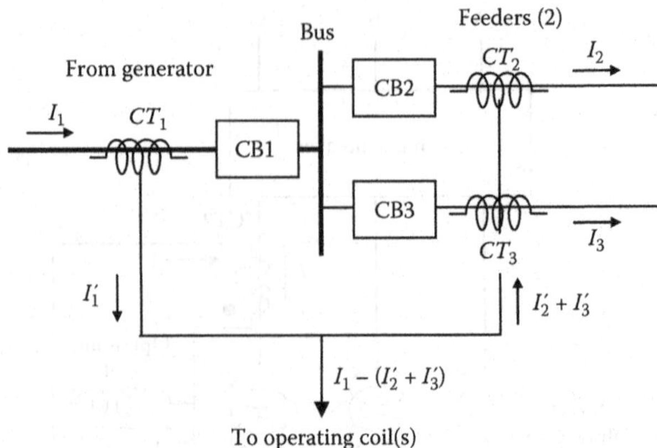

FIGURE 9.16 Differential protection scheme for bus and feeders.

which is zero, and the differential relay does not operate when there is no bus fault. In case of a bus fault, the feeder current balance is lost. The operating coil now see the differential current, activating the trip coil and opening the circuit breaker poles. The scheme can be extended to a large number of feeders from a common bus, although some care is needed to match the CT outputs to avoid nuisance tripping.

9.7 GROUND FAULT CURRENT INTERRUPTER

Like the previous two schemes, the ground fault current interrupter (GFCI) is also a differential protection scheme. A single-phase GFCI widely used in power distribution systems is shown in Figure 9.17. The feeder cable goes through a current sensor—a current transformer (CT) similar to a clamp-on ammeter—that measures

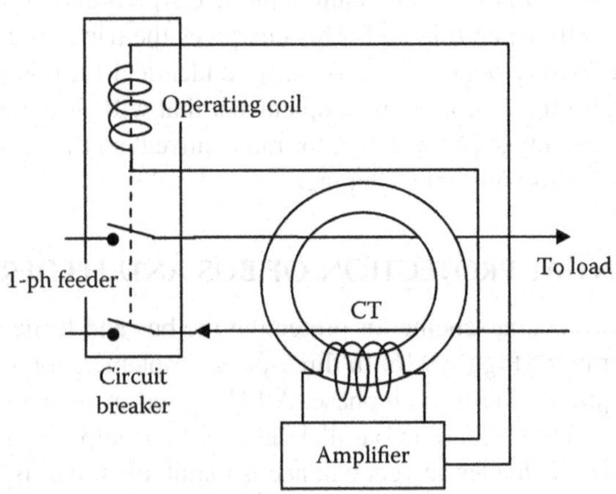

FIGURE 9.17 GFCI working principle.

the net current difference in two wires going through the eye of the CT sensor. Under healthy operation, the lead and return currents are equal, and the sensor output is zero. A short in one line conductor to ground even via a soft fault (high-impedance fault) results in higher-than-rated current in the lead wire with no change in the return current. The net unbalance current is picked up by the sensor, which is usually very small due to a high CT turn ratio. It is, therefore, amplified and then sent to the actuator coil of electromagnetic relay to open the circuit breaker.

The NEC requires that all 120-V single-phase outlets installed outdoors or in bathrooms and kitchens have the ground fault current interrupter (GFCI). Industrial power systems must also use such GFCI in a 600-V *Y*-connected system in each distribution panel board.

9.8 TRANSFORMER PROTECTION

The transformer is generally protected on both primary and secondary sides by a fuse or circuit breaker as shown in Figure 9.18. The continuous current rating of the breaker is selected as follows.

On the secondary side, general-purpose fuse or CB2 rating
= 1.3 × rated secondary current.

On the primary side, because we must allow for the magnetizing inrush current on the switch-on, a time-delay fuse or circuit breaker may be used. Often, the time-delay circuit breaker is not available in small sizes. Then a larger general-purpose circuit breaker or fuse must be used. Thus, the two options are

- Time-delay fuse or CB1 rating = 1.3 × the rated primary current
- General-purpose fuse or CB1 rating = 2 × 1.3 × rated primary current

9.9 MOTOR BRANCH CIRCUIT PROTECTION

The three types of fuse used to protect small motors are illustrated for a 10-hp motor example in Table 9.5. It shows that for a motor with full-load current of 30 A, the continuous current rating requires fuse rating of 90 A for a standard fuse to allow for the starting inrush current or 35 A for a time-delay or current-limiting fuse. If the prospective fault current is 5000 A_{rms} symmetrical, the standard and time-delay fuses will let through a 5000-A peak before interrupting, whereas the current-limiting fuse

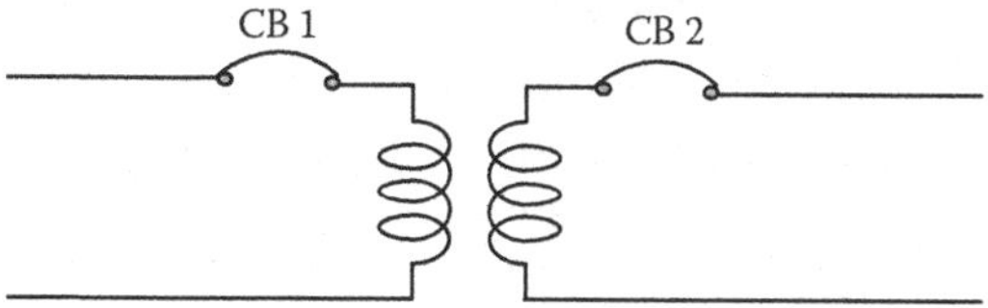

FIGURE 9.18 Circuit breaker protection on both sides of transformer.

TABLE 9.5
Fuse Type Used to Protect 10-hp Motor with Full-Load Current of 30 A

		Let-Through Current in Circuit with Prospective Fault Current (rms Symmetrical) (A)		
Fuse Type	**Fuse Rating Needed**	**3000**	**4000**	**5000**
Standard design	90 A to allow for starting inrush amps	3000	4000	5000
Time delay	35 A	3000	4000	5000
Current limiting	35 A	1700	2000	2500

would let through only 2500 A (one-half as much), thus reducing the mechanical force and heating by a factor of four.

The motor branch circuit is protected by a fuse or a circuit breaker of proper rating in relation to its starting kilovolt-ampere code letter to avoid tripping under a starting inrush current. Typical maximum ratings of these devices are shown in Table 9.6. A motor with the starting kilovolt-ampere code A requires a fuse rating of 300% of the full-load current (FLA) with a standard fuse, but only 150% of FLA if a time-delay fuse is used. If a molded case circuit breaker is used, then its instantaneous magnetic trip should be set to 700% of FLA, and the inverse time thermal trip should be set at 150% of FLA. The table is applicable to all sizes of motors, as the current rating is stated in percentage of the motor full-load ampere rating.

Example 9.3

Determine the continuous current rating and the rms fault current interruption rating of circuit breakers or fuses on both sides of a 100-kVA, 480-V Δ/208-V Y, three-phase transformer that has 5% impedance.

TABLE 9.6
Maximum Rating or Setting of Fuse or Circuit Breaker in Single-Phase or Three-Phase Induction or Synchronous Motor Branch Circuit with Full Voltage or Reduced Voltage Starting

Motor Starting Kilovolt-Ampere Code Letter for AC Motors	Percentage of Full-Load Amperes			
	Fuse Rating		Circuit Breaker Setting	
	Normal Type	**Time-Delay Type**	**Instantaneous Trip**	**Inverse Time Trip**
Code A motors	300	150	700	150
Codes B to F	300	250	700	200
Codes F to V	300	250	700	250
DC motors	200	150	250	150

Solution:

First, we determine the continuous line currents and then the circuit breaker or fuse ratings on both sides (all currents below are continuous current rms values).

Secondary side rated line current = 100 × 1000 ÷ √3 × 208 = 277.6 A
General-purpose CB or fuse rating on secondary side = 1.3 × 277.6 = 361 A
Primary side rated line current = 100 × 1000 ÷ √3 × 480 = 120.3 A
General-purpose CB or fuse rating on primary side = 2* × 1.3 × 120.3 = 313 A
Or, time-delay CB or fuse rating on primary side = 1.3 × 120.3 = 156.4 A
Now, the symmetrical fault current on both sides = 100% V ÷ 5% Z = 20 pu
Fault current interruption rating of secondary side CB or fuse = 20 × 277.6 = 5552 A
Fault current interruption rating of primary side CB or fuse = 20 × 120.3 = 2406 A

9.10 LIGHTNING AND SWITCHING VOLTAGE PROTECTION

The equipment insulation may fail during one of the following overvoltage events:

- Switching overvoltage due to the inductive and capacitive energy of the system overshooting the system voltage that can be as high as 2 × rated voltage. This is analogous to overpressure due to *water hammer* when a water pipe is suddenly shut off (switched off).
- Lightning overvoltage of magnitude IZ_o, where I is a fact-rising lightning current wave entering the system via an outdoor cable with impedance Z_o, which is known by three names—the characteristic impedance, the wave impedance, or the surge impedance of the cable. The lightning voltage entering the equipment can vary from 15 to 1500 kV and may last for about 100 μs.

When a switching or lightning overvoltage occurs in the system that exceeds the insulation strength, the insulation between two live conductors or between one line and the ground can break down in three ways (Figure 9.19):

1. Puncture through solid insulation, which can otherwise withstand the voltage stress of 300–500 V/mil between the conductors (1 mil = 1/1000 in. = 25.4 μm).
2. Strike through air, oil, or gas insulating medium, which can otherwise withstand the voltage stress of 100 to 200 V/mil. Striking an arc through air between HV conductor and grounded metal enclosure is an example of such failure.
3. Creepage (flashover) along the insulation surface, which can start at about 20 V_{rms}/mil for 60-Hz voltage along a clean surface in air and at about 10 V_{rms}/mil along a dusty or greasy surface.

---

* Factor 2 in the general-purpose circuit breaker or fuse rating on the primary side is to avoid nuisance tripping of the circuit under the magnetizing inrush current when the transformer is first connected to the supply lines. A time-delay circuit breaker or fuse on the secondary side is not required because it does not see the magnetizing inrush current. The magnetizing current flows only in the primary side that lines on the instant the transformer connecting to the supply lines.

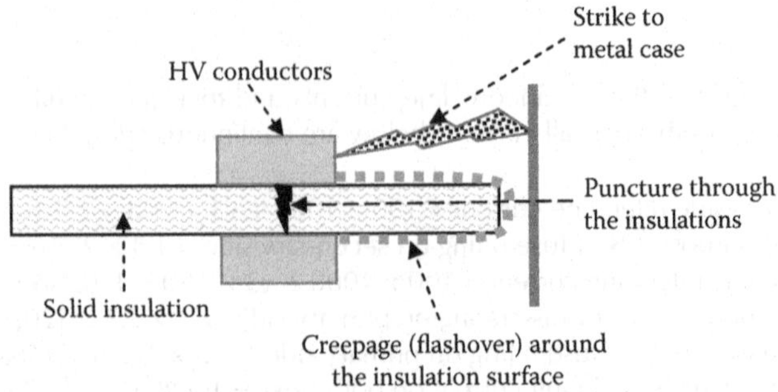

FIGURE 9.19 Three mechanisms of electrical insulation breakdown.

Since the creepage strength of the insulation surface is the least, special insulator stand-offs are used to increase the creepage length between the HV conductor and the ground. Figure 9.20 shows such curvy-shaped porcelain insulator stand-offs between HV lines and grounded metal parts (structure or enclosure). Strings of similar porcelain insulator disks are used to hang HV lines from the transmission towers we often see while driving on highways.

Under normal operating voltage, corona degrades the insulation. Solid void-free insulation typically used in dry-type electrical equipment can withstand about

FIGURE 9.20 Insulator standoffs provide longer creepage distance from HV lines to ground.

80 V_{rms}/mil (2 kV/mm) at 60 Hz for 20–30 years. However, corona (partial discharge) may start at 25 V_{rms}/mil in voids left in the insulation manufacturing process. For this reason, 20 V_{rms}/mil is a good working design limit in most solid insulations with relative dialectic constant around 3.5.

The lightning current can enter the equipment via an incoming line, resulting in overvoltage that can cause damage, particularly at the line connection end of the equipment. The International Standard IEC-1022.1 covers the requirements for protecting equipment against lightning. The generator, transformer, and motor are designed with a certain minimum basic impulse level (BIL) that the equipment can withstand under lightning and switching-type transient voltages consistent with the lightning risk. The risk is proportional to the number of thunderstorms per year in the area. Electrical equipment is further protected from lightning damage by placing a lightning arrester in any one of the three alternative configurations shown in Figure 9.21.

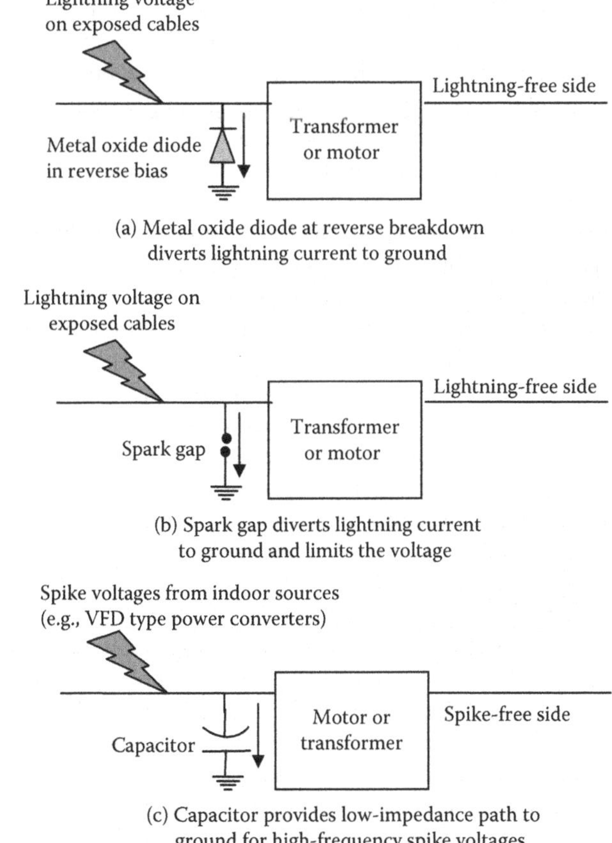

FIGURE 9.21 Lightning arrestor alternatives for protecting electrical equipment from lightning.

(a) *Metal oxide.* Functionally, the arrester works like a diode in the reverse, which breaks down at lightning voltages. It is made from lead peroxide or Thyrite hockey-puck-shaped pellets stacked with series gaps. It consists of a column of diode-type lead peroxide pellets with small series gaps assembled in a porcelain tube placed before the equipment, as shown in Figure 9.21a. The semiconducting pellet material has highly nonliner resistance that behaves like a diode in the reverse bias. The series gap isolates the diode elements from the line voltage until sparked over by a lightning impulse that breaks down the reverse biased diodes. Arresters use zinc oxide and petticoat insulator standoff. The arrester material does not get damaged after the breakdown. It recovers its normal voltage blocking property once the lightning voltage is diverted to the ground. Therefore, it does not need to be replaced after every lightning. It does, however, wear out after a certain number of incidences of lightning and may need to be replaced periodically after several years depending on the type of the material used. The arrester must have proper voltage rating. Too high a rating will not protect the equipment, and too a low rating will cause damage to the arrester itself.

(b) *Spark gap.* A spark gap before the equipment as shown in Figure 9.21b is another way of providing lightning protection. The air in the spark gap under normal operation keeps the ground current zero, but it breaks down under a high lightning voltage, diverting the lightning current to the ground. The air gap recovers its blocking property once it cools down after a strike but may need replacement after numerous lightning strikes causing surface erosion.

(c) *Capacitor.* The best protection of electrical equipment can be achieved by a combination of the lightning arrester and the capacitor placed as close as possible to the equipment as shown in Figure 9.21c. Such a protection scheme is particularly used for motors, where the pf improvement capacitors are also normally used. The lightning voltage that passes onward beyond the lightning arrester reaches the motor and is absorbed by the capacitor. The low impedance of the capacitor for virtually high-frequency lightning voltage smoothens out the steep front part of the voltage surge as it enters the stator winding. The capacitor's approximate ratings found satisfactory in most cases are 1 μF/phase for system voltage 600 V or less, 0.5 μF/phase for $600 < V < 6.6$ kV, and 0.25 μF/phase for 6.6 to 13 kV systems.

In addition to the lightning protection outlined above, a tall metal rod is used to create a lightning protection zone around the area to be protected. Such protection zone has a conical volume (Figure 9.22), starting from the tip of the rod and extending to the floor radius that depends on the rod height and the polarity of the lightning stroke. A round metal ball (solid or a wire mesh) at the tip of the pole eliminates the dielectric stress concentration, thus avoiding the lightning strike in the first place. The tip-ball is a relatively new product in the market that is used to divert the lightning to the neighboring yard.

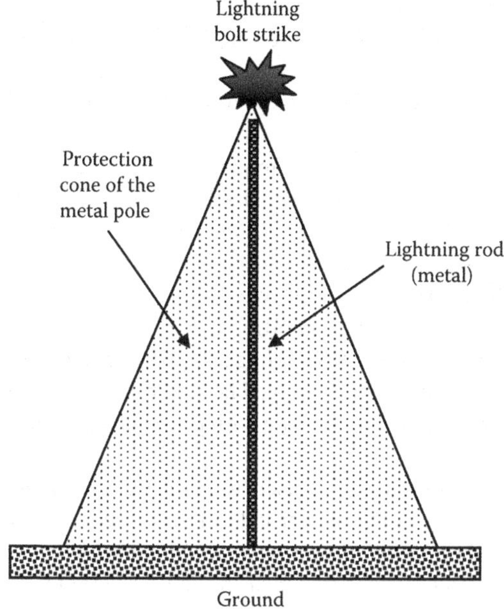

FIGURE 9.22 Protection cone of lightning rod.

9.11 SURGE PROTECTION FOR SMALL SENSITIVE LOADS

The surge protector (suppressor) protects small sensitive loads from damaging voltage surges (spikes). The voltage spikes can come from the utility power lines or be generated from local sources. For example, a large fuse-blow event in the neighborhood can generate large voltage spikes along the power lines, damaging the equipment that is plugged into nearby ac outlets. The surge suppressor protects the connected loads by diverting the voltage spikes to ground through the outlet's ground wire. For this reason, running a surge suppressor from an ungrounded outlet prevents the surge suppressor from providing such protection.

The surge suppressor is connected in front of the load. It typically contains a diode-type material in the reverse bias to the ground. It blocks the ground current under rated line voltage. However, if the line voltage exceeds the diode breakdown voltage, it becomes conducting and diverts the voltage spike to the ground. The highest voltage the load equipment would, therefore, see is the reverse breakdown voltage. The diode is designed to absorb the generated heat in joules without thermal failure. Once cooled down, it heals itself, that is, it recovers the reverse voltage blocking capability. In commercially available surge suppressors, the key components used to reduce or limit the high voltage spikes may include one or more of the following.

Metal oxide varistor (MOV) is the most common protector component, which is typically granular zinc oxide that conducts current only under a voltage above its rated breakdown voltage. It typically limits (clamps) voltage to about three to four times the normal circuit voltage. It degrades in performance when exposed to repeated spikes and hence has a finite life. MOVs may be connected in parallel to increase current capability and life expectancy, providing they are matched sets (MOVs have a tolerance of approximately 20% on voltage ratings). MOVs usually are thermally fused to avoid short circuits and other fire hazards. A failure light indicates a blown thermal fuse on some models. However, even an adequately sized MOV protector eventually degrades beyond acceptable limits without a failure light indication.

Zener diode, also known as the avalanche diode, is another clamping semiconductor similar to the MOV. It provides the best limiting action of protective components but has a lower current capability. Spike voltage can be limited to less than two times the normal operation voltage. Its life expectancy is long if current impulses remain within the device rating. The zener diode may fail short if the rating is exceeded. Since it does not degrade with use, it is often used where spikes are more frequent.

A selenium voltage suppressor is used to limit voltage spikes mostly in high-energy dc circuits such as in the exciter field of an ac generator. It can absorb high energy but does not clamp well. However, it has a longer life than MOV.

There are many rating systems that measure the surge protection:

Joule rating. It is a good parameter for gauging the surge suppressor's ability to absorb surge energy. Since surges commonly occur for microseconds or shorter durations, the energy is typically under 100 J. A well-designed surge protector does not rely on absorbing the surge energy but more on surviving the process of redirecting it. It fails gracefully in reverse bias to divert most of the surge energy to ground, thus sacrificing itself to protect the equipment plugged to it. In general, a 200-J rating gives basic protection and a 400-J surge suppressor provides good protection to most small computer-type equipment. The ampere rating is not a measure of the surge suppressor's ability to protect the load.

Clamping voltage. It specifies the voltage at which the metal oxide varistor inside the suppressor breaks down and diverts the surge to the ground. A lower clamping voltage, also known as the let-through voltage, indicates better protection but a shorter life expectancy. The lowest three levels of protection defined in the UL rating are 330 (suitable for most 120-V ac loads), 400, and 500 V.

Noise filter. Most surge suppressors also include a high-frequency special EMI suppressor (electrical noise filter). The EMI suppression is stated as a decibel level (dB) at a specific frequency (kilohertz or megahertz). The higher decibel rating provides greater attenuation of the incoming noise.

Response time. A short time delay of several nanoseconds exists before the MOV breaks down in reverse. However, surge voltages usually take a few microseconds to ramp up and reach their peak values. The surge suppressor with a nanosecond delay responds fast enough to suppress the most damaging portion of the spike.

9.12 PROTECTION COORDINATION

All circuit breakers and fuses must be properly coordinated. It means that, under a fault downstream, an upstream protection device carrying more load current must interrupt later in time than all downstream devices. Such coordination avoids an upstream feeder from losing power before the downstream feeder. It can be achieved by choosing properly rated fuses and circuit breakers with increasing current versus trip times as we move upstream of the power flow. Recall that the circuit breaker trip time for a given fault current can be adjusted by the user within its design range. Another system design coordination applies to the ring bus or tie sections. Since the power can come from two sides of the fault, the fault must be islanded (isolated or sectionalized) from both sides from where power can flow into the fault.

9.13 HEALTH MONITORING

Health monitoring involves detecting abnormal system operation, which is not a short circuit or other fault that requires tripping the circuit but degrades the system performance. When such an abnormal condition is detected, an alarm is raised—rather than the circuit interrupted. For example, a health monitoring system for a large capacitor bank installed for power factor improvement is shown in Figure 9.23, where each phase of the grounded Y may use multiple capacitor units in parallel. Each capacitor unit is usually fused to clear an internal fault. With all capacitors healthy, the current in all capacitor phases will be balanced with equal magnitudes and 120° out of phase, with their phasor sum equal to zero, and the ground current CT sensor will register zero. However, when one capacitor fuse melts, the current drawn by that phase capacitor becomes zero, and the other two phase currents add to the phasor sum that will have a nonzero value in the neutral. This current is sensed

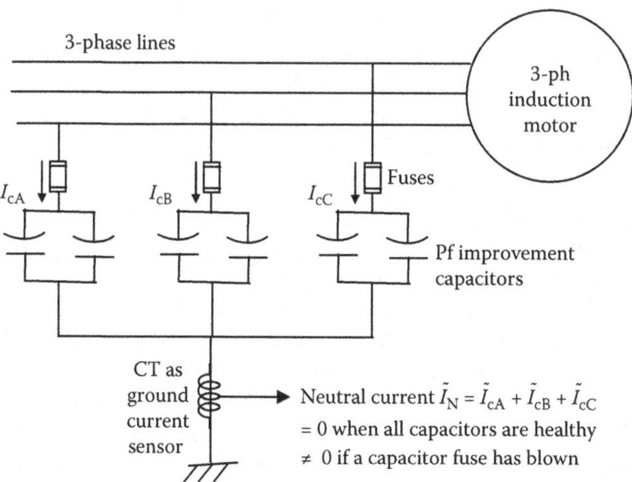

FIGURE 9.23 Health monitoring by neutral ground current sensor.

by the CT and fed to the alarm circuit, while the motor continues working normally with less improved pf.

In an ungrounded system, the *Y*-connected capacitors will have neutral floating. With all capacitors healthy, the insulated neutral-to-ground voltage will be zero because of the balanced operation. However, with one capacitor fuse blown, the unbalanced voltages will cause the neutral-to-ground voltage to be nonzero, which is then detected by PT and fed to the alarm circuit.

Other protections that are generally implemented in most systems are as follows:

- Thermal protection using temperature sensors embedded in the stator slots to protect against blockage of cooling air due to any reason
- Undervoltage protection in case of a brown-out or single-phasing of the lines
- Reverse power protection to keep parallel generators from motoring and overloading other generators

9.14 ARC FLASH ANALYSIS

The electrical power system engineer designs the system that protects itself. However, not much attention has been paid in the past to personnel safety, such as the arc flash safety. With medical cost over a million dollars per person injured by arc flash burn, the emerging industry standards now reflect the increasing importance of conducting the arc flash analysis, such as IEEE-1584 for performing arc flash hazard calculations, NFPA-70(NEC) that governs electrical instillations, NFPA-70B(NEC) that governs the maintenance of electrical equipment, etc.

The system protection analysis and the circuit breaker and fuse selection are based on the bolted fault (hard faults with ground fault impedance equal zero). However, most faults start as soft L-N faults with arcing, which turn into three-phase faults. The arcing generates heat that radiates outside and may injure personnel in the proximity of the faulted equipment. The risk of injury is proportional to the energy released by the arc. Arc flash analysis predicts the amount of the thermal energy in calories per square centimeter area following the worst-case faults, leading to selections of the protective equipment and flash protection boundaries for personnel working around the equipment, as listed in Table 9.7.

Arc flash hazard comes from many factors, such as short circuits due to creepage along the built-up conductive dust or corrosion, accidental shorts due to working with screwdrivers or dropping a tool, and improper work procedures. Although live work is discouraged even for qualified workers, work is often done on live equipment to minimize the downtime cost. For this reason, it is mandated by the National Fire Protection Association (NFPA) that arc flash studies be conducted and warning labels be placed on equipment with flash hazard boundary for carrying out unavoidable work around the energized equipment. Such studies must account for the worst cases in all possible system configurations and conditions. The system modifications and additions since the equipment was first installed must be taken into account. The arc flash studies are often carried out using commercially available software, such as Easy Power, that quantify the following on the arc flash warning labels (Figure 9.24):

TABLE 9.7
Personal Protection Requirement for Various Arc Flash Energy Densities (in cal/cm^2)

NFPA-70E 2004 Equipment Requirements (Proposed)

Category	Energy Level	Typical Personal Protective Equipment required (NFPA-70E)
0	≤2 cal/cm^2	Nonmelting flammable materials
1	4 cal/cm^2	Fire-resistant (FR) shirt and FR pants
2	8 cal/cm^2	FR shirt, FR pants, cotton underwear
3	25 cal/cm^2	Two layers FR clothing, cotton underwear
4	40 cal/cm^2	FR shirt, FR pants, multilayer flash suit, cotton underwear

Source: Adapted from the National Fire Protection Association, NFPA-70E Arc Flash Analysis, Boston, 2009.

Note: Other: face protection face shield and/or safety glasses; hand protection leather over rubber for arc flash protection; leather work boots above 4 cal/cm^2.

WARNING
Arc Flash and Shock Hazard
Appropriate PPE Required
3' - 9" Flash Hazard Boundary
5.8 cal/cm2 Flash Hazard at 18 Inches
#2* PPE Level
FR shirt and FR pants or FR coverall
0.48 kV Shock Hazard when cover is removed
3' - 6" Limited Approach
1' - 0" Restricted Approach - Class 00 Voltage Gloves
0' - 1" Prohibited Approach - Class 00 Voltage Gloves
Equipment Name: MSB-A #3/6/13 (Fed by: MAIN #14)
Date: 4/12/09, IEEE-1584 Calculations, Worst Case AFH Calculations

FIGURE 9.24 Typical arc flash and hazard warning label. (From EasyPower, ESA. With permission.)

- Incident energy in calories per square centimeter at various working distances
- Flash hazard boundaries for limited approach, restricted approach, and prohibited approach
- Personal protection equipment (PPE) class for workers to wear doing live work, such as wearing voltage-rated gloves and cotton underwear with fire-resistant (FR) shirt and FR pants and using properly selected tools
- Shock hazard in kilovolts when the cover is removed for live work

No standards or recommendations exist for arc flash study in dc systems, which is difficult and not well understood at present. However, the dc in today's power system is generally derived by an ac–dc converter (rectifier), such as in dc motor drives and steel mills. In such cases, the arc flash hazard is analyzed on the ac side of the rectifier. It is then assumed that the dc side risk is the same as that on the ac side. The logic here is that the rectifier merely processes the power without bucking or boosting the system energy. This assumption of the equal hazard on the ac and dc sides may not be accurate and may require more than the usual margin in implementing the results.

PROBLEMS

1. Determine the time-delay fuse rating for a three-phase, 208-V, 5-hp motor branch circuit in the engine room, which has an ambient air temperature of 55°C. Assume the motor efficiency of 92% and power factor of 80%. Allow 30% margin.
2. The 4.76-kV circuit breaker in Table 9.4 can be used for a rated fault current interruption of 41 kA at a rated 4.76 kV_{LL} system voltage. Determine the range of system voltage and interruption current it can be used for.
3. Determine the continuous current rating and the rms fault current interruption rating of a circuit breaker or fuse on both sides of a 300-kVA, 600-V Δ/208-V *Y*, three-phase transformer that has 6% impedance.
4. In the figure below, a three-phase, 1000-kVA, 480-V, *Y*-connected generator with ungrounded neutral has armature resistance R_a = 1% and subtransient reactance X''_d = 10%. It feeds a three-phase, 75-kVA, 480/208-V transformer with R = 1% and X = 5%. The cable has R = 0.08 Ω and X = 0.026 Ω per phase per 1000 ft. For an L-L-L fault with no ground involved, 150 ft. away from the transformer terminals, determine the symmetrical rms value and the first offset peak of the fault current in a 150-ft. cable. Repeat the calculations to determine the circuit breaker fault current rating on the 75-kVA transformer secondary terminals.

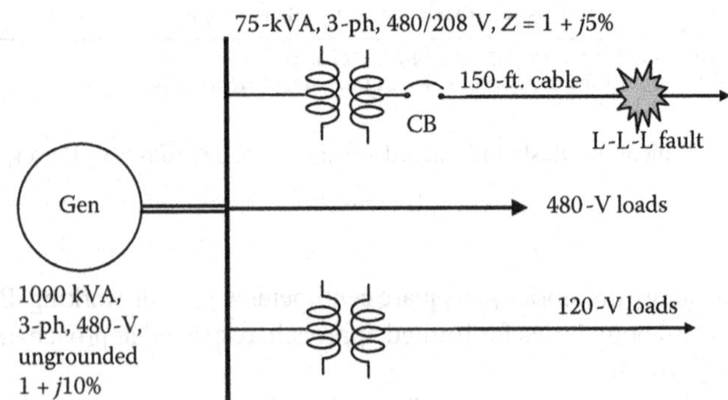

5. In a health monitoring scheme for power factor correction capacitors shown in Figure 9.23, the three capacitor banks in phase A, B, and C draw 12 A_{rms}

each when healthy. Determine the neutral current magnitude (i) when all capacitors are healthy, and (ii) when one capacitor bank gets an internal short causing its fuse to melt.

QUESTIONS

1. Identify equipment and reasons that may require time-delay fuse and fast-acting fuse.
2. Why must each load branch circuit have an overload protection, in addition to the fuse?
3. How is the arcing prevented at the relay contacts while breaking the current?
4. Why do the circuit breaker contacts see arcing when the current is interrupted, and how is it blown away in circuit breakers?
5. What is the differential relay, and how does it work?
6. Describe the working of alternative ways of protecting major electrical equipment from lightning overvoltages.
7. What does the protection coordination mean, and how is it achieved?
8. Briefly describe the arc flash analysis and the end product of it.
9. Identify three alternative ways an insulation between HV and LV conductors can break down, leading to flashover.

FURTHER READING

IEEE Standard-1584. 2008. Guide for Performing Arc Flash Hazard Calculations.
Kimbark, E. 1950. *Power System Stability.* New York: John Wiley and Sons.
National Fire Protection Association. 2009. NFPA-70E Arc Flash Analysis. Boston.
Sleva, A. 2009. *Protective Relay Principles.* Boca Raton: Taylor & Francis/CRC Press.

10 Economical Use of Power

The economical use of power in this chapter means doing the same required work with a minimum expense of energy. Major opportunities in energy savings come from

- Using a high-efficiency motor that runs several hours every day over the year
- Improving the power factor of the equipment or the entire plant to bring it close to unity
- Using variable-frequency drive to match the motor speed with the load requirement
- Storing energy when it is available at lower cost and using it later when it has a higher value
- Converting the kinetic energy of moving mass into electricity when braking, called regenerative braking (RGB)
- Selecting correct rating of the equipment (not oversizing and then using at light loads)

The energy-saving project often requires initial capital investment, which may be recovered in a few years by a continuing stream of savings in the monthly energy bill. We first cover the analytical tools to determine the economic profitability of such projects.

10.1 ECONOMIC ANALYSIS

Making a decision to invest in energy-saving equipment requires an economic analysis of the initial capital investment and monthly savings that can be realized over the life of the proposed equipment. We review here the basics of such analysis and learn how to arrive at the economic decision on whether or not the proposed investment is profitable. Two alternative ways of financing the project are covered below.

10.1.1 CASHFLOW WITH BORROWED CAPITAL

If the initial capital cost is borrowed to finance the project, then the principal and interest must be paid back to the bank or company coffers over a period on a monthly basis. In return, we receive the benefit of a lower utility bill every month over the equipment life. If the project is financed by initial principal P borrowed for M months at an interest rate of i per month, we must make monthly (principal + interest) payments of P_{month} to the lender for M months, which is given by

$$P_{\text{month}} = \frac{P \cdot i}{\left[1 - \dfrac{1}{(1+i)^M}\right]} \text{ \$/month.} \tag{10.1}$$

In return, we save in the monthly energy bill paid to the utility company for the life of the equipment, which is typically 20 to 30 years. If the amount of monthly energy savings is E_{month} kWh/month, and the electrical energy cost is C_{energy} \$/kWh, then the monthly savings in the energy bill is $S_{month} = C_{energy} \times E_{month}$ \$/month. If the monthly savings is greater than the monthly payment, then we have a net positive cash flow of $\text{Net}_{month} = S_{month} - P_{month}$ every month for M months until the borrowed capital is paid off. After M months, the monthly payments stop, and the net positive cash flow jumps to $\text{Net}_{month} = S_{month}$.

The net cash flow of $\text{Net}_{month} = S_{month} - P_{month}$, positive or negative, in the initial months does not impact the profitability of the project. It merely indicates the actual monthly cash flow. One can avoid monthly negative cash flow by borrowing for a sufficiently long time such that the monthly payments to the bank are equal to or less than the energy savings per month, that is, having $P_{month} < S_{month}$.

It is noteworthy that Equation 10.1 is also valid for a home mortgage or car loan. For example, if \$200,000 is borrowed to buy a home with a 30-year mortgage at 6% annual interest rate, then $i = 0.06/12 = 0.005$ and $M = 12 \times 30$, which will require principal-plus-interest payments of

$$P_{month} = \frac{200,000 \times 0.005}{\left[1 - \dfrac{1}{(1+0.005)^{12\times 30}}\right]} = \frac{1000}{\left[1 - \dfrac{1}{6.022575}\right]} = \$1199.10 \text{ per month.}$$

10.1.2 Payback of Self-Financed Capital

When the initial capital investment is self-financed, the net monthly savings is all positive cash flow from month one. In essence, the company paid a big chunk of money on day one to get a series of monthly savings in the future. To be fair in economic analysis in such a case, the future benefits are discounted at a rate equal to the prevailing interest rate to compensate for the lost opportunity of earning interest if the money were deposited in the bank instead. As an example, if the prevailing interest rate is 1% per month ($i = 0.01$ pu), \$2000 coming at the end of the 12th month has the present worth of $2000/(1 + 0.01)^{12} = \$2000 \times 0.8875 = \$1775$. If this amount is invested in the bank at 1% per month interest rate, compounded monthly, it will grow to \$2000 at the end of the 12th month. The factor $1/(1 + i)^{12} = 0.8875$ here is called the *discount factor* for a 12-month waiting period at the discount rate of i per month. Thus, the discount rate shrinks the future fund into its present worth, whereas the interest rate grows the present fund into its future worth. The two rates may be different, but most investors use the same number for both and interchangeably call it the interest rate.

Discounting the future income at the prevailing interest rate i makes savings at the end of the nth month worth less than that at present by the discount factor $1/(1 + i)^n$. Therefore, we convert each future monthly savings of S_{month} at the end of the nth month into its equivalent present worth $PW_n = S_{month}/(1 + i)^n$. The present worth of total savings PW_{Total} during N months is the sum total of all PW_n from $n = 1$ to N:

$$PW_{Total} = \sum_{n=1}^{N} \frac{S_{month}}{(1+i)^n} = \frac{S_{month}}{i}\left[1 - \frac{1}{(1+i)^N}\right]. \tag{10.2}$$

If the present worth of all monthly savings exceeds the initial capital cost, that is, if $PW_{Total} > P$, then obviously the project is profitable. On the other hand, $PW_{Total} = P$ results in a financially break-even project, and $PW_{Total} < P$ results in financial loss.

Many managers and investors think in terms of the payback period, that is, how many months it takes to get the initially invested principal P back from the monthly savings. Figure 10.1a depicts the initial principal P paid (negative) at day 0 and then a continuing string of monthly savings (positive) of equal amount. The present worth of monthly savings, discounted to their present worth, is shown in Figure 10.1b, gradually shrinking more and more as we move farther into future months. The

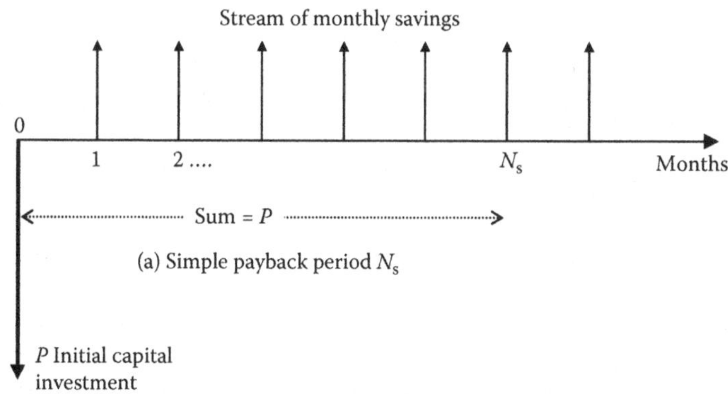

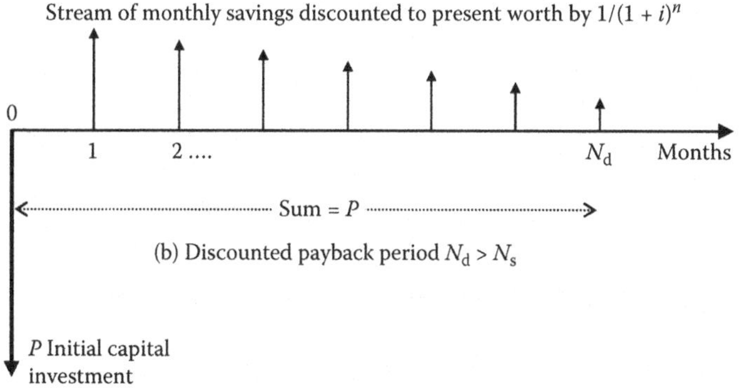

FIGURE 10.1 Monthly savings following an initial capital investment.

TABLE 10.1
Spreadsheet for Determining Discounted Payback Period for Investment

Month (n)	Energy Cost Savings ($)	Discount Factor = $1/1.01^n$	Discounted Savings ($)	Cumulative Savings ($)
1	100	0.9901	99.01	99.01
2	100	0.9803	98.03	197.04
3	100	0.9706	97.06	294.10
4	100	0.9610	96.10	390.20
5	100	0.9515	95.15	485.34
6	100	0.9420	94.20	579.55
7	100	0.9327	93.27	672.82
8	100	0.9235	92.35	765.17
9	100	0.9143	91.43	856.60
10	100	0.9053	90.53	947.13
11	100	0.8963	89.63	1036.76
12	100	0.8874	88.74	1125.51
13	100	0.8787	87.87	1213.37
14	100	0.8700	87.00	1300.37
15	100	0.8613	86.13	1386.51
16	100	0.8528	85.28	1471.79
17	100	0.8444	84.44	1556.23
18	100	0.8360	83.60	1639.83
19	100	0.8277	82.77	1722.60
20	100	0.8195	81.95	1804.56
21	100	0.8114	81.14	1885.70
22	100	0.8034	80.34	1966.04
23	100	0.7954	79.54	2045.58
24	100	0.7876	78.76	2124.34
25	100	0.7798	77.98	2202.32
26	100	0.7720	77.20	2279.52
27	100	0.7644	76.44	2355.96
28	100	0.7568	75.68	2431.64
29	100	0.7493	74.93	2506.58
30	100	0.7419	74.19	2580.77
31	100	0.7346	73.46	2654.23
32	100	0.7273	72.73	2726.96
33	100	0.7201	72.01	2798.97
34	100	0.7130	71.30	2870.27
35	100	0.7059	70.59	2940.86
36	100	0.6989	69.89	3010.75
37	100	0.6920	69.20	3079.95

simple payback period (without discounting the future savings by the interest rate) is given from Figure 10.1a:

$$\text{simple payback period } N_{\text{simple}} = P/S_{\text{month}} \text{ months.} \tag{10.3}$$

If we discount the future savings by a prevailing interest rate—as in Figure 10.1b—the payback period will get a little longer. The discounted payback period, say N months, is obtained by equating P with PW_{Total}, given by Equation 10.2, over N months, that is, discounted payback period N_{disc} is obtained by solving the following equation for N_{disc}:

$$P = \frac{S_{\text{month}}}{i}\left[1 - \frac{1}{(1+i)^{N_{\text{disc}}}}\right]. \tag{10.4}$$

Alternatively, it can be determined by setting up an Excel spreadsheet for adding the discounted monthly savings until the accumulated savings adds up to the initial capital investment. Table 10.1 is an example of such a spreadsheet for \$100 per month energy savings and 1% monthly discount (interest) rate. If the energy-saving equipment has the initial capital cost of \$2000, then the discounted payback period is 22.5 months. If the equipment cost is \$3000, then it will pay back in 36 months.

Typical payback periods for engineering projects are 36–60 months, with 36 months considered attractive by most managers. A project with a payback period longer than 60 months (5 years) hardly gets financed in the industry, except for other considerations such as environmental or public relation reasons.

10.2 POWER LOSS CAPITALIZATION

Utility power engineers often think in terms of the monetary present worth of saving 1-kW power loss in equipment like the transformer or generator, which generally stays connected all the time, 24 h a day and 365 days a year, regardless of whether or not it is delivering power at the output terminals. For example, if the transformer has 5% power loss at rated load, it would have approximately 2% fixed continuous power loss in the magnetic core at all loads (even at no load) and the remaining 3% variable loss in resistance of the coils, which would vary with the square of the actual load current at any given time. The energy cost of 2% fixed continuous loss adds up to a large sum over 25 to 30 years of the equipment life. If the energy cost is C_{energy} \$/kWh and the interest (discount) rate is i per month, every kilowatt reduction in power loss saves 1 kW × 24 h/day × 365 days/year = 8760 kWh per year, which amounts to 8760 × C_{energy} \$/year or 730 × C_{energy} \$/month. Over a 30-year equipment life, the present worth of this annual saving is called the power loss capitalization rate (PLCR), given by

$$\text{PLCR} = \frac{730C_{\text{energy}}}{i}\left[1 - \frac{1}{(1+i)^{30\times 12}}\right] \text{\$/kW.} \tag{10.5}$$

If the equipment saves S_{kW} in power loss and costs ΔP more, it will save money over 30 years only if $\Delta P < S_{kW} \times \text{PLCR}$. The case when $\Delta P = S_{kW} \times \text{PLCR}$ results in breakeven. However, 30 years is too long for most managers who would generally be interested in 1-kW loss saving advantage only if the additional capital cost $\Delta P < ⅓\, S_{kW} \times \text{PLCR}$.

Example 10.1

A 1000-kVA transformer is open for procurement. Vendor A offers 97% efficiency at $90,000 price, and vendor B offers 98% efficiency at $100,000 price. Determine the profitability of buying the high-efficiency transformer. Assume that (1) quoted efficiency is at rated load at 90% pf lagging, (2) core loss = 1/3 total loss at rated load, (3) energy cost = $0.15/kWh, and (4) interest (discount) rate = 9% per year. Frame your answer in terms of the power loss capitalization rate of saving 1-kW power loss in the magnetic core, which remains connected all year round for 30 years.

Solution:

The additional cost of transformer B over A is $10,000. The power loss capitalization rate per kilowatt loss for 30 years (360 months) is given by Equation 10.5:

$$\text{PLCR} = \frac{730C_{\text{energy}}}{i}\left[1-\frac{1}{(1+i)^{360}}\right] = \frac{730\times 0.15}{0.0075}\left[1-\frac{1}{(1+0.0075)^{360}}\right] = \$13{,}618/\text{kW}.$$

The output power at 0.90 pf = 1000 kVA × 0.90 = 900 kW.

The total power loss at 97% efficiency = 900 ÷ [1/0.97 – 1] = 27.8 kW, of which 1/3 × 27.8 = 9.27 kW is in the fixed core loss.

At 98% efficiency, total power loss is 900 ÷ [1/0.98 – 1] = 18.37 kW, of which 1/3 × 18.37 = 6.12 kW is in fixed core loss.

Difference in fixed loss that stays all year around = 9.27 – 6.12 = 3.15 kW.

At the above power loss capitalization rate, 3.15 kW means 3.15 × 13,618 = $42,897 present worth. Since the high-efficiency transformer costs $10,000 more, the simple payback period is about a 1/4 of the full 30 years life, or roughly 7.5 years.

Although many engineers and managers may not be interested in any investment opportunity with longer than a few years' payback period, the utility managers with long-term views may be interested in buying a high-efficiency unit if they have some free capital available for such a low-risk investment.

10.3 HIGH-EFFICIENCY MOTOR

All motors combined consume about 58% of total electrical power generated. Large induction motors of standard design have efficiency in the range of 90% to 95%. Since large three-phase induction motors consume most of the electrical energy, the motor industry offers high-efficiency motors under the class E design at an additional cost to support the national energy conservation plan. However, the economics of using a high-efficiency motor depends on the up-front additional capital cost, energy cost that can be saved every month, and the discount rate for the future savings. Since power loss = input power – output power, and efficiency η = output power/input power, we can write kilowatt power loss in terms of the efficiency and horsepower output of the motor as follows:

$$kW_{loss} = 0.746HP\left(\frac{1}{\eta} - 1\right). \tag{10.6}$$

Power saving by using a motor of higher efficiency instead of a lower efficiency is

$$kW_{saving} = 0.746HP\left(\frac{1}{\eta_{low}} - \frac{1}{\eta_{high}}\right). \tag{10.7}$$

$$\text{Monthly } \$_{monthly.saving} = kW_{saving} \times \text{hours/month used} \times \$/\text{kWh energy tariff.} \tag{10.8}$$

Buying a high-efficiency motor is justified if the cumulative present worth of the monthly savings in the utility bill over the service life of the motor exceeds the additional capital cost. In applications where a large motor runs at full load for 3000 h or more per year (250 h per month on average), the added capital cost may be generally recovered (paid back) typically within a few years. After that, it is all net savings every month over the service life of the motor that is typically 25 to 30 years.

In equipment where the motor runs only a few hours every day at light loads, the energy savings may not add up to the added capital cost of buying a high-efficiency motor in a reasonable number of years. In such lightly used systems, buying the standard-efficiency motor may be economical. However, in large custom-ordered power equipment (motor, transformer, generator, motor drive, etc.) that do not work at full load all the time, we still can save some energy based on Equation 2.9 developed in Section 2.5. It shows that the equipment efficiency varies with load and is the maximum at load point where the fixed loss equals the variable loss. The engineer in charge of buying large custom-ordered equipment may have an option of specifying the % load point where the equipment should be designed to have the maximum efficiency. If the buying engineer specifies the maximum efficiency load point equal to the % load at which the equipment would be operating most of the time, the energy savings can be realized even in standard-efficiency equipment. In any power equipment design—standard or high efficiency—the maximum efficiency is never at full load or at light load; it is typically between 75% and 85% of the full rated load. The engineer can specify the maximum efficiency load point at a lower load if it is used lightly for most of the time.

Since the efficiency is lower at light load, buying a motor or transformer much larger than the actual load it would deliver most of the time results in waste of energy. For example, if the motor is required to deliver 50 hp or less most of the time, and occasionally 60 hp for a couple of hours at a time, then buying any size much greater than 50/0.80 = 62.5 hp is not energy efficient. Other considerations may prevail in some cases, but they must be carefully weighed in making the decision.

Example 10.2

In a large motor drive application under consideration, either a 5000-hp induction motor or a 5000-hp synchronous motor can be used. The estimated system efficiency of the synchronous motor with LCI drive is 96%, and that of the induction motor with CSI drive is 94%. The motor is projected to run 600 h per month. Determine the total energy cost savings over the 30-year expected life of the system using the synchronous motor if the energy cost is 10 cents/kWh. Also, determine the present worth of lifetime savings discounted at 6% annual rate (0.5% per month).

(1) Approximate lifetime savings: approximate difference in input power = difference in % efficiency = 96 – 94 = 2%, and Δenergy cost = 0.02 × 5000 hp × 0.746 kW/hp × 600 h/month × 360 months life × \$0.10/kWh = \$4476 per month × 360 months = \$1,611,360 over 30-year life.

(2) Exact lifetime saving: exact difference in input power = {1/0.94 – 1/0.96} × 100 = 2.2163%, and Δenergy cost = 0.022163 × 5000 hp × 0.746 kW/hp × 600 h/month × 360 months × \$0.10/kWh = \$1,785,600 over 30-year life.

The difference between the exact and approximate power savings is about 10% (2.2163% versus 2%), and the difference in dollar savings of 1,785,600 – 1,611,360 = \$174,240 is also about 10% higher than the approximate savings. The difference is not negligible, suggesting to us to avoid the approximate approach taken in (1).

(3) Discounted present worth of exact savings = \$1,785,600 ÷ $(1 + 0.005)^{360}$ = \$296,485.

The synchronous motor system would be economical if its additional cost were less than \$296,485 over the induction motor system cost.

Example 10.3

A new 10-hp motor is required to replace an old one. It will run 500 h per month, and the energy cost is \$0.15/kWh. You have two options available: (1) a standard motor with 90% efficiency at \$1500 cost or (2) a high-efficiency motor with 92% efficiency at \$2000 cost. Determine the simple and discounted payback periods of investing in the high-efficiency motor at a discount rate of 12% per annum.

Solution:

Power loss savings by using a high-efficiency motor versus standard motor = $10 \times 0.746 \left[\frac{1}{0.90} - \frac{1}{0.92} \right] = 0.18\text{kW}.$

monthly savings = 0.18 × 500 h × \$0.15/kWh = \$13.51

simple payback period = (2000 – 1500) ÷ 13.51 = 37 months.

Discounted payback period at 1% discount rate per month from Equation 10.4 is

$$500 = \frac{13.51}{0.005}\left[1 - \frac{1}{(1+0.005)^{N_{disc}}}\right],$$

which gives N_{disc} = 41 months.

As expected, this is longer than the simple payback period of 37 months.

Alternatively, one can use an Excel spreadsheet similar to Table 10.1 using discount factor = 1 ÷ (1 + 0.005)^n for savings coming after n months. The discounted payback period is when the accumulated present worth of the monthly savings exceeds the additional upfront capital cost. The student is encouraged to set up such a spreadsheet and verify that it also gives the discounted payback period of 41 months.

If the standard motor were damaged and needs repairs that will cost $750, or it can be scrapped with a salvage value of $150, one can trade the repair versus buying a new high-efficiency motor at an additional cost of $2000 – 750 –150 = 1100. Then, the simple payback period would be 1100 ÷ 13.51 = 81.5 months. Although this is relatively long, it may still be attractive to scrap the damaged motor and replace with a new high-efficiency motor.

10.4 POWER FACTOR IMPROVEMENT

In practical power circuits, current $\tilde{I}$ drawn by the load lags the source voltage $\tilde{V}$ by an angle θ in the 30° to 40° range. The real work is done only by the current component $I \times \cos\theta$, which is in phase with the voltage. The quadrature component $I \times \sin\theta$ is orthogonal to the voltage and does no real work. It merely charges the inductor and capacitor in the load circuit for one-half cycle and discharges for the next one-half cycle that follows, contributing zero average work during every cycle. The quadrature component of the current that goes in this charging and discharging effectively leaves less conductor area available for the in-phase component, thus reducing the real power transfer capability of the conductor. In this sense, $I \times \sin\theta$ is like the bad cholesterol in our blood vessels, effectively narrowing the blood artilleries, which is not a healthy situation. Reducing the quadrature component of the current is, therefore, important for the power system health.

With power factor pf = cos θ, where θ = phase difference between the source voltage and the load current, the average real power in ac load is given by watts = $V_{rms} \times I_{rms} \times \cos\theta$, reactive power VAR = $V_{rms} \times I_{rms} \times \sin\theta$, and apparent power VA = $V_{rms} \times I_{rms}$. The power distribution switchboard usually displays kilowatts and kilovolt-amperes and sometimes pf of the aggregate load. The power triangle trigonometry we covered in Section 1.6 gives the following relations:

$$\text{kVA}^2 = \text{kW}^2 + \text{kVAR}^2,\ \text{kW} = \text{kVA} \times \text{pf},\ \text{and kVAR} = \text{kVA} \times \sqrt{(1 - \text{pf}^2)}. \quad (10.9)$$

The pf has the maximum value of 1.0 (best) and the minimum value of zero (worst, delivering no average real power, even while drawing full current at full voltage). Most practical plants draw power at a lagging pf between 0.8 and 0.9. Poor pf results in higher monthly cost to the plant for the following reason—the utility company typically levies two charges every month to the user:

- Energy charge per kilowatt hour consumed to recover the fuel cost, as the amount of fuel used by the generator prime mover depends on the kilowatt hour (real energy) used.
- Demand charge per maximum kilovolt-ampere drawn over the billing period to recover the fixed capital cost of all the equipment (generators, transformers, distribution lines, and cables), which depends on the kilovolt-ampere ratings.

$$\text{Therefore, monthly utility bill } \$_{\text{month}} = C_{\text{energy}} \times \text{kWh} + C_{\text{demand}} \times \text{kVA}, \quad (10.10)$$

where C_{energy} = energy charge rate \$/kWh and C_{demand} = demand charge rate \$/kVA. Average utility tariff rates in the United States in 2012 were about \$0.12/kWh and \$12 per peak kilovolt-ampere drawn over the month (the demand meter is reset every month) for most power users. If pf is poor, the plant draws more kilovolt-ampere for the same power. By improving the pf, we draw the same real power while drawing less kilovolt-ampere, thus reducing the monthly demand charge, although the energy charge would remain the same.

Example 10.4

The electrical power usage in a factory is shown in the table below. The utility contract calls for energy charge of \$0.15/kWh and demand charge of \$15 per peak kilovolt-ampere over any 15-min period during the month. Determine (1) the utility bill in a 30-day month and (2) the new monthly bill if the pf during the first shift is improved to unity.

Shift	Hours Used	Average Kilowatt	Peak Demand Kilovolt-Ampere
1	8	1000	1400
2	8	500	700
3	8	100	150

Solution:

(1) The monthly kilowatt hour energy used = 1000 × 8 + 500 × 8 + 100 × 8 = 12,800 kWh/day

Peak kVA demand during the first shift = 1400 kVA.

Therefore, monthly bill = 12,800 × 30 × 0.15 energy charge + 1400 × 15 demand charge = \$78,600.

(2) With unity pf in the first shift, the kilovolt-ampere demand will be equal to the kilowatt, that is, 1000 kVA, while the energy consumption will remain the same.

Therefore, new monthly utility bill = 12,800 × 30 × 0.15 + 1000 × 15 = \$72,600, which is \$6000 per month less than before.

Example 10.5

A daily power consumption profile of a customer is shown in the table below, where the kilowatt loads are averaged over the period listed and kilovolt-ampere demand meter readings are peak values over 15-min intervals during the period. This pattern repeats for all days in a 30-day month. The utility tariff to this consumer is 12 cents/kWh plus \$15 per peak kilovolt-ampere demand over the month. Determine the monthly utility bill (1) with the present usage pattern and (2) with the pf improved to unity. Also, determine the daily load factor of this consumer as seen by the utility company.

Time Period	Power Usage Kilowatt	Kilovolt-Ampere Demand
12 a.m.–4 p.m.	100	135
4 a.m.–8 a.m.	150	190
8 a.m.–12 a.m.	250	330
12 p.m.–3 p.m.	500	725
3 p.m.–8 p.m.	700	900
8 p.m.–12 p.m.	125	150

Solution:

(1) From the first two columns of the table, we derive daily energy usage = 100 × 4 + 150 × 4 + 250 × 4 + 500 × 3 + 700 × 5 + 125 × 4 = 7500 kWh per day or 7500 × 30 = 225,000 kWh per month. The peak kilovolt-ampere demand for the day is 900 kVA, which occurs over a 15-min period between 3 p.m. and 8 p.m. It does not matter if this happens every day or not, so long as it occurs at least once in the month. The consumer will be charged for this kilovolt-ampere demand for the entire month to recover the capital cost of the equipment capacity installed by the utility company.

Monthly bill = 225,000 × 0.12 energy charge + 900 × 15 demand charge = \$40,500.

(2) If this consumer maintains a unity power factor over every 15-min interval during 3 p.m. to 8 p.m. at least, if not the entire day, the peak kilovolt-ampere demand will be equal to the peak kilowatt demand, which will be 700 kVA instead of 900 kVA. That can save (900 – 700) × 15 = \$3000 per month.

Many consumers in such tariff would indeed install capacitors to maintain at least 0.95 power factor, at least from noon to 8 p.m.

We see that the peak power load is 700 kW, at which rate the total daily consumption would be 700 × 24 = 16,800 kWh per day, whereas the actual energy usage is only 7500 kWh per day.

Daily load factor of this consumer for the utility company = 7500 kWh ÷ (700 kW × 24 h) = 0.4464 or 44.64%.

For improving load pf that is lagging, we first recall that poor pf is caused by large inductance in the load circuit. Therefore, the pf can be improved by compensating the inductance by connecting a capacitor bank in parallel with the load as shown by the one-line diagram in Figure 10.2a. The capacitor is always connected in parallel with the load for better reliability. If the parallel-connected capacitor fails, the load would still run at the old poor pf doing the same work as before, only drawing more kilovolt-ampere until the failed capacitor is replaced. A failure in a series-connected capacitor, on the other hand, would cause complete loss of power to the load.

The current drawn by a capacitor $I_{cap} = V/X_{cap} = C\omega V$, which leads the voltage by 90°. It is added with the motor current in the phasor diagram shown in Figure 10.2b. We note that the resulting line current now is lagging the voltage by a smaller angle than the motor current, and its magnitude is lower, which gives the added secondary benefit of lower I^2R loss in the line cable.

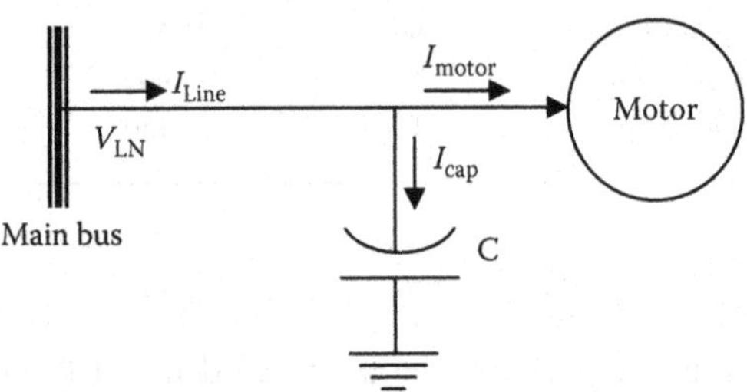

(a) Capacitor connected in parallel with motor

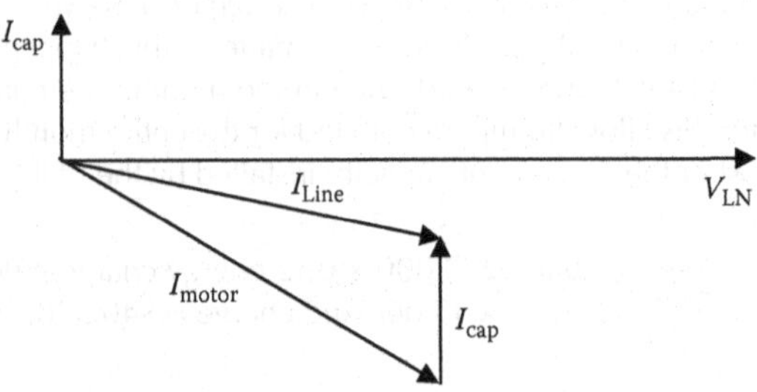

(b) Phasor diagram of the voltage and current per phase

FIGURE 10.2 Power factor improvement capacitor and resulting line current phasor diagram.

The line current drawn from the source is now $\tilde{I}_{\text{Line}} = \tilde{I}_{\text{cap}} + \tilde{I}_{\text{motor}}$, the phasor sum, which is always less than $\tilde{I}_{\text{motor}}$ due to the compensating effect of leading $\tilde{I}_{\text{cap}}$ with lagging motor current. Thus, the motor now produces the same real power with less current drawn from the source, reducing the kilovolt-ampere demand from the line and the monthly demand charge in the utility bill. The energy charge, however, remains the same, except a small reduction in I^2R loss in the line cable.

10.4.1 Capacitor Size Determination

The following analysis determines the capacitor size in farads for improving the pf from power factor angle from θ_1 (old) to θ_2 (new). From the power triangles shown in Figure 10.3 with old and new pf, denoted by suffixes 1 and 2, respectively, we have the following values per phase:

real power in both cases = kW (we get the same work done in both cases)

$$\text{reactive power kVAR}_1 = \text{kW} \tan \theta_1 \quad \text{and} \quad \text{kVAR}_2 = \text{kW} \tan \theta_2.$$

The difference between the two kVAR comes from the capacitor, that is,

$$\text{kVAR}_{\text{cap}} = \text{kVAR}_1 - \text{kVAR}_2. \tag{10.11}$$

The above formula can be rearranged in the following easier form:

$$\text{kVAR}_{\text{cap}} = \text{kW}_{\text{Load}} \left(\sqrt{\frac{1}{\text{PF}_{\text{old}}^2} - 1} - \sqrt{\frac{1}{\text{PF}_{\text{new}}^2} - 1} \right). \tag{10.12}$$

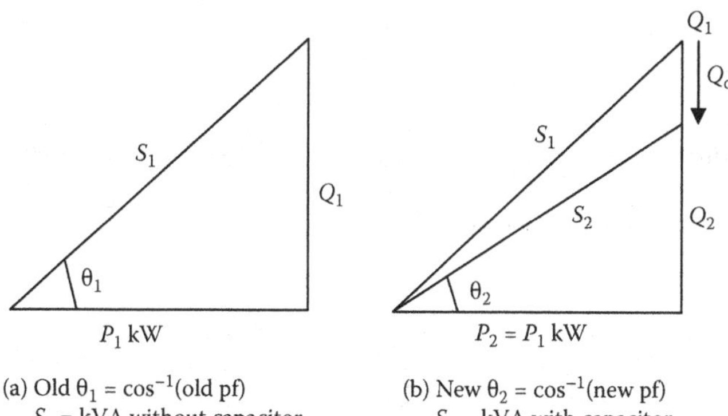

FIGURE 10.3 Power triangles with old and new power factors.

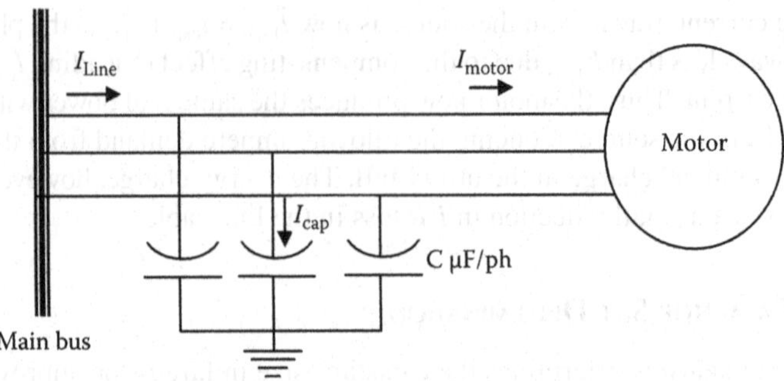

FIGURE 10.4 Three-phase capacitors connected in Y in parallel with load.

The reactive power volt-ampere reactive (VAR) in a reactance (inductive or capacitive) $Q = I^2X = V^2/X$.

$$\text{For capacitor, } X_{cap} = 1/\omega C \quad \text{therefore, } \text{VAR}_{cap} = \frac{V_{LN}^2}{1/\omega C} = \omega C V_{LN}^2.$$

$$\text{Therefore, kVAR}_{cap/ph} = \omega C V_{LN}^2/1000 \text{ and } C = \frac{1000\ \text{kVAR}_{cap/ph}}{\omega V_{LN}^2}\ \text{F/ph.} \quad (10.13)$$

In three-phase systems, three capacitors are generally connected in Y, as shown in Figure 10.4, or in Δ connection if necessary for system considerations.

Example 10.6

Determine the capacitor rating in kVAR and also in farad to improve the load pf from 0.70 lagging (i.e., $\theta_1 = 45.6°$) to 0.90 lagging (i.e., $\theta_2 = 25.8°$) at a three-phase, 60-Hz, 460-V distribution point delivering 1500-kW real power. If the motor runs for 300 h per month, the utility demand charge is \$10/kVA, energy charge is \$0.15/kWh, capacitor cost is \$60/kVAR installed, and cable power loss is 2% per phase, determine the payback period.

Solution:

We will make all calculations per phase.

$$\text{Power delivered} = \frac{1}{3}\,1500 = 500\ \text{kW, and power loss in cable} = 0.02 \times 500 = 10\ \text{kW.}$$

With old pf, $\text{kVAR}_{old} = 500 \tan 45.6° = 511$ kVAR. This must be reduced to new $\text{kVAR}_{new} = 500 \tan 25.8° = 242$ kVAR by capacitor connected in parallel with the motor.

$$\text{Therefore, kVAR}_{cap} = 511 - 242 = 269\ \text{kVAR/ph.}$$

Phase voltage V_{LN} = 460 V_{LL}÷ √3 = 265.6 V/ph, and ω = 2π·60 = 377 rad/s.

Equation 10.13 gives C = 1000 × 269 ÷ (377 × 265.6²) = 10.1 × 10^{-6} F/ph = 10.1 μF/ph.

The Y-connected capacitor bank will have 10.1 μF in each phase of the Y. If the capacitors are connected in Δ for some reason, then X_C in ohms in each Δ-leg must be 3× that in the Y-phase, or C in microfarads in Δ-leg must be 1/3 of that in the Y-phase, that is, 1/3 × 10.1 = 3.367 μF. However, the voltage rating of C in Δ is 460 V instead of 265.6 V in Y, resulting in the same 269 kVAR/ph rating of the capacitor both ways, in Y or in Δ.

With this capacitor added in parallel, kVA_{old} = kW/pf = 500 ÷ 0.70 = 714.28, and kVA_{new} = 500 ÷ 0.90 = 555.56. The differences in kVA demand = 714.28 − 555.56 = 158.72 kVA per phase. Improving the pf from 0.70 to 0.90 reduces the line current to 0.70 ÷ 0.90 = 0.7778 without reducing the real power output of the equipment. This will reduce I^2R loss in the cable to $(0.7778)^2$ = 0.605 pu, saving 0.395 pu or 39.5% from the initial cable power loss of 10 kW per phase, which amounts to 3.95 kW/phase.

Now, for three-phase, kilovolt-ampere saving = 3 × 158.72 = 476.16 kVA, and kilowatt saving in cable loss = 3 × 3.95 = 11.85 kW.

Cost savings = 10 × 476.16 kVA + 0.15 × 11.85 kW × (300 h/month) = 4762 + 533 = $5295 per month.

Total three-phase capacitor cost = $60/kVAR × 3 × 269 kVAR/ph = $48,420.

The capacitor cost will be recovered (paid back) in 48,420 ÷ 5295 = 9.14 months, which certainly is a good investment. This is an extreme example of poor pf to start with, which gives such a short payback period. For a typical plant with a reasonable pf to start with, the payback period may run a few years.

Since the reactive power supplied by the single-phase capacitance C farads/ph is given by $\text{VAR}_{1\text{ph}} = V_{\text{LN}}^2/X_C$, where $X_C = 1/(2\pi f \times C)$, or $\text{VAR}_{1\text{ph}} = V_{\text{LN}}^2 \times 2\pi f \times C$. For a three-phase cap, $\text{VAR}_{3\text{ph}} = 3 \times V_{\text{LN}}^2 \times 2\pi f \times C = V_{\text{LL}}^2 \times 2\pi f \times C$. Thus, the capacitor kVAR depends on both voltage and frequency. The capacitor can be safely operated at voltage and frequency somewhat different than the nameplate rating within a certain range. However, since capacitance C remains constant for a given construction, we can write kVAR relations in the ratio of new to old kVAR at different frequency and voltage:

$$\frac{\text{kVAR}_{3\text{ph}}^{\text{new}}}{\text{kVAR}_{3\text{ph}}^{\text{old}}} = \left(\frac{V_{\text{LL}}^{\text{new}}}{V_{\text{LL}}^{\text{old}}}\right)^2 \cdot \left(\frac{f^{\text{new}}}{f^{\text{old}}}\right). \tag{10.14}$$

Equation 10.14 is used to determine the modified kVAR rating of the capacitor at frequency and voltage different than the rated values.

Example 10.7

A 200-kVAR, 3300-V, 60-Hz, three-phase, American-made capacitor bank is operating at 2400 V, 50 Hz in Europe. Determine its new kVAR capacity in the European power system.

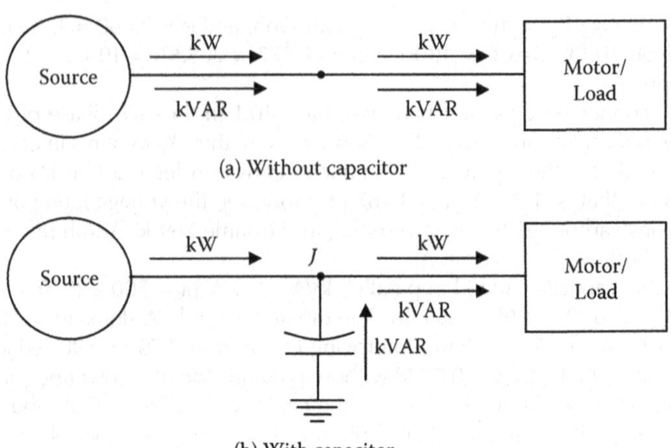

FIGURE 10.5 The kW and kVAR flow with and without capacitors.

Solution:

Using Equation 10.14, we have $\frac{\text{kVAR}_{3\text{ph}}^{\text{new}}}{200} = \left(\frac{2400}{3300}\right)^2 \cdot \left(\frac{50}{60}\right) = 0.4408.$

$$\text{Therefore, kVAR}_{3\text{ph}}^{\text{new}} = 0.4408 \times 200 = 88.16.$$

The kVAR capacity is reduced to 0.4408 pu or 44.08% primarily due to significant reduction in the voltage.

The capacitor improves the pf as explained in Figure 10.5. Without the capacitor in Figure 10.5a, the load kilowatt and kVAR both are supplied by the source. With the capacitor in Figure 10.5b, the capacitor kVAR leads the voltage by 90°, whereas inductive load kVARs lag the voltage by 90°. Thus, they have opposite polarity: when one is going away from the junction node *J*, the other is going toward the node *J*. The net result is that the inductive kVAR is supplied by the capacitor kVAR, and the supply lines are relieved from the kVAR load altogether, reducing the line current.

10.4.2 Parallel Resonance with Source

A concern in placing the pf improvement capacitor C_{pf} is the parallel resonance with the Thevenin source inductance L_{source} at natural undamped resonance frequency:

$$f_r = \frac{1}{2\pi\sqrt{L_{source}C_{pf}}} \text{ Hz.} \tag{10.15}$$

On a bus with a power electronics load drawing harmonic currents, this concern is amplified if any of the current harmonic frequencies matches with this resonance frequency. For example, the 6-pulse power electronics converter draws harmonic currents of the order 5, 7, 11, 13,…., which have frequencies of 300, 420, 660, 780 Hz,…, respectively, in a 60-Hz system. If the resonance frequency f_r = 300 Hz, the total fifth harmonic source impedance will be very high, resulting in high fifth harmonic voltage drop internal to bus. This drop, in turn, results in high bus voltage distortion that will impact all other loads (linear and nonlinear) at the bus. While placing the pf improvement capacitor, such resonance condition should be avoided by all means to maintain quality of power at the bus.

10.4.3 Safety with Capacitors

The medium voltage capacitor ratings generally available for pf improvements are listed in Table 10.2. They are made with two parallel plates enclosed in an insulating liquid and an internal discharge resistor across the terminals to bleed the stored energy after turning the power off. For personnel safety, the NEC code requires sizing the bleed resistor to bring the voltage below 50 V in low voltage (LV) capacitors within 1 min and within 5 min in high voltage (HV) capacitors.

Overcorrecting the pf by placing too many capacitors can make it less than unity in the leading direction, which is as bad as the lagging pf. There is additional danger in overcorrecting the pf. The motor working at leading pf may work as the self-excited induction generator when tripped off the lines and generate high terminal voltage and shaft torque that can damage the motor. The NEMA Safety Standards MG2, therefore, recommends that the pf capacitor value should not exceed that required to correct the no-load pf to unity. Moreover, switching capacitors at the motor terminals when needed is not recommended with the elevator motor, multi-speed motor, motor in jogging or dynamic braking application, and motor used with $Y-\Delta$ starter or reduced-voltage auto transformer. In such cases, the motor manufacturer should be consulted.

TABLE 10.2
Standard Rating for Single-Phase, Low- and Medium-Voltage Capacitors for PF Improvement

Voltage Range (V)	KVAR
240, 480, and 600	Numerous kVAR ratings available
2400–12,470	50
2700–13,280	100
4160–13,800	150
4800–14,400	200
6640–15,125	300
7200–19,920	400
7620–20,800	500

10.4.4 Difference between PF and Efficiency

Improving the pf and improving the efficiency—although both save operating cost—are two different things. They address two different aspects of the power utilization economics. Assuming the same real power output of the equipment at a fixed line voltage and frequency, their differences are as follows:

(1) Pf improvement reduces the power plant capital cost that is based on the kilovolt-ampere load, whereas the efficiency improvement reduces the energy cost.
(2) Pf improvement is done by the plant owner by installing capacitors to lower the kilovolt-ampere demand. The upfront cost of the capacitors is offset by monthly savings in the kilovolt-ampere demand charge.
(3) Efficiency can be improved only by the equipment manufacturer by
- Using larger conductors to reduce I^2R power loss in the conductors
- Using a larger core to reduce magnetic (hysteresis and eddy) losses
- Using better bearings to reduce friction, etc.

The upfront additional cost of high-efficiency equipment is offset by monthly savings in the kilowatt hour energy charge.

Table 10.3 lists the reasons, effects, and remedies of poor pf and poor efficiency.

TABLE 10.3
Reasons and Remedies for Poor PF and Poor Efficiency

	Poor Power Factor	Poor Efficiency
V and I phase relations (visible on oscilloscope)	V and I out of phase (less productive for average real power)	No effect on phase of V and I
Reason	High leakage inductance in equipment and cables	High unwanted resistance in equipment and cables
Effect 1	Draws more current and kilovolt-ampere from the lines	Draws more real power from the source
Effect 2	High losses in the cables	High losses in the equipment
Effect 3	Moderate temperature increase	High temperature increase
Cost	High kilovolt-ampere demand charge in monthly utility bill, or larger generator, transformer, and cable ratings of standalone plant	High kilowatt hour energy charge in utility bill, or greater fuel consumption rate in standalone power plant
Remedy	Use capacitors to neutralize inductance	Use heavier conductors and larger core in transformer and motor
For best performance	Have $\omega L = 1/\omega C$, which gives $X = 0$ with V and I in phase and pf = 1.0 (unity)	[a]Fixed losses in core = variable losses in conductors

[a] Buying large custom-ordered equipment, specify this condition at load point where the equipment is operated for most of the time.

10.5 ENERGY STORAGE

Electrical energy has many benefits over other forms of energy but also has a severe shortcoming. It cannot be easily stored in large quantities and must be used as produced in real time. However, if it could be stored, a new dimension in economical use of power can be realized. For a large power plant as an example, the incremental cost of producing electrical power at night is low when the demand is low. If the electrical power produced at night can be stored, it can be used to meet the peak demand during daytime, when the value of electrical power is high. Such load leveling for the utility power plant results in a lower peak rating of the equipment and lower equipment cost.

Major energy storage technologies—some fully developed and some in early stages—are as follows: (1) electrochemical battery, (2) supercapacitor, (3) superconducting coil, (4) flywheel, (5) compressed air, and (6) pumped hydroplant. The battery is discussed at length in the next chapter.

10.6 VARIABLE-SPEED MOTOR DRIVES

We have seen in Chapter 4 that (1) ac motor speed can be changed by changing the frequency and voltage in the same ratio, and (2) dc motor speed can be changed by changing the terminal voltage and field resistance in the same ratio. The ac motor speed control is achieved by the power electronics converter that converts fixed-frequency, fixed-voltage power into a variable-frequency, variable-voltage power source. Such a converter is called the variable-frequency drive (VFD). The power electronics converter used to control the dc motor speed converts only the voltage level and is called the variable-speed drive (VSD). Both are discussed at length in Chapter 15, but we give a brief summary below.

The VFD consists of a power electronics rectifier and inverter set as shown in Figure 10.6. It is widely used with a large motor driving pump or ventilating fan running long hours every day with a variable fluid pumping requirement. It is desired to maintain the stator flux constant during the motor operation at all speeds in order to produce full rated torque and to avoid magnetic saturation in the core. The ac coil voltage (Equation 2.2) gives $V/f = 4.444 \times \text{turns} \times \text{flux}$. Therefore, maintaining a constant rated flux in an ac motor requires maintaining the constant rated V/f ratio while reducing the frequency for the entire speed range.

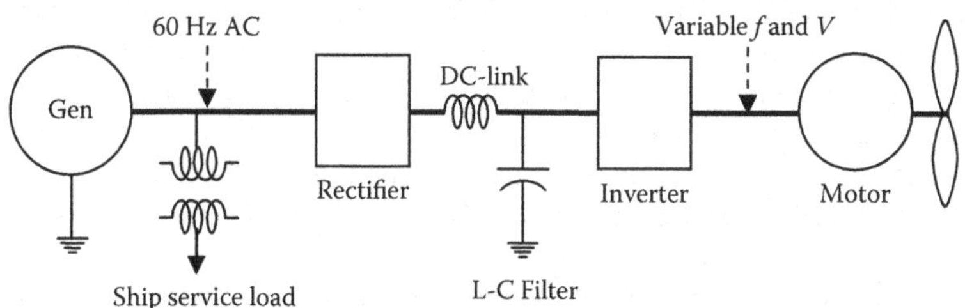

FIGURE 10.6 Variable-frequency motor drive converters with dc link.

The VFD, particularly in a fluid pumping plant such as an oil refinery, saves energy as follows. The old conventional way of reducing the fluid flow rate was to throttle the valve, while the ac motor kept running at full speed. This puts the system in conflict: the motor running at full speed while inserting a hydraulic resistance in the fluid flow, resulting in inefficient utilization of the motor power. The energy-efficient way is to reduce the motor speed in response to reduced flow requirement. The power savings using VFD and no throttle valve can be 20% to 30%. For this reason, the VFD has made significant advances over the last two decades in various industrial and transportation applications.

10.7 REGENERATIVE BRAKING

RGB can significantly save energy in applications where a load with large inertia starts and stops frequently. In RGB, we convert the kinetic energy of moving or rotating mass, or the potential energy of an elevated mass, into electrical energy that is fed back to the source—battery or power grid—from where we again draw the energy to accelerate the system from zero to full speed. At the same time, the use of mechanical brake pads is reduced to about 10%, saving the harmful dust it spreads in the environment. Saving 35% to 50% energy using RGB is possible in some applications.

An ideal candidate for RGB has the following features:

- Runs on an electric source (grid or a battery) where the regenerated electrical energy can be fed back
- Has frequent starts and stops
- Has large inertia or steep grade requiring high energy input to accelerate

Some examples of such candidates are

- City metro trains and buses
- Hybrid cars where the braking energy is stored in batteries for later use
- Electrical cranes and hoists raising and lowering heavy loads
- Elevators in tall buildings
- High-speed, high-inertia lathe machines, drills, mills, etc.

RGB requires the motor to work as the generator during deceleration. With an induction motor, it requires ramping down the supply frequency such that the motor speed becomes and continuously remains supersynchronous. This makes the slip negative, hence the torque also negative by Equation 4.11, forcing the machine to work as the generator. This requires a variable-frequency drive. With a dc motor, RGB requires gradually ramping down the motor terminal voltage, such that the machine back voltage becomes and continues to remain above the terminal voltage. This reverses the current direction, making the machine work as the generator and sending the power back to the supply. The dc RGB system requires a variable voltage converter.

The control system schematic for both ac and dc RGB is shown in Figure 10.7, where the power electronics converter output—frequency in ac and voltage in dc—is controlled by the speed feedback signal.

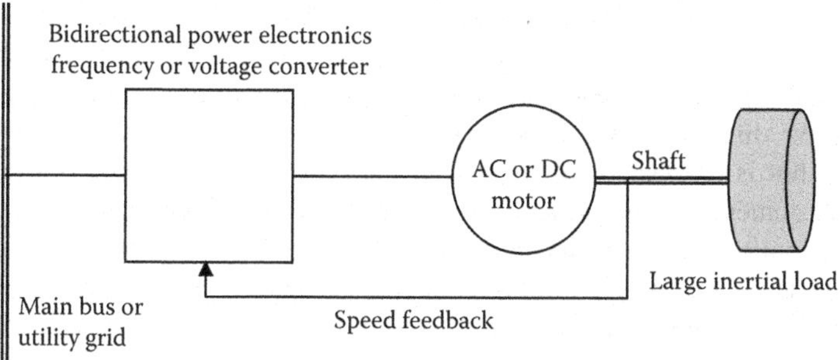

FIGURE 10.7 Control system schematic of RGB.

10.7.1 Induction Motor Torque versus Speed Curve

If the induction motor speed were varied over an extended speed range from synchronous speed backward to twice the synchronous speed forward, the theory we covered in Chapter 4 will give the motor torque versus speed characteristic shown in Figure 10.8. It shows that above the synchronous speed, the motor—although running in the same positive direction—develops negative torque, giving negative mechanical power, meaning it absorbs the mechanical power at the shaft instead of delivering. Since the current also becomes negative (flowing backward to the source), this negative power is converted into the electrical power and fed back to

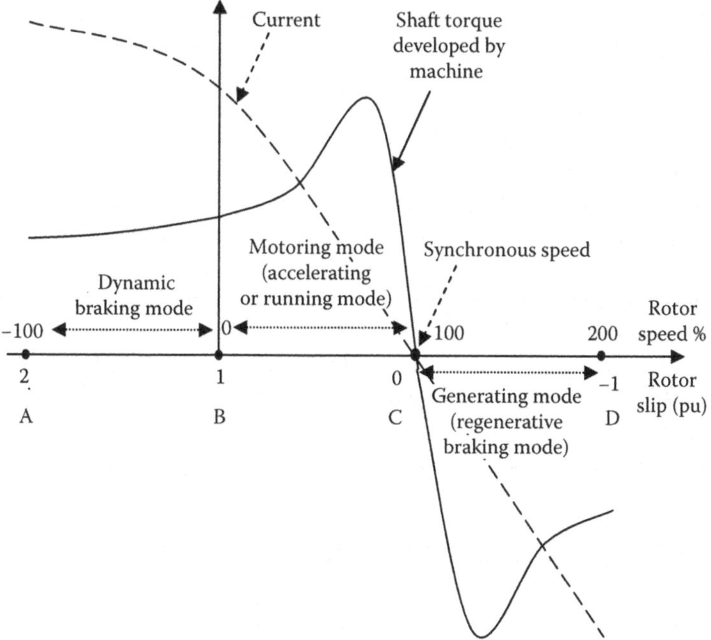

FIGURE 10.8 Induction motor torque and current versus speed over $-n_s$ to $+2n_s$ speed range.

the lines, that is, the machine now works as the generator. On the left-hand side of zero speed, at negative speed (i.e., running backward), the motor torque is still positive, again giving negative mechanical power, but still draws the current in the same positive direction from the source. This power, therefore, is still drawn from the source but is dissipated in I^2R heat in the rotor. The additional power loss in this region comes from the kinetic energy of the rotor, which slows down, as if a brake was applied, hence the name *dynamic braking.* It is commonly used in many diesel-electric locomotives to limit wear on the mechanical brakes. Thus, in dynamic braking, the power flows backward from the shaft but does not come out as electrical power at the machine terminal for outside use. It is rather dissipated in rotor resistance—often external resistance inserted via slip rings, which are cooled by forced air that can be used to heat the passenger compartments in winter.

Thus, RGB is not to be confused with dynamic braking. RGB means that the braking energy is converted into electrical energy that can be reused. Dynamic braking is not really regenerative, since it does not reuse the kinetic energy but instead dissipates it as heat through the resistor. We discuss only RGB in the section that follow.

Thus, the induction machine operating regions are as follows:

- Region A to B: dynamic braking
- Region B to C: motoring
- Region C to D: generating

10.7.2 Induction Motor Braking

We illustrate the working principle of RGB in further detail in Figure 10.9. We consider a four-pole, 60-Hz, 1750-rpm induction motor, which has the synchronous speed $n_s = 120\,f/P = 120 \times 60/4 = 1800$ rpm. If this motor is running at full rated speed of 1750 rpm, and we suddenly lower the frequency to 50 Hz in one step, the new synchronous speed will be $120 \times 50/4 = 1500$ rpm. However, the motor will still keep running near 1750 rpm for a while due to moving or rotating mass inertia. The operating point shifts from O_1 on a 60-Hz curve to O_2 on a 50-Hz curve, generating the machine torque equal to $-T_G$ at that instant. The negative sign here means that the shaft absorbs the torque and converts it into electrical power, depleting the kinetic energy and subsequently braking the motion of moving mass attached to the shaft (train, car, etc.). If we want to stop the motor with continued RGB, then we gradually ramp down the frequency from 60 Hz to a low value, simultaneously ramping down the voltage to keep the V/f ratio constant.

In SI units, power = torque × ω_{mech}, where $\omega_{mech} = 2\pi \times \text{rpm} \div 60$ rad/s and rpm = actual running speed of the machine.

$$\text{Therefore, regenerative power fed back to source } P_{RG} = \frac{T_G \times 2\pi \times \text{rpm}}{60}. \quad (10.16)$$

As the speed comes down, the frequency and voltage are lowered further down, and down, and down… to keep the machine torque negative in order to continuously

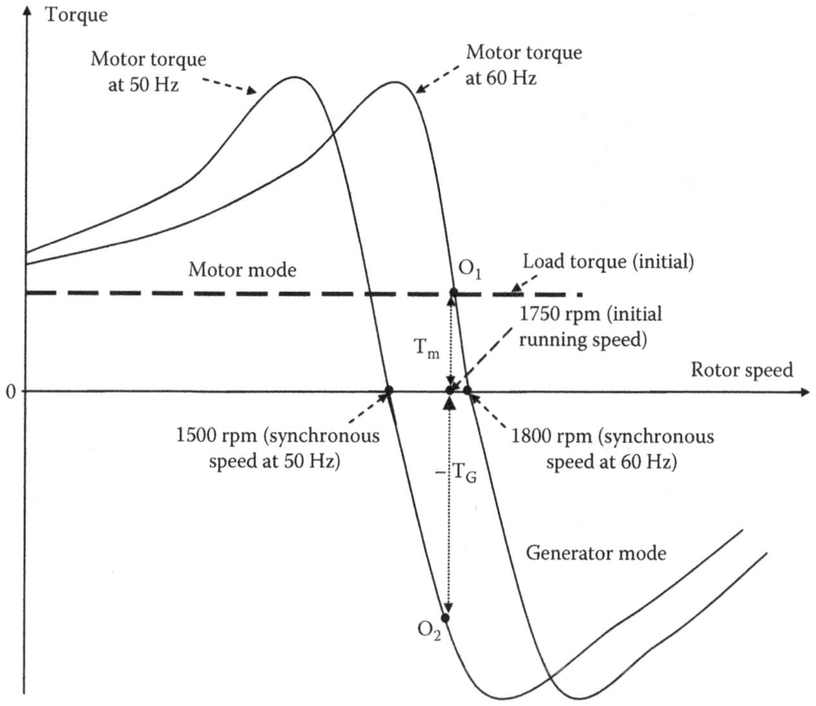

FIGURE 10.9 RGB of induction motor by changing the supply frequency from 60 to 50 Hz in the first step.

recapture most of the kinetic energy of the system as shown in Figure 10.10. When negligible kinetic energy is left in the system at very low speed, the RGB is turned off, and the mechanical brakes are applied to bring the vehicle to a full stop. The regenerative power fed back to the source in this ramp down is given by Equation 10.16.

A simpler alternative explanation of the induction motor braking follows. The motor typically runs 3%–5% below the synchronous speed with positive slip = (synchronous speed − rotor speed), and I_{rotor} = constant × slip. If the frequency is suddenly reduced, the synchronous speed becomes lower than the rotor speed, which cannot change suddenly due to inertia. The slip momentarily becomes negative, and the current and torque become negative, converting the kinetic energy of the rotor inertia into the electrical energy and feeding back into the source (grid or battery).

Example 10.8

A three-phase, four-pole, 60-Hz induction motor is running at 1746 rpm under full load. Determine the regenerative braking torque at the instant of frequency suddenly reduced to 95% using variable frequency drive while maintaining the *V/f* ratio constant. Develop your answers in percentage of rated values.

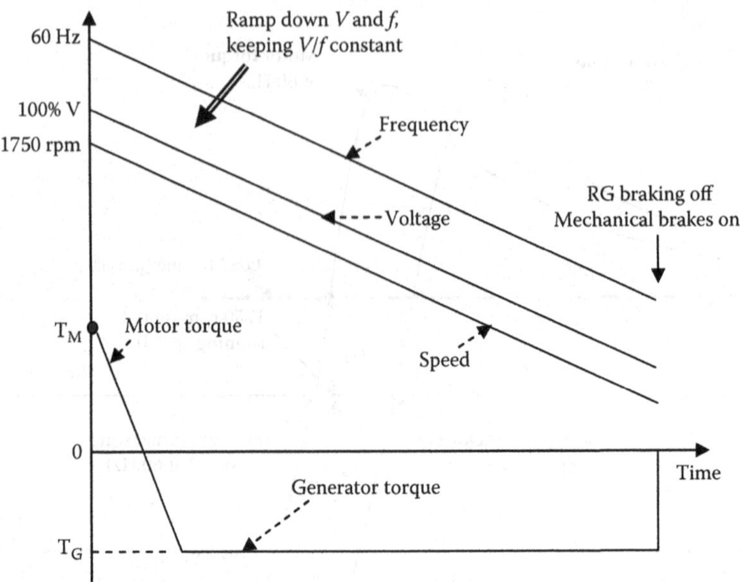

FIGURE 10.10 Frequency and voltage ramped down for induction motor braking from full to near-zero speed.

Solution:

From the motor fundamentals, we know the following, where all *K*'s are proportionality constants.

$$\text{Electromechanical torque of the motor } T_{em} = K_1 \cdot \text{stator flux} \cdot \text{rotor current} \quad \text{(a)}$$

$$\text{With constant } V/f \text{ ratio, stator flux amplitude remains constant} \quad \text{(b)}$$

$$\text{Rotor current} = K_2 \cdot d\phi/dt = K_3 \cdot \text{rotor slip speed} \quad \text{(c)}$$

Equations (b) and (c) in (a) give $T_{em} = K_4 \times$ rotor slip speed.

At 60 Hz, torque = rated value, synch speed = 1800 rpm for a four-pole motor, rotor speed = 1746 rpm, and slip speed = 1800 – 1746 = 54 rpm (positive).

At suddenly reduced frequency to 95%, that is, at a new frequency of 0.95 × 60 = 57 Hz, new synchronous speed = 1710 rpm, but the motor would still continue to run at 1746 rpm for a while due to inertia, so new slip speed = 1710 – 1746 = –36 rpm (negative).

$$\text{New torque} \div \text{old torque} = \text{new slip speed} \div \text{old slip speed} = -36 \div 54 = -0.667$$

New torque = –0.667 or –66.7% of old torque.

The negative sign signifies the regenerative (not motoring) torque, converting the inertial energy of the rotor into electrical energy and feeding back to the source, simultaneously slowing down the motor speed, hence called the *regenerative braking torque*.

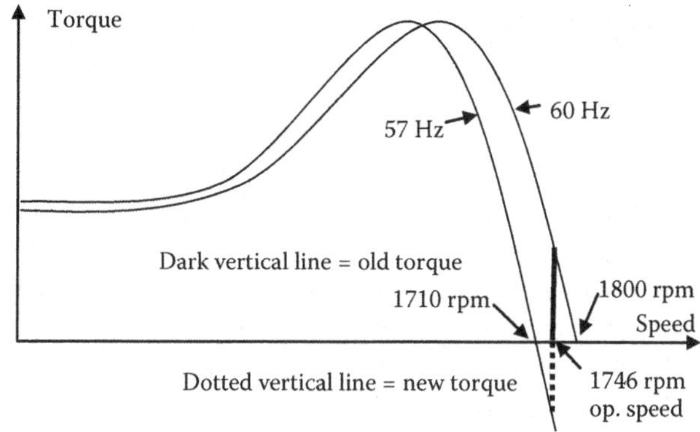

10.7.3 DC Motor Braking

The RGB scheme with a dc motor is shown in Figure 10.11. In the motoring mode, with notations shown in the figure, the armature current going into the machine is given by

$$I_a = \frac{V_T - E_a}{R_a} = \frac{V_T - KI_f n}{R_a}. \tag{10.17}$$

For RGB, we lower the machine terminal voltage V_T by a power electronics voltage controller such that V_T is lower than the counter emf, that is, $V_T < KI_f n$. The armature current I_a—given by Equation 10.17—now becomes negative, flowing

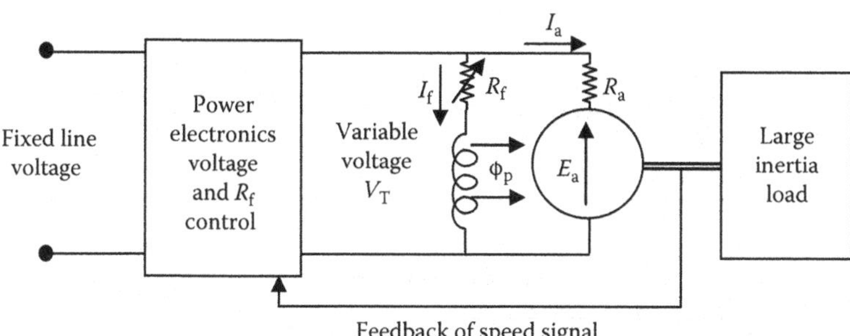

FIGURE 10.11 Variable-speed drive configuration for dc motor.

backward from the machine terminals. The machine now works as the generator, and the load's kinetic energy gradually depletes as it gets converted into electrical power fed back to the source. To extract most of the kinetic energy of the load, we keep lowering V_T so that I_a remains negative as the speed gradually drops to a low value, when we switch to the mechanical brakes. The power fed back to the source in this process is given by

$$P = E_a \cdot I_a = (K \cdot I_f \cdot n)\left(\frac{V_T - K \cdot I_f \cdot n}{R_a}\right). \tag{10.18}$$

In a simpler explanation, the motor back voltage is always less than the terminal voltage, making the current flow into the rotor. If the terminal voltage is suddenly reduced before the back emf (function of speed) can adjust, the rotor current reverses and so does the torque and power flow. This depletes the kinetic energy of the moving mass, which slows down. In dc machines, $I_f = V_T/R_f$ ratio is maintained constant while reducing the terminal voltage during the entire RGB operation as shown in Figure 10.12. A constant V_T/R_f ratio essentially keeps the flux constant at a rated value to maintain the rated torque on the shaft.

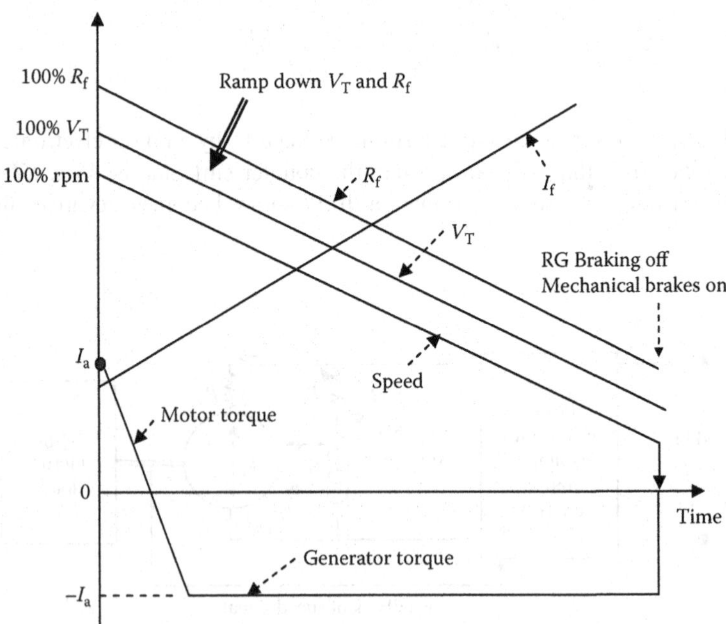

FIGURE 10.12 Voltage and field resistance ramped down for dc motor braking from full to near-zero speed.

10.7.4 New York and Oslo Metro Trains

For estimating possible energy savings with RGB in metro trains as a study project, we first review the Long Island–New York City metro trains in operation for decades and the newly installed metro trains (T-Bane) in Oslo City, Norway.

The Long Island trains use nominal 600-V_{dc} power drawn from the third rail running parallel to two traction tracks. Each car is 10 ft. wide × 10 ft. tall × 51 ft. long on 4 axles, sits 44 passengers, and is individually propelled by its own motors. Each car weighs 40 tons empty and 50 tons full. A train may use 4 to 12 cars depending on the route and time of the day. It can travel up to 55 mi./h (mph) on the standard-gage tracks. All cars in a train are controlled from a central control station. Acceleration is controlled at a rate of 2.5 mph/s up to full speed of 55 mph. Deceleration is at a rate of 2.5 mph/s by RGB from 55 down to 10 mph, and then by pneumatic air brakes to full stop.

The dc motors are four-pole compound wound with commutating poles, using both series and shunt field coils, which are changed as required during the acceleration and deceleration periods. Each motor is rated 100 hp, 300 V, 280 A, and 1175 rpm. Each car has four motors and gear assemblies. All four motors are connected in series for reduced voltage starting to control the initial starting inrush current, and the motors are connected (two in series × two in parallel) during full-speed operation.

The total resistance (drag) R on each rail car is a function of the static and dynamic rolling frictions and the aerodynamic drag. On a straight track without curvature or grade, it is given by the Davis formula:

$$R_1 = 1.3 + (29/w) + bV + (cAV^2/wn) \quad \text{lbf per ton of car weight,} \qquad (10.19)$$

where w = car weight on each axle (tons per axle of car); b = dynamic friction coefficient (0.04); c = aerodynamic drag coefficient (0.0025 for the first car and 0.00035 for the trailing cars); A = cross section perpendicular to travel (~100 ft.^{2}); n = number of axles per car; and V = speed of travel (mph).

The total resistance to each car is given by

$$R_{car} = R_1 wn. \qquad (10.20)$$

Total resistance R_{Train} to the entire train is then the sum of R_c's of all cars, including the first car, which sees greater aerodynamic resistance, that is, $R_{Train} = \sum R_{car}$. If η = motor shaft to traction efficiency (~85%), then the horsepower required for the train to overcome this drag is

$$\text{HP} = R_{Train}V \div (375\ \eta) \quad \text{where} \quad R_{Train} = \sum R_{car}. \qquad (10.21)$$

The city of Oslo in Norway during the 2008–2010 time period replaced its 65 old T-bane metro trains with new three-car trains from Siemens that are about 35% more energy efficient than the old trains. This is achieved by using RGB in the new trains, which converts up to 45% of the kinetic energy of the moving train into electric energy when braking.

The environment benefits are a major side benefit of RGB. By one estimate, the average metro train or tram contribute about 25 g of CO_2 to the atmosphere per kilometer traveled. RGB can reduce it to about 15 g. The T-bane trains made from lightweight aluminum require less energy to accelerate in the first place. Using RGB, it recaptures about 45% of the electrical energy back into the grid. Using the balance of the energy (55%) mainly from hydroelectric power, Oslo City has one of the smallest carbon footprints for a city of its size in the world.

Two design challenges with RGB in the metro train system are as follows:

(1) The power generation occurring in one decelerating train has to match in time with power demand from another nearby train that is accelerating, so that the metro authority has to draw less net power from the power grid at a given time.

(2) The distribution line voltage drops when the train is drawing power during acceleration and rises when the train is decelerating with RGB. The voltage drop and rise are equal in magnitude. If the drop is, say, 7% (relatively high since a train at low voltage draws heavy current), the voltage fluctuation is $2 \times 7 = 14\%$, which is not easy to work with in a practical system. For this reason, power distribution stations for the train power where the voltage can be closely regulated need to be closely spaced, so that 7% in this example can be reduced to say 2.5%, giving a manageable $2 \times 2.5 = 5\%$ voltage fluctuation. This may not be easy in many large cities, where the right of way to build new distribution stations may not come easy. This situation is currently faced by New York City subway trains; many of them are still not using RGB.

PROBLEMS

1. A 2-MVA transformer is open for procurement. Vendor A offers 96.5% efficiency at $160,000 price, and vendor B offers 97.8% efficiency at $180,000 price. Determine the present worth of buying the high-efficiency transformer. Assume that (1) quoted efficiency is at rated load at 90% pf lagging, (2) core loss = 1/3 total loss at rated load, (3) energy cost = $0.12/kWh, and (4) discount rate = 12% per year. Answer in terms of the power loss capitalization rate.
2. A 5000-hp variable-frequency motor drive is under consideration using either an induction motor or a synchronous motor. The estimated system efficiency of the synchronous motor with drive is 92%, and that of the induction motor with drive is 88%. The motor is projected to run 600 h per month on average. Determine the total energy cost savings over a 30-year expected life of the system using the synchronous motor drive if the energy cost is 12 cents/kWh. Also determine the present worth of lifetime savings discounted at 0.75% per month.

3. A new 25-hp motor is required to replace an old one. It will run 500 h per month, and the energy cost is \$0.15/kWh. You have two options available: (1) a standard motor with 90% efficiency at \$3000 cost or (2) a high-efficiency motor with 93% efficiency at \$4000 cost. Determine the simple and discounted payback periods of the additional cost of a high-efficiency motor at a discount rate of 9% per annum.
4. The table below shows a factory's electrical power usage pattern. The utility contract calls for energy charge of \$0.12/kWh and demand charge of \$14 per peak kilovolt-ampere over any 15-min period during the month. Determine (1) the utility bill in a 30-day month and (2) the new monthly bill if the pf during the first shift is improved to unity.

Shift	Hours Used	Average Kilowatt	Peak Demand Kilovolt-Ampere
1	8	1200	1800
2	6	800	1000
3	4	600	800

5. The daily power consumption profile of a customer is tabulated below, where the kilowatt loads are average over the period listed and kilovolt-ampere demand meter readings are peak over 15-min intervals during the period. The utility tariff to this consumer is 15 cents/kWh plus \$18 per peak kilovolt-ampere demand over the month. In a 30-day month, determine the monthly utility bill (1) with the present usage pattern and (2) with the pf improved to unity.

Time Period	Power Usage Kilowatt	Kilovolt-Ampere Demand
12 a.m.–3 a.m.	200	270
3 a.m.–8 a.m.	200	240
8 a.m.–10 a.m.	500	700
10 a.m.–3 p.m.	900	1200
3 p.m.–9 p.m.	600	800
9 p.m.–12 p.m.	200	250

6. Determine the capacitor rating in kVAR and also in farads to improve the load pf from 0.75 lagging to 0.95 lagging at a three-phase, 60-Hz, 480-V distribution point delivering 2400-kW real power. Assume that the motor runs for 400 h per month, the utility demand charge is \$15/kVA, the energy charge is \$0.15/kWh, the capacitor cost is \$70/kVAR installed, and the cable power loss is 1% per phase.
7. A 200-kVAR, 2400-V, 50-Hz, three-phase European-made capacitor bank is connected to 3300-V, 60-Hz lines in the United States. Determine its new kVAR capacity in the United States.

8. A three-phase, 100-hp, four-pole, 60-Hz induction motor is running at 1740 rpm under full load. Determine the RGB torque at the instant of reducing frequency to 96% using a variable-frequency drive while maintaining the *V/f* ratio constant.
9. What is the present worth of $1000 coming at the end of (1) the 6th month, (2) the 24th month, and (3) 30 years at the discount rate of 1% per month?
10. Your plant needs to buy a 1000-hp motor for a new conveyor belt. You have the following information: (1) a standard motor having 90% efficiency costs $96,000; (2) a high-efficiency motor having 95% efficiency costs $145,000; (3) the motor will be operating 6 days a week, 16 h per day, all year round; (4) the energy cost to your plant is $0.10/kWh; (5) the cost of money to your company is 1% per month; and (6) the life of the motor is 30 years. As the project manager, you are considering buying a high-efficiency motor. Ignoring the savings in the demand charge, first determine the savings per month in the energy cost. At that monthly savings, determine the discounted payback period for the additional cost of a high-efficiency motor. Would you buy the high-efficiency motor? Why? Be quantitative in your answer.
11. The plant you are working uses three-phase, 30-MW, 60-Hz power at 6.6 kV, 0.80 pf lagging. Determine (1) the size of a three-phase *Y*-connected capacitor bank that will improve the pf to 0.95 lagging, (2) the current drawn from the utility mains with the old pf and the new pf, and (3) the simple payback period for recovering the capital cost of the capacitor. Assume that (i) the power loss reduction in the cables is negligible, (ii) the capital cost of capacitors is $40/kVAR installed, (iii) the discount rate is 1% per month, and (iv) the monthly utility tariff is $10/kVA demand + $0.10/kWh energy.

QUESTIONS

1. Explain the difference between the terms *interest rate* and *discount rate* used in the economic analysis for determining a profitability of investment in a capital project.
2. What is the difference between simple payback period and discounted payback period?
3. Why is the power loss capitalization rate more important for the transformer and generator than for the motor?
4. Differentiate clearly between poor power factor and poor efficiency, and their causes and remedies.
5. Explain the difference between regenerative braking and dynamic braking and at what speed they take place.
6. Identify a few more candidates for RGB (other than those listed in the book).
7. Why does RGB of the dc motor require variable voltage and variable field resistance, whereas the ac motor requires variable frequency and variable voltage?

8. What is a rough estimate of % energy that can be saved using RGB in city metro trains with stations only 1 mi. apart?
9. Name the types of energy storage systems you can use for large-scale energy storage, say, for a wind farm or a solar park.

FURTHER READING

Ahmed, A. 1999. *Power Electronics for Technology.* Upper Saddle River: Prentice Hall.

Bose, B. 2006. *Power Electronics and Motor Drives*. Burlington: Academy Press.

Capehart, B.L., Turner, W.C., and Kennedy, W.J. 2008. *Guide to Energy Management.* Lilburn: Fairmount Press.

Jordan, H.E. 1983. *High Efficiency Electric Motors and Their Applications.* New York: Van Nostrand Reinhold.

Mohan, N., Undeland, T.M., and Robbins, W. 2003. *Power Electronics.* Hoboken: John Wiley & Sons.

NEMA Standard Publication No. MG10. 1977. *Energy Management Guide for Selection and Use of Polyphase Motors*. Washington, DC.

11 Electrochemical Battery

Electrical power, although a very convenient form of energy to distribute and use, cannot be easily stored on a large scale. Almost all electrical power generated by utility power plants is consumed simultaneously in real time. However, various technologies are presently available to store energy on a relatively small scale in electrical, mechanical, chemical, and magnetic form. The energy storage densities in these alternatives and their typical durations of use are compared in Table 11.1.

The electrochemical battery stores energy in electrochemical form for a wide variety of applications in consumer products and industrial plants. Its energy conversion efficiency from electrical to chemical or vice versa is about 85%. There are two basic types of electrochemical battery.

Primary battery. It converts the chemical energy into electrical energy, in which the electrochemical reaction is nonreversible, and the battery after a full discharge is discarded. It has high energy density, both gravimetric (Wh/kg) and volumetric (Wh/L). For this reason, it finds applications where high energy density for one-time use is required.

Secondary battery. It is also known as the rechargeable battery. Its electrochemical reaction is reversible. After a discharge, it can be recharged by injecting a direct current from an external source. In the charge mode, it converts the electrical energy into chemical energy. In the discharge mode, it is reversed, converting the chemical energy into electrical energy. In both charge and discharge modes, about 15% of energy is converted into heat each way, which is dissipated to the surrounding medium. Therefore, the round-trip energy conversion efficiency is 0.85×0.85, that is, 70% to 75%, depending on the electrochemistry.

The rechargeable battery is used in industrial plants for (1) emergency power for essential loads, (2) control circuits, and (3) starting power for the generator prime mover. At the utility power grid level, large batteries have applications for (1) energy storage in wind farms and solar parks, (2) smart grid with communication and cyber security lines, (3) improving transmission system reliability to prevent voltage collapse and blackout, (4) improving quality of power in the utility and end-user electrical systems, and (5) programs to reduce emissions, promote more efficient energy use, and improve environment quality.

The internal construction of a typical electrochemical cell, primary or rechargeable, is shown in Figure 11.1. It has positive and negative electrode plates with insulating separators and a chemical electrolyte in between. The two groups of electrode plates are connected to two external terminals mounted on the casing. The cell stores electrochemical energy at a low electrical potential, typically 1.2 to 3.6 V depending

TABLE 11.1
Electrical, Mechanical, and Magnetic Energy Storage Technologies

Energy Storage Technology	Energy Form in Storage	Storage Duration in Typical Applications	Relative Energy Density (kWh/kg)
Capacitor	Electrical field	< Seconds	Low
Inductor	Magnetic field	< Seconds	Moderate
Battery	Electrochemical	Days, months	Moderate
Flywheel	Kinetic energy	Days, months	High
Superconducting magnet (coil)	Magnetic field	Days, months	Very high

on the electrochemistry. The cell's charge-holding capacity is denoted by C, which is measured in ampere-hours (Ah) it can deliver. Using the basic definition of 1 A = 1 C/s, we have 1 Ah = 1 C/s × 3600 s = 3600 C of charge. A cell of capacity C can deliver C amperes for 1 h or C/n amperes for n hours. The ampere-hour capacity depends linearly on the electrode plate area. The cell voltage, on the other hand, depends solely on the electrochemistry and is independent of the plate area.

The battery is made of numerous electrochemical cells connected in a series–parallel combination to obtain the desired battery voltage and current. A higher voltage requires a greater number of cells in a series, whereas a higher current requires a greater number of cells in parallel. The overall battery rating of all cells combined is stated in terms of the average voltage during discharge and the ampere-hour capacity it can deliver before the voltage drops below the specified limit. The product of the average voltage and the ampere-hour rating forms the watt–hour energy rating of the battery that it can deliver to the load from fully charged condition.

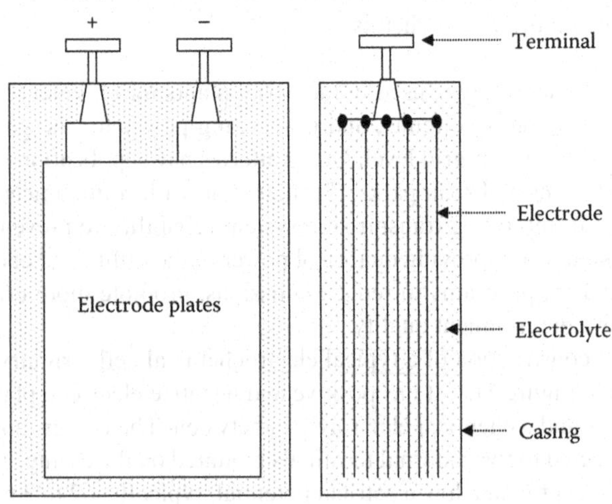

FIGURE 11.1 Electrochemical cell construction.

11.1 MAJOR RECHARGEABLE BATTERIES

Several battery technologies have been fully developed and are in use in the industrial and consumer markets. New electrochemistries are being continuously developed by the U.S. Advanced Battery Consortium for a variety of applications, such as electric vehicles, spacecraft, utility load leveling, and, of course, wind and solar power systems with inherently intermittent power generation. These electrochemistries are

- Lead–acid (Pb–acid)
- Nickel–cadmium (NiCd)
- Nickel–metal hydride (NiMH)
- Lithium ion (Li ion)
- Lithium polymer (Li poly)
- Sodium battery

The average voltage during discharge depends on the electrochemistry, as listed in Table 11.2. The energy densities of various batteries, as measured by the watt–hour capacity per kilogram mass and watt–hour per liter volume, are compared in Figure 11.2. The selection of the electrochemistry for a given application is a matter of performance, size, and cost optimization. Some construction and operating features of the above electrochemistries are presented below.

11.1.1 Lead–Acid

This is the most common type of rechargeable battery in use today because of its maturity and high performance per unit cost, even though it has the least energy density by weight and volume. In the lead–acid battery under discharge, water and lead sulfate are formed, the water dilutes the sulfuric acid electrolyte, and specific gravity of the electrolyte decreases with the decreasing state of charge. The recharging reverses the reaction, in which the lead and lead dioxide are formed at the negative and positive plates, respectively, restoring the battery into its originally charged state.

TABLE 11.2
Average Cell Voltage during Discharge in Various Rechargeable Batteries

Electrochemistry	Typical Applications	Cell Voltage	Remarks
Lead–acid	Most widely used in industry, cars, boats	2.0	Least-cost technology
Nickel–cadmium	Portable tools	1.2	Exhibits memory effect
Nickel–metal hydride	Automobiles	1.2	Temperature sensitive
Lithium ion	Computers, cell phones, spacecraft	3.6	Safe, contains no metallic lithium
Lithium polymer	Computers, cell phones, spacecraft	3.0	Contains metallic lithium

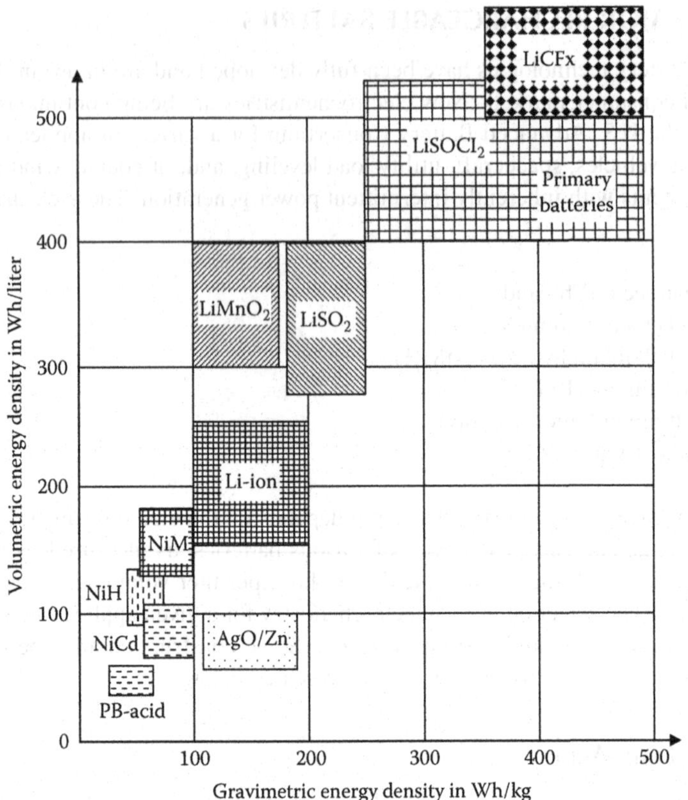

FIGURE 11.2 Energy density comparison between various battery cells.

The lead–acid battery comes in various versions. The shallow-cycle version is used in automobiles, where a short burst of energy is drawn from the battery to start the engine. The deep-cycle version, on the other hand, is suitable for repeated full charge and discharge cycles. Most energy storage applications require deep-cycle batteries. The lead–acid battery is also available in a sealed *gel–cell* version with additives, which turns the electrolyte into a nonspillable gel. The gel–cell battery, therefore, can be mounted sideways or upside down. The high cost, however, limits its use in military avionics.

11.1.2 Nickel–Cadmium

NiCd is a matured electrochemistry, in which the positive electrode is made of cadmium and the negative electrode is made of nickel hydroxide. The two electrodes are separated by Nylon separators and placed in potassium-hydroxide electrolyte in a stainless steel casing. With sealed cell and about 50% weight of the conventional lead–acid, the NiCd battery has been widely used in the past to power most rechargeable consumer products. It has a longer deep cycle life, and is more temperature tolerant than the lead–acid battery. However, this electrochemistry has a memory effect (explained in Section 11.3.5), which degrades the capacity if not used for a long time. Moreover, cadmium has recently come under environmental regulatory scrutiny. For these reasons, the NiCd is being replaced by NiMH and Li-ion batteries in laptop computers and other similar high-priced consumer electronics.

11.1.3 Nickel–Metal Hydride

NiMH is an extension of the NiCd technology, which offers an improvement in energy density and has a negligible memory effect. Its anode is made of a metal hydride that eliminates the environmental concerns of cadmium. Compared to NiCd, it has a lower peak power delivering capability, has a higher self-discharge rate, and is susceptible to damage due to overcharging. It is also expensive at present, although the future price is expected to drop significantly. This expectation is based on current development programs targeted for large-scale applications of this technology in electric vehicles.

11.1.4 Lithium Ion

The Li-ion technology is a recent development, which offers about three times the energy density over that of lead–acid. Such a large improvement in the energy density comes from lithium's low atomic weight of 6.9 versus 207 for lead. Moreover, the lithium ion has a higher cell voltage, 3.5 V per cell versus 2.0 V for lead–acid and 1.2 V for other electrochemistries. This requires fewer cells in series for a given battery voltage, thus reducing the assembly cost.

On the negative side, the lithium electrode reacts with any liquid electrolyte, creating a sort of passivation film. Every time the cell is discharged and then charged, the lithium is stripped away, a free metal surface is exposed to the electrolyte, and a new film is formed. This is compensated for by using thick electrodes, or else the electrode life would be shortened. For this reason, Li ion is more expensive than NiCd. Moreover, the lithium-ion electrochemistry is vulnerable to damage from overcharging or other shortcomings in the battery management. Therefore, it requires more elaborate charging circuitry with adequate protection against overcharging.

New companies in the United States have started to develop new Li-ion batteries for the electric cars and to compete with the present Asian suppliers such as Sanyo and Hitachi. Companies like A123 and Sakti3 are focusing on key performance requirements for eclectic car batteries, which (1) provide enough power for acceleration, (2) store enough energy to guarantee the required range, (3) survive numerous charge/discharge cycles over 8 to 10 years for an average user, and (4) keep cost below the level acceptable to electric car buyers. The Li-ion battery has high power and energy density compared with other batteries at present. This results in concentrated heat in a small volume, causing high temperature rise that can explode the battery, as has been experienced recently in some laptop computers. At about $1000/kWh cost in 2010, the GM Volt's 16-kWh battery alone would cost about $16,000. This is rather high at this time but is expected to fall to one-half in several years based on the ongoing developments.

11.1.5 Lithium Polymer

This is a lithium battery with solid polymer electrolyte. It is constructed with a film of metallic lithium bonded to a thin layer of solid polymer electrolyte. The solid polymer enhances the cell's specific energy by acting as both the electrolyte and the separator. Moreover, the metal in solid electrolyte reacts less than it does with a liquid electrolyte. A few companies are developing the lithium polymer battery for cars that may double the range compared with other batteries.

11.1.6 Sodium Battery

The sodium battery is the newest electrochemistry that is entering the market. The General Electric Company makes this advanced battery in a new plant near Albany, NY, for hybrid locomotives, heavy service vehicles, and backup storage in wind and solar energy farms. The sodium electrochemistry is based on a sodium–metal halide technology, which is better suited for short bursts of intense power required to start a vehicle moving and also for high power over a long period of time.

11.2 ELECTRICAL CIRCUIT MODEL

The battery—in the first approximation—works like a constant internal voltage source V_i with a small internal resistance R_i as shown in Figure 11.3, both of which vary with ampere-hour discharged as follows:

$$V_i = V_o - K_1 \cdot \text{DoD and } R_i = R_o - K_2 \cdot \text{DoD}, \tag{11.1}$$

where

$$\text{depth of discharge, DoD} = \frac{\text{Ah drained from battery}}{\text{rated Ah capacity}} \tag{11.2}$$

$$\text{state of charge, SoC} = \frac{\text{Ah remaining in battery}}{\text{rated Ah capacity}} = 1 - \text{DoD} \tag{11.3}$$

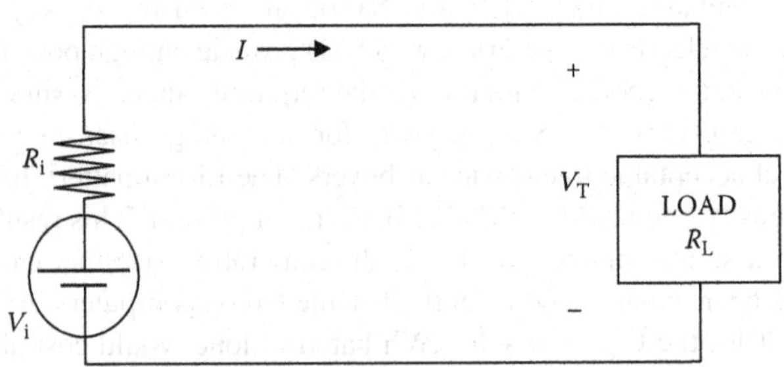

FIGURE 11.3 Electrical model of battery for steady-state performance.

and V_i = open-circuit (or electrochemical) voltage that decreases linearly with DoD; R_i = internal resistance that increases linearly with DoD; V_o and R_o = values of V_i and R_i in fully charged state (DoD = 0); and K_1 and K_2 = constants determined by curve fitting through the battery test data.

Thus, the value of V_i is lower, and R_i is higher in a partially discharged state (i.e., when DoD > 0). The terminal voltage of a partially discharged battery delivering load current I is less than V_i by the internal voltage drop $I \cdot R_i$, that is,

$$V_T = V_i - IR_i = V_o - K_1\text{DoD} - IR_i. \tag{11.4}$$

The power delivered to the external load is I^2R_L, and the internal power loss is I^2R_i, which is dissipated as heat. As the battery is discharged, its internal resistance R_i increases, which progressively generates more heat.

Example 11.1

A battery with an internal resistance of 0.01 Ω/cell needs to deliver 10-A current at 120 V at the load terminals. Determine the number of cells in series if the electrochemical voltage is 1.25 V/cell.

Solution:

First, we recognize that every cell in series must carry the load current, 10 A in this example. The battery terminal voltage = number of series cells × (internal electrochemical voltage less the voltage drop in the internal resistance per cell).

With N series cells, $120 = N(V_i - I \times R_i) = N(1.25 - 10 \times 0.01)$

Therefore, $N = 120 \div 1.15 = 104.35$ cells in series.

We then choose 105 cells to have slightly higher than 120 V at the load terminals at the beginning of battery life, which would decrease slightly with age.

11.3 PERFORMANCE CHARACTERISTICS

As stated earlier, the cell capacity is measured in ampere-hours (Ah). The cell of capacity C ampere-hours can deliver C amperes for 1 h or C/n amperes for n hours. Each cell delivers the rated ampere-hour at the terminal voltage equal to the cell voltage (1.2 to 3.6 V), which depends on the electrochemistry. The battery charge and discharge rates are stated in a unit of its capacity C. For example, charging a 100-Ah battery at $C/10$ rate means charging at 100/10 = 10 A, at which rate the battery will be fully charged in 10 h. Discharging a 100-Ah battery at $C/2$ rate means drawing 100/2 = 50 A, at which rate the battery will be fully discharged in 2 hours.

The battery voltage varies over time with the charge level, but the term *battery voltage* in practice refers to the average voltage during discharge. The battery capacity is defined as the ampere-hour charge it can deliver before the voltage drops below a certain limit (typically about 80% of the average voltage). Energy storage capacity is the product of voltage and ampere-hour capacity, that is, V × Ah = (VA) × h = Wh.

The battery performance is characterized in terms of (1) charge/discharge (c/d) voltages, (2) charge/discharge ampere-hour ratio (c/d ratio), (3) self-discharge and trickle charge rates, (4) round-trip energy efficiency, and (5) life in number of charge/discharge cycles (c/d life).

11.3.1 Charge/Discharge Voltages

Figure 11.4 shows the charge/discharge voltages of a nominal 1.2-V cell, such as NiCd and NiMH. We notice that the battery voltage has a flat plateau while discharging and another plateau during charging. The average values of these plateaus are known as the discharge voltage and the charge voltage, respectively. The c/d voltages also depend on temperature and on how fast the battery is charged or discharged.

11.3.2 c/d Ratio and Charge Efficiency

It is defined as the ratio *ampere-hour charged/ampere-hour discharged* with no net change in the state of charge. The c/d ratio—always greater than unity—depends on the charge and discharge rates and the temperature as shown in Figure 11.5. For example, if a battery has a c/d ratio of 1.1 at a certain temperature, it means that the battery needs 10% more ampere-hour charge than what was discharged in order to restore it back to the fully charged state. The *charge efficiency*—also known as the *coulombic efficiency*—is the inverse of the c/d ratio.

11.3.3 Round-Trip Energy Efficiency

The energy efficiency in one round trip of discharge and charge is defined as the ratio *energy output/energy input* at the terminals, which is given by

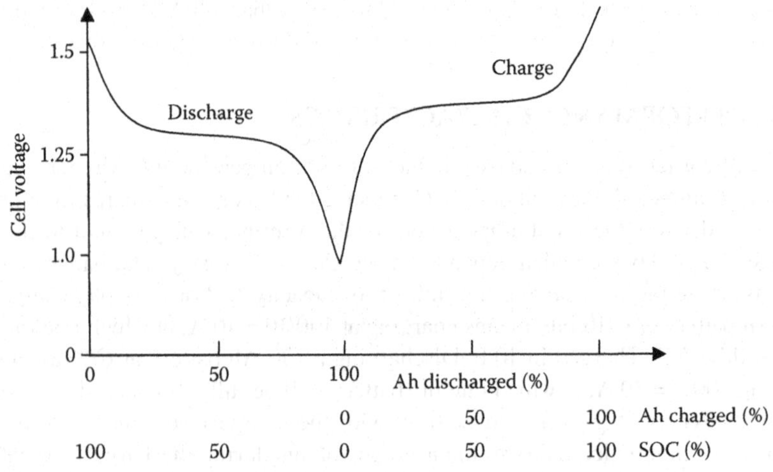

FIGURE 11.4 Discharge and charge voltages of 1.2-V cell.

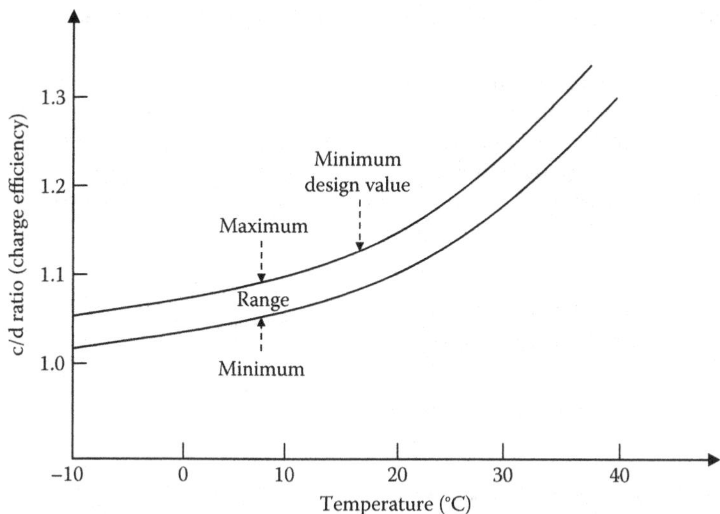

FIGURE 11.5 c/d ratio (charge efficiency) versus temperature.

$$\text{round-trip energy efficiency} = \frac{\text{average discharge voltage} \times \text{Ah capacity}}{\text{average charge voltage} \times \text{c/d ratio} \times \text{Ah capacity}}. \tag{11.5}$$

For example, for a cell of capacity C with an average discharge voltage of 1.2 V, an average charge voltage of 1.45 V, and a c/d ratio of 1.1, the round-trip energy efficiency at the cell terminals is

$$\eta_{\text{energy}} = \frac{1.2 \times C}{1.45 \times 1.1C} = 0.75 \text{ or } 75\%. \tag{11.6}$$

Thus, in this battery, 25% energy is wasted (turned into heat) in a round trip of full discharge and then full charge. The battery room, therefore, is generally hot and hazardous with combustible fumes, especially if nonsealed cells are used.

Example 11.2

You have two options in procuring 100-Ah battery cells for a large battery assembly in an industrial plant, which requires frequent charging and discharging. Cell A has an average cell voltage of 1.2 V during discharge and 1.4 V during charge and a c/d ratio of 1.1. Cell B has an average cell voltage of 1.3 V during discharge and 1.5 V during charge and a c/d ratio of 1.15. Both cells A and B cost the same per ampere-hour rating. Select the favorable cell.

Solution:

From the given data, both battery cells appear to be of the same electrochemistry. The small difference in the performance may be due to different cell design and construction methods—such as electrode thickness and electrolyte concentration—used by two manufacturers.

With equal cost, the decision has to be based on the round-trip energy efficiency. Using Equation 11.5, we obtain

$$\text{round-trip energy efficiency of cell A} = \frac{100 \times 1.2}{100 \times 1.1 \times 1.4} = 0.779$$

$$\text{round-trip energy efficiency of cell B} = \frac{100 \times 1.3}{100 \times 1.15 \times 1.5} = 0.754.$$

Cell A is preferred since it has a higher efficiency of 77.9% versus 75.4% for cell B.

11.3.4 Self-Discharge and Trickle Charge

The battery slowly self-discharges—even in open-circuit condition—at a rate that is typically less than 1% per day. To maintain full charge, the battery must be continuously trickle-charged to counter the self-discharge rate.

After the battery is fully charged, the energy storage stops rising, and any additional charge is converted into heat. If overcharged at a higher rate than the self-discharge rate for an extended period of time, the battery overheats, posing a safety hazard of potential explosion. Overcharging also produces internal gassing, which scrubs and wears down the electrode plates, shortening the life of the battery.

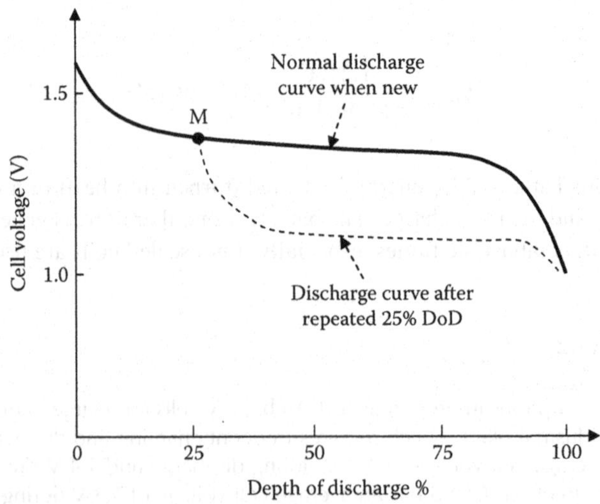

FIGURE 11.6 Memory effect in NiCd battery.

Therefore, the battery charger should have a regulator that detects full charge and cuts back the charge rate to trickle rate after the battery is fully charged.

11.3.5 Memory Effect in NiCd

The NiCd battery remembers its c/d pattern and changes the performance accordingly. After it is repeatedly charged and discharged to, say, 25% of its full capacity to point M in Figure 11.6, it remembers point M. Subsequently, if it is discharged beyond point M, the cell voltage drops much below its original normal value shown by the dotted line, resulting in a loss of full capacity. This phenomenon is like losing our body muscle due to lack of use over a long time. A remedy for restoring the full capacity after many shallow discharge cycles is to *exercise* the battery, also called *reconditioning* the battery. It means to fully discharge it to almost zero voltage once every few months and then fully charge it to full voltage per cell. The memory effect is unique to the NiCd battery; most other types of electrochemistries do not have such memory effect.

11.3.6 Temperature Effects

The operating temperature significantly influences the battery performance in three ways: (1) the charge storage capacity and charge efficiency decrease with increasing temperature, (2) the self-discharge rate increases with increasing temperature, and (3) the internal resistance decreases with increasing temperature. Various battery characteristics affecting the thermal design are listed in Table 11.3.

11.4 BATTERY LIFE

The battery life is determined by its weakest cell in the long chain of cells. The battery cell can fail short or open. A shorted cell loses the voltage and ampere-hour capacity completely and works as a load on healthy cells in the battery. Charging the battery with a shorted cell may result in heat-related damage to the battery or to the charger. The cell can also fail open, disabling the entire battery of series-connected cells.

TABLE 11.3
Battery Characteristics Affecting Thermal Design

Electrochemistry	Op. Temperature Range (°C)	Overcharge Tolerance	Heat Capacity (Wh/kg K)	Mass Density (kg/L)	Entropic Heating on Discharge (W/A cell)
Lead–acid	−10 to 50	High	0.35	2.1	−0.06
Nickel–cadmium	−20 to 50	Medium	0.35	1.7	0.12
Nickel–metal hydride	−10 to 50	Low	0.35	2.3	0.07
Lithium ion	+10 to 45	Very low	0.38	1.35	Negligible
Lithium polymer	+50 to 70	Very low	0.40	1.3	Negligible

Charging two parallel batteries by one common charger—especially with a shorted cell in one battery—can result in a highly uneven current sharing and may result in overheating both batteries and shortening their life. Even worse, the healthy battery can end up charging the defective battery, in addition to providing full load by itself. Three remedies to avoid this are as follows: (1) charge both batteries with individual chargers at their own rated charge current, (2) replace the shorted cell immediately (this can sometimes be impractical), and (3) use an isolation diode in both batteries.

In general, an individual charger for each battery is the best strategy, as shown in Figure 11.7. It incorporates isolation diodes to prevent one battery charging the other in case of an internal short in one of the batteries. It may also allow replacement of any one of many parallel batteries with somewhat different age and load-sharing characteristics. Batteries are usually replaced several times during the economic life of the plant.

Even without a random failure, the cell electrodes eventually wear out and fail due to repeated c/d cycles. The battery life is measured by the number of cycles it can be discharged and recharged before the electrodes wear out. The cycle life depends strongly on the electrochemistry and also on the depth of discharge and temperature, as shown in Figure 11.8 for the NiCd battery. Other electrochemistries will have a different but similar pattern. A very important observation from Figure 11.8 is that the life at a given temperature is inversely related to the depth of discharge. For example, at 20°C in Figure 11.8, the battery has 10,000 cycles life at 30% DoD but only about 6000 cycles at 50% DoD. That is, the lower the DoD, the longer the cycle life, giving the following approximate relation:

$$\text{number of cycles until failure} \times \text{DoD} = \text{constant K}. \quad (11.7)$$

Equation 11.7 holds well for most electrochemistries. However, the constant K decreases with increasing temperature. This means that the battery at a given

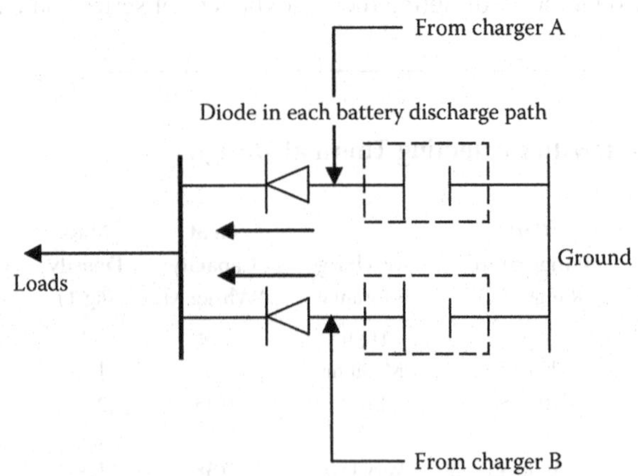

FIGURE 11.7 Two parallel batteries with separate chargers and isolation diodes.

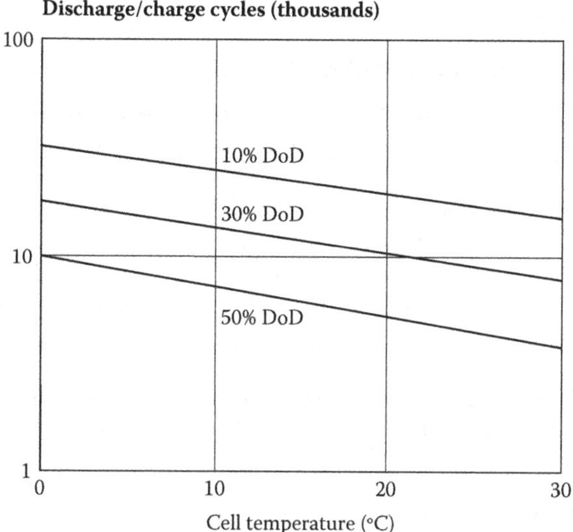

FIGURE 11.8 Charge–discharge cycle life of NiCd battery versus DoD and temperature.

temperature can deliver a certain number of equivalent full (100% DoD) cycles of ampere-hour charge and discharge regardless of the depth of discharge. Expressed differently, the total ampere-hours it can deliver over its life is approximately constant. Such observation is useful in comparing the cost of various batteries for a given application.

Life consideration is a key design driver in determining the battery ampere-hour rating. Even when the load may be met with a smaller capacity, the battery is oversized to meet the life requirement as measured in the number of c/d cycles. For example, with the same watt–hour load, the battery that must deliver twice as many cycles over its life must have double the ampere-hour capacity for the same life in calendar years.

Example 11.3

A battery lasts 2000 charge/discharge cycles of 100% DoD. Determine its approximate cycle life at 40% DoD and at 25% DoD.

Solution:

The battery's approximate cycle life is inversely related with the cyclic DoD; it lasts longer at lower DoD. Therefore, we expect the battery to last longer than 2000 c/d cycles at 40% DoD and even longer at 25% DoD.

Life at 40% DoD = 2000 × 100 ÷ 40 = 5000 cycles of charge and discharge

Life at 25% DoD = 2000 × 100 ÷ 25 = 8000 cycles of charge and discharge

TABLE 11.4
Life and Cost Comparison of Various Batteries

Electrochemistry	Cycle Life in Full Discharge Cycles	Calendar Life in Years	Self-Discharge % per Month at 25°C	Relative Cost ($/kWh)
Lead–acid	500–1000	5–8	3–5	200–300
Nickel–cadmium	1000–2000	10–5	20–30	1500
Nickel–metal hydride	1000–2000	8–10	20–30	400–600
Lithium ion	1500–2000	8–10	5–10	500–800
Lithium polymer	1000–1500	n/a	1–2	>2000

We must remember that these are just the first-order approximations, and the exact life estimate comes from the manufacturer's extensive life test data.

11.5 BATTERY TYPES COMPARED

Life and cost of various batteries are compared in Table 11.4. We note that the Li-ion battery has a relatively longer c/d cycle life and lower self-discharge rate but higher cost. The Li-poly battery is still too new to meaningfully compare with other types at present. The lead–acid battery—because of its least cost per kilowatt hour delivered over life—has been the workhorse of the commercial industry for a long time. It is available in small to large ampere-hour capacities in various terminal voltages, such as 6, 12, 24, 36 V, etc.

11.6 MORE ON LEAD–ACID BATTERY

This battery is most widely used for its overall performance at low cost. Nominal cell voltage of a lead–acid battery is 2.0 V, and the internal resistance of a 50-Ah cell is around 1 mΩ. The cycle life is 500–1000 full charge–discharge cycles for medium-rate batteries. The self-discharge rate is typically less than 0.2% per day at normal room temperature. As in all batteries, its ampere-hour capacity is sensitive to temperature. For example, its ampere-hour capacity at −20°F is only about 20% of that at 100°F. For this reason, the car is hard to start in the winter.

Table 11.5 shows the effect of state of charge on the electrolyte specific gravity, freezing point, and cell voltage. The electrolyte in a fully charged battery has a

TABLE 11.5
Effects of SoC on Specific Gravity and Freezing Point of Lead–Acid Battery

State of Charge	Specific Gravity	Freezing Point	120-V Battery Voltage
1 (fully charged)	1.27	− 65°F (−54°C)	128
75%	1.23	− 40°F (−40°C)	124
50%	1.19	− 10°F (−23°C)	122
25%	1.15	+5°F (−15°C)	120
0 (fully discharged)	1.12	+15°F (−9°C)	118

high specific gravity and freezes at −65°F. On the other hand, a fully discharged battery freezes at +15°F. This explains the importance of keeping the battery fully charged on cold days. The table also shows that we can estimate the battery state of charge by measuring the specific gravity of the electrolyte. Since this is not possible in sealed cells, other means must be used, such as monitoring the voltage or the ampere-hour going in and out (ampere-hour bookkeeping). The ampere-hour bookkeeping can be done numerically or by an electronic integrating circuit. In either case, the error accumulated over a long time makes the end result less reliable and often useless.

While the unsealed lead–acid battery is an old technology, it has one distinct advantage over modern sealed batteries. The measurement of specific gravity gives an accurate state of charge of each cell. Moreover, over a period of use, the engineer is able to feel the trend in each cell and often can detect a weak cell before it fails. If a weak cell is dragging other healthy cells down, it can be bypassed from the battery. For this reason, many engineers prefer the unsealed lead–acid battery. It is economical and offers an easy way to track the health of individual cells.

11.7 BATTERY DESIGN PROCESS

The battery is an essential part of power systems to store energy for starting the engine or for emergency power for essential loads. Other alternative energy storage technologies can also be used where beneficial. For example, compressed air in a submarine can blow the ballast tanks dry quickly due to rising to the surface in a hurry. However, compressed air energy storage, although more reliable, requires much more space to store the same energy.

Battery design for a given application proceeds in the following eight steps:

1. Select the electrochemistry suitable for the overall system requirements.
2. Determine the number of series cells required to meet the voltage requirement.
3. Determine the ampere-hour discharge required to power the load (i.e., load current × duration).
4. For the required number of c/d cycle life, determine the maximum allowable depth of discharge.
5. The ampere-hour capacity of the battery is then determined by dividing the ampere-hour discharge needed to power the load by the allowable depth of discharge (i.e., step 3 result ÷ step 4 result).
6. Determine the charge and discharge rates and necessary controls to protect from overcharging or overdischarging.
7. Determine the temperature rise and the thermal cooling requirement.
8. Determine the ventilation need of the battery room.

Being highly modular (i.e., built from numerous cells), there is no fundamental technological size limitation on the battery system that can be designed and operated. The world's largest 40-MW peak power battery was built and commissioned in

2003 at a $30-million cost. The system uses 14,000 sealed NiCd cells manufactured from recycled cadmium by Saft Corporation at a total cell cost of $10 million. The cells will be recycled again after its 20-year calendar life. The battery system is operated by Golden Valley Electric Association in Fairbanks for an Alaskan utility company. The spinning energy reserve of the battery provides continuous voltage support to the utility and lowers blackout possibility.

As stated earlier, when two or more batteries are placed in parallel, it is important that they are as identical as possible for equal load sharing for the reason explained in Section 2.8.1. Also, isolation diodes must be placed in each battery as shown in Figure 11.7 to avoid the stronger battery (with higher internal voltage) charging the weaker battery (with lower internal voltage).

Example 11.4

Design the battery for an industrial application that requires 10-kWh energy in 1.5 h for a dc load at 120 V three times daily for 250 days every year. The battery must be completely recharged in 2.0 h after each discharge. Use Li-ion cells in this example, but the design process will be applicable to other electrochemistries as well. Use the following Li-ion cell parameters: average discharge voltage 3.6 V, average charge voltage 4.2 V, c/d ratio 1.1, trickle charge rate C/100, and average life 2000 cycles at 100% DoD. For the purpose of reliability, design the battery with two strings in parallel. To avoid one string charging the other, place an isolation diode with 1-V drop in the discharge path of each string as shown in the figure below. Assume that each cell has 10-mΩ internal resistance at the operating temperature. If the battery must last five calendar years before replacement, determine (1) the number of cells in series in each parallel string, (2) the discharge current, (3) the ampere-hour ratings of each cell, (4) the charge current, (5) the charge voltage, and (6) the trickle charge current.

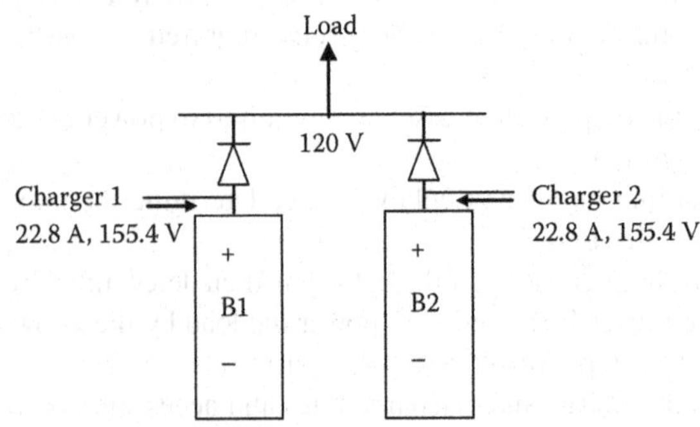

Solution:

Total ampere-hours discharged to load = 10,000 Wh ÷ 120 V = 83.33 Ah.
For two strings in parallel, each string must deliver 83.33 ÷ 2 = 41.47 Ah.
Current discharged from each string = 41.47 Ah ÷ 1.5 h = 27.65 A for 1.5 h.

If N = number of series cells in each string, allowing for voltage drops in the internal resistance and the isolation diode, we must have

$120 = N(3.6 - 0.010 \times 27.65) - 1$, which gives $N = 121 \div 3.32 = 36.42$ cells.

Therefore, we use 37 cells in series, each requiring 42-Ah capacity.

The number of discharge and charge cycles in 5 years = 3 per day × 250 × 5 = 3750 cycles, which is more than 2000 cycles—its life at 100% DoD. To survive this life duty, the required ampere-hour rating of the battery = 42 × 3750 ÷ 2000 = 78.75 Ah.

The next standard cell of 80 Ah is selected, which will last 3750 cycles at DoD = 41.47 ÷ 80 = 0.518 or 51.8%.

Charge ampere-hour required to restore full state of charge after each discharge = 1.1 × 41.47 = 45.62 Ah in 2 h, that is, at $C/2$ rate. Therefore, charge current = 45.62 ÷ 2 = 22.8 A.

Required charger voltage = 37 series cells × 4.2 V/cell = 155.4 V.

The trickle charge rate to counter the self-discharge = $C/100$, that is, 80/100 = 0.8 A.

The charger must be designed as a constant current source that delivers 22.8 A for 2 h and then shifts to 0.8 A for the rest of the time.

11.8 SAFETY AND ENVIRONMENT

Batteries use toxic chemicals such as lead, sulfuric acid, lithium, cadmium, etc. In unsealed batteries, the acidic fumes pose various hazards to the personnel working around. For example, the battery-charging line officer aboard a nuclear submarine often has to crawl in a small space on top of a large unsealed lead–acid battery to take specific gravity readings during the charge until the battery is fully charged. Line officers often see their clothes developing small holes, especially in their cotton undershirts, after a few washes. Also, the tools left in the battery room quickly corrode. This is undoubtedly due to the sulfuric acid fumes. Therefore, eye protection (and also protection for other body parts) must be worn when working closely with the batteries. It is more of an environmental safety and health issue than an environmental contamination issue.

Since the battery generates ignitable gas when overcharged, and some may leak out from an unsealed or defective battery, the battery room has to be provided with adequate air ventilation. As a guideline, the required fresh air ventilation Q (in m^3/h) is given by

$$Q_{m3/h} = 0.0055 \times \text{Ah capacity/cell} \times \text{no. of elementary cells in a room.} \quad (11.8)$$

The 12-V lead–acid battery has six elementary cells of 2 V each inside the case. Therefore, if there are 10 12-V lead–acid batteries in the room, then the number of elementary cells is 10 × 6 = 60 in Equation 11.8.

Example 11.5

A large 240-V battery is assembled from 12-V, 110-Ah battery modules commercially available off the shelf. Determine the air ventilation requirement for each room.

Solution:

For the number of cells in the room in Equation 11.8, we assume the battery module of lead–acid electrochemistry, which has 2 V/cell, so a 12-V module has six elementary cells in series inside the case.

The number of series cells in battery room = 240 V battery voltage ÷ 12 V module voltage = 20 modules × 6 series cells/module = 120 elementary cells in series for the 240-V battery.

Ampere-hour capacity of each series cell = 120 Ah.

Therefore, required air ventilation $Q = 0.0055 \times 110 \times 120 = 72.6$ m³/h.

Besides the personnel safety issues discussed above, the most important safety consideration for protecting the battery is not to overcharge. Any overcharge above the trickle charge rate is converted into heat and pressure buildup, which can explode the battery if allowed to overheat beyond a certain limit. For the lead–acid battery, the trickle charge rate is typically below $C/20$ amperes (5% of the rated capacity). Another caution is never to overcharge above the float charge voltage of 2.25 V/cell at 25°C in a lead–acid battery. Above this voltage, all charging should be terminated.

Environmentally, the battery can be both safe and unsafe, depending on how we use it. It is safe until the time comes to dispose of it with many toxic materials inside. The battery is not *green* by itself, but it supports the energy sources that are safe for the environment. Proper disposal of the toxic materials can minimize the environmental impact but still leaves harmful waste products. Moreover, the batteries are costly to recycle. Not everyone recycles small cells at the consumer level. The lead–acid batteries use about 90% of the lead in the United States, and a large percentage of it is recycled. The recycled lead and plastic are used for new batteries, and the sulfuric acid converted to water or to sodium sulfate is used in detergent, textile, and glass manufacturing. Recycling NiCd batteries needs advanced recovery methods, where cadmium is reused for making new batteries, and nickel is used in steel production. The liquid potassium sulfate may be used in some wastewater treatment plants, but this use is not widespread as yet. Lithium batteries discarded from consumer products, cell phones, and personal electronic devices are recycled to a lesser extent at about 65%. The military recycles about 90% of the batteries. All these data suggest that the batteries pose some potential dangers but are not necessarily unsafe if used and recycled properly.

PROBLEMS

1. A battery with an internal resistance of 0.015 Ω/cell needs to deliver 20-A current at 240 V to the load. Determine the number of cells in series if the electrochemical voltage is 2.0 V/cell.
2. Select the favorable cell from two options available in procuring 100-Ah battery cells. It is for a large battery assembly in an industrial plant, which

requires frequent charging and discharging. Cell A has an average cell voltage of 1.25 V during discharge and 1.45 V during charge and a c/d ratio of 1.15. Cell B has an average cell voltage of 1.3 V during discharge and 1.5 V during charge, and a c/d ratio of 1.2. Both cells A and B cost the same per ampere-hour rating.

3. A battery lasts 3000 charge/discharge cycles of 100% DoD. Determine its approximate cycle life at 30% DoD and at 20% DoD.
4. Design the battery for an industrial application that requires 20-kWh energy in 2 h for a dc load at 240 V four times daily for 300 days every year. The battery must be completely recharged in 2.5 h after each discharge. Use NiMH cells with the following cell parameters: average discharge voltage 1.2 V, average charge voltage 1.45 V, c/d ratio 1.15, trickle charge rate *C*/80, and average life 1500 cycles at 100% DoD. For the purpose of reliability, design the battery with two strings in parallel. To avoid one string charging the other, place an isolation diode with 1-V drop in the discharge path of each string. Assume that each cell has 15-mΩ internal resistance at the operating temperature. If the battery must last 7 calendar years before replacement, determine for each parallel string (1) the number of cells in series, (2) the discharge current, (3) the ampere-hour ratings of each cell, (4) the charge current, (5) the charge voltage, and (6) the trickle charge current.
5. A large 240-V battery is assembled from 2-V, 160-Ah lead–acid battery cells commercially available off the shelf. The maximum charging rate is *C*/4, and the charging rate during gas formation is *C*/20. Determine the air ventilation required in the battery room.
6. A new electrochemical cell has an average charge voltage of 4.5 V, an average discharge voltage of 3.0 V, and a c/d ratio of 1.15. Determine its round-trip energy efficiency.
7. The winter temperature of your car battery when parked outdoors on a street in Alaska can fall to −40°F at midnight. Determine the minimum SoC you must maintain in your battery to avoid electrolyte freezing.
8. A battery in your plant lasts 7 years when repeatedly discharged to 50% DoD. Determine its approximate life if the repeated DoD is increased to 75%.
9. Design a NiCd battery for a 10-year calendar life, which is required to (1) power 10-kW load at 240 V_{dc} for 1 h three times every day (once every 8 h) and (2) charge it back in 6 h, allowing 2-h margin for the battery to cool down after being fully charged and before the next load cycle starts. Assume average discharge voltage 1.2 V/cell, internal resistance 0.02 Ω/cell, and operating temperature 20°C.

   You must use three batteries in parallel to share full load and one dormant spare battery for reliability. With one failed battery, the other three must power full load. For each of the four batteries, determine the number of cells in series, cell capacity in ampere-hours, and the required charge current if the c/d ratio is 1.1 and the charge voltage is 1.45 V. Use a diode with a 1-V drop in each battery to prevent a stronger battery charging a weaker battery in case of an internal cell short or for other reasons. See Figure 11.8.

QUESTIONS

1. If a 100-Ah battery is being charged at *C*/5 rate, how long will it take to get fully charged?
2. If you have used or designed any type of battery, would you replace it with another electrochemistry, and for what reason?
3. Identify the battery type that has the highest volumetric and gravimetric energy densities (1) in the secondary battery group and (2) in the primary battery group.
4. What is the difference between charge efficiency and energy efficiency?
5. Describe the memory effect in the NiCd battery and the remedy for minimizing it.
6. Discuss the key performance requirement for a battery for an electric car.
7. When using two or more batteries in parallel, why should each battery have a separate charger and diode in the discharge path?
8. Name alternatives to a battery for large-scale energy storage, say, for a wind farm or a solar park.
9. Argue for and against the electric car in solving (1) the environment problem and (2) the energy problem.
10. Give your thoughts on the impact of the battery on the energy shortage problem many nations face today. Will it solve it or aggravate it?
11. Identify alternatives to the battery in (1) the marine industry, (2) the commercial industry, and (3) consumer electronics.

FURTHER READING

Nazri, G.A. and Pistoia, G. 2009. *Lithium Batteries Science and Technology*. Kluwer Publishing.

Reddy, T.B. and Linden, D. 2010. *Linden's Handbook of Batteries*. New York: McGraw Hill.

Part B

Power Electronics and Motor Drives

No other technology has brought a greater change in the electrical power industry, while still holding the potential of bringing future improvements, than power electronics. The power electronics equipment prices have declined to about 1/10 since the 1990s, fueling an exponential growth in their applications throughout the power industry.

The subject of *electronics* is broad, covering the following five classes of functionally different electronics:

(1) Power electronics, where solid-state semiconducting devices are used primarily as *on* or *off* switches to process and condition power—change voltage, current, or frequency—in order to increase the system's energy efficiency.
(2) Audio and video electronics in radio and TV circuits, where three-terminal semiconducting devices (transistors) are primarily used to amplify weak signals from the airwaves before feeding to the speaker coil, TV, or the display monitor screen.
(3) Operational amplifiers, which are used extensively in feedback control systems (servo controls).
(4) Digital electronics in computer-type equipment, where the semiconducting devices are used to register the presence or absence of signals (1 or 0 digit).
(5) Programmable logic controllers (PLC), which use digital electronics and relays to program discrete steps of operation in a logical sequence that is set in the program.

The above five branches of electronics are so different in operation and analysis that a course focusing on one branch would generally not cover the other four branches. In Part B of the book, we cover only power electronics.

Power electronics saves energy while offering great flexibility in electrical equipment operation. This is achieved by using semiconductor junction devices with two terminals (diode) or three terminals (controlled diode and transistor) as switches to turn the power on and off in the desired direction and at the desired time. The term *switch* in power electronics means a semiconducting device periodically switched at line frequency (60 or 50 Hz) or at high frequency up to hundreds of kilohertz. In theory—and only in theory—one can achieve the same results by using ordinary wall switches if one can switch them on and off by hand at the desired frequency for a long time. Even then, the wall switches would wear out (erode) under unavoidable sparking at the moving metal contacts. The power electronics switches, on the other hand, can switch power on and off at hundreds of kilohertz frequency for decades without wearing out. Since the semiconducting switches are solid in form, and were invented to replace hollow vacuum tubes, they are often called solid-state switching devices.

Part B includes Chapters 12 through 15 and starts with power electronics switches and then covers the dc–dc converters, ac–dc–ac converters, and frequency converters used in variable frequency motor drives. Chapter 16 discusses the quality of power, the importance of which has increased in modern power systems with many power electronics loads of relatively large size.

The equipment's power handling capability is limited by the temperature rise that must remain within the allowable limit. Therefore, understanding the cooling design is important to maximize the equipment power rating. Chapter 17 covers the power converter cooling by air and also by water, taking the power engineer into some important interdisciplinary topics.

12 Power Electronics Devices

The power electronics devices in electrical power systems basically perform the following functions by periodically switching the current *on* and *off* at a desired frequency: (1) convert ac into dc, (2) convert dc into ac, (3) convert frequency, (4) control ac and dc voltages, and (5) control power without converting voltage or frequency.

The direct or indirect benefit of a power electronics controller is system efficiency improvement. Figure 12.1 is a simple example in a process heat control, which can be achieved in two ways: (1) by inserting a control resistance (rheostat) or (2) by a control switch that is repetitively turned on for duration T_{on} and off for duration T_{off}. In the rheostat control, the average power going in the load is fraction $R_L \div (R_L + R_c)$ of that coming out of the source, and the remaining power is wasted in R_c, giving the energy efficiency of $R_L \div (R_L + R_c)$. In the switch control with an ideal switch, all energy coming from the source goes to the load, and the energy efficiency is 100%, although the average power going in the load is lower by the duty ratio $D = T_{on} \div (T_{on} + T_{off})$. Thus, the power electronics switch controls the load power without wasting energy in the control (dummy) resistor.

Example 12.1

A 240-V dc generator is powering a chemical process heater that has 20-Ω load resistance and produces 2880-W heat when fully on. A certain operation requires a partial heat of only 2160 W on average. Determine (1) the required series resistance (rheostat) value and the process energy efficiency and (2) switching duty ratio *D* and the energy efficiency if an ideal power electronics control is employed to reduce the heat.

Solution:

1. With the rheostat control of power, the ratio of the reduced to full powers gives $2160 \div 2880 = R_L \div (R_L + R_c) = 20 \div (20 + R_c)$, which leads to R_c = 6.667 Ω. Then, the energy efficiency = $20 \div (20 + 6.667) = 0.75$, which means 25% power is wasted in the rheostat.
2. With the power electronics control, the power ratio $2160 \div 2880$ = duty ratio $D = T_{on} \div (T_{on} + T_{off})$. The actual values of T_{on} and T_{off} can depend on the mass (thermal inertia) of the load. For small mass to be heated uniformly in time, T_{on} and T_{off} could be in seconds. For large mass, they can be longer. The energy efficiency of the switch control is always 100%, regardless of the switching times, meaning no energy is wasted in the switch or anywhere else.

Obviously, the power electronics control is energy efficient, saving 25% energy in this example.

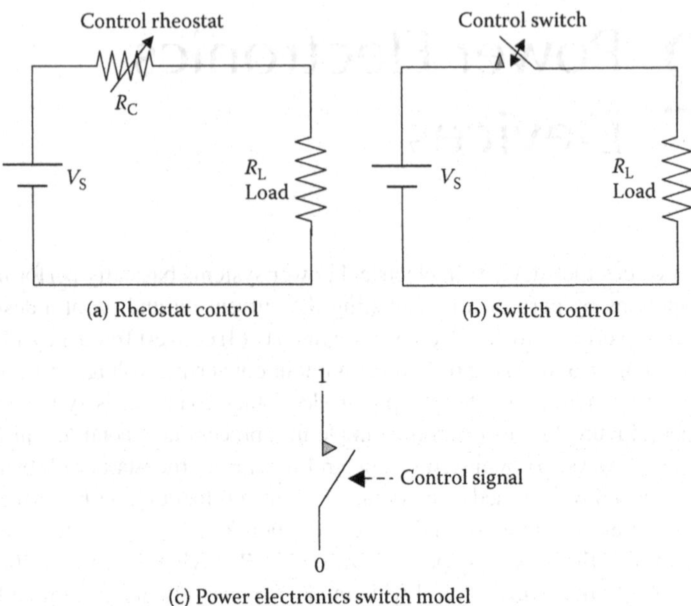

FIGURE 12.1 Power control by variable series resistor versus on–off switch.

The U.S. Department of Energy (DoE) estimates that over 100 billion kWh can be saved every year in the United States by using power electronics motor drives. At $0.10/kWh energy cost, that would save about $10 billion per year and would eliminate about eleven 1000-MW power plants in the country. A significant fraction of electrical power in the United States is already supplied through power electronics and perhaps higher in other countries where the energy cost is higher.

Semiconducting devices have discrete conducting and nonconducting states that depend on the direction of the prospective current flow and the bias signal—voltage or current—at its control terminal. They are made with one or more junctions of two differently doped semiconducting materials, known as p-type and n-type. The semiconducting devices when used to process high power (as opposed to signals) are called power electronics devices. In such applications, they are generally used only in two operating states, either *on* with near-zero resistance or *off* with a very large resistance across the main power terminals.

The diode is a two-terminal device, which automatically turns on (offering virtually zero resistance) when the p-terminal voltage is positive (higher) over the n-terminal voltage and is off (offering virtually infinite resistance) under the negative (lower) voltage at the p-terminal. The switchover from the *on* to *off* state is automatic and is not controllable by any signal. Other devices covered in this chapter use the diode as the building block, in which a third terminal is added to control (trigger) the on and off states by injecting a control signal at the control terminal. Thus, they virtually work as variable resistance that can be controlled ideally between zero (*on*) and infinity (*off*) by a small control signal applied at the third terminal. For this reason, the power electronics device in the electrical circuit diagram is often shown by a simple switch symbol between two power terminals and the third control terminal, as shown in Figure 12.1c. A variety of switching devices are available for use as controlled switches. However, a few basic switching devices are as follows:

- Thyristor or silicon-controlled rectifier (SCR)
- Gate-turn-off thyristor (GTO)
- Bipolar junction transistor (BJT)
- Insulated gate bipolar transistor (IGBT)
- Metal–oxide semiconducting field-effect transistor (MOSFET)
- Junction field-effect transistor (JFET)

A power electronics switching device of any type is triggered periodically on and off by a train of control signals of suitable frequency. The control signal may be of rectangular, triangular, or other wave shape and is generated by a separate triggering circuit, which is often called the firing circuit. Although it has a distinct identity with many different design features, it is generally incorporated in the main power electronics assembly.

Figure 12.2 shows the external construction features of these devices: (a) the diode with two power terminals; (b) the thyristor with three terminals—p-terminal with a large eye, n-terminal with threads to go on a grounded assembly plate, and a control (gate) terminal with a small eye; (c) the IGBT with three terminals; and (d) the gate triggering integrated circuit chip. The generally used circuit symbols of the six switching devices covered in this chapter are shown in Figure 12.3. A

(a) Diode (b) Thyristor (c) IGBT

(d) Trigger IC (e) Pressed-pack and studded thyristors

FIGURE 12.2 Power electronics devices' external construction features.

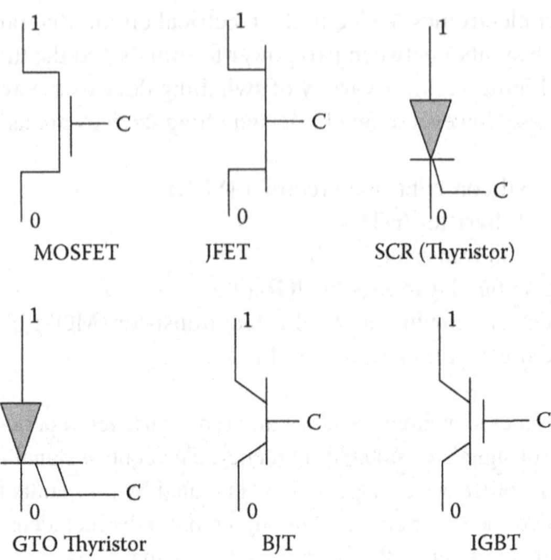

FIGURE 12.3 Circuit symbols of power electronics devices used as controlled switches.

common feature among these devices is that all are three-terminal devices. The two power terminals 1 and 0 are connected in the main power circuit. The control gate terminal C is connected to the auxiliary control circuit, called the firing circuit. In a normal conducting operation, terminal 1 is generally at a higher voltage than terminal 0. Since the device is primarily used for switching power on and off as required, it is functionally represented by a gate-controlled switch as shown in Figure 12.4a. In the absence of the gate control signal, the device resistance between the power terminals is large—the functional equivalence of an open switch. When the control signal is applied at the gate, the device resistance approaches zero, making the device function like a closed switch. The device in this state lets the current flow freely through its body. With the control signal on and off periodically, the device

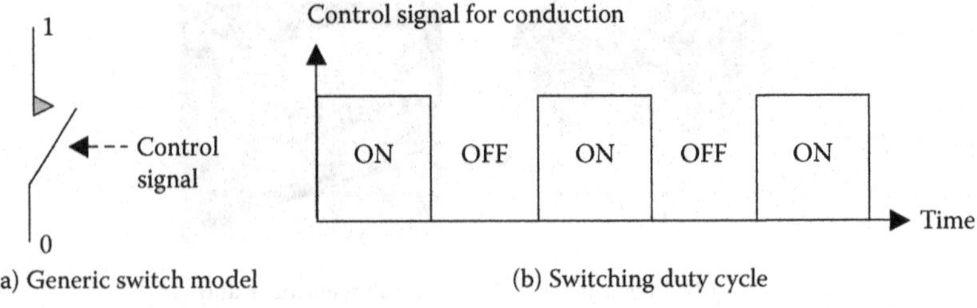

FIGURE 12.4 Power electronics device operation as periodic on–off switch.

conduction can be turned on and off at high frequency as shown in Figure 12.4b. With no moving contacts, such a virtual switch can be operated on and off trillions of times per day for decades with no wear at all, requiring no maintenance over the entire operating life.

The term *voltage versus current (v–i) characteristic* of the device is often used to describe the device performance at its own two power terminals, although the device makes only a part of a long circuit loop. To be more specific, it is the plot of voltage drop versus current at the device power terminals measured locally, as shown in Figure 12.5. The current would flow only if the power loop is complete (not shown here) and the device is conductive, or else it will have zero current even in a closed loop. The current in any closed loop is calculated by setting up the Kirchhoff voltage law (KVL) equation and solving it for the loop current. The KVL states that *in any closed loop, the sum of source voltages equals the sum of voltage drops in the loads.* The voltage drop v across the device at a given current i in formulating the KVL equation is that given by the v–i characteristic relation of the device, which is linear (Figure 12.5a) for the resistor and nonlinear (Figure 12.5b) for the power electronics devices. For this reason, the power electronics devices are called nonlinear devices, and the loads with power electronics devices are called nonlinear loads.

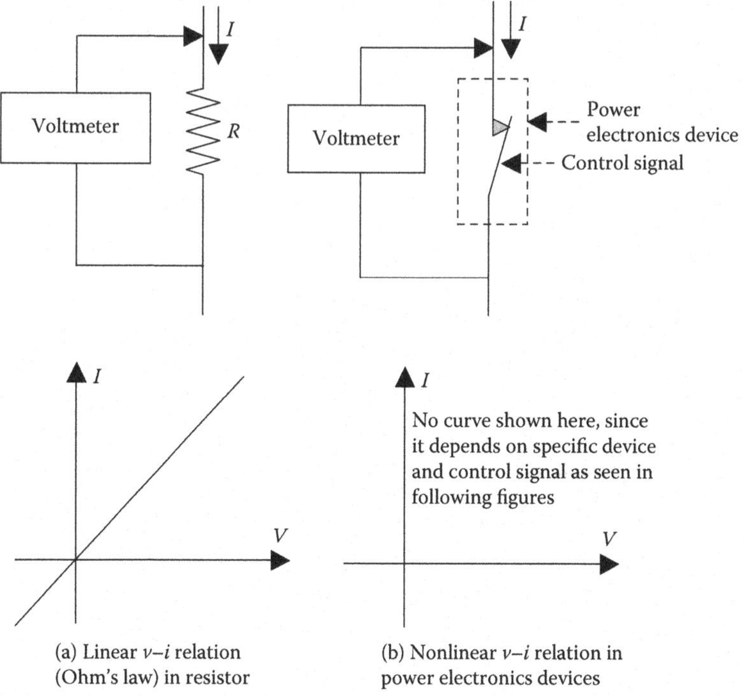

FIGURE 12.5 Local voltage drop versus current characteristic at device terminals.

12.1 DIODE

A diode is often used to rectify (convert) ac into dc and hence is also known as the *rectifier.* Although not a controllable switching device, the diode is the most basic semiconducting device that forms the building block for other devices. Made of p- and n-type semiconductors forming a p–n junction, it conducts current from P (anode) to N (cathode) terminals but not in the reverse direction. Figure 12.6 depicts the circuit symbol and the *v–i* characteristics of the ideal and practical diodes. It shows that the voltage drop in the ideal diode conducting current in the forward direction is zero regardless of the current magnitude. The practical diode, however, has a small voltage drop typically < 1 V. If the voltage of negative polarity is applied at the power terminals, that is, if the diode is reverse biased, the current in reverse direction is negligible until the junction breaks down at what is known as the *reverse breakdown voltage,* which is typically in hundreds of volts. After the breakdown, a large current flows with the reverse voltage drop approximately constant.

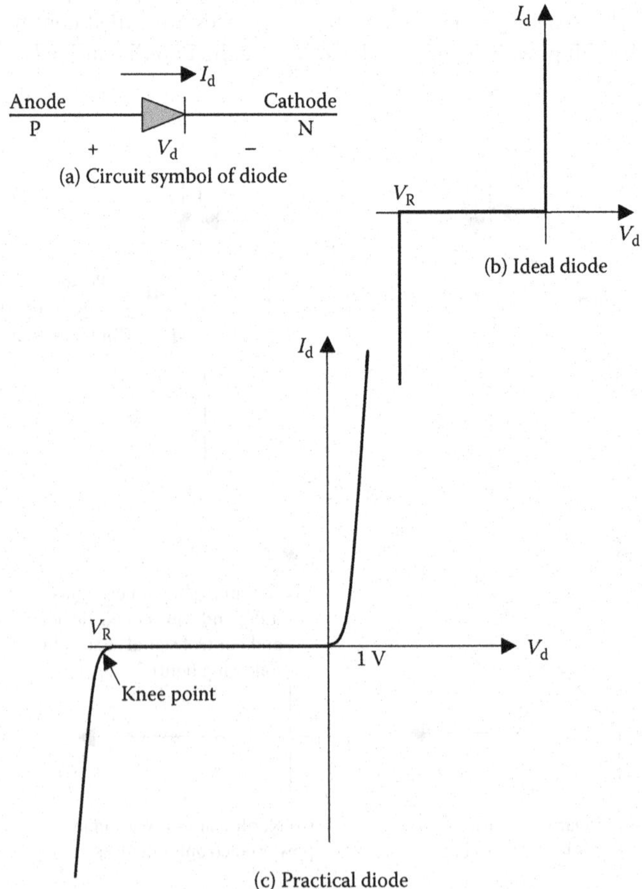

FIGURE 12.6 Ideal and practical diode voltage drop versus current characteristics.

The diode is rated in terms of the reverse voltage it can withstand without breakdown and the forward current it can carry without overheating. The diode is also rated by the switching time from the off state to the on state and vice versa. Large power diode ratings can be as high as 5000 V, 5000 A with small leakage current in the off state. However, high-power diodes with a large charge stored at the junction take a longer time to switch and hence are limited to relatively low-frequency applications, such as for rectifying up to 400-Hz ac into dc. Diodes in power electronics converters operating at switching frequency of 10 to 20 kHz require special high-speed diodes.

Fast switching diodes. For fast switching operations, the reverse recovery time of the diode must be short to minimize the power loss. Fast recovery diodes for high-frequency applications are available, but their power handling capability is lower due to higher power loss at higher switching frequency. The power-frequency (60 or 50 Hz) diodes are available up to 7 kV and 10 kA ratings, a few-kilohertz diodes up to 3000 V and 1500 A, a few-hundreds-of-kilohertz diodes up to 1000 V and 100 A, and Schottky diodes are available for high-frequency applications up to 1 MHz in ratings below 150 V and a few amperes.

Zener diode. This special diode permits current in the forward direction like a normal diode but also in the reverse direction if the voltage is greater than the reverse breakdown voltage, which is called the *zener knee voltage* or *zener voltage* V_z. The device is named after Clarence Zener who first discovered it. The conventional diode draws negligible current in reverse bias below the reverse breakdown voltage V_R. However, when the applied reverse bias voltage exceeds V_R, the conventional diode draws high current due to avalanche breakdown and generates large power loss equal to $V_R \times I$. The resulting heat would permanently damage the diode junction unless the avalanche current is limited by external circuit elements. The zener diode exhibits almost the same characteristic, except that the device is specially designed to have much reduced breakdown voltage (*zener voltage*). Moreover, it exhibits a controlled breakdown and allows the current just enough to maintain constant voltage drop across the terminals equal to the zener voltage V_z without thermal damage. For example, a zener diode with $V_z = 3.2$ V in Figure 12.7 will maintain a constant voltage of 3.2 V at the output terminals even if the input voltage (reverse-bias voltage applied across the zener) varies over a wide range above the zener voltage. In this way, the zener diode is typically used to generate a constant reference

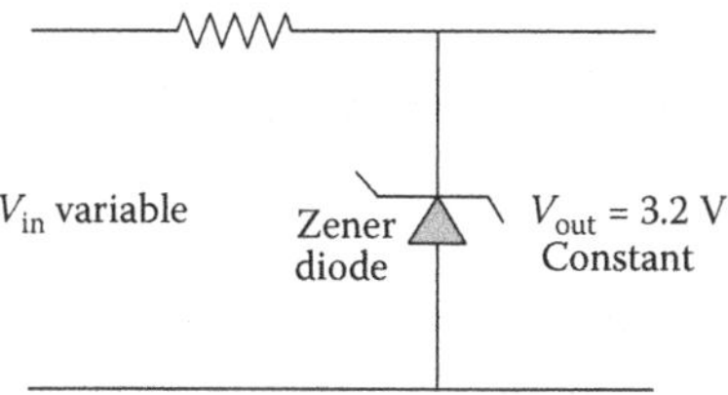

FIGURE 12.7 Zener diode circuit to derive constant 3.2-V reference voltage.

voltage for an amplifier stage or as a voltage stabilizer for low-current applications. The zener breakdown voltage V_z can be controlled quite accurately in the semiconductor doping process. The most widely used voltage tolerance is 5%, although close tolerances within 0.05% are available for precise applications.

Trigger (PNPN) diode. It blocks the current in both directions but can be made conducting with about 1-V drop in the forward direction if the applied forward voltage exceeds the forward breakdown value V_{BO} as shown in Figure 12.8. It turns off when the diode current falls below the holding current I_H, which is typically in several milliamperes.

Diac. The diac is made of two trigger diodes placed back to back. It conducts when the voltage exceeds the breakdown voltage in either direction, as shown in Figure 12.9, and remains conducting with small voltage drop across it until the current falls below the holding current I_H.

Example 12.2

Determine the current, voltage, and power loss in the diodes of circuits (a), (b), and (c) shown below, where all diodes have a forward voltage drop of 1 V and reverse breakdown voltage of 100 V.

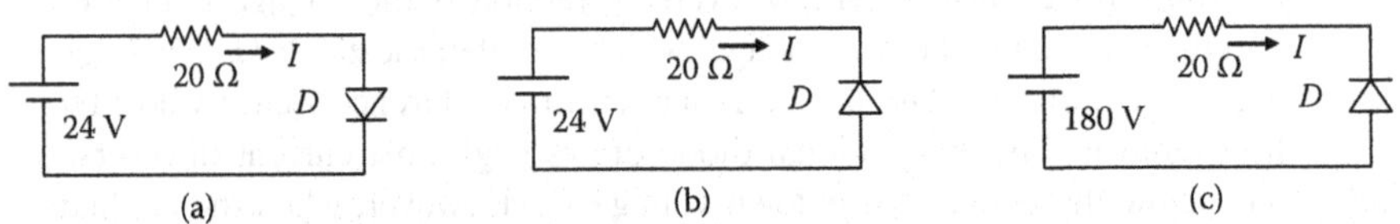

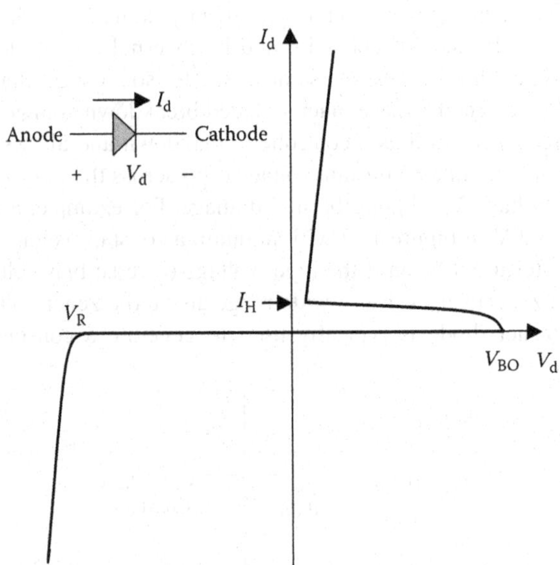

FIGURE 12.8 Trigger diode voltage drop versus current characteristic.

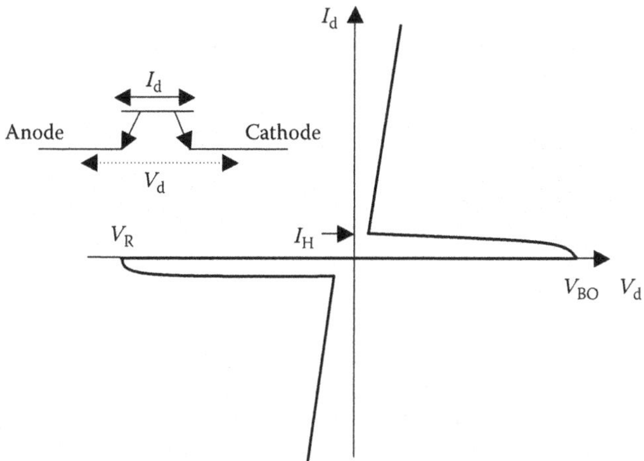

FIGURE 12.9 Diac voltage drop versus current characteristic.

Solution:

We first recall that (1) the voltage drop in the diode when it conducts in the forward or reverse direction is independent of the current magnitude and (2) the power loss in diode = voltage drop in diode × current through the diode.

In circuit (a), the diode is forward biased (in the conducting direction) and hence will conduct since the applied voltage exceeds the prospective voltage drop of 1 V. Using KVL, we derive the current $I = (24 - 1)\ \text{V} \div 20\ \Omega = 1.15\ \text{A}$. The voltage across diode = 1 V (regardless of the current magnitude), and power loss in diode = 1 × 1.15 = 1.15 W.

In (b), the diode is reverse biased, with the applied voltage less than the reverse breakdown voltage. Hence, the diode will not conduct ($I = 0$), the voltage across diode = 24 V (full source voltage since the voltage drop across the resistor is zero), and the power loss in diode = 0 × 24 = 0.

In (c), the diode is reverse biased, with the applied voltage greater than the reverse breakdown voltage. Hence, the diode will break down and conduct the current in the reverse direction (against the diode's symbolic arrow) with voltage drop of 100 V (again, regardless of the current magnitude). That gives $I = (180 - 100)\ \text{V} \div 20\ \Omega = 4\ \text{A}$, voltage drop across diode = 100 V, and the power loss in diode = 4 × 100 = 400 W. Such high power loss in a diode conducting in the reverse direction—as opposed to 1.15 W in the forward direction—will burn the junction almost instantly. Therefore, the circuit diodes are normally selected with sufficient margin to avoid reverse breakdown under normal operation and transient conditions.

Example 12.3

For the circuit shown below with ideal diodes (zero forward voltage drop and high reverse breakdown voltage), determine the average power absorbed by the resistor.

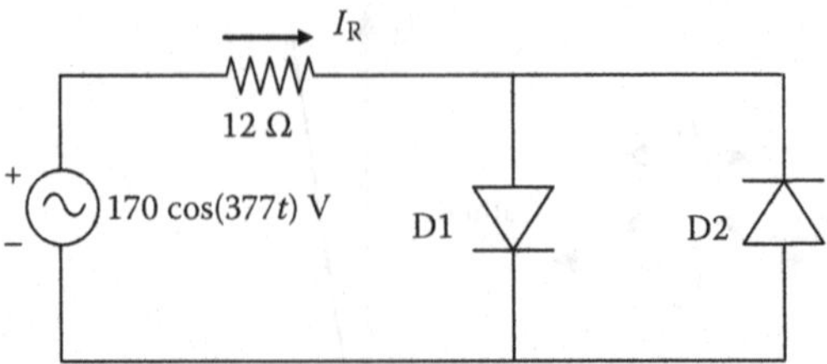

Solution:

When the ac voltage has positive polarity at the top as shown, diode D1 is forward biased and conducts, but D2 in the reverse bias does not conduct. With zero voltage drop in the ideal diode D1, the current in the resistor will be 170 cos(377*t*) ÷ 12 A, left- to right-hand side.

In the following 1/2 cycle, the ac voltage will have positive polarity at the bottom, and diode D2 will conduct but diode D1 will not. The current in the resistor will be – 170 cos(377*t*) ÷ 12 A, now right- to left-hand side.

Therefore, for the whole cycle, the resistor current will be a complete ac sine wave, as if two diodes do not exist in this circuit. The average power absorbed by the resistors is then the same as that in any ac circuit, that is,

$$\text{Pavg} = I_{rms}^2 R = \left(\frac{170}{\sqrt{2}\times 12}\right)^2 12 = 1204\ \text{W}.$$

12.2 THYRISTOR (SCR)

The thyristor is a three-terminal device derived from the diode by adding a control terminal. It is made of four PNPN layers with two power terminals—A (anode) and C (cathode)—and a control terminal G (gate). Figure 12.10 shows its circuit symbol and the terminal *v–i* characteristic. The thyristor is normally nonconducting in both directions but can be made conducting in the forward direction by applying a gate current. A larger gate current starts the conduction sooner. Once it is triggered in the forward conduction mode, its voltage drop is low as in the diode (<1 V), and it remains conducting until the current falls below the holding current I_H. In the reverse direction, it remains nonconducting until the reverse breakdown voltage is reached and then conducts current that is determined by the entire circuit loop. The thyristor is available in ratings from a few amperes to about 3000 A.

The thyristor is also known as the SCR since it is usually made using silicon diode (rectifier) and can control the conduction by a gate signal. The third terminal offers the control on making the device conducting or nonconducting, or delaying the start of conduction until reaching a desired phase angle (firing angle) on a sinusoidal voltage wave. A few other variations to the thyristor are available with somewhat varying operating characteristics as described below.

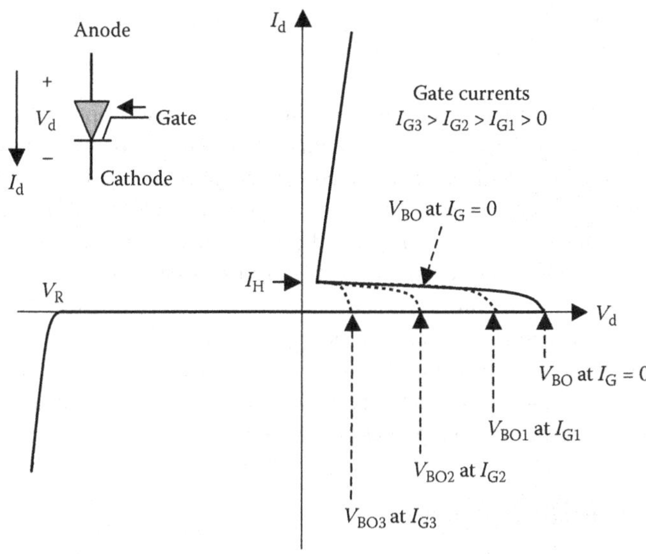

FIGURE 12.10 Thyristor (SCR) voltage drop versus current characteristic.

Triac. It has five semiconducting layers (PNPNP) and behaves like two PNPN thyristors placed back to back. Its circuit symbol and the *v*–*i* characteristic are shown in Figure 12.11. Because a single triac can conduct in both directions, it can replace a pair of two thyristors in many ac voltage control circuits and is widely used to step down the ac voltage by delaying the firing

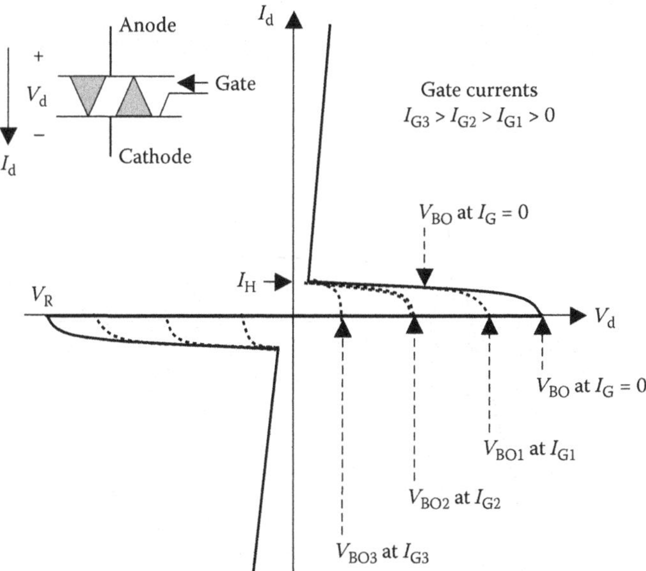

FIGURE 12.11 Triac voltage drop versus current characteristic.

angle. However, since its switching speed is slower than the thyristor, it is limited in use in low to medium power at 60 or 50 Hz, such as in lighting circuits.

GTO. The GTO is a recent introduction in the high-power applications. The thyristor, once triggered into the conduction, remains conducting until the control voltage is removed and the current is brought back below the holding current I_H (i.e., when the thyristor is commutated or quenched). That requires some special circuit—called the commutating circuit—that adds to the complexity and cost. The GTO eliminates the need for the commutating circuit. It can be turned off even from current exceeding I_H by applying a large-enough negative pulse (about 20% of the load current) at the gate terminals for about 20–30 μs. Many large GTOs have a hockey-puck (pressed-pack) shape and are water-cooled to remove large power loss produced internally while processing high power.

12.3 POWER TRANSISTOR

The power transistor (PTR) also has three terminals: C (collector), E (emitter), and B (base). The collector and emitter terminal names have continued from the vacuum

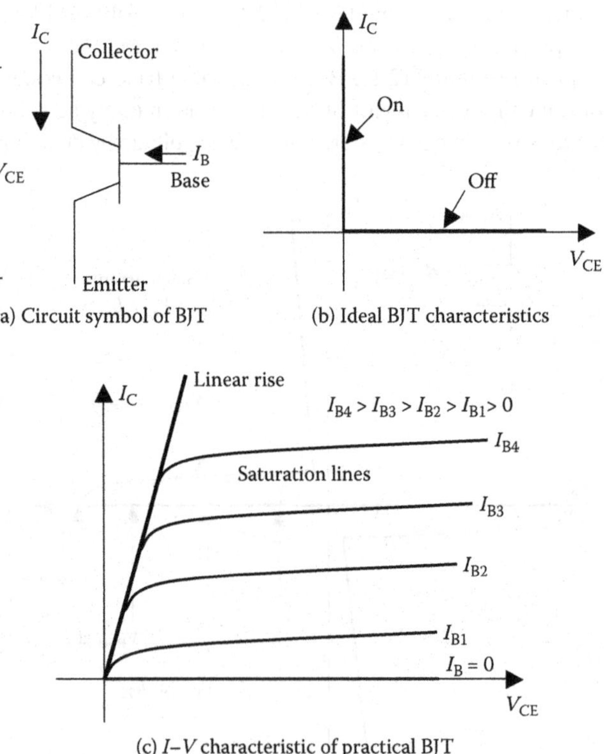

FIGURE 12.12 BJT voltage drop versus current characteristic.

tube that was replaced by the transistor. Among a variety of transistors available in the market, some commonly used devices are as follows.

BJT. The circuit symbol and the *v–i* characteristic of the BJT are shown in Figure 12.12. Even in the forward direction, the transistor remains nonconducting if the base current is zero. A base current triggers the transistor into the conduction mode and builds up the current until a certain value, beyond which the current remains constant (called the saturation current limit). A larger base current drives the transistor into a higher saturation current limit. Since a small base current signal can produce a large collector current, the transistor is widely used as an amplifier. However, in the power industry, it is used as a controlled switch that can be effectively turned on or off by the base current signal.

The BJT comes with NPN or PNP junctions. The NPN transistor has higher voltage and current ratings and hence is better suited for power electronics. It has low gain (I_C/I_B ratio less than 10), meaning that the base current must be more than 10% of the current to be switched, which may require a large base circuit. Bipolar transistors have lower power handling

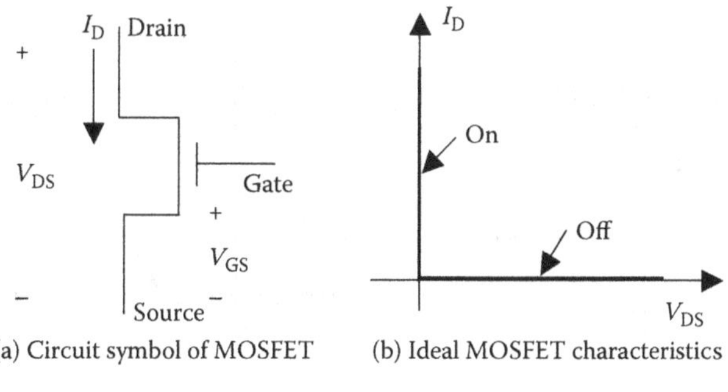

(a) Circuit symbol of MOSFET (b) Ideal MOSFET characteristics

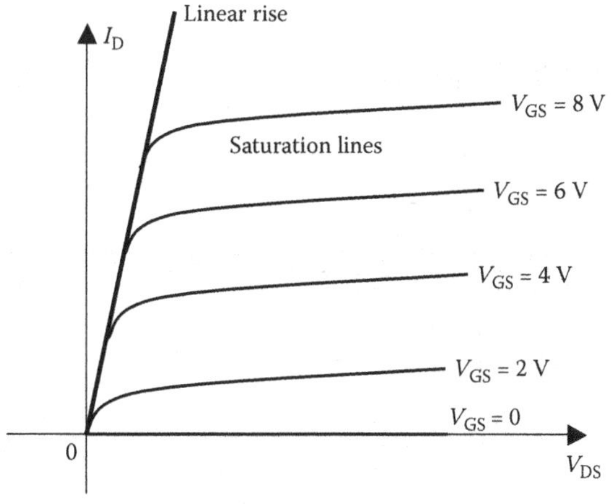

(c) *I–V* characteristic of practical N-channel MOSFET

FIGURE 12.13 MOSFET voltage drop versus current characteristic.

capabilities than thyristors in general but have good control characteristics and high frequency switching capabilities. Turning on and off the current in BJT is inherently slow due to its own internal stray inductance. Since BJTs are difficult to operate in parallel at high frequency above 10 kHz, they have been traditionally used in low-power applications in the past. However, the BJT has been largely replaced by MOSFET in low-voltage applications and by IGBT in high-voltage applications.

MOSFET. This device is controlled by voltage (<15 V in power MOSFET) rather than current, as shown in Figure 12.13. Its high gate impedance and low gate current make the drive circuit simple. It can operate above 100 kHz. The power capability is low in a few kilowatts, but it can be easily paralleled for greater power capability. MOSFETs are available in ratings up to 600 V with lower current or up to 200 A with lower voltage.

IGBT. The IGBT combines good characteristics of MOSFET and BJT. It is a relatively recent device that works on the voltage signal applied at the base and does not require the commutating circuit. Its circuit symbol and *v–i* characteristics are shown in Figure 12.14. Since turning on and off the voltage signals in the IGBT is faster, it has been used in high-power,

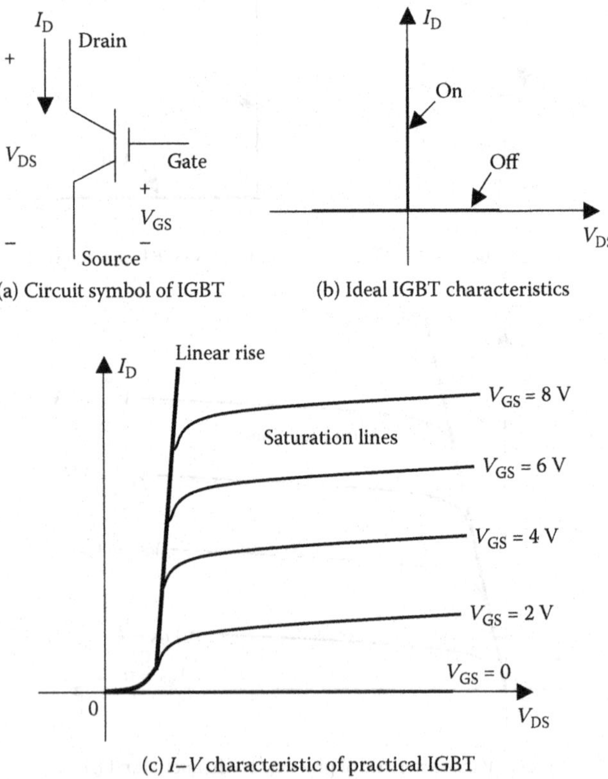

FIGURE 12.14 IGBT voltage drop versus current characteristic.

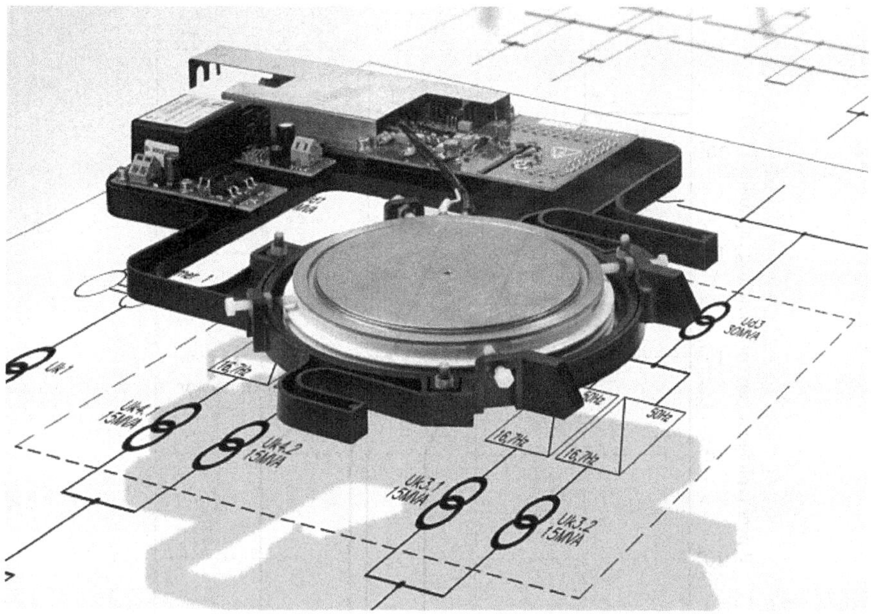

FIGURE 12.15 Large IGBT in hockey-puck (pressed-pack) shape assembled in base frame. (From Converteam Inc. With permission.)

high-frequency applications in motor drives up to several hundreds of horsepower. IGBTs are available from small to high ratings up to 1200 A at 330 V, 900 A at 4500 V, and some even higher ratings. Like GTOs, large IGBTs have a hockey-puck (pressed-pack) shape as shown in Figure 12.15. Large IGBTs are water-cooled to remove large internal power loss produced while processing high power.

12.4 HYBRID DEVICES

The thyristor technology has advanced into a variety of hybrid devices. The GTO and the static induction thyristor (SITH) turn on by a positive pulse to the thyristor gate and turn off by applying a negative pulse at the gate. Thus, they offer good forced-commutation techniques but require high firing power. The GTOs are available up to 4500 V/3000 A, which makes multimegawatt inverters possible for motor drives. SITHs are available up to 1200 V/300 A. Both GTOs and SITHs have high flexibility in design and can be easily controlled. SITHs have high frequency switching capability, way above GTOs that are limited to about 10 kHz. The SCRs and IGBTs, on the other hand, are limited to 100 Hz at present.

The gate commutated thyristor (GCT) and the integrated GCT (IGCT) are based on the GTO. They conduct like GTO but turn off like a transistor, which reduces switching power loss to about one-half compared to that in the GTO. Commercially

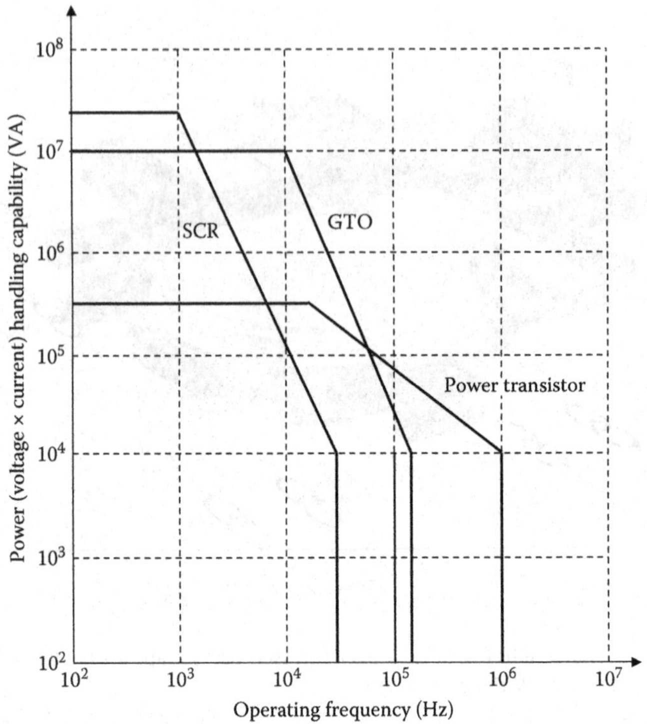

FIGURE 12.16 Power handling capability versus switching frequency of various power electronics devices.

developed around 2000, the GCT has started to replace GTO in high-power applications.

The MOS-controlled thyristor (MCT), reverse-conducting thyristor (RCT), gate-assisted turn-off thyristor (GATT), and light-activated SCR (LASCR) are other examples of specialty devices, each having niche advantage and application.

Figure 12.16 shows the power handling versus switching frequency capabilities of the most commonly used power electronics devices available to the design engineer at present, and higher rated devices are emerging in an evolutionary manner.

12.5 d*i*/d*t* AND d*v*/d*t* SNUBBER CIRCUIT

The voltage drop across two terminals of a current carrying inductor L is $v = L \times \mathrm{d}i/\mathrm{d}t$. Therefore, switching the current on and off in a short time in a high frequency switching circuit gives high d*i*/d*t* that causes high voltage stress on the junction, which may damage the device unless suppressed by a snubber circuit.

The voltage drop across two terminals of a current carrying capacitor C is $v = \frac{\int i \cdot \mathrm{d}t}{C}$, the differentiation of which leads to $i = C \times \mathrm{d}v/\mathrm{d}t$. Therefore, switching the voltage on and off in a short time in a high frequency switching circuit gives high dv/dt that causes high current stress on the junction, which may damage the device.

Moreover, both high di/dt and high dv/dt can cause high electromagnetic interferences (EMI) in the neighboring circuits.

When a firing signal is applied at the gate terminal of the thyristor and gate-turn-off thyristor-type (GTO-type) devices under forward voltage bias, the conduction across the junction starts in the immediate neighborhood of the gate connection and spreads from there across the whole area of the junction. A high rate of rise in anode current—that is, high di/dt—results in high current density around the junction area. This may cause a local hot spot, thermally damaging the junction and making the thyristor inoperative. The maximum di/dt a thyristor can withstand without damage is specified by the manufacturer. The converter design engineer must limit the di/dt on the anode current by inserting a small inductor in series with the thyristor, as shown in Figure 12.17. Recall that the inductor slows down the di/dt in order to keep its magnetic energy constant.

If a sinusoidal voltage is applied to the thyristor, the minimum required inductor to limit the di/dt below the maximum permissible value can be derived from the $v = L \times \mathrm{d}i/\mathrm{d}t$ relation, which gives

$$L_{\min} = \frac{V_{\text{peak(anode)}}}{(\mathrm{d}i/\mathrm{d}t)_{\text{max(allowed)}}} \text{ henrys.} \tag{12.1}$$

The thyristor dv/dt rate is controlled by placing a capacitor in parallel with the thyristor, as shown in Figure 12.17, where a small series resistance R_s is placed to protect the capacitor from high inrush charging current. Under high dv/dt, which is virtually a high-frequency voltage, the snubber capacitor branch offers much lower impedance, which draws most of the inrush current diverting away from the

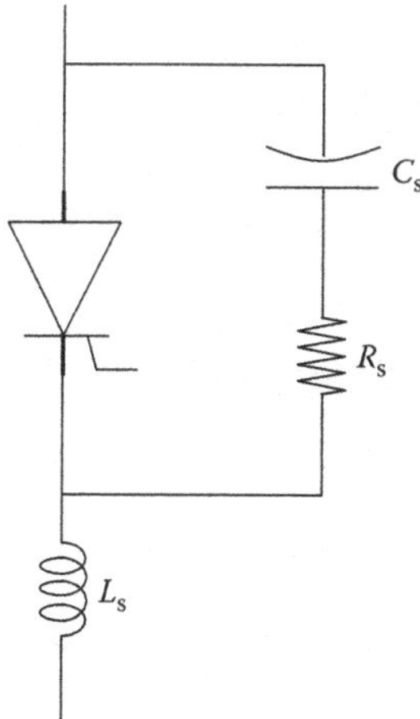

FIGURE 12.17 Snubber circuit for di/dt and dv/dt suppression as part of thyristor power circuit.

thyristor. The required value of such a capacitor is determined from the $i = C \times dv/dt$ relation. It is also related with L_s.

Figure 12.17 is a typical snubber circuit for the thyristor or a transistor. Its main purpose is to protect the device from a large rate of change in voltage and current. The series inductor limits d*i*/d*t*, and the parallel capacitor branch limits d*v*/d*t*. Typical values of L_s, R_s, and C_s are in a few millihenrys, a few ohms, and a few tenths of microfarads, respectively. The power loss in R_s decreases the efficiency of the converter, and hence, the efficiency is traded for the level of d*i*/d*t* and d*v*/d*t* protection desired.

Another problem associated with the thyristor and gate-turn-off thyristor-type (GTO-type) devices is the high rate of rise in the forward voltage. High d*v*/d*t* applied at the anode terminal turns on the thyristor even in absence of the gate firing signal. Possibility of such anomalous conduction due to stray d*v*/d*t* signals from neighboring power equipment's radiated electromagnetic interference (EMI) is one reason why thyristors are not used in spacecraft with an extremely high reliability requirement, where the probability of malfunction in each device must be less than one in billions.

Example 12.4

A thyristor in a certain 440-V converter design can see a maximum forward voltage of 254 V_{LNrms}. Its data sheet specifies a maximum allowable d*i*/d*t* of 100 A/μs. Determine the snubber inductor you must include in the converter design to limit the d*i*/d*t* stress to 70 A/μs to allow for a desired margin.

Solution:

To limit the d*i*/d*t* to 70 A/μs, that is, 70 × 10⁶ A/s, we use Equation 12.1 to obtain

$$L_{min} = 254 \times \sqrt{2} \div (70 \times 10^6) = 5.13 \times 10^{-6} \text{ H or } 5.13\ \mu\text{H}.$$

Example 12.5

A power electronics device turns off 50 A in 200 μs and builds up the voltage from zero to 110 V in 220 μs. Determine the d*i*/d*t* and d*v*/d*t* stresses on the device.

Solution:

We must first be clear about the definition of change, which is defined as

$$\text{change } \Delta = (\text{new value} - \text{old value})$$

$$\text{50-A turn-off means } \Delta i = (0 - 50) = -50 \text{ A in } \Delta t = 200\ \mu\text{s}$$

$$\text{100 V buildup means } \Delta v = (110 - 0) = +110 \text{ V in } \Delta t = 220\ \mu\text{s}$$

Therefore, d*i*/d*t* = – 50/200 = – 0.25 A/μs and d*v*/d*t* = +110/220 = + 0.50 V/μs.

12.6 SWITCHING POWER LOSS

The power electronics switch when turned on or off encounters some power loss. The converter design engineer chooses a device with low power loss and provides adequate cooling to maintain the junction temperature below the allowable limit. The power loss in any switch can be estimated from a generic diagram, shown in Figure 12.18, where the switch is turned on by the control signal in time T_{swon} and off in time T_{swoff}. On the *on* signal, the switch current takes time t_{ir} to rise to full value, and the voltage takes time t_{vf} to fall to zero, making $T_{swon} = t_{ir} + t_{vf}$. On the *off* signal, the voltage rises to full value in time t_{vr}, and the current falls to zero value in time t_{if}, making $T_{swoff} = t_{vr} + t_{if}$. The power loss in the switch is the product of the voltage across the switch and the current through the switch, and the energy loss = power × time duration, which is shown by the shaded area. To sum up the total energy loss, we have the following.

Switching energy loss in two hump areas:

$$E_{sw} = 1/2\ V_{oc} \times I_{on} \times T_{swon} + 1/2\ V_{oc} \times I_{on} \times T_{swoff}. \tag{12.2}$$

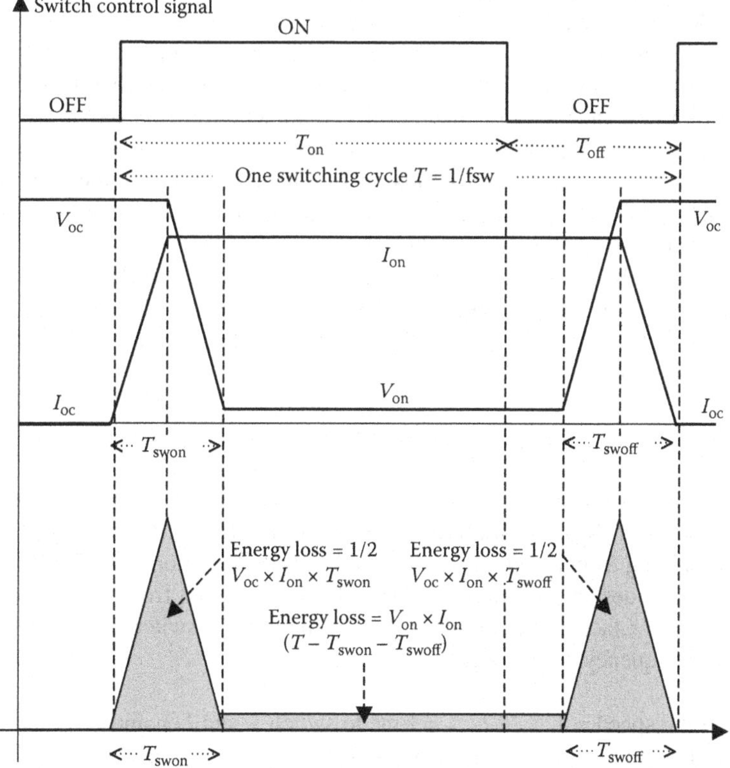

FIGURE 12.18 Energy losses during rising voltage and falling current in one switching period of fast switch (for worst-case analysis).

Switching repetition period $T = T_{on} + T_{off}$, and switching frequency $f_{sw} = 1/T$. (12.3)

Switching power loss in fast switch:

$$P_{sw} = E_{sw}/T = \frac{1}{2} V_{oc} \times I_{on} \times (T_{swon} + T_{swoff}) \times f_{sw}. \tag{12.4}$$

Other power losses occur during conduction on and off. The power loss during the on-state, shown in the narrow long shaded area, amounts to the average value of

$$P_{on} = V_{on} \times I_{on} \times (T - T_{swon} - T_{swoff}) \div T = V_{on} \times I_{on} \times D \tag{12.5}$$

where D = duty ratio, the fraction of cycle time the conduction is on. Here, we have ignored the switching on and off times since they are very short compared to the switching repetition period T.

$$\text{Power loss during off-state } P_{off} = V_{oc} \times I_{leak} \times (1 - D). \tag{12.6}$$

The value of P_{off} is usually negligible and is ignored in most practical calculations.

We note that P_{sw} and P_{on} vary directly with the switching frequency and the switching times. Therefore, the device with short switching time and low on-state voltage V_{on} minimizes the switching power loss that allows high-frequency operation.

Equation 12.2 and the subsequent results are for a fast turn-on and turn-off switch with short T_{swon} and T_{swoff} times. In slow switches, the rise time and fall time overlap. The voltage falls as the current builds up simultaneously during the turn-on time, and the current falls as the voltage builds up simultaneously during the turn-off time. The hump area, therefore, has height only one-half of that in a fast switch and the shape of one-half sine wave, which has a shaded area equal to Peak × 2/π. The net result is reduced switching energy loss that is given by $E_{sw} = (2/\pi)\ \{1/2 V_{oc} \times 1/2\ I_{on} \times T_{swon} + 1/2\ V_{oc} \times 1/2\ I_{on} \times T_{swoff}\}$, which is about 1/3 of that in a fast switch. Therefore, for a slow switch

$$P_{sw} = E_{sw}/T = 0.16\ V_{oc} \times I_{on} \times (T_{swon} + T_{swoff}) \times f_{sw}. \tag{12.7}$$

The switch with switching speed between the two extreme cases—fast and slow—discussed above would have the factor in the front of switching power loss formula ranging from a high of 0.50 to a low of 0.16. Equation 12.4, therefore, gives a conservative design until the actual switching speed data from the device vendor support a lower factor. It is important to distinguish the switching speed from the switching frequency:

switching speed = $T_{swon} + T_{swoff}$ = time to switch V and I on and off = duration of the humps (usually in microseconds)

switching frequency = $f_{sw} = 1/T$ = number of times the voltage and current get switched on and off in 1 s (usually line frequency to 500 kHz in some high-frequency converters).

The switching times come from the device vendor data sheet, whereas the switching frequency is set by the design engineer, and the switching power loss depends on both.

Example 12.6

In the circuit below, the power electronics switch is operating at 1000-Hz switching frequency with a 60% duty ratio. The switch has on-state voltage drop of 1 V, off-state leakage current of 1.5 mA in reverse bias, switch-on time of 1.2 μs, and switch-off time of 1.5 μs. Determine the average power loss in the switch.

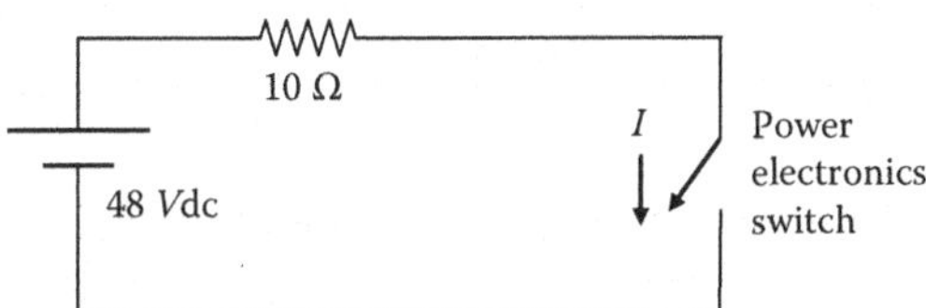

Solution:

We use formulas developed in Section 12.6. One cycle takes 1/1000 = 0.001 s or 1 ms. With 60% duty ratio, the switch conducts for 0.6 ms and is off for 0.4 ms. When it conducts, the current will be (48 – 1) ÷ 10 = 4.7 A with a forward voltage drop of 1 V. In the off state, the voltage across the switch will be equal to the applied voltage of 48 V and the leakage current 1.5 mA.

Since 1000 Hz is a relatively low switching frequency, we use Equation 12.7 to obtain

switching power loss $P_{sw} = 0.16 \times 48 \times 4.7 \times (1.2 + 1.5)10^{-6} \times 1000 = 0.0975$ W.

Using Equation 12.5, power loss during on-time $P_{on} = 1 \times 4.7 \times 0.6 = 2.82$ W.

Power loss during off-time $P_{off} = 48 \times 0.0015 \times 0.4 = 0.0288$ W (negligible, 1% of P_{on}).

Power loss averaged over the cycle = 0.0975 + 2.82 + 0.0288 = 2.95 W.

If this switch were operated at very high frequency, say, 100 kHz, Equation 12.4 would apply for

$$P_{sw} = 1/2 \times 48 \times 4.7 \times (1.2 + 1.5)10^{-6} \times 100{,}000 = 30.5 \text{ W},$$

which is about 300 times 0.0975 W loss at 1000 Hz. This is clearly high, which makes this switch unsuitable for such high switching frequencies.

The on-state power loss in the transistor is always on as long as the power terminals are under voltage, although it becomes conducting only when the base signal is present. It does not multiply by the duty ratio (i.e., conducting/nonconducting time ratio).

12.7 DEVICE APPLICATION TRENDS

The device selection for a specific application depends on the voltage, current, and frequency requirements of the system to be designed. The characteristics of the switching devices presently available to the design engineer are listed in Table 12.1. The maximum voltage and current ratings with unique operating features of thyristors and transistors widely used in high-power applications are listed in Table 12.2.

The thyristor used as the switching device requires the current commutation (turn off the device) when its anode–cathode voltage is reversed. The thyristor current in a line-frequency (60 or 50 Hz) load-commutated converter is naturally reverse-biased and turns off when the next thyristor is gated on. However, in other—such as pulse width modulated (PWM)—converters, it requires a separate forced-commutation

TABLE 12.1
Characteristics of Power Electronics Semiconductor Device

Type	Function	Voltage (V)	Current (A)	Upper Frequency (kHz)	Switching Time (μs)	On-State Resistance (mΩ)
Diode	High power	6500	8000	1	100	0.1–0.2
	High speed (50 kHz)	3000	1500	10	2–5	1
	Schottky (1 MHz)	< 150	< 100	20	0.25	10
Forced turned-off thyristor	Reverse blocking	5000	5000	1	200	0.25
	High speed	1200	1500	10	20	0.50
	Reverse blocking	2500	400	5	40	2
	Reverse conducting	2500	1000	5	40	2
	GATT	1200	400	20	8	2
	Light triggered	6000	1500	1	200–400	0.5
TRIAC	Back-to-back thyristors	1200	300	1	200–400	3–4
Self-turned-off thyristor	GTO	4500	3000	10	15	2–3
	SITH	1200	300	100	1	1–2
Power transistor	Single	400	250	20	10	5
		400	40	20	5	30
		600	50	25	2	15
	Darlington	1200	400	10	30	10
Power MOSFET	Single	500	10	100	1	1
		1000	5	100	1	2
		500	50	100	1	0.5
IGBTs	Single	1200	400	100	2	60
MCTs	Single	600	60	100	2	20

TABLE 12.2
Maximum Voltage and Current Ratings of Power Electronics Switching Devices

Device	Voltage Rating (V)	Current Rating (A)	Operating Features
BJT	1500	400	Requires large current signal to turn on
IGBT	1200	400	Combines the advantages of BJT, MOSFET, and GTO
MOSFET	1000	100	Higher switching speed
SCR	6000	3000	Once turned on, requires heavy turn-off circuit

circuit, which adds to the converter cost and power loss. High-power BJT, IGBT, and GTO, which do not require commutation circuits, have replaced the switchmode thyristor converter in new designs. The GTO is a new technology that has the commutation capability by a gate signal in the reverse. It has been implemented successfully in many high power applications. One particular attraction of this technology is the consequent ability to select the induction motor in favor of the synchronous motor where beneficial. GTOs are currently available for power levels over 10 MW and can be connected in parallel for even higher power levels. However, the controlling electronics for device turn-on and turn-off cost more.

The IGBT is now widely used in new designs due to its faster switching frequency, better controllability, lower power losses, and simpler and more compact control circuit than possible with GTOs. However, most are available for relatively small prepackaged drive applications (up to a few megawatts) with principal limitation in the voltage rating up to about 1500 V rms.

12.8 DEVICE COOLING AND RERATING

The device junction temperature must be below the allowable limit to avoid junction failure. This is difficult considering that the operating heat is generated in a small wafer-thin volume around the junction way inside the device. The design power rating of any device is limited by the permissible temperature rise of the junction, which depends on the cooling method and the temperature of the ambient air where all the switching power loss is ultimately dissipated. Lower ambient air temperature, therefore, allows greater power handling capability, and vice versa. The device vendor provides the nominal rating at the industry-standard ambient air temperature of 40°C. If the device is operating in a higher temperature ambient, it cannot handle the vendor-rated power and must be derated accordingly. On the other hand, it can handle higher than the vendor-rated power in ambient cooler than 40°C. The device rerating (derating or uprating) factor for the ambient air deviating from 40°C is shown in Figure 12.19.

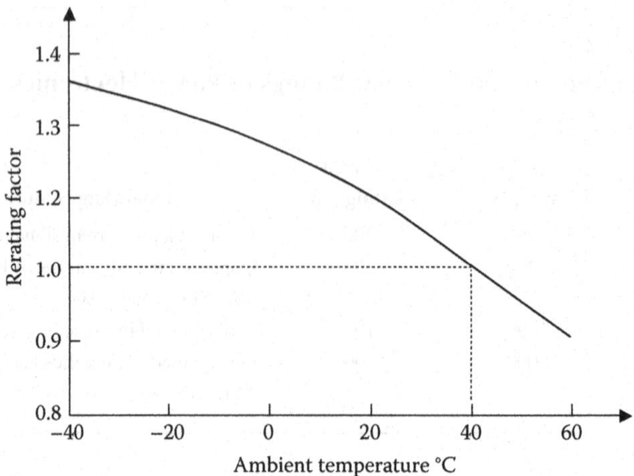

FIGURE 12.19 Device rerating factor versus ambient air temperature.

PROBLEMS

1. An industrial process heater delivers power from a 480-V dc generator to a 30-Ω load resistance that produces 7680-W heat when fully on. A reduced operation requires only 4224 W of average heat generation rate. Determine (1) the required series rheostat resistance and the process energy efficiency and (2) the switching duty ratio D and the energy efficiency if an ideal power electronics control is employed.
2. Determine the current, voltage, and power loss in the diodes of circuits (a), (b), and (c) shown in the figure below, where all diodes have a forward voltage drop of 0.8 V and reverse breakdown voltage of 50 V.

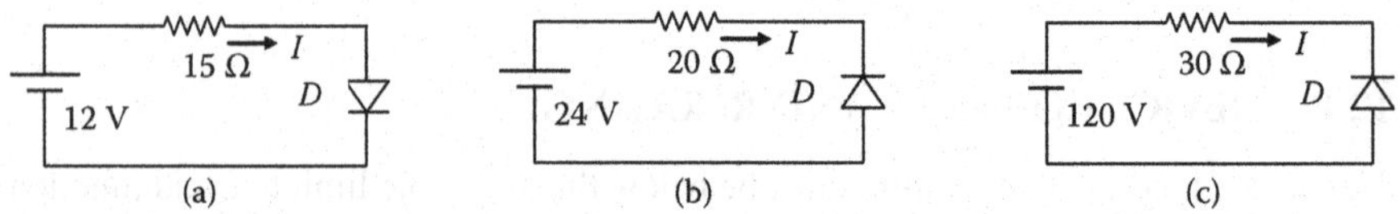

3. With ideal diodes (zero forward voltage drop and high reverse breakdown voltage) in the figure below, determine the average power absorbed by the resistor.

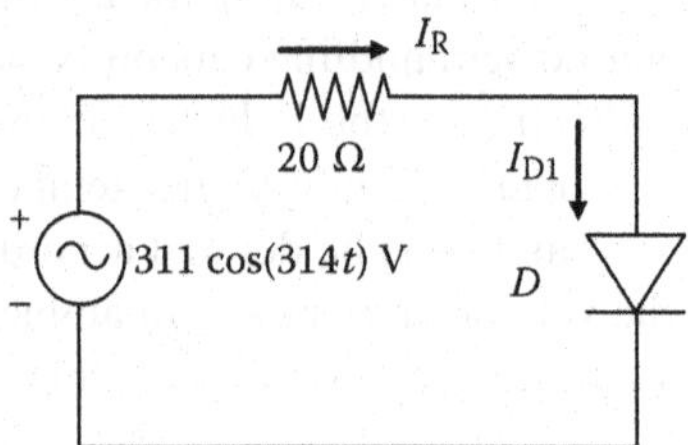

4. A thyristor in a certain 240-V converter design can see a maximum forward voltage of 120 V_{LNrms}. Its data sheet specifies a maximum allowable di/dt of 80 A/μs. Determine the snubber inductor you must include in the converter design to limit di/dt to 60 A/μs to allow for the desired margin.
5. A power electronics device builds up the voltage from 0 to 120 V in 180 μs and turns off 100 A in 250 μs. Determine the dv/dt and di/dt stresses on the device.
6. In the circuit shown below, the power electronics switch is operating at 2000-Hz switching frequency with a 50% duty ratio. The switch has an on-state voltage drop of 0.8 V, off-state leakage current 2 mA in reverse bias, switch-on time 1.5 μs, and switch-off time 2.0 μs. Determine the average power loss in the switch.

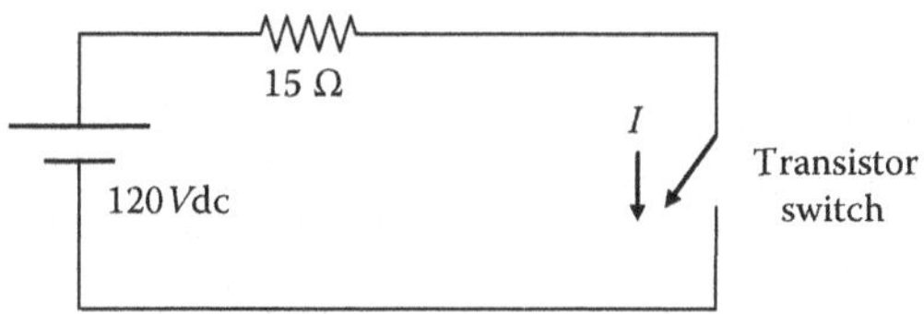

QUESTIONS

1. Explain the difference between the diode and the thyristor in construction and in operation.
2. In what two different ways can the transistor be used in electronic circuits?
3. Differentiate the terms *switching speed* and *switching frequency*.
4. High di/dt causes what, high voltage stress or high current stress? Why?
5. High dv/dt causes what, high voltage stress or high current stress? Why?
6. When would you need to rerate (derate or uprate) a device and for what reason?

FURTHER READING

Ahmed, A. 1999. *Power Electronics for Technology.* Upper Saddle River: Prentice Hall.

Ang, S. and Oliva, A. 2010. *Power Switching Converters.* Boca Raton: CRC Press/Taylor & Francis.

Hart, D. 2011. *Power Electronics.* New York: McGraw Hill.

Mohan, N., Undeland, T.M., and Robbins, W. 2003. *Power Electronics Converters, Applications, and Design.* New York: John Wiley & Sons/IEEE Press.

13 DC–DC Converters

The power electronics devices covered in Chapter 12 make the building blocks of a variety of power electronics converters for changing ac and dc voltage level or changing frequency at the interface of the power mains and the load. This chapter covers dc–dc converters, whereas ac–dc–ac converters are covered in the next chapter. The dc–dc converter is needed to change voltage for load equipment or for charging and discharging a battery in the dc power distribution system. Since such a converter achieves the desired function by switching the dc power *on* and *off* (chopping the power) at high frequency, it is also known as the *dc chopper* or *switchmode power converter.* The following sections describe the circuit topology and the operational characteristics of various types of the dc–dc converter.

13.1 BUCK CONVERTER

The buck converter steps down the input voltage and hence is also known as the *step-down converter.* Figure 13.1a is the most widely used buck converter circuit. The switching device used in this converter may be the bipolar junction transistor (BJT), metal–oxide semiconducting field-effect transistor (MOSFET), or insulated gate bipolar transistor (IGBT). The switch is turned on and off periodically at high frequency, typically in tens of kilohertz as shown in Figure 13.1b. The duty ratio D of the switch is defined as

$$\text{duty ratio } D = \frac{\text{time on}}{\text{period}} = \frac{T_{\text{on}}}{T} = T_{\text{on}} \cdot \text{switching frequency}\cdot \tag{13.1}$$

Since the buck converter is widely *used* to charge the battery, it has yet another name, the *battery charge converter.* It is required to buck (step down) the dc bus voltage to the battery voltage during charging. Its operation during one on and off period triggering signal is shown in Figure 13.2. During the on-time, the switch is closed and the circuit operates as in Figure 13.2a. The dc source charges the inductor and capacitor, in addition to supplying power to the load. During the off-time, the switch is open, and the circuit operates as in Figure 13.2b. The power drawn from the dc source is zero. However, full load power is supplied by the energy stored in the inductor and the capacitor, with the diode carrying the return current. Thus, the inductor and the capacitor provide short-time energy storage to ride through the off period of the switch. The load current during this period is known as the freewheeling current, and the diode is known as a freewheeling diode. The voltage and current waveforms over one complete cycle are displayed in Figure 13.3. The line voltage is either fully turned on or off. The inductor current decays during off-time and rises during on-time, and so do the switch current and the diode current. Either the switch or the diode carries the inductor current as shown in the last two waves. A suitable

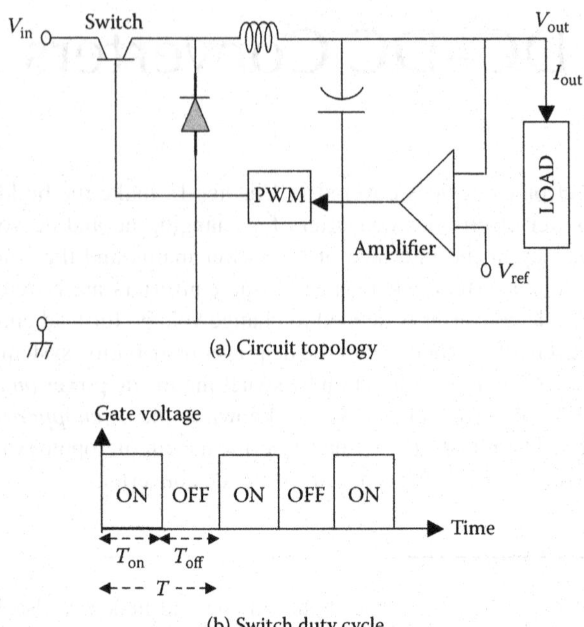

FIGURE 13.1 Buck converter circuit and switch duty ratio.

bleeding (dummy) resistor at the load terminals is sometimes incorporated in the design to keep the converter working without the load.

The analytical principles detailed below for the buck converter are applicable also for other converters described in this chapter. The analysis is primarily based on the energy balance over one switching period, that is,

energy supplied to load over total period T = energy drawn from source during on-time, which is the only energy fed to the circuit

energy supplied to load during off-time = energy discharged from inductor and capacitor during off time, which equals the energy deposited in the inductor and capacitor during on-time

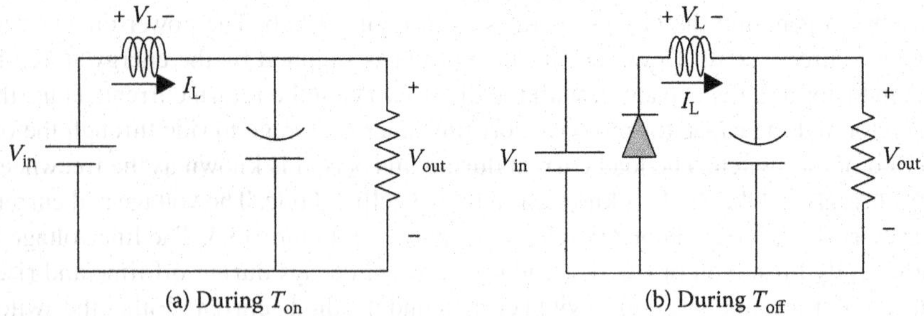

FIGURE 13.2 Buck converter operation during on and off periods of switch.

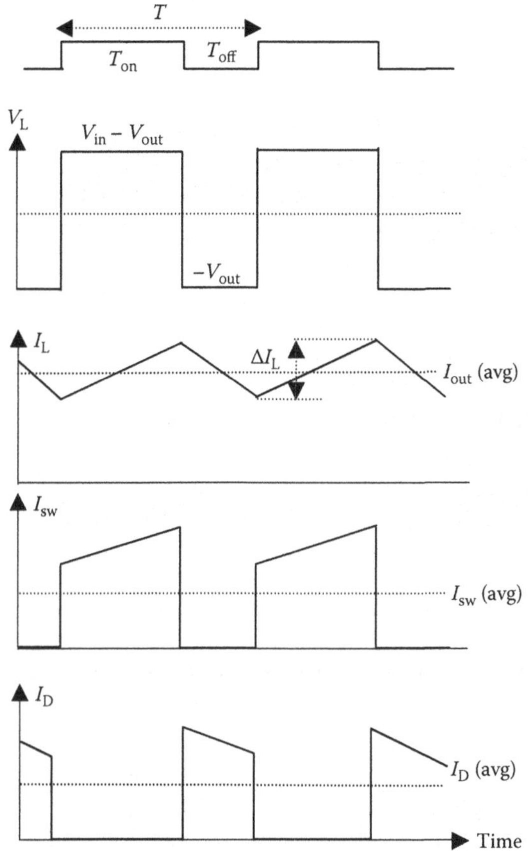

FIGURE 13.3 Current and voltage waveforms in buck converter operation.

The freewheeling diode circulates the load current through the inductor, which depletes the energy during off time and then refurbishes the energy during on-time. Therefore, the performance analysis is based on the energy balance in the inductor during one cycle, since the inductor draws power from the source during on-time. The inductor stores energy in the magnetic flux and generally uses magnetic core with a lumped or distributed air gap. The flux density in the core has to be kept below the magnetic saturation level. For this reason, the net change in flux over one cycle of switching has to be zero. Otherwise, the core would eventually *walk away* to the magnetic saturation on the high side or get depleted of energy on the low side. The magnitude of the voltage drop V_L across the inductor is given by Faraday's law, that is, $V_L = N\, d\varphi/dt$, where N = number of inductor turns, φ = flux in the inductor core, and t = time. It can also be written as $V_L \cdot dt = N \cdot d\varphi$. Since the inductor flux must maintain balance over one cycle in the steady state, we must have $d\varphi = 0$ over one complete cycle, or

$$\int_{0}^{T} V_L \cdot dt = 0. \tag{13.2}$$

This requires that the (voltage × time) products during on and off periods must be equal and algebraically opposite to maintain the energy balance over one complete switching cycle. Therefore, equating the two is called the volt-second balance method of analysis. Since the voltage is a measure of energy per unit charge, the $v \times dt$ product balance during on and off periods essentially gives the energy balance in the inductor over one steady-state cycle. Since $v = L \times di/dt$ or $v \times dt = L \times di$, the $v \times dt$ balance also requires the $L \times di$ balance. The steady-state current in the inductor may rise and fall during one cycle, but the net change in one cycle has to be zero for it to be in a steady-state operation.

We now apply the volt-second or the $L \times di$ balance in the voltage and current waveforms shown in Figure 13.3, where I_L = inductor current:

$$\text{during on-time,} \quad L \times \Delta I_L = \text{voltage across inductor} \times T_{on} = (V_{in} - V_{out}) \times T_{on} \tag{13.3}$$

$$\text{during off-time,} \quad L \times \Delta I_L = \text{voltage across inductor} \times T_{off} = V_{out} \times T_{off}. \tag{13.4}$$

Equating Equation 13.3 with Equation 13.4 and rearranging result in the following output-to-input voltage relation, where D = duty ratio:

$$\frac{V_{out}}{V_{in}} = \frac{T_{on}}{T_{on} + T_{off}} = \frac{T_{on}}{T} = D. \tag{13.5}$$

Ignoring the internal power loss, the balance of input power and output power gives $V_{in} \times I_{in} = V_{out} \times I_{out}$, which leads to the current relation:

$$\frac{I_{out}}{I_{in}} = \frac{1}{D}. \tag{13.6}$$

Since the duty ratio D is always less than unity, the output voltage of the buck converter is always less than the input voltage. Thus, the buck converter can only step down the voltage, hence the name.

Example 13.1

A buck converter operating at 50-Hz switching frequency has T_{on} = 5 ms. Determine the average source current if the load current is 40 A.

Solution:

In the buck converter, we expect the source side at higher voltage to draw lower current than the load side at lower voltage.

$$\text{Switching period } T = 1/50 = 0.02 \text{ s} = 20 \text{ ms}$$

$$\text{Duty ratio } D = 5/20 = 0.25$$

$$\text{Average source current} = 40 \times 0.25 = 10 \text{ A.}$$

Example 13.2

A buck converter is operating at 1-kHz switching frequency from a 120-V dc source. The inductance is 50 mH. If the output voltage is 60 V and the load resistance is 12 Ω, we determine the following:

$$\text{Duty ratio } D = 60 \text{ V} \div 120 \text{ V} = 0.50 \quad \text{period } T = 1/1000 = 0.001 \text{ s} = 1 \text{ ms.}$$

$$T_{on} = 0.50 \times 1 = 0.5 \text{ ms} \quad T_{off} = 1 - 0.5 = 0.5 \text{ ms.}$$

$$\text{Average load current} = 60 \text{ V} \div 12\ \Omega = 5 \text{ A} \quad \text{Output power} = 60 \times 5 = 300 \text{ W.}$$

$$\text{Average source current} = 5 \times 0.50 = 2.5 \text{ A} \quad \text{Input power} = 120 \times 2.5 = 300 \text{ W.}$$

Equation 13.4 gives peak-to-peak ripple current:

$$\Delta I_L = V_{out} \times T_{off} \div L = 60 \times 0.0005 \div 0.050 = 0.6 \text{ A.}$$

If the switching frequency were 10 kHz, T = 1/10,000 = 0.1 ms, and T_{off} = 0.05 ms, the inductance required for the same ripple current = 60 V × 0.05 ms/0.6 A = 5 mH, which is 1/10 the size of that required at 1 kHz. This illustrates the benefit of high switching frequency.

The inductor current I_L rises during the switch on-time (charging) and decays during the off-time (discharging), with an average value equal to the output current I_{out}. The capacitor charging and discharging does not affect the average value of I_{out}. If the inductor is large enough, as is usually the case in practical designs, the change in the inductor current is small, and the peak value of the inductor current is given by

$$I_{peak} = I_{out} + \frac{1}{2}\Delta I_L \tag{13.7}$$

where the average load current $I_{out} = V_{out}/R_{load}$ = average value of the inductor current, and ΔI_L = ripple current in the inductor, as shown in Figure 13.3.

This converter can charge the battery only if V_{out} is greater than the battery voltage, which rises as the battery approaches full charge. Therefore, V_{out} must also rise with the battery state of charge. It is seen from Equation 13.5 that varying the converter duty ratio controls the output voltage, which is done in a feedback control loop with the battery voltage or charge current as the feedback signal. Since the duty ratio is controlled by modulating the pulse width of T_{on}, such converter adds one more name, the *pulse width modulated (PWM) converter.*

The relations 13.5 and 13.6 between the output and input voltages and currents (all dc) are analogous to the ac transformer relations. The duty ratio D in the dc–dc converter effectively works like the turn ratio in an ac transformer. This converter is therefore often incorrectly called the dc transformer, although it works on a fundamentally different principle than the ac transformer as we know it.

We have assumed in the above analysis that the buck converter is operating in the continuous conduction mode, which requires a minimum output current. If $I_{out} < 1/2\ \Delta I_L$, the load current would fall to zero and remain there until the switch is turned on again. This is known as the discontinuous mode of operation. To establish the boundary between the continuous and discontinuous conduction modes of the buck converter, we again use the basic relation $v = L\, di/dt$ or $di = v \times dt \div L$, which leads to

$$I_{out.boundary} = \frac{1}{2}\frac{(V_{in} - V_{out})T_{on}}{L}. \tag{13.8}$$

The converter operates in continuous conduction mode if the load current $I_{out} > I_{out.boundary}$.

Another design approach is to establish the minimum required inductance that will store the energy to keep the output current continuous even during the off-time. That requires $I_{pk\text{-}pk} = 2 \times I_{out} = T_{off}V_{out} \div L$, which gives $L_{min} = T_{off}V_{out} \div (2 \times I_{out})$. Since $V_{out}/I_{out} = R_L$, we obtain

$$L_{min} = \frac{T_{off} \times R_L}{2}. \tag{13.9}$$

With a capacitor filter, in addition to the inductor, the output voltage ripple magnitude is

$$\Delta V_{out} = \frac{\Delta Q}{C} = \frac{1}{C}\frac{1}{2}\frac{\Delta I_{out}}{2}\frac{T}{2} = \frac{T}{8C}\frac{V_{out}}{L}(1-D)T = \frac{T^2 V_{out}}{8CL}(1-D). \tag{13.10}$$

Example 13.3

A buck converter has 120-V input voltage, $R_L = 12\ \Omega$, switching frequency of 1 kHz, and on-time of 0.5 ms. If the average source current is 2 A, we determine the following:

For 1-kHz switching frequency, $T = 1/1000$ s = 1 ms, duty ratio = 0.5/1 = 0.5, T_{on} = 0.5 ms, and T_{off} = 1 – 0.5 = 0.5 ms

average output voltage = 0.5 × 120 = 60 V

average output current = 2/0.5 = 4 A

average output power = $V_{out} \times I_{out}$ = 60 × 4 = 240 W.

For continuous conduction, using Equation 13.9, we have L_{min} = 0.5 ms × 12 ÷ 2 = 3 mH.

Example 13.4

For a buck converter with V_{out} = 5 V, f_s = 20 kHz, L = 1 mH, C = 470 μF, V_{in} = 12.6 V, and I_{out} = 0.2 A, determine the peak-to-peak ripple ΔV_{out} in the output voltage.

Solution:

Switching period T = 1/20,000 = 0.00005 s, and $D = V_{out}/V_{in}$ = 5/12.5 = 0.4. Using Equation 13.10, we obtain

$$\Delta V_{out} = \frac{0.00005^2}{8 \times 470 \times 10^{-6}} \times \frac{5}{0.001} \times (1 - 0.40) = 0.002\,\text{V}.$$

This is quite a low ripple voltage, giving a smooth dc voltage output in this converter.

The converter efficiency is calculated as follows. During the on-time, the input voltage less the voltage drop in the switch equals the circuit voltage. The energy transfer efficiency is therefore $(V_{in} - V_{sw})/V_{in}$, where V_{in} = input voltage and V_{sw} = voltage drop in the switch. During the off period, the circuit voltage equals the output voltage plus the diode drop. The efficiency during this period is therefore $V_{out} \div (V_{out} + V_{diode})$, where V_{out} = output voltage and V_{diode} = voltage drop in the diode. In addition, there is some loss in the inductor, the efficiency of which we denote by η_{ind}. The losses in the capacitor and wires are relatively small and can be ignored for simplicity. The overall efficiency of the buck converter is, therefore, the product of the above three efficiencies, namely,

$$\eta_{conv} = \left(\frac{V_{in} - V_{sw}}{V_{in}}\right)\left(\frac{V_{out}}{V_{out} + V_{diode}}\right)\eta_{ind}. \tag{13.11}$$

With commonly used devices in such coveters, the transistor switch and diode voltage drops are typically 0.6 V each, and the inductor efficiency can approach 0.99. Therefore, the overall efficiency of a typical 28-V output converter is around 95%. The losses in the wires, capacitors, and magnetic core may further reduce the efficiency by a couple of percents, making it in the 92%–94% range. Equation 13.11 indicates that the buck converter efficiency is a strong function of the output voltage. It decreases with decreasing output voltage, since the switch and diode voltage drops remain constant. For example, a 5-V output converter would have efficiency in the 75%–80% range, and a 3-V converter would have even lower efficiency, around 65%, unless special low-voltage devices are used in the design.

13.2 BOOST CONVERTER

The boost converter steps up the input voltage to a higher output voltage. The converter circuit is shown in Figure 13.4. When the transistor switch is on, the inductor is connected to the input voltage source, the inductor current increases linearly and the capacitor supplies the load current while the diode is reverse biased. When the switch is off, the diode becomes forward biased, the inductor current flows through the diode and the load, and the inductor voltage adds to the source voltage to increase the output voltage. The output voltage of the boost converter is derived again from the volt-second balance principle in the inductor. With duty ratio D of the switch, it can be shown, in a manner similar to that in the buck converter analysis, that the output voltage and current are given by the following expressions:

$$\frac{V_{out}}{V_{in}} = \frac{T_{off} + T_{on}}{T_{off}} = \frac{1}{1-D} \tag{13.12}$$

$$\frac{I_{out}}{I_{in}} = 1 - D. \tag{13.13}$$

Since D is always a positive number less than 1 ($0 < D < 1$), the output voltage is always greater than the input voltage. Therefore, the boost converter can only step up the voltage, hence another name: *step-up converter*. It is widely used for discharging the battery to feed loads at constant voltage, hence, it has yet another name: the *battery discharge converter*. Since the battery voltage under discharge sags with increasing depth of discharge, a voltage converter with feedback controlled duty ratio is required to continuously boost the battery voltage to a regulated output voltage for the loads.

The efficiency of the boost converter can be shown to be

$$\eta_{conv} = \left(\frac{V_{in} - V_{sw}}{V_{in}}\right)\left(\frac{V_{out}}{V_{out} + V_{diode}}\right)\eta_{ind}, \tag{13.14}$$

which is generally higher than in the buck converter due to higher output voltage.

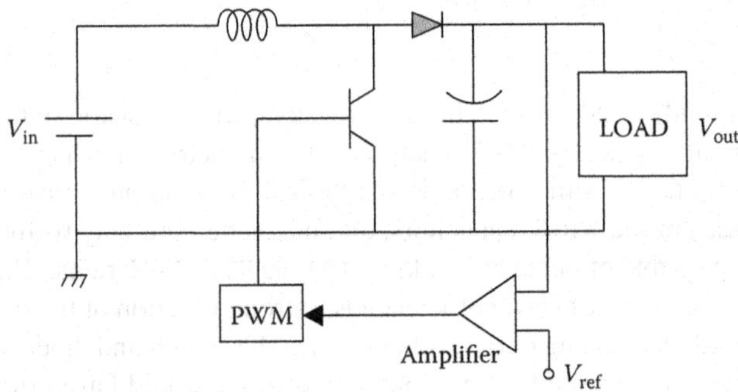

FIGURE 13.4 Boost converter circuit.

Example 13.5

A boost converter powers a 4-Ω resistor and 1-mH inductor load. The input voltage is 60 V, and the output load voltage is 80 V. If the on-time is 2 ms, we determine the following.

Using Equation 13.12, 80 ÷ 60 = 1 ÷ (1 – D), which gives the duty ratio D = 0.25.

Therefore, T_{on} = 0.25 T, that is, 2 ms = 0.25 T, which gives T = 8 ms, where T = T_{on} + T_{off}.

Switching frequency = 1/T = 1/0.008 = 125 Hz.

Output current = 80/4 = 20 A Input current = 20 (1 – 0.25) = 15 A.

Example 13.6

For a boost converter with switching frequency of 500 Hz, input of 50 V dc, output of 75 V dc, inductor of 2 mH, and load resistance of 2.5 Ω, we determine the following:

period T = 1/500 = 0.002 s = 2 ms

V_{out}/V_{in} = 75/50 = 1/(1 – D), which gives D = 0.333

T_{on} = 0.333 × 2 ms = 0.666 ms T_{off} = 2 – 0.666 = 1.334 ms

I_{out} = 75/2.5 = 30 A I_{in} = 30 ÷ (1 – 0.333) = 45 A

$\Delta I_{Ripple.pk\text{-}pk} = V_{in} \times T_{on} \div L$ = 50 V × 0.666 ms ÷ 2 mH = 16.65 A (a large ripple).

13.3 BUCK–BOOST CONVERTER

Combining the buck and boost converters in a cascade (Figure 13.5a) gives a buck–boost converter, which can step down or step up the input voltage. A modified buck–boost converter often used for this purpose is shown in Figure 13.5b. The voltage and current relations in either circuit are obtained by cascading the buck and boost converter voltage and current relations as follows:

$$\frac{V_{out}}{V_{in}} = \frac{D}{1-D} \tag{13.15}$$

$$\frac{I_{out}}{I_{in}} = \frac{1-D}{D}. \tag{13.16}$$

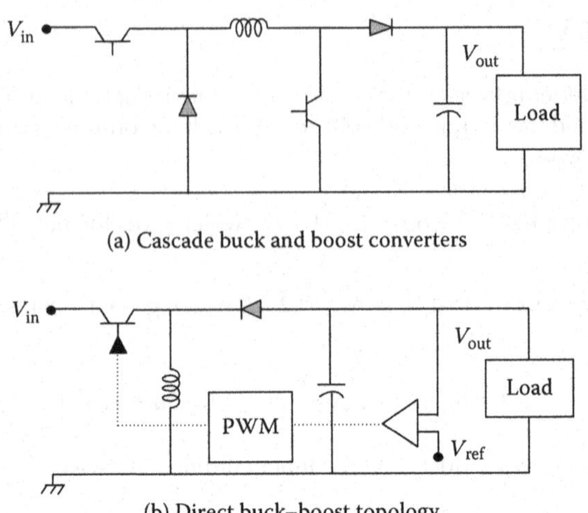

FIGURE 13.5 Buck–boost converter alternative circuits.

Equation 13.15 shows that the output voltage of this converter can be higher or lower than the input voltage depending on the duty ratio D (Figure 13.6), and the voltage ratio equals 1.0 when $D = 0.5$.

The buck–boost converter is capable of a four-quadrant operation (i.e., both V and I can be positive or negative), making it suitable in variable-speed drives for a dc motor with regenerative braking. The converter is in the step-up mode during the generating operation and in the step-down mode during the motoring operation.

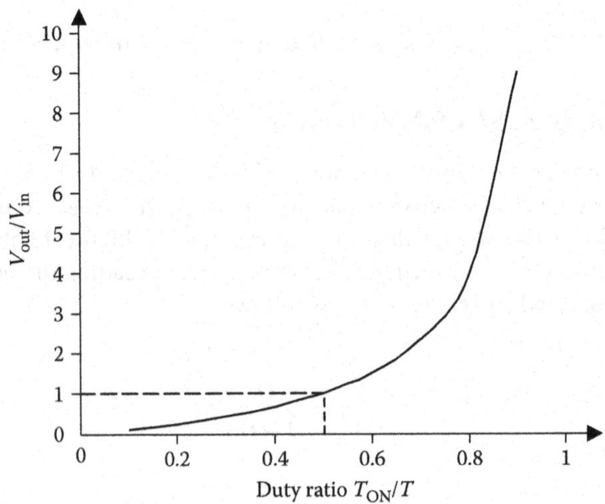

FIGURE 13.6 Buck–boost converter voltage ratio versus duty ratio.

13.4 FLYBACK CONVERTER (BUCK OR BOOST)

Figure 13.7 shows the topology of the flyback converter, which can buck or boost the voltage depending on the coupled inductor's turn ratio N_{in}/N_{out}. The energy is stored in the primary side of the inductor during on-time. When the switch is turned off, the primary side energy gets inductively transformed—gets kicked or flyover—to the secondary side and delivered to the load via diode. The polarity marks of the two coils are of essence in this circuit.

The magnetic flux in air only—as opposed to flux in a magnetic core—can store magnetic energy. Therefore, the energy storage inductor must have air in its magnetic flux path. The inductor core in the converter is, therefore, made with an air gap in one lump or distributed throughout the core, such as in a potted magnetic metal powder core with nonmagnetic binder interleaved that is magnetically equivalent to air. The energy is stored in the distributed air gap of the core.

The ideal transformer core cannot store magnetic energy since it has zero air gap and magnetically soft steel. Therefore, the magnetic flux in the core is established with negligible magnetizing current I_{mag} in the inductance L_{mag}, and the energy stored in the magnetic flux = $1/2\, L_{mag} \times I_{mag}^2 = 0$ since $I_{mag} = 0$. The mechanical analogy of this is an infinitely soft spring. When compressed, it cannot store potential energy in the spring compression.

The voltage ratio of the flyback converter is the cascade of the duty ratio and the turn ratio

$$\frac{V_{out}}{V_{in}} = D \cdot \left(\frac{N_{out}}{N_{in}} \right). \tag{13.17}$$

This gives great flexibility in the output voltage without limitation of the duty ratio. It can buck or boost depending on whether the turn ratio is less than or greater than 1.

A flyback converter uses a dual-purpose two-winding inductor with N_{in}/N_{out} turn ratio to step up or down the voltage like a transformer and also to store energy like an inductor.

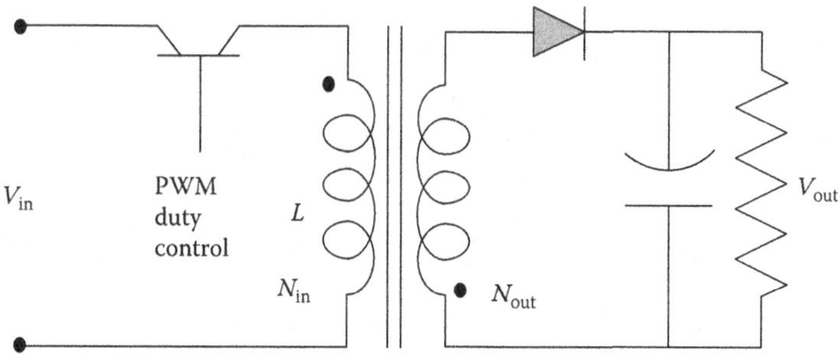

FIGURE 13.7 Flyback converter circuit with inductor.

Denoting the efficiency of coupled inductors by η_{ind} (typically 96%–98%), the overall converter efficiency derived in a similar manner as before results in the converter efficiency

$$\eta_{conv} = \left(\frac{V_{in} - V_{sw}}{V_{in}}\right)\left(\frac{V_{out}}{V_{out} + V_{diode}}\right)\eta_{ind}. \tag{13.18}$$

An advantage of the flyback converter over the classical buck–boost converter is that it electrically isolates the two sides, thus minimizing the conducted EMI and enhancing safety. Moreover, it stores the inductive energy and simultaneously operates like a classical forward buck converter with a transformer, stepping up or down the input voltage as needed.

13.5 TRANSFORMER COUPLED FORWARD CONVERTER

The converter circuit shown in Figure 13.8 employs a transformer with suitable turn ratio N_{in}/N_{out} in the conventional buck converter such that the output voltage can be below or above the input voltage. The polarity marks are of essence. A bleeding resistor is required in parallel with the load to keep the converter working with a certain minimum load. Its voltage relation is identical to that for the buck converter modified by the transformer turn ratio. That is,

$$\frac{V_{out}}{V_{in}} = D \cdot \left(\frac{N_{out}}{N_{in}}\right) \tag{13.19}$$

and the converter efficiency

$$\eta_{conv} = \left(\frac{V_{in} - V_{sw}}{V_{in}}\right)\left(\frac{V_{out}}{V_{out} + V_{diode}}\right)\eta_{ind}\eta_{trfr}, \tag{13.20}$$

where η_{trfr} = transformer efficiency, which is typically 96%–98%.

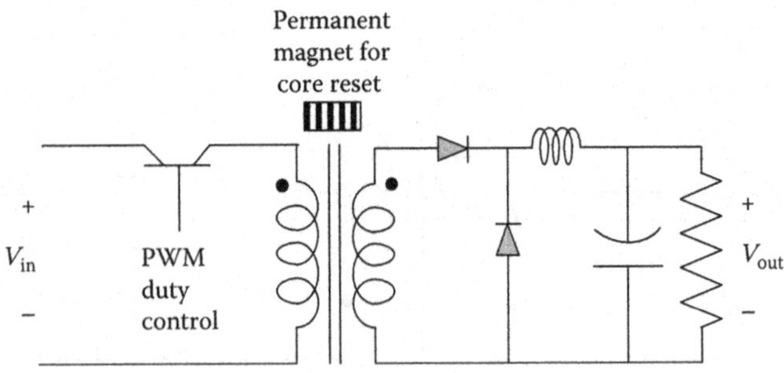

FIGURE 13.8 Forward converter circuit with transformer.

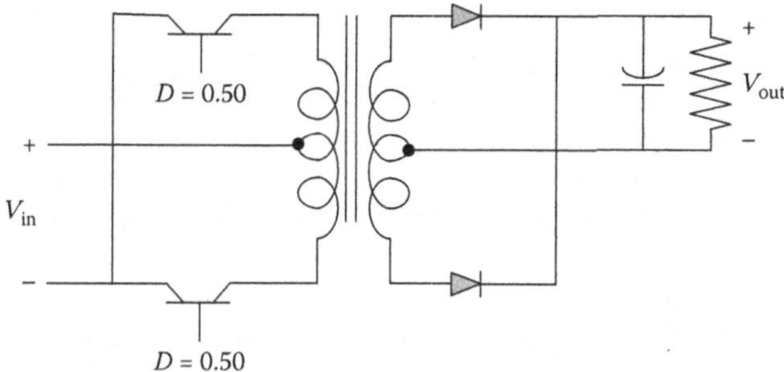

FIGURE 13.9 Push–pull converter with center-tapped transformer.

13.6 PUSH–PULL CONVERTER

This is again a buck–boost converter as shown in Figure 13.9. The converter topology uses a center-tapped transformer, which gets square-wave excitation. It is seldom used at low power levels but finds applications in systems dealing with several tens of kilowatts and higher power. It can provide desired output voltage by setting the required transformer turn ratio. The voltage ratio and the efficiency are the same as those given by Equations 13.19 and 13.20.

13.7 INDUCTOR COUPLED BUCK CONVERTER

This converter is also known as the *Cuk converter* after its inventor. All converters presented above need an L-C filter to control the ripple in output voltage. The buck converter needs a heavier filter than the boost converter. Since ripples are present in both the input and output sides, it is possible to cancel them out by coupling the ripple current slopes of two sides with matched inductors as shown in Figure 13.10. The ripple magnitudes are matched by using the required turn ratio, and the polarities are

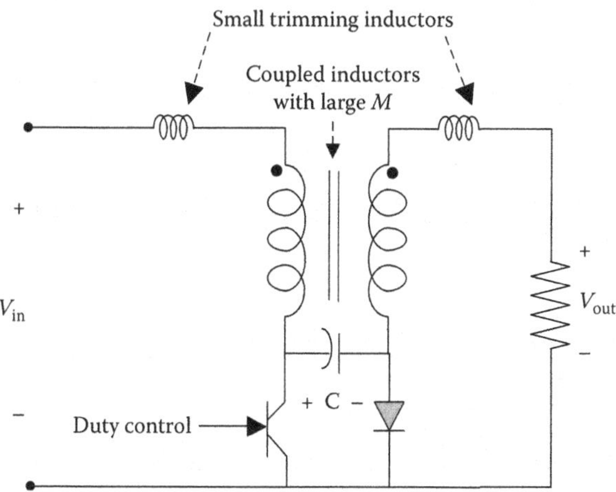

FIGURE 13.10 Cuk converter with coupled inductors.

inverted by winding the two inductors in magnetically opposite directions. The net ripple on the output side is zero at a certain air gap, below which it changes the polarity. A gap-adjusting screw is needed to tune the gap for a precise match. The output capacitor C_o (not shown) is not really needed in a perfectly coupled design, but a small value of C_o improves the performance. The coupling capacitor C is needed for the performance. The total value of the two capacitors is about the same as that needed in the classical buck converter. On the output side, some ripple may be left due to the winding resistances. The input side ripple is the same as that without the coupled inductor. The load current changes do not affect the ripple cancellation.

13.8 DUTY RATIO CONTROL CIRCUIT

The duty ratio that was shown in Figure 13.1b can be controlled by a feedback control in the triggering circuit as shown in Figure 13.11. In the functional schematic (Figure 13.11a), the actual output voltage of the converter is compared with the desired voltage, and the difference is fed to the operational amplifier. The amplified error signal—called the control voltage—is subtracted from a sawtooth voltage by a comparator, whose output is applied to the gate terminals of the power electronics switch of the converter. If the power electronics device is a transistor, it conducts as long as the ($V_{sawtooth}$ − $V_{control}$) is positive, or else it blocks the conduction. The

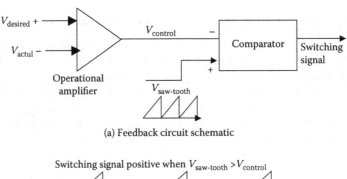

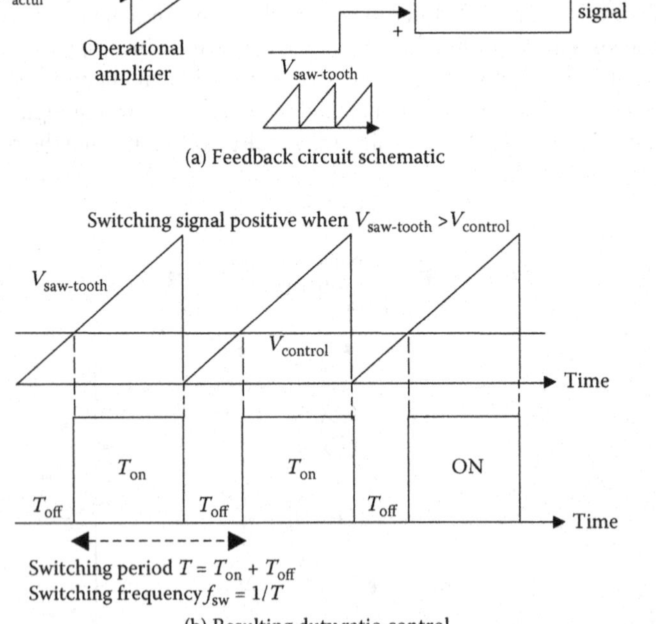

FIGURE 13.11 Feedback duty ratio control circuit and operation.

resulting duty ratio is shown in Figure 13.11b. The device turns on when $V_{sawtooth} > V_{control}$, which happens when angle $\alpha = \frac{V_{control.dc}}{V_{sawtooth.peak}} T$, and remains on for $(180 - \alpha)°$. Therefore, it yields the duty ratio

$$D = \frac{T - \alpha}{T} = 1 - \frac{V_{control.dc}}{V_{sawtooth.pk}}. \tag{13.21}$$

Obviously, $D = 1$ if $V_{control.dc} = 0$, and $D = 0$ if $V_{control.dc} = V_{sawtooth.pk}$. The value of D can be adjusted to any value in between by varying the value of $V_{control.dc}$.

13.9 LOAD POWER CONVERTER

The load power converter (LPC) is connected between the power bus and the load, with its output suitable for a specific load. Its design requirements come primarily from the user requirements on the load and output regulation (such as constant voltage, constant current, constant power, quality of power, etc.). The necessary control circuit to regulate the output is generally incorporated in the LPC design package.

Loads having stringent ripple requirement may require a prohibitively large filter capacitor at the power supply output. The inductor stores much greater energy per unit mass and therefore significantly reduces the filter size. Alternatively, an L-C filter, shown in Figure 13.12a, at the load interface may be more cost effective. When using such an L-C filter, the transient analysis becomes important on switch-on, as both the current and voltage may overshoot to damaging levels. If the LPC is disconnected from the bus using a mechanical relay, the energy stored in the filter's L and C elements may cause sparks at the relay contacts. Moreover, large negative voltage transients could occur at the user interface when the relay contact is opened. This can

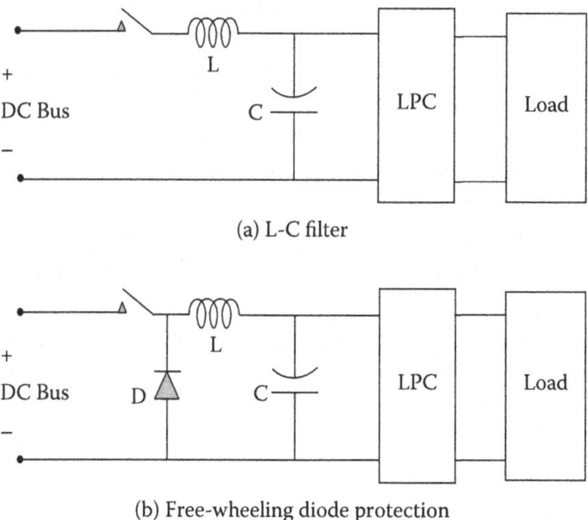

FIGURE 13.12 L-C ripple filter between power bus and LPC.

result in component damage. A freewheeling diode D shown in Figure 13.12b can eliminate any such sparking and negative voltage transients by providing a circulating current path when the switch is opened.

The basic converter—called the main power train—in the LPC box constitutes the bulk of the LPC design, around which many bells and whistles are added to make the LPC suitable for the user load. The input voltage may vary over a wide range in some poor-quality power buses. If the LPC output voltage is required to be constant with variable voltage input, a feedback voltage regulator is used in the design with a PWM converter with duty ratio control. If the input voltage variation is small, a linear dissipative series voltage regulator may be used. The linear resistive control is more reliable than a switching converter because of fewer parts and no high-voltage switching stress, although it has lower efficiency and runs hotter. The duty ratio-controlled buck converter is simple and dynamically stable in operation. All other converters are relatively complex and may become unstable, as they can have poles on the right-hand side of the Laplace plane.

If a current-limiting feature is desired, the converter control circuit regulates and limits the current. In such an LPC design, the current is limited to some high value during overloads, or even under a load fault during which the bus voltage is not driven to zero. It just folds back the current to some low value.

Thus, the LPC control loop can be designed to maintain constant voltage, constant current, or constant power. It can also be designed for a combination of two or more performance parameters. For example, the converter may maintain constant voltage up to a certain current limit and then maintain constant current at the set limit, or even fold back a little in current. It is all in the control system, rather than in the basic power train design.

13.10 POWER SUPPLY

Power supply is a converter having one or more regulated output voltages such that the output is suitable for a range of loads. The output may be electrical power isolated from the input side by a transformer for safety or EMI reasons. The power supply design is focused primarily on the user requirements for the load and the output regulation (such as a constant voltage, constant current, constant power, quality of power, etc.). The necessary control circuit to regulate the output is generally incorporated in the power supply packaging (Figures 13.13 and 13.14). Thus, the converter's main power train constitutes the bulk of the power supply design around which the controls are added to make it suitable for various loads. The output-to-input relations for the power supply are essentially the same as those for the main power train but with finer controls. Thus, we can say that power supply = power train converter in the basic form + fine controls for finished output with desired regulation. Typical power supply requirements are as follows:

- Regulated voltages within a few percent (regardless of wide variations in input voltage, temperature, etc.)
- Ripple limited to a low specified value
- Isolations with n:1 or 1:1 turns ratio transformer

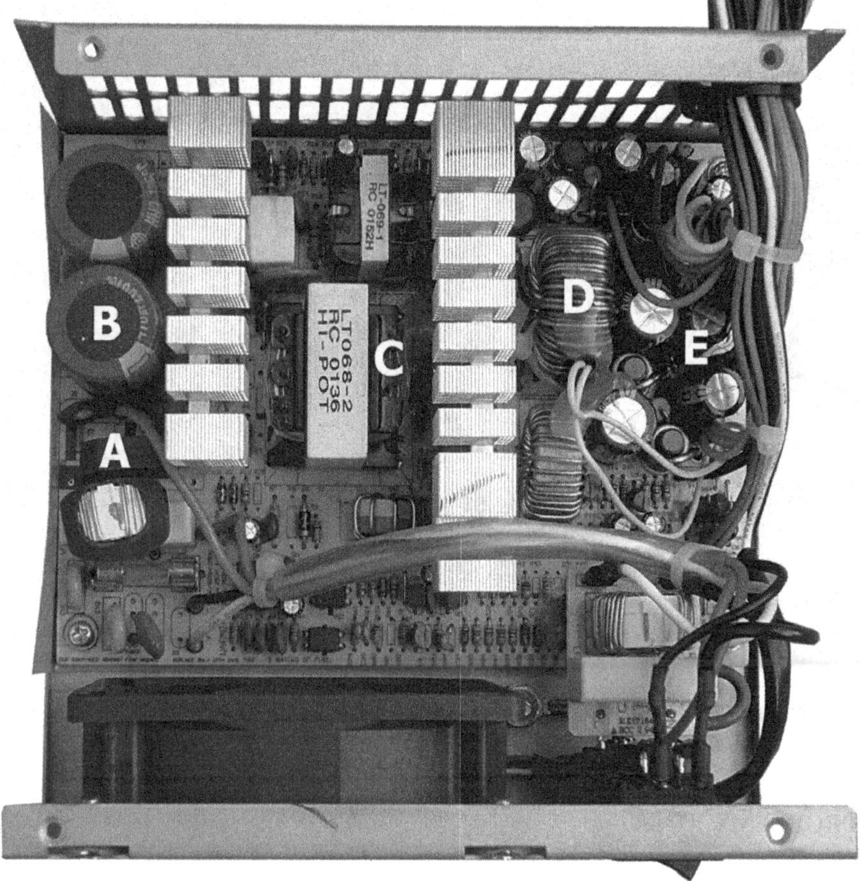

FIGURE 13.13 Switchmode power supply interior packaging (ATX Inc.).

- Multiple output (using three-winding transformer, if needed)
- Protection (overvoltage, overcurrent, current limiting, etc.)

The power supply uses a buck, boost, or buck–boost converter with a transformer inserted to step up or down the voltage or just for the electrical isolation of the output side from the input side.

In all types of converters covered in this chapter, there is a trend to increase the switching frequency to reduce the size of the power train inductor and the output capacitor filter. Frequencies up to a few hundred kilohertz are common at present but are approaching 1 MHz in some high-density power converters. The design challenge increases as the switching frequency exceeds 500 kHz, as the lead wire inductance and capacitance become significant due to stray and parasitic effects, which are difficult to analyze.

FIGURE 13.14 Switchmode power supply exterior view (Voltcraft 4005). (Courtesy of Nuno Nogueira, Wikimedia Commons.)

PROBLEMS

1. A buck converter operating at 60-Hz switching frequency has $T_{on} = 7$ ms. Determine the average source current if the load current is 12 A.
2. A buck converter is operating at 2-kHz switching frequency from a 220-V dc source. The inductance is 30 mH. If the output voltage is 110 V and the load resistance is 15 Ω, determine (1) the switch *on-time* and *off-time*, (2) the load current and power output, (3) the source current and power input, and (4) the peak-to-peak ripple current.
3. A buck converter has 220-V dc input voltage, $R_L = 18$ Ω, switching frequency of 5 kHz, and on-time of 80 μs. If the average source current is 10 A, determine (1) the average output voltage, current, and power and (2) the minimum inductance needed to assure continuous conduction.
4. For a low-voltage buck converter with $V_{out} = 3.5$ V, $f_s = 50$ kHz, $L = 200$ μH, $C = 500$ μF, $V_{in} = 12$ V, and $I_{out} = 1.5$ A, determine the peak-to-peak ripple ΔV_{out} in the output voltage.
5. A boost converter has an input voltage of 115 V and an output voltage of 230 V dc connected to a 10-Ω resistor and 2-mH inductor load. If the on-time is 3 ms, determine (1) the switching frequency, (2) the output current, and (3) the input current.

6. A boost converter has an input of 60 V dc, output of 100 V dc, inductance of 3 mH, and load resistance 4 Ω. If its switching frequency is 2 kHz, determine (1) the output and input currents and (2) the peak-to-peak value of the ripple current.

QUESTIONS

1. Explain the principle on which the buck converter analysis is based.
2. As the duty ratio changes from 0.1 to 0.9, how does the buck–boost converter voltage ratio change?
3. At what duty ratio does the buck–boost converter change from buck to boost?
4. What does the term *electrical isolation* mean? How is it achieved in a converter design?
5. Identify the similarities and differences between the dc–dc converter and the ac transformer.
6. Identify the construction and functional differences between the transformer and the inductor.
7. What flies back in the flyback converter?
8. Identify and explain the benefit of using high frequency switching in the switchmode power converter design.

FURTHER READING

Ahmed, A. 1999. *Power Electronics for Technology.* Upper Saddle River: Prentice Hall.
Fang, L.L. and Hong, Y. 2003. *Advanced DC/DC Converters.* Boca Raton: CRC Press/Taylor & Francis.
Fang, L.L. and Hong, Y. 2005. *Essential DC/DC Converters.* Boca Raton: CRC Press/Taylor & Francis.
Kazimierczuk, M. 2008. *Pulse-Width Modulated DC–DC Power Converters.* Hoboken: John Wiley & Sons.
Kularatna, N. 2011. *DC Power Supplies.* Boca Raton: CRC Press/Taylor & Francis.
Wu, K.C. 1997. *Pulse Width Modulated DC/DC Converters.* New York: Chapman and Hall.

14 AC–DC–AC Converters

This chapter covers the following power electronics converters widely used in electrical power systems, with their alternative names given in the parentheses:

- AC–DC converter (rectifier)
- AC–AC voltage converter (phase-controlled converter)
- DC–AC converter (inverter)
- AC–DC–AC (dc-link) frequency converter
- AC–AC direct frequency converter (cycloconverter)

Converters perform their functions using power electronics (solid-state semiconductor) devices periodically switched on and off at certain frequency. Numerous circuit topologies have been developed for these converters, and new ones are evolving every year. A number of computer software programs, such as PSPICE and SABER, are available to analyze such converters in great detail. However, the following sections present the basic circuits and the voltage, current, and power relationships at the input and output terminals of some widely used converters, leaving details for more advanced books dedicated to the subject of power electronics.

14.1 AC–DC RECTIFIER

14.1.1 Single-Phase Full-Wave Rectifier

The single-phase full-wave rectifier shown in Figure 14.1 is built with four thyristors. It is typically used to charge a battery or drive a dc motor, often at variable speed. Its generalized circuit includes the ac side source inductance L_{ac}, the dc side inductance L_{dc}, and the back emf E_{dc}. E_{dc} could be the internal emf of a battery or the counter emf of a dc motor. R_{dc} may be the motor armature resistance or the battery internal resistance.

The inductance L_{ac} may be the cable inductance or the filter inductance purposely placed in the circuit. The cable inductance is due to the leakage flux between the cable conductors, the value of which comes from the cable manufacturer's data sheet. The internal source inductance in the Thevenin source model is generally negligible if the source is much larger than the converter rating, which is often the case in practice. If significant, it is included in L_{ac}.

In the converter operation, when the upper terminal of the ac source is positive, the thyristors T1 and T4 are fired to conduct downward current in the dc load. After one-half cycle, when the lower ac terminal becomes positive, T2 and T3 are fired, again conducting downward current in the dc load. Thus, two pulses of positive voltage are applied to the load over one full ac wave (hence the name), and the load current is always positive downward (rectified dc). The dc side current would have

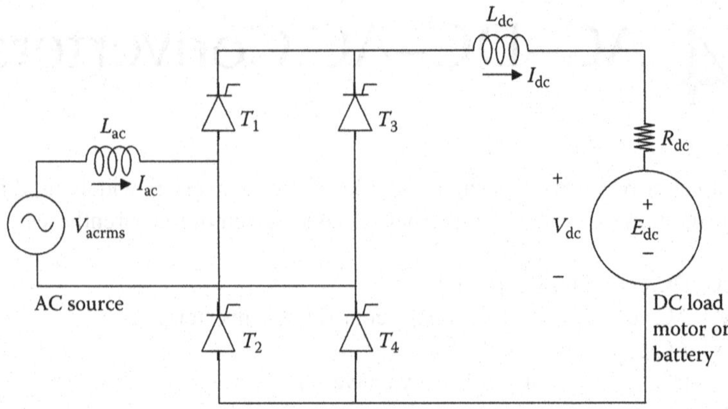

FIGURE 14.1 Single-phase full-wave rectifier circuit for dc motor drive or battery charging.

some ripples but can be made smooth by using a large dc side filter inductor L_{dc}, as is usually done in practical rectifier designs. The dc side voltage can be controlled by delaying the thyristors' firing angle α as shown in Figure 14.2.

When no dc side load is connected, the no-load dc voltage V_{dc0} at the converter terminals is given by the average value of the dc side voltage over the conducting pulse of the rectified voltage shown in Figure 14.2. Mathematically, it is the average value of the sine function from α to 180° over the 180° span. With $v(t) = \sqrt{2}\, V_{rms} \sin \omega t$ for ac voltage and the angle measured in radians, we have

$$V_{dc0} = \frac{\frac{1}{\omega}\int_{\alpha}^{\pi} v(t)\cdot d(\omega t)}{\pi} = \frac{\sqrt{2}}{\pi} V_{rms}(1+\cos\alpha) = 0.45 V_{rms}(1+\cos\alpha). \qquad (14.1a)$$

However, with inductive load, the output current continues in the negative side even after the converter voltage has come to zero, posing a negative voltage on the

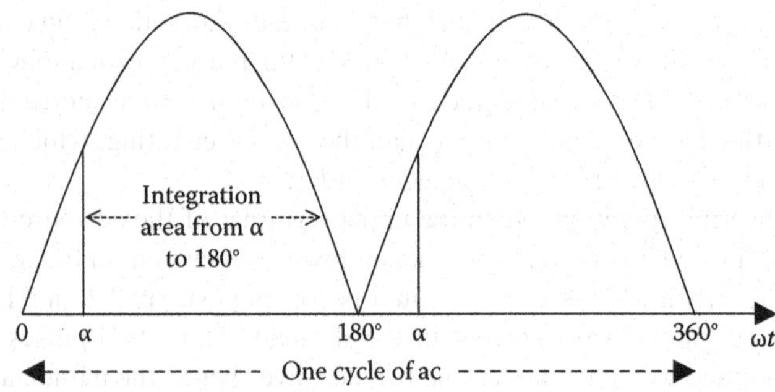

FIGURE 14.2 Single-phase full-wave rectifier output with firing delay angle α.

load that subtracts from the above integral. This results in a slightly reduced output voltage at the converter terminals:

$$V_{dc0} = \frac{\frac{1}{\omega}\int_{\alpha}^{\pi} v(t)\cdot d(\omega t)}{\pi} = \frac{2\sqrt{2}}{\pi} V_{rms}\cos\alpha = 0.90 V_{rms}\cos\alpha. \tag{14.1b}$$

The average value of dc is not a smooth constant value but is superimposed with high-frequency ripples. When the dc side delivers a constant load current I_{dc} via a large ripple filter inductor L_{dc}, the smooth dc voltage at the load terminals is given by

$$V_{dc} = 0.90 V_{rms}\cos\alpha - \frac{2\omega L_{ac}}{\pi} I_{dc} = E_{dc} + I_{dc}R_{dc}. \tag{14.2}$$

L_{dc} has no effects on V_{dc}; it merely smoothes out the ripples from I_{dc}, as is assumed in the analysis.

The dc side voltage can be controlled by controlling the firing angle α as desired. If the diodes were used instead of thyristors, then α is always zero as the diodes automatically get conducting immediately on becoming positively biased. Such a diode rectifier is called an *uncontrolled rectifier*, for which Equation 14.2 holds true with $\alpha = 0$, giving a maximum dc side voltage of 0.90 V_{rms}.

Figure 14.3 shows how the thyristor gate firing (triggering) signal pulses are generated by comparing a reference sawtooth wave voltage with a dc control voltage. When $V_{sawtooth} > V_{control}$, a positive voltage results on the thyristor gate terminal, starting the conduction (i.e., firing the thyristor). The firing angle is controlled from 0° to 180° by varying the value of $V_{control}$. A higher $V_{control}$ results in later firing. Thus, the firing delay angle α is proportional to $V_{control}$ as given by

$$\alpha = \left(\frac{V_{control(dc)}}{V_{sawtooth.peak}}\right)180°. \tag{14.3}$$

Varying the $V_{control}$ signal varies the ac output voltage. A rectifier with $0° < \alpha < 90°$ can be easily made an inverter by increasing the firing delay angle α to above 90° (i.e., $90° < \alpha < 180°$), when cos α becomes negative. Then, Equation 14.2 would give negative I_{dc}, taking power from the dc side to the ac side, that is, inverting the dc into ac.

The voltage across the thyristor is near zero when it is conducting. The reverse voltage equals the full source voltage V_s when the thyristor is not conducting, as there are no voltage drops anywhere else in the circuit due to zero current. With an ac source, the peak value of the reverse voltage is 1.414 $V_{ac.rms}$. This can be visualized by thinking of a circuit with only a thyristor and a resistor in the loop.

The output voltage of the rectifier is not a purely smooth dc, as is the characteristic of all converters we will cover in this book. For comparing the output performance of various converters, we develop in the following two sections a

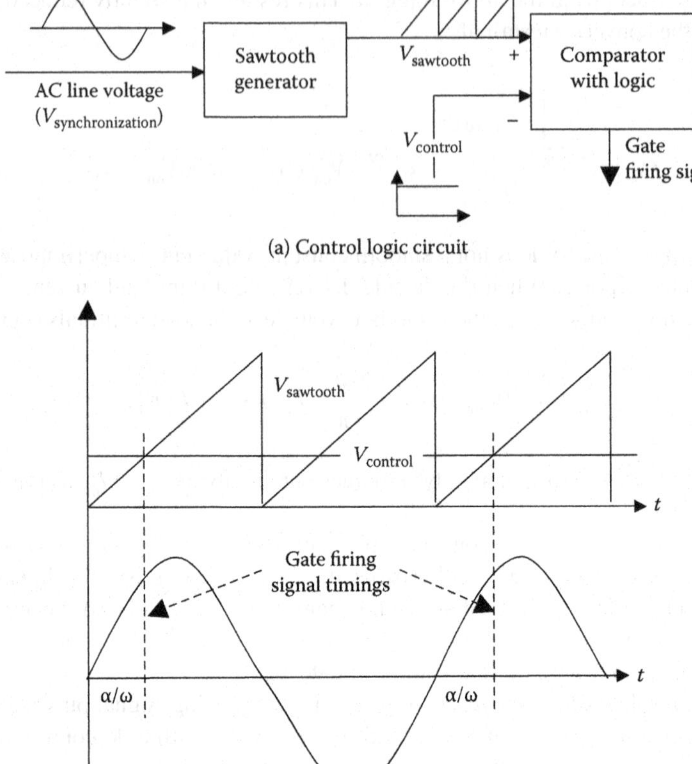

FIGURE 14.3 Thyristor firing angle control logic and operation.

method of taking into account the amount of deviation from the ideally smooth dc or sinusoidal ac.

Example 14.1

A single-phase full-wave rectifier shown in Figure 14.1 is built using diodes and a large inductor. Determine the average and rms values of the current in each diode.

Solution:

The current in each diode is the load side current I_{dc}, which is constant as long as that particular diode is conducting and is zero all other times. Since each diode conducts only for 1/2 cycle or $1/2T$, where T = period of the cycle, we have

$$\text{Diode current} = I_{dc} \text{ for } 1/2T$$
$$= 0 \text{ for the rest of the cycle}$$

$$\text{Therefore, } I_{diode(avg)} = \frac{I_{dc} \times \frac{T}{2} + 0 \times \frac{T}{2}}{T} = \frac{I_{dc}}{2} \text{ and}$$

$$I_{diode(rms)} = \sqrt{\frac{I_{dc}^2 \frac{T}{2} + 0^2 \frac{T}{2}}{T}} = \sqrt{\frac{I_{dc}^2 \cdot \frac{T}{2}}{T}} = \frac{I_{dc}}{\sqrt{2}}.$$

The quantity in the first square root sign above is the mean of the square over period T of the cycle.

Example 14.2

A single-phase full-wave bridge rectifier using phase-controlled thyristors with a large L_{dc} converts 60 Hz, 120 V rms to dc for a dc side load resistance of 10 Ω. For the firing delay angle $\alpha = 60°$, determine the dc side voltage and the power absorbed by the load, ignoring the cable inductances and ripples.

Solution:

We use Equation 14.2 to obtain $V_{dc} = 0.9 \times 120 \times \cos 60° = 54$ V.

Then, $I_{dc} = 54/10 = 5.4$ A and $P_{dc} = 54 \times 5.4 = 291.6$ W.

The dc side ripple current has zero average value, but its rms value will add into the power absorbed by the 10-Ω resistor. However, we ignore the ripple effect here.

Example 14.3

The rectifier of Example 14.2 with the ac side cable inductance of 2 mH and the dc side cable resistance of 0.5 Ω is charging a battery at 5.4-A charge rate instead of powering the 10-Ω resistance. Determine the voltage available at the battery terminal.

Solution:

From the dc voltage of 54 V calculated in Example 14.2, we deduct the voltage drops due to the ac side inductance and the dc side resistance. With $\omega = 2\pi \times 60 = 377$ rad/s for 60 Hz, using Equation 14.2 leads to

$$E_{dc} = 54 - 2 \times 377 \times 0.002 \times 5.4 \div \pi - 5.4 \times 0.5 = 54 - 2.6 - 2.7 = 48.7 \text{ V}.$$

The dc side voltage has now dropped from 54 V in Example 14.2 to 48.7 V due to the inductance and resistance of the cables on both sides.

14.1.2 Ripples in DC and Ripple Factor

The output voltage of the above rectifier is not a smooth dc. The term *ripple* means periodic rise and fall in dc voltage or current output of the rectifier circuit. Ripples can be filtered out by either capacitor or inductor, or both in some combinations. Even then, small rises and falls—exponential in theory but practically linear over a small range—remain superimposed on the average dc output voltage or current. The ripple factor is a measure of the deviation from the smooth dc output of a dc converter and is defined as

$$\text{ripple factor } r = \frac{V_{\text{ripple.rms}}}{V_{\text{dc}}}. \tag{14.4}$$

The dc output with ripples has an rms value greater than its average dc value as given by

$$V_{\text{out.rms}} = \sqrt{V_{\text{dc}}^2 + V_{\text{ripple.rms}}^2} = V_{\text{dc}}\sqrt{1 + \frac{V_{\text{ripple.rms}}^2}{V_{\text{dc}}^2}} = V_{\text{dc}}\sqrt{1+r^2}. \tag{14.5}$$

We can also write Equation 14.5 as

$$\left(\frac{V_{\text{out.rms}}}{V_{\text{dc}}}\right)^2 = 1 + r^2 \quad \text{or} \quad \%\,\text{ripple } r = \sqrt{\left(\frac{V_{\text{out.rms}}}{V_{\text{dc}}}\right)^2 - 1} \times 100. \tag{14.6}$$

14.1.3 Harmonics in AC and Root Sum Square

On the ac input side of the converter in Figure 14.1, although the ac voltage input is purely sinusoidal, the ac input current will have high-frequency sine waves superimposed to keep balance with the high-frequency ripples in the dc output current. The term *harmonics* means high-frequency sinusoidal components in ac voltage or current at the input or output of a power electronics converter. Thus, the terms *harmonics* and *ripples* are similar, except that ripples are a deviation from the dc, whereas harmonics are a deviation from pure sine-wave ac. Like ripples, the harmonics can also be filtered out by L-C filters, which can be tuned to specific harmonic frequencies.

Using the Fourier series theory, any *periodic* wave can be resolved into a series of orthogonal functions. The sine and cosine functions of different frequencies are orthogonal since the product of any two over the period is zero. Accordingly, a nonsinusoidal voltage (or current) that is not a pure sinusoidal wave can be resolved into a series of high-frequency sinusoidal harmonic functions as follows:

$$V(t) = V_1 \cos \omega t + \sum_{h=2}^{\infty} V_h \cos(h\omega t + \alpha_h). \tag{14.7}$$

The first component on the right-hand side of the above equation is the fundamental component, whereas all other higher-frequency terms ($h = 2, 3\ldots\infty$) are called harmonics. The nonsinusoidal current $i(t)$ can be represented by a similar series of harmonic currents. The frequency of the hth harmonic current is $h \times$ fundamental frequency. For example, the frequency of the 7th harmonic in a 60-Hz power system is $7 \times 60 = 420$ Hz.

It can be shown that the rms value of the total nonsinusoidal ac voltage $v(t)$ with harmonics is given by the *root sum square (rss)* value of the individual harmonic rms values, that is,

$$V_{\text{ac.rms}} = \sqrt{V_{1\text{rms}}^2 + V_{2\text{rms}}^2 + V_{3\text{rms}}^2 + \ldots\ldots. V_{\text{hrms}}^2} = \sqrt{V_{1\text{rms}}^2 + V_{\text{Hrms}}^2} \tag{14.8a}$$

where V_{hrms} = rms harmonic voltage of frequency ($h \times$ fundamental frequency) for $h = 2$, $3, \ldots, \infty$, and the total harmonic rms voltage $V_{\text{Hrms}} = \sqrt{V_{2\text{rms}}^2 + V_{3\text{rms}}^2 + \ldots\ldots. V_{\text{hrms}}^2} = \sqrt{\Sigma V_{\text{hrms}}^2}$.

Equation 14.8a can be stated in words as

rms value of total voltage = root sum square (rss) value of rms harmonic voltages. (14.8b)

For the *root mean square* value of the total, we do not take the mean (average) of the rms values of the component squares but the *root sum of the component squares*. This is analogous to what we know from vector algebra: any three-dimensional vector V can be resolved into three orthogonal components V_x, V_y, V_z on x-y-z axis, and the magnitude of the vector V is the root sum square of the component magnitudes, that is, $V = \sqrt{V_x^2 + V_y^2 + V_z^2}$. The same applies to the harmonics, which are mathematically orthogonal functions.

If a harmonic is not present, then its value is taken as zero in the above formula. Practical voltage and current waves are symmetrical about the time axis, for which the Fourier series always results in all even harmonics equal to zero.

The performance of the system with harmonic voltages can be calculated by calculating the system performance under each harmonic voltage separately and then superimposing the individual performances to get the total performance of the system.

14.1.4 Harmonic Distortion Factor

Power engineers express ac voltage and current in customarily implied rms values, generally omitting the suffix rms. The total harmonic content normalized with the fundamental value as base is called the *total harmonic distortion* (THD) factor. The THD in voltage is then defined with implied rms values as

$$\mathrm{THD_v} = \frac{\text{total harmonic voltage rms}}{\text{fundamental voltage } V_1 \text{ rms}},$$

that is,

$$\mathrm{THD_v} = \frac{\sqrt{V_{3\mathrm{rms}}^2 + V_{5\mathrm{rms}}^2 + \ldots\ldots V_{h\mathrm{rms}}^2}}{V_{1\mathrm{rms}}}. \quad (14.9)$$

Usually the fundamental voltage $V_{1\mathrm{rms}}$ is taken as the base for $\mathrm{THD_v}$. Recalling that V_H represents *the root sum square of all harmonics*, Equation 14.8a can then be rewritten as

$$V_{\mathrm{ac.rms}} = \sqrt{V_{1\mathrm{rms}}^2 + V_{\mathrm{Hrms}}^2} = V_{1\mathrm{rms}}\sqrt{1 + \frac{V_{\mathrm{Hrms}}^2}{V_{1\mathrm{rms}}^2}} = V_{1\mathrm{rms}}\sqrt{1 + \mathrm{THD_v^2}}. \quad (14.10)$$

We can also write Equation 14.10 as

$$\left(\frac{V_{\mathrm{ac.rms}}}{V_{1\mathrm{rms}}}\right)^2 = 1 + \mathrm{THD_v^2} \quad \text{or} \quad \%\,\mathrm{THD_v} = \sqrt{\left(\frac{V_{\mathrm{ac.rms}}}{V_{1\mathrm{rms}}}\right)^2 - 1} \times 100. \quad (14.11)$$

Similarly, the total harmonic rms current $I_{\mathrm{Hrms}} = \sqrt{I_{3\mathrm{rms}}^2 + I_{5\mathrm{rms}}^2 + I_{7\mathrm{rms}}^2 + \ldots}$, and the THD factor in current with no even harmonic is defined as

$$\mathrm{THD_i} = \frac{\sqrt{I_{3\mathrm{rms}}^2 + I_{5\mathrm{rms}}^2 + \ldots\ldots I_{h\mathrm{rms}}^2}}{I_{1\mathrm{rms}}} = \sqrt{\left(\frac{I_{3\mathrm{rms}}}{I_{1\mathrm{rms}}}\right)^2 + \left(\frac{I_{5\mathrm{rms}}}{I_{1\mathrm{rms}}}\right)^2 + \left(\frac{I_{7\mathrm{rms}}}{I_{1\mathrm{rms}}}\right)^2 \ldots} \quad (14.12)$$

The $\mathrm{THD_i}$ can be derived from the harmonic current ratios often given for the equipment. Alternatively, we can also write the total ac rms current:

$$I_{\mathrm{ac.rms}} = \sqrt{I_{1\mathrm{rms}}^2 + I_{\mathrm{Hrms}}^2} = I_{1\mathrm{rms}}\sqrt{1 + \left(\frac{I_{\mathrm{Hrms}}}{I_{1\mathrm{rms}}}\right)^2} = I_{1\mathrm{rms}}\sqrt{1 + \mathrm{THD_i^2}} \quad (14.13)$$

or

$$\%\,\mathrm{THD_i} = \sqrt{\left(\frac{I_{\mathrm{ac.rms}}}{I_{1\mathrm{rms}}}\right)^2 - 1} \times 100. \quad (14.14)$$

Example 14.4

A single-phase half-wave diode rectifier draws power from a single-phase, 240-V, 60-Hz source and delivers to a 2-kW load on the dc side at constant voltage V_{dc} with a large filter capacitor as shown in the figure below. The ac side harmonics to fundamental current ratios I_h/I_1 in percentage are 73.2, 36.6, 8.1, 5.7, 4.1, 2.9, 0.8, and 0.4 for h = 3, 5, 7, 9, 11, 13, 15, and 17, respectively. Determine the rms ripple current in the filter capacitor. Ignore all losses.

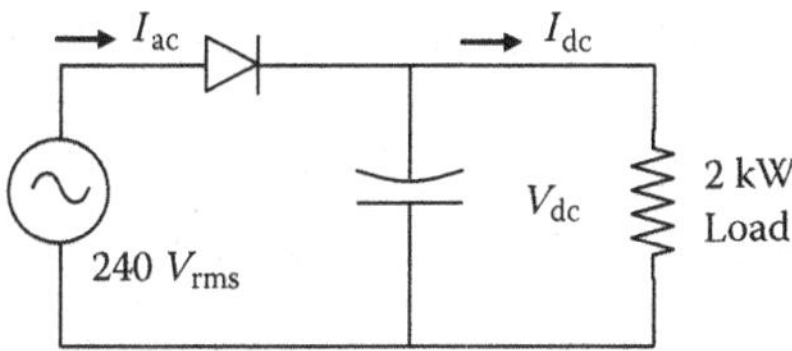

Solution:

With a large filter capacitor, the dc bus voltage V_{dc} = peak of the ac input voltage.

$$\text{Therefore, } V_{dc} = \sqrt{2} \times 240 = 339 \text{ V}.$$

For 2-kW dc output, $I_{dc(avg)}$ = 2000 W/339 V = 5.9 A.

With no firing delay angle in the diode rectifier, DPF = 1.0; ignoring all losses, we must have $P_{ac} = P_{dc}$.

Therefore, fundamental ac side rms current $I_{s1(rms)}$ = 2000 W/240 V = 8.33 A.

Since there is no transformer in this circuit, we must have total rms current continuity on both sides, that is,

$$I_{ac(rms)} = I_{dc(rms,\ total)} = \sqrt{I^2_{dc(avg)} + I^2_{dc(ripple,rms)}}$$

$$\text{Therefore, } I_{dc(ripple,rms)} = \sqrt{I^2_{ac(rms)} - I^2_{dc(avg)}}\,.$$

From the given ac side harmonics, Equation 14.12 gives

$$\text{THD}_i = \sqrt{0.732^2 + 0.366^2 + \ldots\ldots + 0.004^2} = 0.82 \text{ or } 82\%.$$

Equation 14.13 gives the total $I_{ac(rms)} = I_{ac1} \times \sqrt{1^2 + 0.82^2} = 8.33 \times 1.2932 = 10.8\,\text{A}$

$$\text{Therefore, } I_{dc(ripple)} = \sqrt{10.8^2 - 5.9^2} = 9\,\text{A}.$$

As seen here, the ac side THD and dc side ripples are rather large in the half-wave rectifier. That is why it is seldom used in practical applications.

14.1.5 Three-Phase Six-Pulse Rectifier

The circuit diagram of the full-wave, three-phase ac-to-dc rectifier is shown Figure 14.4 with thyristors (silicon-controlled rectifiers) as the power switches. It has two thyristors (T5 and T6) added in the single-phase full-wave rectifier of Figure 14.1. Keeping the same notations for the ac and dc side parameters, its operation over one ac cycle is as follows. At any instant of time, one of T1, T3, and T5 thyristors and one of T2, T4, and T6 thyristors are fired in proper sequence as the ac phase voltage goes from the positive to negative cycle. For example, when the phase voltage V_a is most positive and the phase voltage V_b is most negative, thyristors T1 and T4 are fired to conduct I_{dc}. This lasts for 60° of the full 360° cycle. For the next 60°, when V_a is most positive and V_c is most negative, thyristor T6 is fired, and I_{dc} flows through T1 and T6. The current switchover from T4 to T6 is called the commutation, which may be delayed by a few electrical degrees due to the circuit inductance. For the next four 60° intervals, T3 + T6, T3 + T2, T5 + T2, and T5 + T4 in pairs conduct I_{dc}. The operation is like an orchestra of thyristors conducted by the triggering circuit. Each of the three phases contributes two pulses, making a total of six pulses in the output side (hence the name). Each thyristor conducts for 120° every cycle, so it carries an average current of $I_{dc}/3$ and an rms current of $I_{dc}/\sqrt{3}$.

Example 14.5

A three-phase, full-bridge, six-pulse rectifier shown in Figure 14.4 is built with diodes and a large inductor. Determine the average and rms values of the current in each diode.

Solution:

The current in each diode is the load side current I_{dc}, which is constant as long as that diode is conducting and is zero all other times. Since each diode conducts for only 1/3 cycle or 1/3 T, where T = period of the cycle, we have

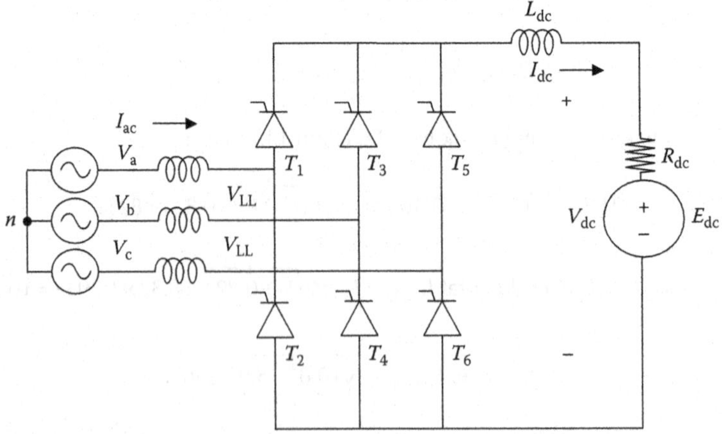

FIGURE 14.4 Three-phase full-wave six-pulse rectifier circuit.

diode current = I_{dc} for $1/3T$

= 0 for the rest of the cycle

$$\text{Therefore, } I_{\text{diode(avg)}} = \frac{I_{dc} \times \frac{T}{3} + 0 \times \frac{2T}{3}}{T} = \frac{I_{dc}}{3} \quad \text{and}$$

$$I_{\text{diode(rms)}} = \sqrt{\frac{I_{dc}^2 \frac{T}{3} + 0^2 \left(\frac{2T}{3}\right)}{T}} = \sqrt{\frac{I_{dc}^2 \cdot \frac{T}{3}}{T}} = \frac{I_{dc}}{\sqrt{3}}.$$

The quantity in the first square root sign above is the mean of the square over period T of the cycle.

The rectified output voltage has six pulses as shown in Figure 14.5. With a firing delay angle of α, the average V_{dc} value is given by

$$V_{dc} = \frac{3\sqrt{2}}{\pi} V_{\text{LLrms}} \cos\alpha - \frac{3\omega L_{ac}}{\pi} I_{dc} = E_{dc} + I_{dc} R_{dc}. \tag{14.15}$$

Again, V_{dc} is controlled by controlling α, and if $\alpha > 90°$, cos α becomes negative, making I_{dc} negative. The three-phase rectifier then becomes a three-phase inverter, taking power from the dc side and delivering it to the ac side. With no dc side load in the rectifier mode,

$$V_{dco} = \frac{3\sqrt{2}}{\pi} V_{\text{LLrms}} \cos\alpha = 1.35 V_{\text{LLrms}} \cos\alpha. \tag{14.16}$$

If diodes were used, $\alpha = 0$, and the maximum dc side voltage with no load is 1.35 V_{LLrms}.

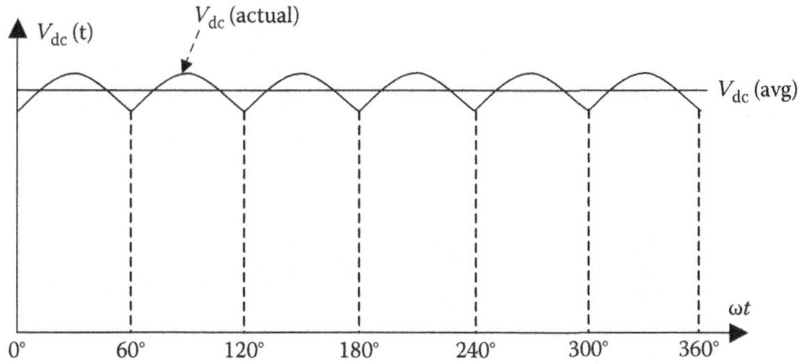

FIGURE 14.5 Three-phase full-wave six-pulse rectifier output voltage.

If L_{ac} is negligible, then it can be shown that the total and fundamental components of the ac side current have the following relations:

$$I_{rms} = \sqrt{\frac{2}{3}} I_{dc} \quad I_{1rms} = \frac{3}{\pi}\sqrt{\frac{2}{3}} I_{dc} \quad \text{and} \quad \frac{I_{1rms}}{I_{rms}} = \frac{3}{\pi} = 0.955. \tag{14.17}$$

The total harmonic current is therefore $I_{Hrms} = \sqrt{I_{rms}^2 - I_{1rms}^2} = 0.2966 I_{rms}$ and

$$\text{THD}_i = \frac{I_{Hrms}}{I_{1rms}} = \frac{0.2966}{0.955} = 0.31 \text{ perunit (pu) or } 31\%. \tag{14.18}$$

With line-frequency ac side voltage, only the fundamental ac side current can produce average real power. The average power is, therefore, reduced by the current ratio I_{1rms}/I_{rms}, called the *harmonic power factor*, which is derived from Equation 14.17 to be equal to $3/\pi$ for this rectifier. The firing delay angle α introduces the phase difference $\theta_1 = \alpha$ between the voltage and the fundamental component of the ac side current. This is known as the *displacement power factor*. When L_{ac} is not negligible, it causes further delay in transferring (commutating) the conduction from one thyristor to another by additional angle u, called the *commutating angle*, given by the following relation:

$$\cos(\alpha + u) = \cos\alpha - \frac{2\omega L_{ac}}{\sqrt{2} V_{LLrms}} I_{dc}. \tag{14.19}$$

This equation gives the commutating angle u for given L_{ac} and I_{dc}.

Example 14.6

A three-phase six-pulse thyristor rectifier with a large dc side filter inductor delivers 500-kW power at 525 V dc and draws 60-Hz power from 460 V_{LL} via a cable with leakage inductance of 25 μH. Determine the thyristor's commutation angle.

Solution:

From the given dc side power parameters, we have

$$I_{dc} = P_{dc}/V_{dc} = 500{,}000/525 = 952.4 \text{ A}.$$

Using Equation 14.15, $V_{dc} = \dfrac{3\sqrt{2}}{\pi} 460 \cos\alpha - \dfrac{3 \times 2\pi \times 60 \times 25 \times 10^{-6}}{\pi} 952.4.$

That gives $\cos\alpha = 0.8589$ and $\alpha = 30.8°$.

Using Equation 14.19, $\cos(\alpha+u)=0.8589-\frac{2\times 2\pi\times 60\times 25\times 10^{-6}}{\sqrt{2}\times 460}952.4.$

That gives $\cos(\alpha + u) = 0.831$ and $\alpha + u = 33.8°$.

Therefore, the commutation angle $u = 33.8 - 30.8 = 3°$.

With some approximation, it can be shown that the effective phase difference between $V_{1\text{rms}}$ and $I_{1\text{rms}}$ results in a virtual power factor, called the total *displacement power factor*, which is given by

$$\text{DPF} = \cos\left(\alpha + \frac{u}{2}\right). \tag{14.20}$$

Ignoring small commutation angle u, from Equations 14.17 and 14.20, we have the following:

total ac side power factor = harmonic power factor × displacement power factor.

This equation leads to the following:

$$\text{total ac side power factor} = \frac{3}{\pi}\cos\alpha. \tag{14.21}$$

The harmonics superimposed on the average dc voltage in the output of this converter are of the order

$$h = 6k \pm 1, \text{ where } k = 1, 2, 3,\ldots. \tag{14.22}$$

Smooth dc can be obtained by a filter made of an inductor in series or a capacitor in parallel with the rectified output voltage. The load determines the dc side current and power, that is,

$$\text{DC side load current } I_{\text{dc}} = \frac{V_{\text{dc}}}{R_{\text{Load}}} \quad \text{and} \quad \text{DC side power } P_{\text{dc}} = V_{\text{dc}}I_{\text{dc}}. \tag{14.23}$$

In the steady-state operation, the balance of power must be maintained on both ac and dc sides. That is, the power on the ac side must be equal to the sum of the dc load power and the losses in the rectifier circuit. Moreover, the three-phase ac power is given by $P_{\text{ac}} = \sqrt{3}V_{\text{LL}}I_L pf$. The balance of ac and dc powers, therefore, gives the following from which we obtain the ac-side line current I_{L} for a given I_{dc} load current:

$$\sqrt{3}V_{\text{LL}}I_{\text{L}}\text{pf} = \frac{V_{\text{dc}}I_{\text{dc}}}{\text{converter efficiency}}. \tag{14.24}$$

14.2 AC–AC VOLTAGE CONVERTER

14.2.1 Single-Phase Voltage Converter

The single-phase ac voltage converter steps down a fixed line voltage to a variable voltage on the load side. It uses two back-to-back thyristors or a triac in the circuit shown in Figure 14.6a. A firing delay angle α controls the output voltage by effectively chopping out a portion of each half-cycle of the line voltage. The thyristor T1 is forward-biased during the positive half-cycle and is turned on at an angle α. It then conducts from α to 180° and supplies power to the load. The thyristor T2 is turned on after 1/2 cycle at α + 180°, which then conducts up to 360° supplying power to the load. With a purely resistive load, the load current follows the output voltage in phase as shown by the lower wave in Figure 14.6b. Both the voltage and current on

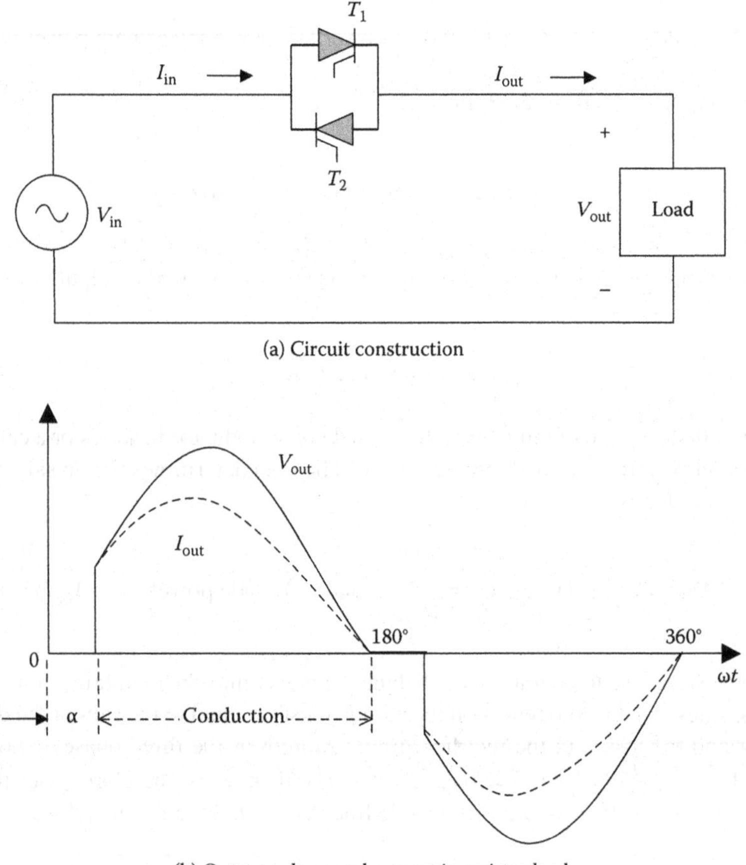

FIGURE 14.6 Single-phase ac voltage converter.

the load side are alternating at the same frequency, although chopped off for angle α in both halves of each cycle. The output rms voltage and current can be shown to be

$$V_{\text{out.rms}} = V_{\text{in.rms}}\sqrt{1-\frac{\alpha}{\pi}+\frac{\sin 2\alpha}{2\pi}} \quad \text{and} \quad I_{\text{out.rms}} = \frac{V_{\text{out.rms}}}{R_{\text{Load}}} \tag{14.25}$$

$$\text{output power } P_{\text{out.avg}} = V_{\text{out.rms}} \times I_{\text{out.rms}} = I^2_{\text{out.rms}} \times R_{\text{Load}}. \tag{14.26}$$

On the source side, the input rms current is the same as that on the load side, that is, $I_{\text{in.rms}} = I_{\text{out.rms}}$. The apparent input power $S_{\text{out}} = V_{\text{in.rms}} \times I_{\text{in.rms}}$, and the load power factor due to delayed current conduction even with a resistive load is

$$\text{load power factor} = \frac{P_{\text{out}}}{S_{\text{out}}} = \sqrt{1-\frac{\alpha}{\pi}+\frac{\sin 2\alpha}{2\pi}}. \tag{14.27}$$

The output voltage is controlled by varying α. With large α, the output voltage and effective power factor decrease even for a purely resistive load, and both become zero when α = π (only in theory, never done in practice).

Since each thyristor carries $I_{\text{out.rms}}$ for one-half cycle, its rms current for self-heating calculations is $I_{\text{Thy.rms}} = I_{\text{out.rms}}/\sqrt{2}$. If a triac were used instead, it would carry current for the whole cycle with $I_{\text{triac.rms}} = I_{\text{out.rms}}$.

With small load-side inductance with a load power factor around 0.85 (as in most practical loads), the load current takes some time to build up initially but also takes about the same time to come to zero due to the inductive inertia of the circuit. The net result is that the current conduction time remains essentially the same as in a resistive circuit. The above equations, therefore, hold approximately true. For purely inductive loads, α must be greater than 90° for the circuit to perform, and the converter operation changes significantly such that the above equations do not hold true even in an approximate sense.

14.2.2 Three-Phase Voltage Converter

The three-phase voltage converter shown in Figure 14.7 is made of three single-phase converters of Figure 14.6a placed in each phase. It reduces the voltage at the output terminals by delaying the thyristor conduction in each phase. The analysis gets complex, as the converter's operating mode changes with the value of α in the ranges of 0°–30°, 30°–60°, 60°–90°, 90°–120°, 120°–150°, and 150°–180°. The power electronics design engineer generally uses simulation software, such as PSPICE, for the performance analysis. However, when the thyristor delay angle α and the load side inductance are small, the single-phase relations presented in Section 14.2.1 hold approximately true with values per phase on the input and output sides. The output voltage will have six pulses, with harmonics of the order $h = 6k \pm 1$, where $k = 1, 2, 3, \ldots$

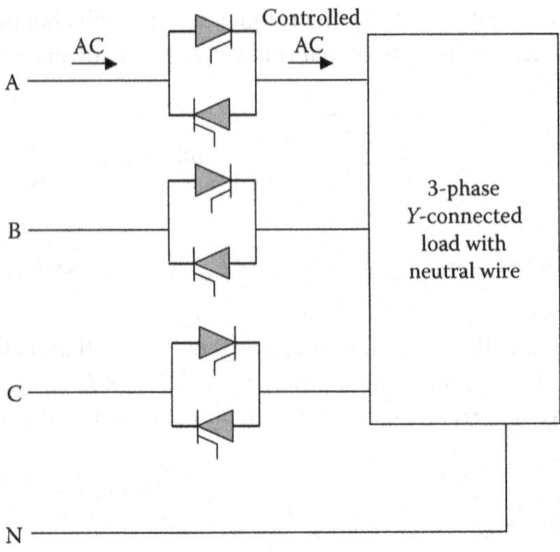

FIGURE 14.7 Three-phase ac voltage converter.

This converter can be used for controlling the power input to a three-phase static load, which gets power in proportion to V_{out}^2. It can also be used for speed control of a single-phase or three-phase induction motor, particularly driving a pump or a fan. With reduced voltage applied at the motor terminals, the torque–speed curve shrinks vertically in V_{out}^2 proportion, keeping the same synchronous speed. The new operating speed is then at the intersection of the motor torque and the load torque characteristics.

14.3 DC–AC INVERTER

The power electronics circuit used to convert dc into ac is known as the *inverter*, although the term *converter* is often used to mean either the rectifier or the inverter. The dc input to the inverter can be from a rectifier, battery, photovoltaic panel, or fuel cell.

14.3.1 Single-Phase Voltage Source Inverter

In its simplest form, the single-phase full-wave voltage source inverter (VSI) is made by using four thyristor switches, shown in Figure 14.8a. The four thyristors are fired on and off sequentially in diagonal pairs for equal duration, that is, each pair for one-half cycle. This results in the load being alternatively connected to + and – polarities of the dc source, thus making the load voltage alternate between + and – polarities, as shown by rectangular pulses in Figure 14.8b. For the first one-half cycle, from time 0 to 1/2 *T*, thyristors T1 and T4 are fired to conduct the load current downward. For the second one-half cycle, from time 1/2 *T* to *T*, the thyristors T3 and T2 are fired to conduct the load current upward. This way, although the source voltage is pure V_{dc}, the load voltage alternates between $+V_{dc}$ and $-V_{dc}$ in a rectangular wave shown in

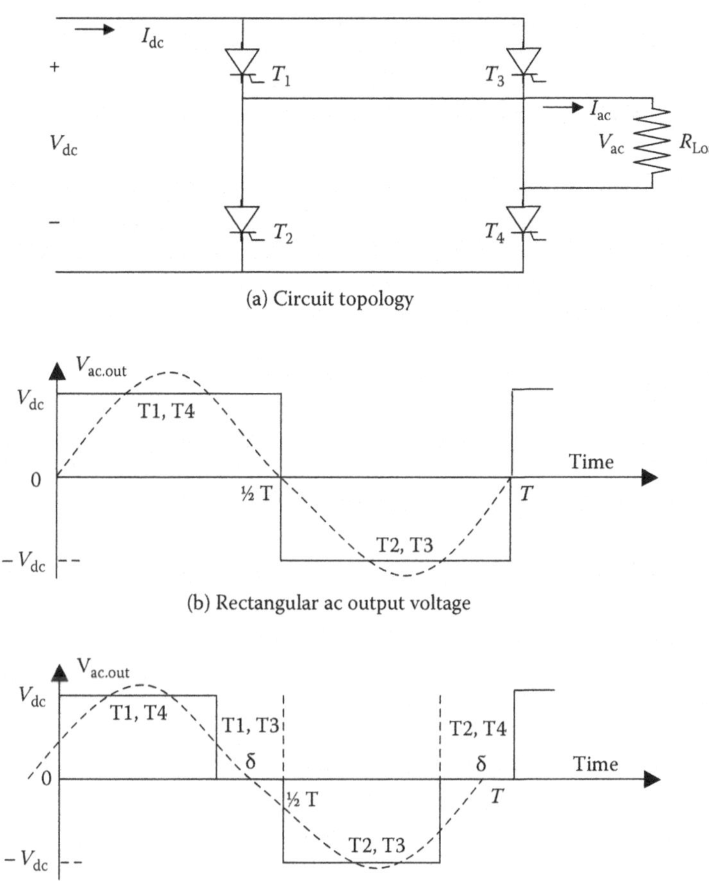

FIGURE 14.8 Single-phase full-wave inverter with resistive load.

Figure 14.8b, with frequency $f = 1/T$, where T = switching period. This rectangular wave ac output voltage can be resolved into the Fourier series of a fundamental sine wave (dotted curve) plus a third harmonic ac and a series of higher-order harmonics. Such output has $V_{1peak} > V_{dc}$, but $V_{1rms} < V_{dc}$. Since the input side rms value is the same as V_{dc}, the total rms values on both sides must be equal, that is,

$$V_{ac.rms} = V_{dc}. \tag{14.28}$$

The Fourier series harmonic analysis would give the fundamental and harmonic voltages as

$$V_{1rms} = \frac{4V_{dc}}{\pi\sqrt{2}} \quad \text{and} \quad V_{hrms} = \frac{4V_{dc}}{h\pi\sqrt{2}} \quad \text{for} \quad h = 3, 5, 7, 11, 13, \ldots, \tag{14.29}$$

and the total voltage harmonic distortion factor, which is rather large, is

$$\mathrm{THD_v} = \frac{V_{\mathrm{Hrms}}}{V_{\mathrm{1rms}}} = \frac{\sqrt{V_{\mathrm{acrms}}^2 - V_{\mathrm{1rms}}^2}}{V_{\mathrm{1rms}}} = 0.483\mathrm{pu} \quad \text{or} \quad 48.3\%. \tag{14.30}$$

V_{ac} can be controlled either by varying V_{dc} or by introducing an additional thyristor switching state in which the output is shorted ($V_{ac} = 0$) by switching either T1 and T3 simultaneously or T2 and T4 simultaneously in turn for angle δ in each one-half cycle, as shown in Figure 14.8c. With such an inverter, the total rms value of the ac output voltage is given by

$$V_{\mathrm{ac.rms}} = V_{\mathrm{dc}}\sqrt{1 - \frac{2\delta}{T}}. \tag{14.31}$$

For inductive loads, we must add freewheeling diodes D1, D2, D3, and D4 to modify the previous circuit in Figure 14.8a to that shown in Figure 14.9a. The function of

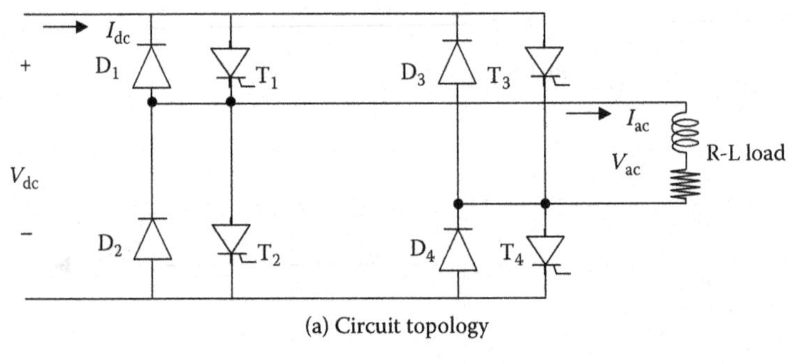

(a) Circuit topology

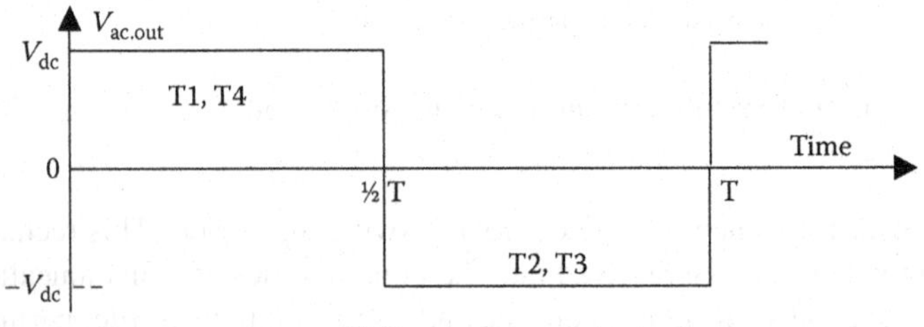

(b) Rectangular ac output voltage

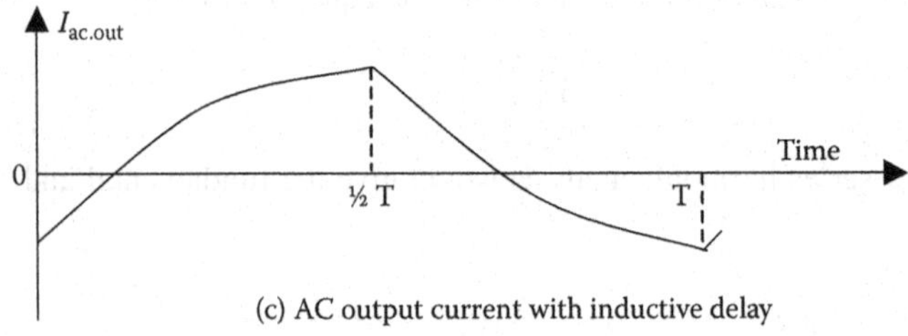

(c) AC output current with inductive delay

FIGURE 14.9 Single-phase full-wave VSI with freewheeling diode for inductive load.

the diodes is to provide a continuing path for the load current soon after the thyristors are turned off until the current naturally comes back to zero. The diodes get turned on and off automatically as required to carry the freewheeling current. The voltage in the new circuit is the same as before, but the current now becomes continuous due to the inductive inertia, as shown in Figure 14.9b and c, respectively.

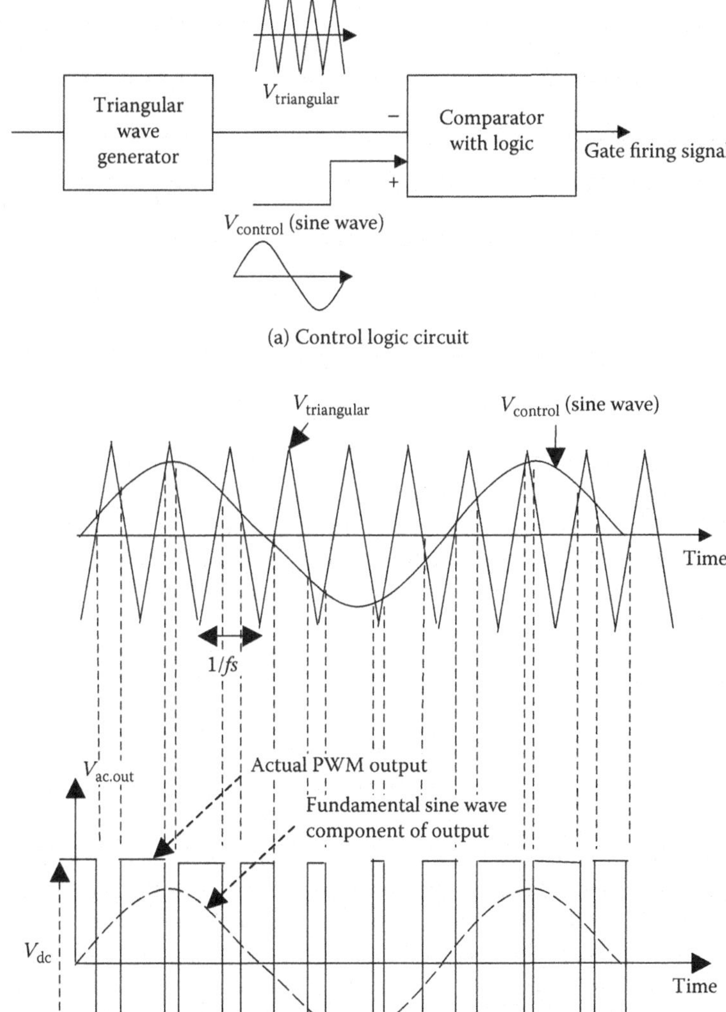

FIGURE 14.10 PWM inverter switching scheme and voltage waves.

14.3.2 Single-Phase Pulse with Modulated Inverter

The large 48.3% voltage harmonics distortion factor in Equation 14.30 with the line frequency switching scheme above can be reduced by using a high-frequency switching scheme, shown in Figure 14.10, known as the *pulse width modulated (PWM) VSI.* Two VSIs shown in Figure 14.8 are stacked vertically in a bipolar (center tap) configuration to share a common dc voltage source V_{dc}, each stack taking +1/2 V_{dc} and −1/2 V_{dc}. The PWM scheme can be implemented by comparing the sinusoidal reference and triangular carrier signal as shown in the middle part of Figure 14.10. The diagonally opposite thyristors T1 and T2 are turned on when $V_{sine.control} > V_{triangular}$, giving $V_o = +1/2\ V_{dc}$. Thyristors T3 and T4 are turned on when $V_{sine.control} < V_{triangular}$, giving $V_o = -1/2\ V_{dc}$. Modulating the switch duty ratio this way varies the dc voltage output in a sine wave. The reference frequency of the sine-control voltage is the desired output frequency of the inverted voltage, and the triangular carrier frequency is the switching frequency at which the input dc voltage polarity is switched with varying duty ratio. Since the output voltage alternates between +1/2 V_{dc} and −1/2 V_{dc} as shown in the lower part of Figure 14.10, such an inverter is called the bipolar switching inverter. The following important definitions apply to such a PWM switching scheme:

$$\text{frequency modulation ratio } M_f = \frac{f_{carrier}}{f_{reference}} = \frac{f_{triangular}}{f_{sine\ control}} = \frac{f_{switching}}{f_{desired}}. \tag{14.32a}$$

Increasing the carrier frequency increases the switching power loss and the harmonic frequencies. The value of M_f generally ranges from 10 to 100 or even much greater. In the PWM converter, note the following differences carefully:

frequency of triangular wave = switching frequency for turning devices *on* and *off*

frequency of sine wave control voltage = desired ac output frequency. (14.32b)

The switching frequency f_{sw} of the triangular control voltage is 20 kHz or higher in modern converters, and the output power frequency set by the sinusoidal control voltage is typically 60 or 50 Hz or even lower in ac motor drives. The higher switching frequency results in lower L-C filter energy storage requirement to power the load during off-time from the source, leading to lower filter weight and cost. For this reason, the switching frequency in some modern low-power converters is 200 to 500 kHz and often approaches 1000 kHz for achieving high power density in some converters for defense aircrafts and satellites, where the component weight has a high premium.

Another important definition in the PWM switching is the following:

$$\text{amplitude modualtion ratio } M_a = \frac{V_{\text{peak-reference}}}{V_{\text{peak-carrier}}} = \frac{V_{\text{peak-sinecontrol}}}{V_{\text{peak-triangular}}}. \tag{14.33}$$

If $M_a < 1$, the peak amplitude of the fundamental frequency output voltage V_1 is linearly proportional to M_a, that is,

$$V_{1\text{peak}} = M_a \cdot V_{dc} \text{ and } V_{1\text{rms}} = M_a V_{dc}/\sqrt{2}. \tag{14.34}$$

Thus, V_1 is linearly controlled by M_a. Such control is used for varying V_1 under fixed V_{dc} or holds the output V_1 constant under fluctuating V_{dc}.

If $M_a > 1$, V_1 increases with M_a, but not linearly, and remains between V_d and $4V_{dc}/\pi$.

If the inverter output load with some inductance draws a sinusoidal current of $I_{\text{out.rms}}$, then each thyristor switch must have a peak current rating of $\sqrt{2}\, I_{\text{out.rms}}$ and a peak voltage rating of V_{dc}. The inverter kilovolt-ampere output is $V_{1\text{rms}} \times I_{\text{out.rms}}$.

The rectangular pulses of varying width in $V_{\text{ac.out}}$ shown in Figure 14.10 can be resolved into the fundamental sine wave and a series of high-frequency harmonics. The harmonic amplitudes in this inverter depend on M_a because each pulse width depends on the relative amplitudes of the sine and triangular waves. The harmonic frequencies are around the integer multiples of the modulation frequency M_f, as listed in Table 14.1. For $h = 2M_f \pm$ few and $h = 3M_f \pm$ few, the $V_{\text{h.peak}}/V_{dc}$ ratio remains around 0.20 and 0.15, respectively, and 0.100 for all higher-order harmonics up to $9M_f$, beyond which they are not a function of M_f.

The PWM switching scheme can also be applied to the three-phase inverter, covered in the following sections. The harmonics in a PWM inverter modulated with 12 and 24 pulses per cycle are listed in Table 14.2. The PWM converter has no fifth and seventh harmonics but has much greater higher-order harmonics. It is important to note that, although the lower-order harmonics are eliminated in the PWM inverter, the higher-frequency harmonics of much higher amplitudes are present as seen in Figure 14.11. This can be damaging to the motor performance and can increase the motor noise in the audible range.

TABLE 14.1

Harmonic Voltages in PWM Inverter for Large M_f (Normalized to V_{dc})

Harmonic Order	$M_a = 0.1$	$M_a = 0.4$	$M_a = 0.7$	$M_a = 1.0$
Fundamental	0.10	0.40	0.70	1.0
$h = M_f$	1.27	1.15	0.92	0.60
$h = M_f \pm 2$	0	0.06	0.17	0.32
$h = 2M_f \pm 3$	0	0.02	0.10	0.20

TABLE 14.2
Percent Harmonics Content of 12- and 24-Pulse PWM Inverters

Harmonic Order	12-Pulse per Cycle PWM	24-Pulse per Cycle PWM
1	100	100
5	0	0
7	0	0
11	40	0
13	40	0
17	5	0
19	11	0
23	13	40
25	13	40
29	18	0
31	9	0

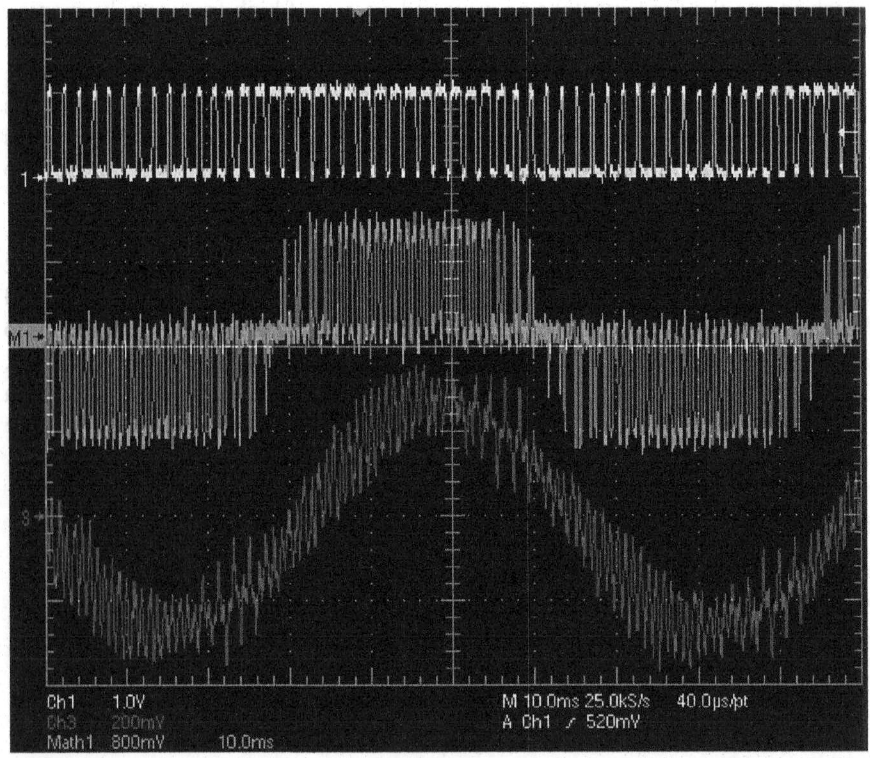

FIGURE 14.11 High-frequency harmonics in PWM converter. (Courtesy of Bill Veit, U.S. Merchant Marine Academy.)

14.3.3 Three-Phase Six-Pulse Voltage Source Inverter

Figure 14.12a depicts a six-pulse VSI fed with constant voltage V_{dc} with a large capacitor. The thyristors are fired in square wave mode for 180° at a time as shown in Figure 14.12b, 120° out of phase in each phase. The resulting step wave alternating line voltage has the fundamental and harmonic voltages as follows:

$$V_{LL1rms} = \frac{\sqrt{3}}{\sqrt{2}}\frac{4}{\pi}\frac{V_{dc}}{2} = \frac{\sqrt{6}}{\pi}V_{dc} = 0.78V_{dc}. \tag{14.35}$$

The output ac voltage magnitude can be controlled only by V_{dc}, and the frequency can be controlled by the switching frequency. The voltage is independent of the load and has harmonics of the order $h = 6k \pm 1$ (k = 1, 2, 3, ….) with line-to-line rms magnitudes given by

$$V_{LLhrms} = \frac{\sqrt{6}}{\pi h}V_{dc} = \frac{0.78V_{dc}}{h}. \tag{14.36}$$

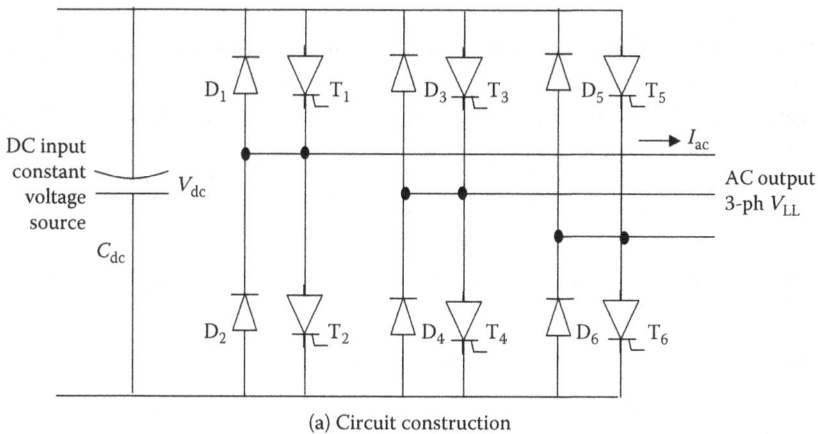

(a) Circuit construction

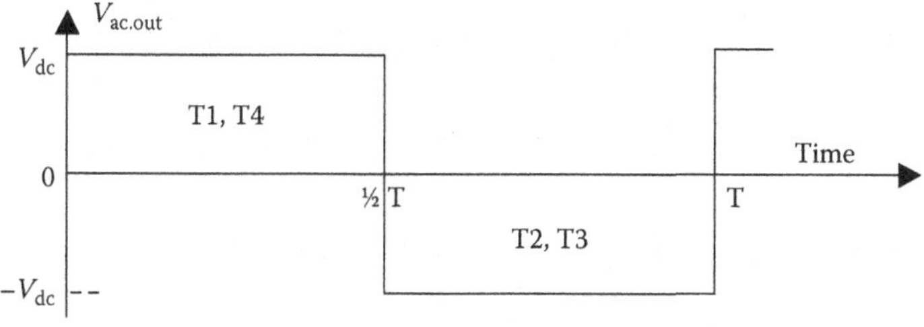

(b) Rectangular ac output voltage

FIGURE 14.12 Three-phase VSI.

Example 14.7

A three-phase six-pulse voltage source inverter (VSI) is fed with 240 V dc. Determine the ac side rms fundamental line-to-line voltage and the first four harmonic voltages.

Solution:

Using Equation 14.35, $V_{LL1rms} = 0.78 \times 240 = 187.2$ V.

Harmonics in the six-pulse VSI are of the order $h = 6k \pm 1$, where $k = 1, 2, 3, \ldots$.

Therefore, the first four harmonics are 5, 7, 11, and 13, with the rms voltage magnitudes 187.2/*h*.

That gives $V_5 = 187.2/5 = 37.44$ V, $V_7 = 187.2/7 = 26.74$ V, $V_{11} = 187.2/11 = 17.02$ V, and $V_{13} = 187.2/13 = 14.4$ V, all rms values because 187.7 V is an rms value.

14.3.4 Three-Phase Six-Pulse Current Source Inverter

The output of a photovoltaic panel at a constant solar intensity is a constant dc current proportional to the panel area. Therefore, the solar panel works like a constant current source, which must be inverted to 60 or 50 Hz ac for local use or for feeding back to the grid. In many variable-frequency motor drives operating with a large dc filter in a dc link, the inverter power source is also a constant current source. We therefore study here, the three-phase current source inverter (CSI) circuit shown in Figure 14.13 that is used in solar photovoltaic power plants and in variable frequency motor drives. The circuit is essentially the same as in Figure 14.12, except that a large input side capacitor is replaced with a large series inductor, which maintains a constant current at the input terminal of the inverter. At any given instant, only two thyristors, as shown in Figure 14.12b by their numbers, are turned on in sequence. Each thyristor conducts for 120°. The three-phase output current is a 120° wide rectangular step wave of magnitude I_{dc}, and the output frequency equals the switching frequency of the thyristors. The current and voltage relations are as follows:

$$\text{In each phase (leg) of the inverter, } I_{out.rms(phase)} = \frac{\sqrt{2}}{3} I_{dc}. \tag{14.37}$$

$$\text{In each output line of the inverter, } I_{out.rms(line)} = \frac{\sqrt{2}}{\sqrt{3}} I_{dc}. \tag{14.38}$$

The output voltage per phase (line to ground) is

$$V_{out.rms} = \frac{V_{dc}}{\sqrt{6}\cos\theta}, \tag{14.39}$$

where $\cos\theta$ = load power factor, that is, $\theta = \tan^{-1}(X_{Load}/R_{Load})$.

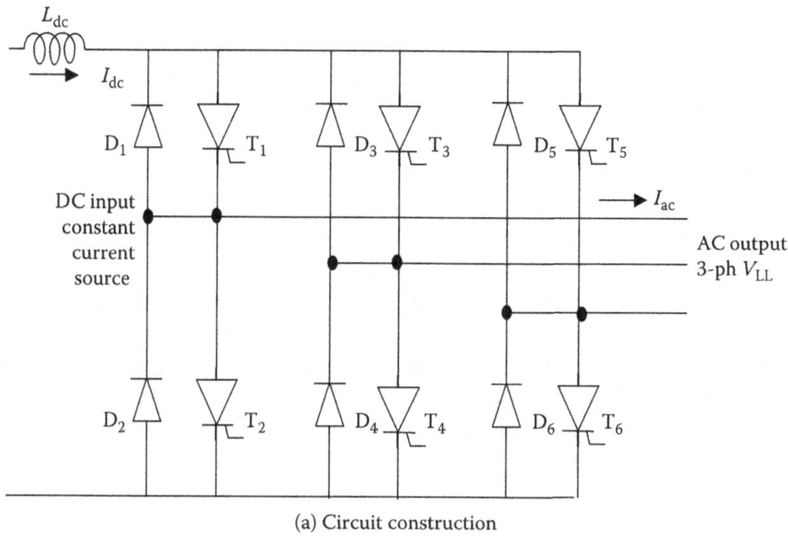

(a) Circuit construction

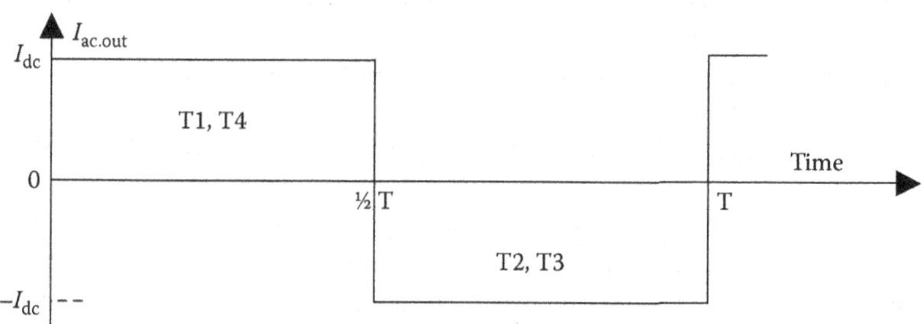

(b) Rectangular ac output current

FIGURE 14.13 Three-phase CSI.

Example 14.8

A three-phase six-pulse CSI is fed from a 240-V dc bus with a dc-link current of 80 A. The ac side load has 0.85 power factor lagging. Determine the ac side rms phase and line currents and output phase and line voltages.

Solution:

Using Equations 14.37 and 14.38,

$I_{ac.rms.phase} = 80\sqrt{2} \div 3 = 37.71$ A and $I_{ac.rms.phase} = \sqrt{3} \times 37.71 = 65.32$ A.

Using Equation 14.39, the ac side phase voltage is

$V_{ac.rms.phase} = 240 \div (\sqrt{6} \times 0.85) = 115.27$ V line-to-ground,

which gives $V_{ac.rms.line} = \sqrt{3} \times 115.27 = 200$ V line-to-line.

Unlike in bipolar junction transistor (BJT), metal–oxide semiconducting field-effect transistor (MOSFET), and insulated gate bipolar transistor (IGBT), the thyristor current, once switched on, must be forcefully switched off (commutated) to cease conduction. Therefore, a converter using thyristors as the switching devices must incorporate an additional commutating circuit—a significant part of the inverter—for the converter performance. There are two main types of inverters: the line commutated and the force commutated.

Line-commutated inverter. This type of inverter must be connected to the ac system into which it feeds power. The design method is matured and has been extensively used in high-voltage dc transmission line inverters. The inverter is simple and inexpensive and can be designed in any size. The disadvantage is that it acts as a sink of reactive power and generates a high content of high-frequency harmonics. Therefore, its output needs a heavy harmonic filter to improve the quality of power at the ac output. This is done by an inductor connected in series and a capacitor connected in parallel of the inverted output voltage, just like it was done in the rectifier covered earlier.

The poor power factor and high harmonic content of the line-commutated inverter significantly degrade the quality of power at the grid interface. This problem has been recently addressed by a series of design changes. Among them is the 12-pulse inverter circuit and an increased harmonic filtering. These new design features have resulted in today's inverter operating near unity power factor and less than 3% to 5% THD. The quality of power at the utility interface at many modern solar power sites now exceeds that of the grid they interface.

Force-commutated inverter. This inverter type does not have to be supplying load and can be free-running as an independent voltage source. The design is relatively complex and expensive. The advantage is that it can be a source of reactive power, and the harmonics content is low.

The power electronics device prices dictate the most economical design for a given system. Present inverter prices are about $1500/kW for ratings below 1 kW, $1000/kW for 1–10 kW, $600/kW for 10–100 kW, and $400/kW for ratings approaching 1000 kW. The efficiency of a dc–ac converter with transformer at full load is typically 85% to 90% in small ratings and 92% to 95% in large ratings.

In addition to the high efficiency in dc–ac conversion, it must have low harmonic distortion, low EMI, and high power factor. The inverter performance and testing standards are covered by IEEE 929-2000 and UL 1741 in the United States, EN 61727 in the European Union, and IEC 60364-7-712 for the international standards. The THD generated by the inverter is regulated by the international standard IEC 61000-3-2. At full load, it requires $THD_i < 5\%$ and the voltage $THD_v < 2\%$ for the harmonics spectra up to the 49th harmonic. At partial loads, the harmonic distortions are usually much higher.

14.4 FREQUENCY CONVERTER

Two main types of frequency converters are (1) the dc-link converter, which first converts ac of one frequency into dc voltage or current using a rectifier, and then inverts dc into another frequency ac using the voltage source or current source inverter, and (2) the cycloconverter, which converts one frequency ac directly into another frequency ac by ac–ac conversion. In the past, the cycloconverter was more economical than the dc-link inverter. With decreasing prices of power semiconductors, it has been gradually replaced by the dc-link converter, which gives lower harmonics in the output. However, recent advances in fast-switching devices and microprocessor controls are increasing the efficiency and power quality of the cycloconverter, which may come back in large power applications.

14.4.1 DC-Link Frequency Converter

A dc-link frequency converter consists of a phase-controlled (α-controlled) three-phase rectifier feeding a dc link with a large filter inductor and a three-phase inverter as shown in Figure 14.14. The large inductor in a dc link works like a constant current source for the inverter. The performance of this converter can be analyzed by cascading the performances of the three-phase rectifier and the three-phase current source inverter (CSI) covered in Sections 14.1 and 14.3. The advantage of a dc-link converter is a common design that can be developed for both the induction motor and the synchronous motor of comparable ratings. For this reason, it is widely used in variable-frequency drives for ac motors, as discussed further in Chapter 15.

The analytical formulation of the instantaneous dc-link current is complex. However, it can be simplified by the conservation of power relation, which equates the instantaneous dc power $V_{dc} \times I_{dc}$ with the sum of ac power in three phases, leading to the following simple relation:

$$i_{dc} = \frac{v_a i_a + v_b i_b + v_c i_c}{V_{dc}}. \tag{14.40}$$

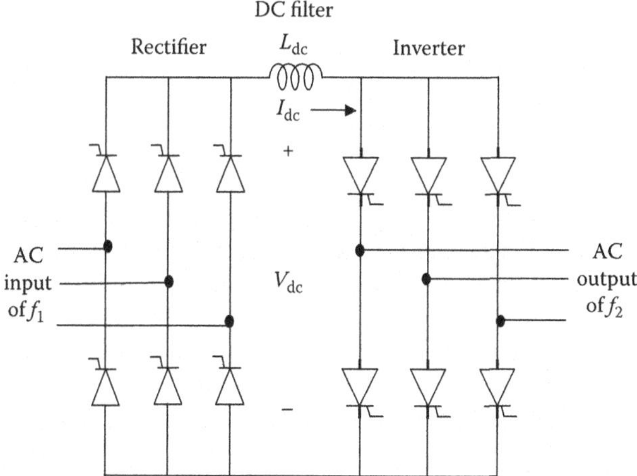

FIGURE 14.14 Three-phase dc-link frequency converter.

In the same manner, $I_{dc.rms} = \sqrt{I^2_{dc.avg} + I^2_{ripple.rms}}$, which gives the ripple current:

$$I_{ripple.rms} = \sqrt{I^2_{dc.rms} - I^2_{dc.avg}}\,. \tag{14.41}$$

The dc-link filter inductor L_{dc} is designed to smooth out ripples caused by the source and load side converters while keeping the ripple current low. It should have the current rating equal to I_{dc} and low resistance to keep the I^2R loss low. If $I_{ripple(p\text{-}p)}$ = permissible ripple current (peak-to-peak), $V_{ripple.in}$ and $V_{ripple.out}$ = ripple voltage on the line and load sides, respectively, and Δt = time increment between the ripple current peaks, then

$$L_{dc} = (V_{ripple.in} - V_{ripple.out})\, \Delta t \div I_{ripple(p\text{-}p)}. \tag{14.42}$$

A smaller ripple current requirement results in a larger inductor. The ripple current is generally worse at the upper edge of the constant-torque region of the motor drive operation. The size of inductor L_{dc} is usually much larger in the load-commutated inverter than in the high-frequency Change to pulse width modulated (PWM) inverter. In construction, L_{dc} is usually an iron-core inductor with an air gap to store energy and prevent magnetic saturation.

Example 14.9

Determine the dc-link voltage for a variable frequency drive (VFD) to drive a three-phase 460-V induction motor.

Solution:

Using Equation 14.35, $V_{LL.rms} = \sqrt{6} \times V_{dc} \div \pi$, from which we obtain

$$V_{dc} = \pi \times 460 \div \sqrt{6} = 590 \text{ V}.$$

14.4.2 Single-Phase Cycloconverter

The circuit shown in Figure 14.15a is similar to the three-phase, three-pulse, half-wave, phase-controlled rectifier fed from ac source voltage V_{rms} of frequency f_1. Its dc output voltage, as was analyzed in Section 14.1.1, is given by Equation 14.1b, that is, $V_{dc0} = 0.90V_{rms} \cos\alpha$, where α = conduction delay angle. With fixed V_{rms}, the ratio $V_{d0}/V_{rms} = 0.90 \cos\alpha$ changes with the delay angle α. It becomes zero at $\alpha = 90°$ and negative for $\alpha > 90°$. Such variations in V_{dc} polarity can be used to change frequency by the conduction delay angle control as shown below:

Conduction delay angle $\alpha°$	0	30	60	90	120	150	180
V_{dc}/V_{rms} ratio	0.90	0.78	0.45	0.00	–0.45	–0.78	–0.90

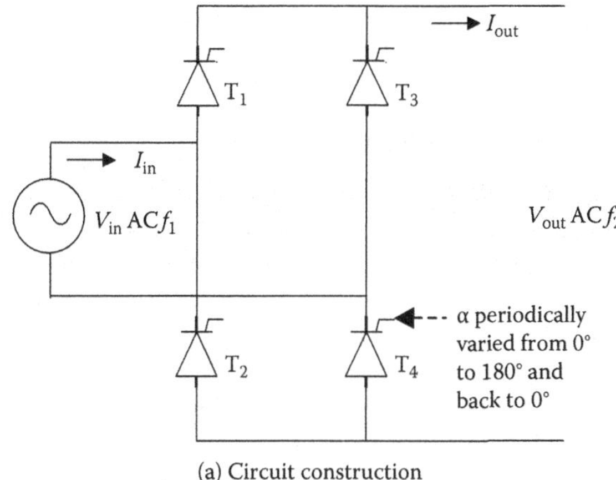

(a) Circuit construction

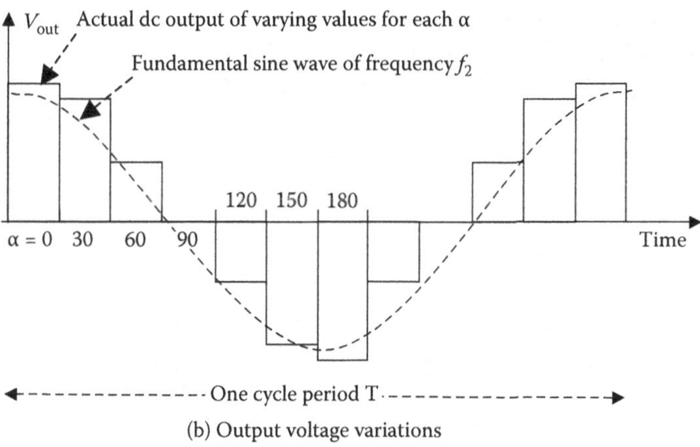

(b) Output voltage variations

FIGURE 14.15 Single-phase cycloconverter.

As seen in the above table, if the firing delay angle is controlled over one-half cycle from 0° to 180°, V_{dc} as the multiple of V_{rms} varies in discrete steps from a maximum value of 0.90 to 0 to –0.90. Such variations are plotted in Figure 14.15b, which represents one-half cycle of the ac output made of discrete steps of various V_{dc} values. In such cycling of firing angle α, V_{dc} at any given instant depends on α in force at that instant. If the variations in α continue in small time intervals with a repetition frequency f_2, then the converter output is also the ac of frequency f_2. The superimposed harmonics can be made small by modulating α in small steps.

Notice that we have used the single-phase rectifier of Section 14.1.1 to rectify ac by cycling the firing delay angle α between 0° and 180° to obtain the output of desired frequency. This gives an inherent limitation in the cycloconverter operation; its output frequently can only be lower than the input frequency, with the high end of the output frequency about one-third of the input line frequency in practical designs.

The firing delay angle α varying over a wide range makes it a harmonic-rich converter, requiring a heavy filter.

14.4.3 Three-Phase Cycloconverter

The high harmonic content in the single-phase cycloconverter can be reduced by using the three-phase cycloconverter circuit shown in Figure 14.16. Each output phase consists of a single-phase cycloconverter, shown earlier in Figure 14.15. The conduction delay angle α in each rectifier and inverter in each phase is cyclically controlled to yield a low-frequency sinusoidal output voltage. Again, due to the harmonic limitations, the maximum output frequency is limited to about one-third of the input line frequency. For this reason, the three-phase cycloconverter is generally used with a low-speed, high-power induction motor or synchronous motor.

The same thyristor commutating techniques are used in the cycloconverter as in the ac–dc rectifier. Some high-power cycloconverters use six thyristors for the positive wave voltages and another six thyristors for the negative wave voltage in each phase, thus 36 thyristors in total for the three-phase design, which is many more than the dc-link frequency converter. The conduction angle cycling scheme controls the ac output voltage, and the firing frequency controls the output frequency. Both the α-cycling and frequency controls are complex. The cycloconverter has been used in

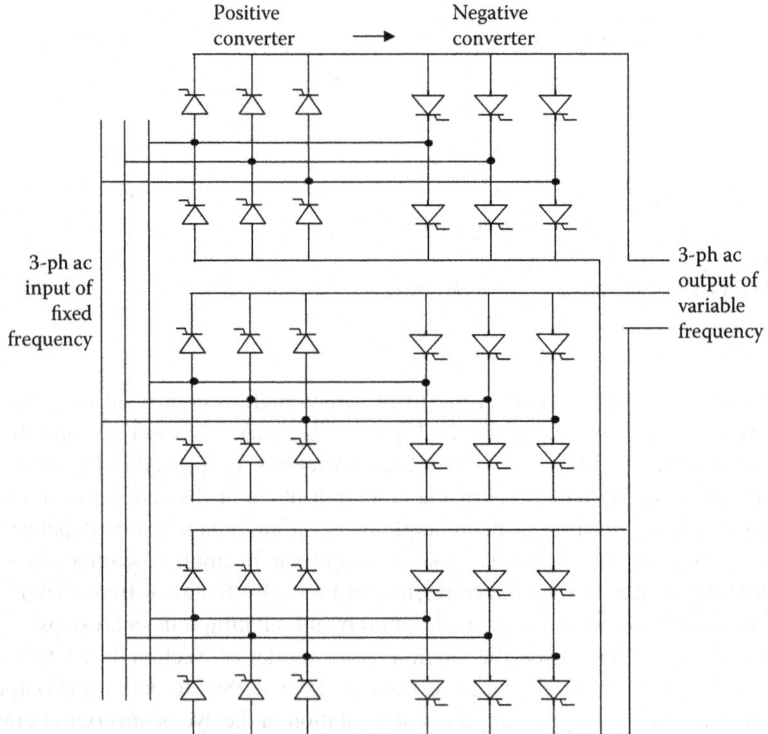

FIGURE 14.16 Three-phase cycloconverter.

the past but was generally out of favor for many years due to complex controls and many more switching devices, which makes it more expensive, less reliable, and less efficient. The new power electronics devices and new cycloconverter designs, however, are gradually bringing this old technology into a new light.

Example 14.10

A 3-ph, 5000-hp, 12-pole, 60-Hz synchronous motor uses a cycloconverter to vary speed. Determine the maximum speed at which it can be operated to limit the harmonics to the generally acceptable level.

Solution:

The maximum output frequency of the cycloconverter must be limited to one-third of the input frequency, that is, 60/3 = 20 Hz in 60-Hz drives. At this frequency, the 12-pole motor can run at a maximum speed of $120 \times 20 \div 12 = 200$ rpm.

14.5 THYRISTOR TURNOFF (COMMUTATION) CIRCUITS

The thyristor can be turned off (commutated) only by reducing the current below the holding current I_H and applying a reverse bias voltage for a certain minimum duration of time called the turnoff-time t_{off}. Only then does it recover the forward blocking characteristic for continuing operation. A variety of commutating circuits can be designed for this purpose as discussed below.

14.5.1 Line Commutation

In the line frequency converter, the thyristor current is naturally turned off when the thyristor is reverse-biased periodically under a sinusoidal voltage. For example, in a simple circuit shown in Figure 14.17a, the thyristor current naturally comes to zero every one-half cycle, when it gets turned off without an additional commutation circuit. When there are multiple thyristors, as in the full-wave rectifier, the thyristor current is naturally commutated, and the device turns off when the next thyristor in the sequence is fired on. This is not true in the switchmode converter, where the thyristor current needs to be forcefully turned off using an additional commutation circuit.

14.5.2 Forced (Capacitor) Commutation

In the switchmode converter, a forced commutating circuit, shown in Figure 14.17b, is needed to turn off the thyristor. It uses the capacitor energy to send the counter-current to bring the thyristor current to zero. The + and – polarities are those of the precharge voltage on the capacitor before the closure of the auxiliary switch to turn off the main thyristor. The inductor controls the di/dt stress on the switch, and the diode provides the freewheeling path for the current when needed.

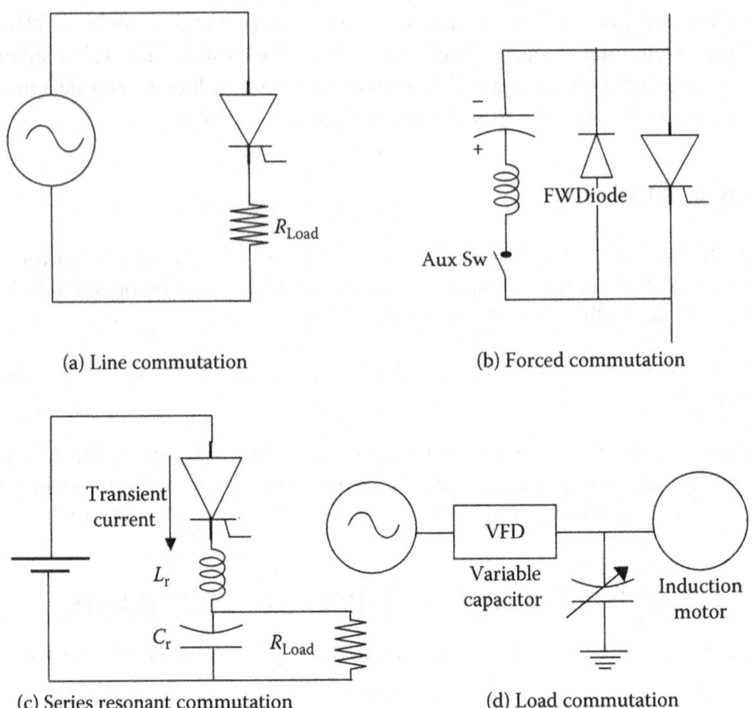

FIGURE 14.17 Thyristor turn-off (commutation) circuits.

14.5.3 Resonant Commutation

The resonant converter uses additional inductance L_r and capacitor C_r to cause the converter circuit to resonate locally at the switch as shown in Figure 14.17c. Such a resonant circuit results in transient dc current and voltage oscillating between zero and peak values, and the switching is done at the natural zero of current or voltage. The resonant converter is typically a full-wave converter that includes the inductor and the capacitor as the resonant elements.

14.5.4 Load Commutation

For an inductive (R-L) load, static or dynamic, as shown in Figure 14.17d with induction motor load, the thyristors in the inverter require forced commutation. However, with a parallel capacitor as shown, a resonance is created with the load circuit, and the load current commutation is possible. The inverter frequency should be slightly above the resonance frequency, so that the effective load power factor is leading. The capacitor should be variable to match with the load parameters for satisfactory commutation. This is especially true with the induction motor with variable frequency drives, where the motor frequency can vary over a wide range, and the leading power factor can be maintained at all frequencies only with a variable capacitor.

The synchronous motor is naturally suitable for the load commutation without the capacitor bank, as the motor itself can be operated at a leading power factor with an overexcited rotor field coil. For that reason, the synchronous motor is better suited to work with a VFD.

14.5.5 ZCS and ZVS

Some high-power converters with high switching frequency employ zero current switching (ZCS) or zero voltage switching (ZVS). The switching is done when the voltage or current is passing through natural zero on a sinusoidal cycle, so that the current or voltage is not abruptly interrupted from a high value to zero in a very small time (in microseconds or even less). Thus, the ZVS and ZCS converters eliminate high d*i*/d*t* and d*v*/d*t* stresses, giving a *soft operation*. They result in higher efficiency, smaller size, lighter weight, and better dynamic response.

14.6 OTHER POWER ELECTRONICS APPLICATIONS

14.6.1 Uninterruptible Power Supply

Certain essential loads need uninterruptible power supply (UPS) to continue critical operations even when the main power source is down following a system fault. The UPS comes in two types:

Type 1: provides critical power until a backup generator comes online in a short time (in minutes). This type has small kilowatt-hour ratings.

Type 2: provides specified power for specified duration (in hours). This type has large kilowatt-hour ratings (e.g., 10 kW for 1 h).

In either type, the required backup energy is stored in a battery or in a flywheel in some newly developed UPS units. The stored energy is discharged when needed. The system schematic is shown in Figure 14.18. In normal operation, the line frequency power is rectified to charge the battery, first at a full charge rate and then tapered down to the trickle charge rate. In case of the ac line outage, the automatic bus transfer (ABT) switch disconnects the battery from the rectifier and connects it to the inverter. The battery power is then inverted to the line frequency ac and supplied to the critical loads via the same distribution cables. The harmonics in the output power of the rectifier and the inverter are filtered to improve the quality of power.

14.6.2 Static kVAR Control

As we recall from the electrical machines theory, the unloaded synchronous motor with overexcited rotor field draws the armature current at a leading power factor. Such a capacitor-type machine—called the synchronous condenser—is often used in large systems for power factor improvement. The synchronous condenser can provide variable kVAR to match with the system need. When the vibration and noise coming from a rotating machine are objectionable, large static capacitors are used

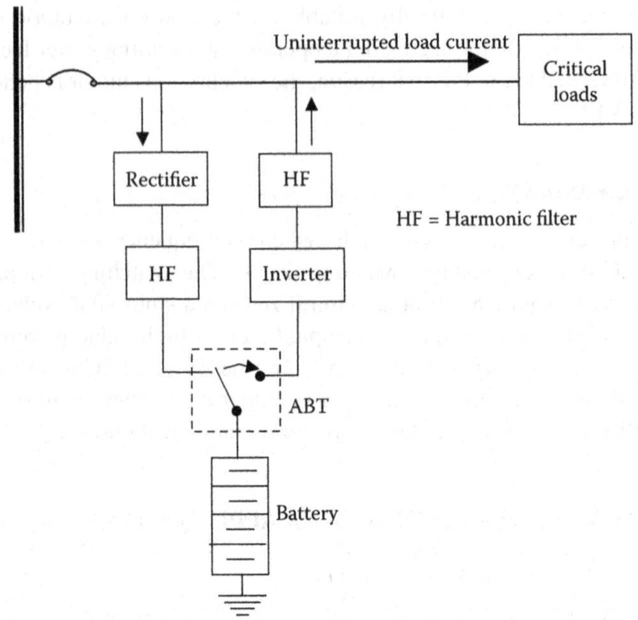

FIGURE 14.18 UPS functional diagram.

with a scheme to switch some capacitors on and off. That varies the capacitor kVAR with changing load power factor to keep the main line power factor near unity at all times. The power electronics kVAR compensation shown in Figure 14.19 is often used to provide variable kVAR to the load. A fixed base capacitor is connected to provide a fixed kVAR, and the variable kVAR is obtained by the thyristors with variable firing delay angle α necessary to maintain the unity power factor at the load terminals. The capacitors with the thyristors can even be switched off if both thyristors in each phase were not fired at all. A small inductor (not shown) is placed in series with each thyristor, or else damaging inrush current may result when the thyristor is

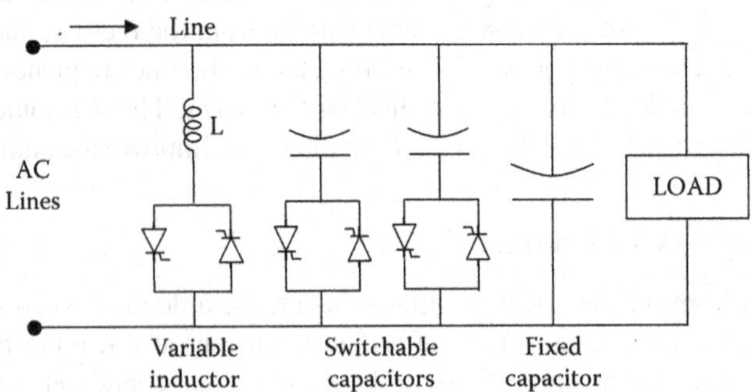

FIGURE 14.19 Static variable kVAR control.

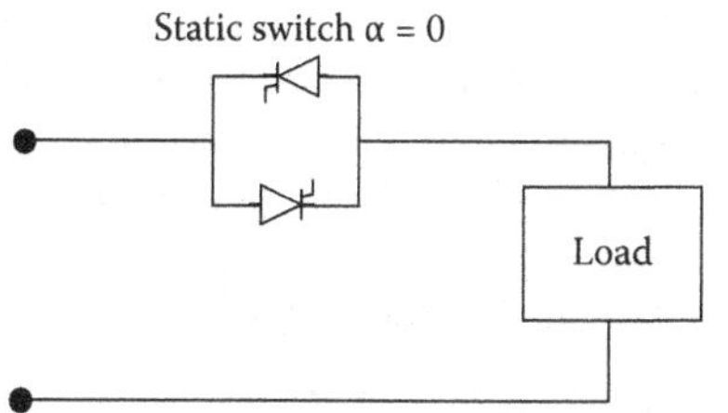

FIGURE 14.20 Static (solid-sate) switch and relay.

switched directly across the fixed capacitor. The inductor slows down the d*i*/d*t* rate in the thyristor to its design limit. In case the load power factor becomes leading, it can be corrected to unity by introducing a variable inductor, shown in the left-hand side in Figure 14.19. Such static kVAR control has infinitely fine resolution on kVAR compensation, as opposed to discrete resolution with capacitors switched on and off. Therefore, it is increasingly used in modern power systems.

14.6.3 Static Switch and Relay

The power electronics device can be used to turn on or off high power without using mechanical contacts, which may spark and wear out after a certain number of operations. The static switch or static relay—also known as the solid-state switch or solid-state relay (SSR)—has no moving parts, and hence has no wear-out failure mode, and lasts billions of operations over decades. It is made of back-to-back thyristors or a triac in ac power circuits, as shown in Figure 14.20. The SSR finds increasing applications for switching a variety of loads, such as motors, transformers, lighting, heating, etc. It can control a large power by a small signal from the control circuit. The light-emitting diode with an optical coupler can provide electrical isolation for safety. The SSR is usually designed for single-pole applications with triac normally off (i.e., normally open contacts).

14.7 COMMON CONVERTER TERMS

Widely used terms in discussing the power electronics converters are summarized below.

Switchmode converter (SMC): converter in which the power electronics devices are used as switches (as opposed to amplifiers).

PWM converter: converter in which the output voltage and frequency are controlled by adjusting the pulse width of the gate signal (i.e., *on* and *off* time). A switchmode converter can also be a PWM converter.

Voltage source inverter (VSI): it maintains a constant voltage on the inverter input side by using a large capacitor. If the dc side capacitor C_{dc} in Figure 14.12 were made infinitely large, it would become a VSI.

Currents source inverter (CSI): it maintains a constant current on the inverter input side by using a large inductor. If the dc side inductor L_{dc} in Figure 14.13 were made infinitely large, it would become a CSI.

Commutation: it refers to turning off one thyristor and shifting the current to another. It can be natural or forced by additional circuits.

Flyback converter: the dc–dc converter in which the stored magnetic energy in one inductor coil jumps to another inductor coil—flying from the source side to the load side on the back—when the switch opens. The magnetically coupled flyback inductor coils essentially form an air-core transformer with turn ratio N_{in}/N_{out} that can be less than or greater than one to buck or boost the voltage as required in the design.

Forward converter: the dc–dc buck converter with a transformer on the load side (forward in the power flow direction). The transformer turn ratio N_{in}/N_{out} can be less than or greater than one to buck or boost the voltage as required in the design.

14.8 NOTES ON CONVERTER DESIGN

Fuse protection. Since the power electronics device has a very small thermal capacity, it must be protected from heavy fault current. This is commonly done in thyristor circuits by using the fast-acting fuse. The thyristor has two types of current ratings: (1) the maximum continuous rms current and (2) the nonrepetitive peak current for one-half cycle. It also has a subcycle I^2t rating for time duration between 1.5 and 8.35 or 10 ms (one-half cycle of 60 or 50 Hz). As a rule, both the peak current and I^2t ratings of the fuse must be lower than those of the device. A similar protection principle applies to the diode as well. However, since the fast fuse is expensive, a new trend is to design a fuseless thyristor converter. For insulated gate bipolar transistor (IGBT), the fuse protection is not necessary because its gate blocking capability is lost at high fault current. The same is true for metal–oxide semiconducting field-effect transistor (MOSFET).

Converter with both sides inductive or capacitive. In the phase-controlled thyristor converter, both sides can be inductive since the current is commutated from the outgoing to the incoming set of devices. However, the thyristor converter cannot operate if both sides are capacitive due to the capacitor charging inrush current problem. In general, a pulse width modulated (PWM) converter requires one side to be inductive and the other side capacitive.

Device power rating and heat sink size. The converter rating is determined by two limiting parameters of the power electronics devices: the peak current I_{pk} and the maximum junction temperature T_{jmax}. The converter power losses are primarily made of the ohmic conduction loss and the switching loss. At high power, the losses and the junction temperature increase. At rated power, the steady-state junction temperature T_j must be below T_{jmax}. If the cooling is improved, T_j will decrease, and the steady-state power loading can be increased up to the limit of I_{pk} and T_{jmax}. Or, with the same cooling (i.e., with the same heat sink), the short-time power rating can be higher (overload) within the peak current limit of the device for a certain time duration within the T_{jmax} limit. Thus, improved cooling increases the converter steady-state rating up to the peak current limit. It also improves the overload capability for a certain time duration limited by the thermal inertia and T_{jmax} of the device.

PWM switching frequency. The switching frequency in a motor drive design is selected to minimize the total power loss in the converter and the motor combined. The converter switching loss increases with the switching frequency. However, the motor harmonic copper loss decreases at higher switching frequency due to lower harmonic current magnitudes, although the harmonic iron loss increases by a relatively small amount. Switching speed of the device also enters in the design consideration. Faster devices have low switching loss, and vice versa. The total losses in the converter and motor combined are plotted with the switching frequency to determine the optimum switching frequency for the minimum total loss. The following factors may also contribute in determining the optimum switching frequency.

- Switching losses in IGBT and IGCT are lower than those in GTO. Therefore, the IGBT and IGCT generally operate at higher switching frequencies.
- Harmonic filter weight, volume, and cost get smaller at higher switching frequencies.
- Acoustic noise is another consideration in selecting the switching frequency. In general, a higher switching frequency results in lower acoustic noise.

Electrolyte capacitor size. The large dc-link filter capacitor size in the VSI is determined by the harmonic current filtering requirement while keeping the temperature rise due to the capacitor's equivalent series resistance (ESR) within the safe design limit—otherwise, the capacitor may explode. For the three-phase inverter in UPS with a battery, most of the harmonic currents must be filtered out by the capacitor to keep the battery life from degrading due to the harmonic current heating.

IGBT versus IGCT for high-power applications. IGBT PWM converters have been used in few-megawatts applications and are moving into the higher-power applications. The IGCT—introduced in 1996—is suited to even higher power than the IGBT. The two are similar in construction but have the following operational characteristics:

- IGBT is a voltage-controlled device, whereas IGCT is a current-controlled device.
- Both devices are suitable for PWM frequency below 1 kHz.
- Both devices can be used in series–parallel combination for high voltage and high current as needed in high-power applications.
- Neither device needs commutating circuit in operation.
- IGCT has a lower conduction drop, giving higher efficiency.

In general, the IGCT gives higher efficiency and needs fewer devices, so it is better suited for high-power applications.

Double-sided PWM converter versus phase-controlled cycloconverter. The phase-controlled cycloconverter has been used in tens of megawatts four-quadrant motor drives for icebreaker ships, steel rolling mills, and mining applications.

However, the recent trend is to use a double-sided PWM converter because the cycloconverter has the following disadvantages: (1) the current commutation often fails if the line voltage dips, (2) it generates complex harmonic currents on the line and load sides, which are difficult to filter and may cause excessive losses and magnetic saturation in the motor and line side transformer, and (3) it results in poor line side displacement power factor.

The cycloconverter, however, being a direct frequency converter without a dc link, does not need an intermediate energy storage element.

EMI in PWM converter. Modern PWM converters need many control signal circuits in close proximity with the power circuit. This creates a difficult EMI problem, particularly at high switching frequencies. Analyzing and solving the EMI problem is often difficult as it does not lend itself to a clean analysis. The common-mode EMI generated by an IGBT snubber circuit is particularly difficult to filter. The following design practices generally control the EMI.

- Twisted and shielded signal wires with shield grounded to a low-resistance plane
- Low leakage inductance in the power cables
- Greater physical distance between the power and signal cables
- Using differential and common-mode EMI filters
- Lower switching frequency
- Grounding the metal frame of transformer, converter, and the motor with a low-resistance ground wire

VSI versus CSI converter for main frequency power. The VSI is a matured and less expensive design, available in 6-pulse, 12-pulse, and 18-pulse configurations. It provides voltage to the motor in a manner typical of the normal power lines, except at a variable frequency and variable voltage. The CSI, on the other hand, acts as a constant current source to the motor and hence is not a popular converter in the power industry. One shortcoming of the CSI is the induction motor crawling at low speed (one-fifth of synchronous speed) due to the magnetic phase belt harmonics in space, which rotates at main synchronous speed ÷ space harmonic order. For example, the fifth space harmonic in a two-pole motor flux rotates at $3600 \div 5 = 720$ rpm, causing the motor to run around that speed even at 60 Hz. Since this can be a problem, the range of speed control with CSI is not as large as other drives. The large dc filter required in the dc link can be bulky, often as large as the motor. The advantage of the CSI is that it is a bidirectional converter, which can be used in the regenerative braking scheme to convert the kinetic energy of the load inertia into electrical power and feed back to the source. That can be simply accomplished by having the rectifier delay angle greater than 90°.

VSI versus CSI converter for 10 to 20 kHz power. The U.S. Navy often uses high-frequency power converters to meet the low noise standards. For this application, both the CSI and VSI can be considered with the trade-offs listed in Table 14.3.

Power electronics is a very wide and developing subject. The intent of this chapter has been to present some basic converters used in electrical power systems. Further details can be obtained from books dedicated to power electronics.

TABLE 14.3
Performance Trade-Off between VSI and CSI for 10–20 kHz Power

Performance Parameter	Voltage Source Converter	Current Source Converter
Circuit configuration	Series resonant	Parallel resonant
Inverter frequency	Tracks the load resonance frequency	Tracks the load resonance frequency
Load power factor required for soft switching to eliminate switching losses	Lagging	Leading
Overall comparison	Lower cost, higher efficiency, and higher frequency possible	Higher cost, lower efficiency, and frequency limitation

PROBLEMS

1. A single-phase full-wave rectifier shown in Figure 14.1 is built using a thyristors. Determine the average and rms values of the current in each thyristor when the firing delay angle $\alpha = 0$ and the input voltage is (1) 240-V, 50-Hz ac, or (2) 120-V, 60-Hz ac.
2. A single-phase full-bridge bridge rectifier using phase-controlled thyristors with a large L_{dc} converts 50-Hz, 240-V rms to dc for a dc side load resistance of 12 Ω. For the firing delay angle $\alpha = 30°$, determine the dc side voltage and the power absorbed by the load, ignoring the cable inductances and ripples.
3. The rectifier of Example 14.2 with the ac side cable inductance of 3 mH and the dc side cable resistance of 1.0 Ω is charging a battery at a 10-A charge rate instead of powering the 10-Ω resistance. Determine the voltage available at the battery terminal.
4. A half-wave diode rectifier draws power from a single-phase, 240-V, 50-Hz source and delivers to a 3-kW load on the dc side at constant voltage V_{dc} with a large filter capacitor as shown in the figure below. The ac side harmonics to fundamental current ratios I_h/I_1 in percentage are 80, 40, 10, 6, 4, 3, 1, and 0.5 for the harmonic orders h = 3, 5, 7, 9, 11, 13, 15, and 17, respectively. Determine the rms ripple current in the filter capacitor. Ignore all losses.

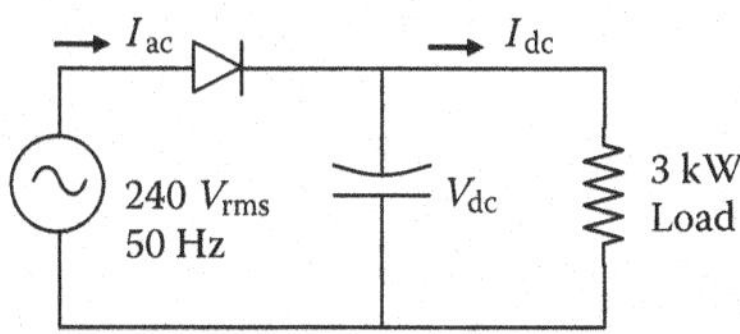

5. A 460-V, three-phase, full-bridge, six-pulse rectifier shown in Figure 14.4 is built with diodes and a large inductor. Determine the average and rms

values of the current in each diode if the output load current is (1) 50-A dc and (2) 135-A dc.

6. A three-phase, six-pulse thyristor rectifier with a large dc side filter inductor delivers 200-kW power at 240-V dc and draws 60-Hz power from 220 V_{LL} via a cable with leakage inductance of 30 μH. Determine the thyristor's firing delay angle and the commutation angle.
7. A three-phase six-pulse VSI is fed with 480-V dc. Determine the ac side rms fundamental line-to-line voltage and the first four harmonic voltages.
8. A three-phase six-pulse CSI is fed from a 120-V dc bus with dc-link current of 60 A. The ac side load has 0.90 power factor lagging. Determine the ac side rms values of (1) phase and line currents and (2) phase and line voltages.
9. Determine the dc-link voltage for a variable frequency drive (VFD) to drive a three-phase 6600-V induction motor for ship propulsion.
10. A three-phase, 1000-hp, eight-pole, 50-Hz synchronous motor uses a cycloconverter to vary speed. Determine the maximum speed it can be operated at to limit the harmonics to the generally acceptable level.

QUESTIONS

1. Fill in the table below with the key performance features of controllable power electronics switches covered in this chapter.

TABLE Q14.1
Key Performance Features of Controllable Power Electronics Switches

Switch Name	SCR	BJT	MOSFET	GTO	IGBT
Power terminal name					
Control terminal names					
Latching or nonlatching					
Saturating or nonsaturating					
Control signal, voltage, or current					
Control signal, pulse, or continuous					
Relative power handling capability					
Relative switching frequency					

2. What is the difference between the terms *ripples* and *harmonics*?
3. Explain the difference between the firing delay angle α and the commutation delay angle u.
4. Identify the source of the load power factor, displacement power factor, and harmonic power factor.
5. What is the principal difference between CSI and VSI?
6. Explain the difference between the snubber circuit and the commutation circuit.
7. Decades ago, large uninterruptible power supply design in ac systems used

an ac electrical machine coupled to a dc machine connecting to a battery to convert the ac into dc and vice versa. Draw such an electromechanical UPS schematic incorporating the ac bus and the user loads. Today, the electrical machines have been replaced by bidirectional power electronics converters. Discuss the pros and cons of using the electrical machines versus power electronics in UPS.

FURTHER READING

Ahmed, A. 1999. *Power Electronics for Technology.* Upper Saddle River: Prentice Hall.

Ang, S. and Oliva, A. 2010. *Power Switching Converters.* Boca Raton: CRC Press/Taylor & Francis.

Bose, B. 1996. *Power Electronics and Variable Frequency Drives.* New York: IEEE Press.

Hart, D. 2011. *Power Electronics.* New York: McGraw Hill.

Luo, F.L. and Ye, H. 2010. *Power Electronics Advanced Conversion Technologies.* Boca Raton: CRC Press/Taylor & Francis.

Mohan, N., Undeland, T.M., and Robbins, W. 2003. *Power Electronics Converters, Applications, and Design.* New York: John Wiley & Sons/IEEE Press.

15 Variable-Frequency Drives

The power electronics motor drive comes in two major types: (1) the servo motor drive to control the precise speed and position of the rotor and (2) the motor drive for varying motor speed in response to a desired change in the load output, such as the air ventilation and fluid pumping rate. The response time and precision in position are important in the servo drive but not in the variable-speed motor drive. Our interest in this chapter is primarily focused on the variable-speed motor drives.

The variable-frequency drives (VFDs) for ac motors were initially developed in the 1970s for oil refineries, where numerous motors are used to pump oil all year around. Although costly at the time, the significant energy saving at the end of every month paid back the drive cost within a few years. Then onward, for the rest of the drive's life of 20–30 years, the VFDs added to the corporate profits. Most power installations now use VFDs with pumps for water, oil, and refrigeration, and with fans for air ventilation. Most cruise ships use VFDs for propulsion motors as well.

The VFD primarily controls the speed, torque, acceleration, deceleration, or direction of rotation of an ac motor. Unlike the dual-pole—hence dual-speed—motor, the VFD allows a continuous resolution of the speed control within its operating range. The benefit of the infinitely variable-speed motor is to increase the energy efficiency—doing the same work at a minimum expenditure of kilowatt-hours—or to optimize the product quality or the production process. In addition to their use with large motors in oil refineries and ship propulsion, the following are a few other examples in medium power ranges:

- A pump supplying water in a high-rise building may run at low speed during night time and high speed in the afternoon to provide the required flow rate while maintaining the system pressure.
- Lathes or machine tools can run small-diameter jobs at high speed and large-diameter jobs at low speed to increase the production rate.
- A printing press can operate at a speed that optimizes the print quality depending on the type of paper and ink.
- Provide smoothly controlled acceleration in a spinning mill to avoid yarn breakage.

The induction motor is the most economical and reliable motor widely used in the industry. However, it runs essentially at constant speed with a few percent slip below the synchronous speed. The slip—not the speed—is proportional to the load torque. This is a disadvantage in applications where the motor speed needs to change in response to the load requirement. The motor driving (1) an oil pump in a refinery or an oil tanker and (2) a large ventilator fan in a commercial or an industrial building are just two examples. The fluid-pumping system can use a throttle valve to reduce the flow rate while the motor runs essentially at a constant speed. Turning the valve partially closed adds resistance to the fluid flow, which lowers the energy efficiency of the system. Reducing the speed to reduce the fluid flow rate is more energy efficient. The oil refineries in the past sometimes used gears or dual-speed motors to change the pump speed. However, the backlash and resulting vibrations following a sudden speed change often broke the shaft and reduced the equipment life.

An energy-efficient solution is to reduce the fluid flow rate by reducing the motor speed. We recall that the induction and synchronous motor speeds depend directly on the frequency. Therefore, the motor speed can be changed by varying the frequency of the motor input in response to the flow rate requirement. If a motor larger than 500 hp is running at least 8 h every day—about 2500 h per year—the extra investment in the drive cost is generally paid back in a few years by the energy savings every month. The energy-saving potential of the variable-speed drive (VSD) in a fluid pumping system is first analyzed below.

15.1 PUMP PERFORMANCE CHARACTERISTICS

We consider a fluid pumping system that moves fluid through plumbing that has internal friction in the pipelines. The pump is the pressure source that moves the fluid, and the pipeline friction is the pressure load on the system. A stable system operation is achieved at the flow rate when the pressure rise in the pump is equal to the pressure drop in the pipelines. Figure 15.1 depicts a typical pump pressure versus

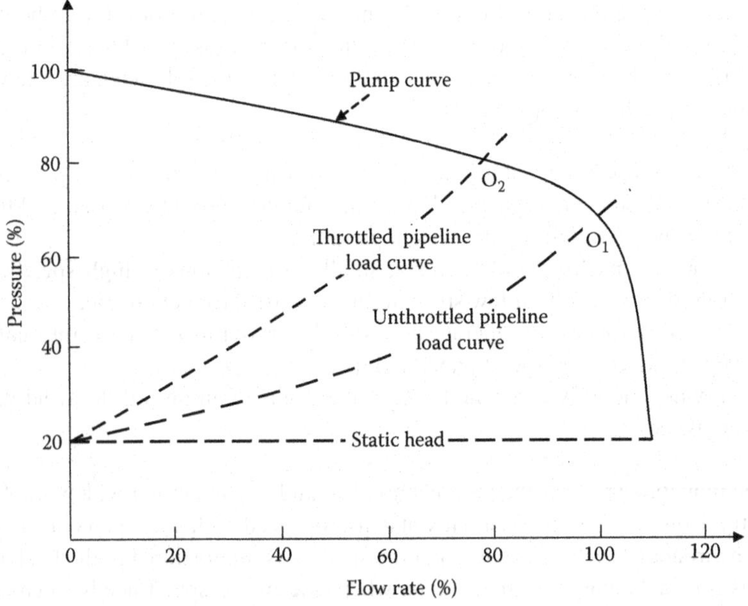

FIGURE 15.1 Pump pressure versus fluid flow rate with and without throttle in pipeline.

fluid flow rate (heavy solid line). When there is no resistance to the pump, the flow is the maximum and the pump pressure is minimum, which is just equal to the static head value. At the other extreme, if the pump outlet is blocked, the flow is zero, and the pump pressure is maximum (100%). On the load side, the pressure drop in the pipeline varies approximately as a square function of the fluid flow rate, that is,

$$\text{pipeline friction pressure load} = k \times (\text{flow rate})^2 \tag{15.1}$$

where k is the proportionality constant that depends on the friction in the pipeline and throttle valve combined. With no throttle valve or valve fully open (unthrottled system), the load pressure versus flow rate is shown by the heavy dotted curve. The system will operate at point O_1 where the pump curve and the load curve intersect. If we throttle the valve, the load curve shifts upward (more pressure drop and less flow), and the operating point moves to O_2 and further to the left with increasing throttle. The power required to drive the pump is given by one of the following two equations:

$$\text{pump motor output power} = (\text{pressure}) \times (\text{flow rate}) = k \times (\text{flow rate})^3 \tag{15.2}$$

$$\text{propeller or fan motor output power} = k \times \text{speed}^3. \tag{15.3}$$

Ignoring the internal losses in the motor and the pump for simplicity here, the motor power input equals the pump power output. The powers at operating points O_1 and O_2 in Figure 15.1—the products of pressure and flow rate at those points—translate to points O_1 and O_2 in Figure 15.2, respectively. From point O_1 (100% flow rate)

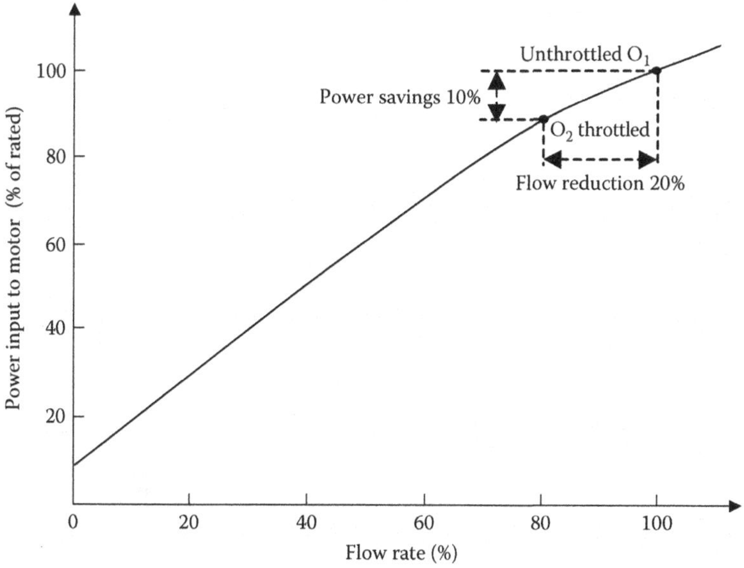

FIGURE 15.2 Motor power input versus fluid flow rate with throttle.

to point O_2 (throttled to 80% flow rate), we see that the power is reduced by 10% for 20% reduction in the flow rate. Thus, the throttled system uses more energy per cubic meter of fluid pumped, which is an uneconomical use of power.

15.2 PUMP ENERGY SAVINGS WITH VFD

Now consider the pump operation with variable-speed motor drive and no throttle valve in the system. In Figure 15.3, the top solid curve at 100% pump speed and dotted unthrottled load curve are the same as in Figure 15.1. At a reduced pump speed, say, at 50%, the solid pump curve in Figure 15.3 shifts downward, but the load curve still remains the same—it is still the same pipe and still unthrottled. This results in the operating point moving from point O_1 to O_2. The power input to the motor is now shown by points O_1 and O_2 in Figure 15.4. It shows the power dropping to 30% (70% saving) for 50% reduction in the flow rate. This results in less energy required per cubic meter of the fluid pumped. This is expected since the pipeline pressure drop varies with the flow rate squared.

The power savings that can be realized by using a variable-speed motor at various flow rates is shown in Figure 15.5. It shows zero savings at 100% speed (obviously true), the maximum savings at around 50% speed, and small savings at lower speeds (because of some fixed losses regardless of the speed). The actual power savings will be somewhat lower than the chart indicates because of the internal power losses in the variable-frequency drive (VFD), the motor, and the pump.

The speed of the pump using the induction motor can be varied by varying the frequency of the motor input. Figure 15.6 depicts the torque-speed characteristics

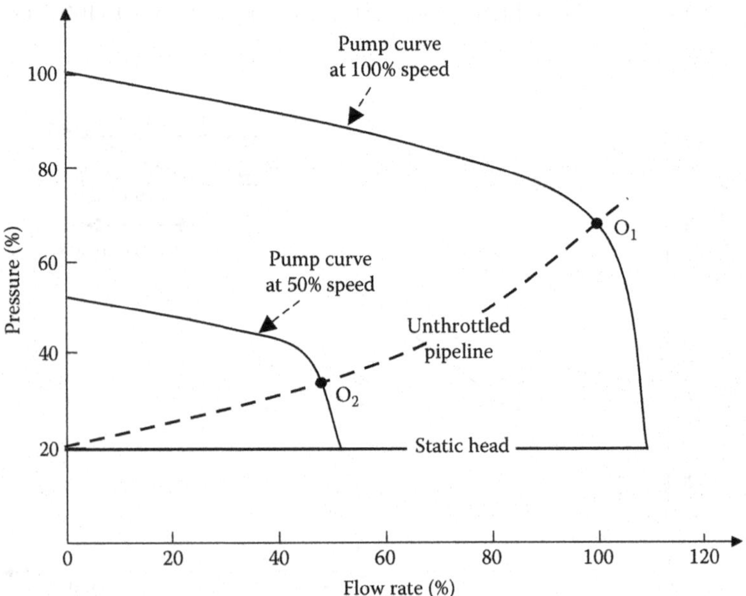

FIGURE 15.3 Pump pressure versus fluid flow rate at 100% and 50% motor speed without throttle.

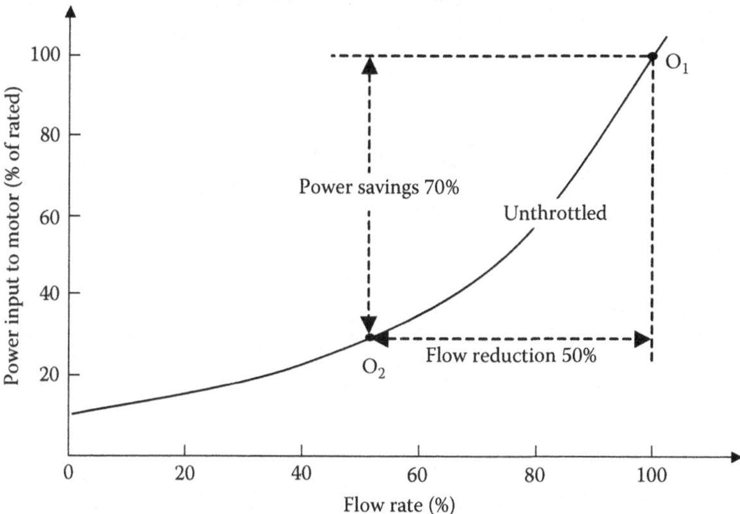

FIGURE 15.4 Motor power input versus fluid flow rate without throttle.

(the humped curves) of the induction motor at various frequencies, where $f_4 < f_3 < f_2 < f_1$ (60 or 50 Hz). The unthrottled pump load torque varies as the speed squared, as plotted by the dotted curve. The operating points (shown by heavy dots at the intersection) shift to lower speeds to n_2, n_3, n_4 as the frequency is lowered.

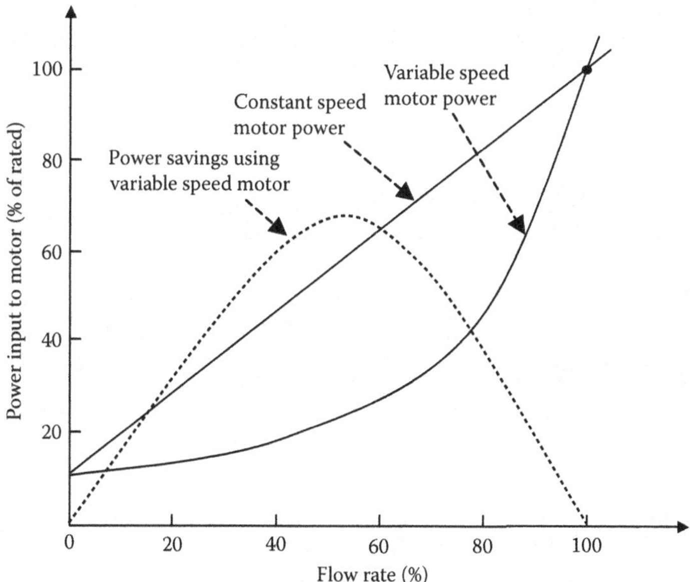

FIGURE 15.5 Motor power saving with variable-frequency drive (VFD) for fluid flow rate variation from 0% to 100%.

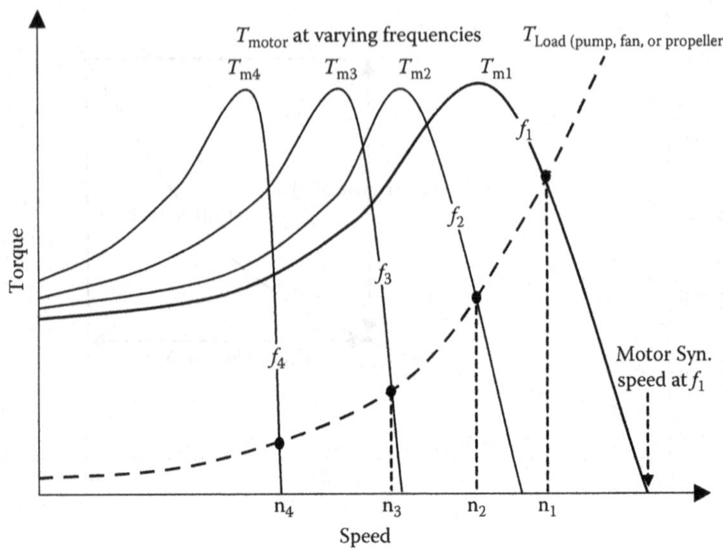

FIGURE 15.6 Induction motor torque versus speed characteristics at various frequencies with pump, fan, or propeller load.

Example 15.1

A water pump driven by a 1000-hp induction motor pumps 100,000 m^3/day at rated speed. If the motor were run at slower speed using VFD to reduce the flow rate to 80% over a longer time to pump the same quantity of water per day, determine the percent savings in kilowatt-hour energy consumed per day.

Solution:

Using Equation 15.2, power at 80% flow rate = $0.80^3 \times$ old power.
Hours to pump the same water quantity = old hours/0.80.

Therefore, new energy consumption to pump the same quantity of water = ($0.80^3 \times$ old power) × (old hours/0.80) = $0.80^2 \times$ old kWh = 0.64 × old kWh per day.

This gives energy savings of 0.36 or 36% of the old energy consumption for pumping the same quantity of water per day. This clearly justifies investing in VFD for saving energy.

Example 15.2

An existing cruise ship New Horizon was designed to travel at 20 knots with a dedicated 6.6-kV, 46-MW_e propulsion power plant consisting of generators, transformers, VFDs, and motors rated accordingly. A new cruise ship is being designed with essentially the same hull as New Horizon, except for a higher speed of 26 knots. Determine the rating of the new propulsion power plant and the recommended voltage level.

Solution:

Using Equation 15.3 in ratio, we have $\frac{\text{new power}}{\text{old power}} = \left(\frac{\text{new speed}}{\text{old speed}}\right)^3$.

Therefore, new propulsion power plant rating = 46 × $(26/20)^3$ = 101 MW.

As for the new voltage level, the industry experience indicates the following,

$$\frac{\text{new voltage}}{\text{old voltage}} = \sqrt{\frac{\text{new power}}{\text{old power}}}$$

which gives

$$\text{new voltage} = \sqrt{\frac{101}{46}} \times 6.6 = 9.78\ \text{kV},$$

which can be rounded to the nearest standard voltage of 11 kV.

Example 15.3

A cargo ship travels 5000 nautical miles transatlantic at 25 knots speed, taking 200 h and using 520,000 gal. of fuel oil. If the speed were reduced by 10%, determine the percent change in total gallons of fuel consumption.

Solution:

New speed = 0.9 × 25 knots, and new power = 0.9^3 = 0.729 × old power.

The fuel consumption rate per hour is proportional to power and hence will reduce to 0.729 × old rate.

New journey time = old journey time ÷ 0.9 = 200 × 1.111 = 222.2 h.

Gallons of fuel used = (0.729 × old rate) × (1.111 × old journey time) = 0.81 × old gallons.

Therefore, reduction in fuel consumption = (1 – 0.81) = 0.19 or 19%.

We note that gallons of fuel consumption = $k \times \text{speed}^2$. In this example, the fuel consumption has decreased since the speed has decreased. At a higher speed, the total fuel consumption for the same journey will increase in the squared proportion of the speed (true for a car trip also). This is why the high-speed ships are not only a design challenge; they also consume more fuel for the same journey. On the benefit side, higher speed lowers the manpower and capital costs per journey since the ship can make more trips per year.

15.3 VFD ON SHIPS AND IN OIL REFINERIES

Large ships such as cruise liners and ice breakers traditionally use large pumps and propellers driven by motors via VFDs. About 80%–85% of the installed drives on ships are for auxiliary systems such as fans, pumps, compressors, etc. Only about 15%–20% of the drives are used for ship propulsion. However, the propulsion drives are much larger in ratings (about 20 MW each), making a significant contribution in the energy efficiency. Some common applications of VFD on ships are as follows:

- Deck equipment, such as under-deck cranes for pallets handling, gantry cranes for container handing, cargo winches, windlass, and constant tension mooring winches
- Oil pumps and water pumps
- Compressor pumps in refrigerators and air-conditioners
- Air ventilation fans
- Dynamic positioning of ships or floating platforms
- Propulsion motors (synchronous, induction, or other types)
- Compressor fans in the antiheeling system

Oil refineries use numerous heavy motors to pump oil through the process. The energy-saving potential of VFD is so great here that VFD research in the 1970s and 1980s was heavily funded by oil refineries. They were the early pioneers in using VFD for large pump motors.

15.4 VFD FOR MEDIUM-SIZE MOTOR

The widely used VFD configuration for most medium-size motors—other than large propulsion motors—is the dc-link frequency converter shown in Figure 15.7. It can use the current source inverter (CSI) or the voltage source inverter (VSI) design depending on the type of motor used for the overall system level optimization. The VFD takes

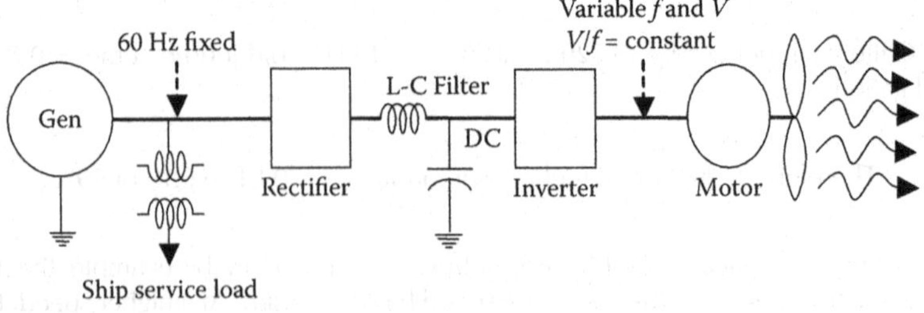

FIGURE 15.7 DC-link VFD schematic for ac motors.

60 or 50 Hz power from the generator, rectifies it into dc, and then inverts it into a lower-frequency, lower-voltage output, which is applied to the induction or synchronous motor. In general, the high-frequency harmonics produced by the switchmode rectifier and inverter are filtered by an L-C filter. However, the induction motor VFD using a constant-current source inverter needs a large inductor, whereas the synchronous motor VFD using a constant-voltage source inverter needs a large capacitor. In high-power propulsion motor VFDs, the design drivers are the torque and speed characteristics as they relate to the propeller efficiency, in addition to the size, weight, total systems cost, and availability of the high-power devices. Beginning in the late 1980s, high-power VFDs used high-voltage (5000 V) gate turn-off thyristors (GTOs) until the advent of the insulated gate bipolar transistors (IGBTs) along with the gate commutated thyristors (GCTs), both of which are now gradually replacing the GTOs.

Figure 15.8 is an interior view of a 300-kW induction motor drive, whereas Figure 15.9 shows an exterior view of VFDs in various kilowatt ratings. The VFD designs at present use various configurations shown in Figure 15.10. Their alternate names and key features are as follows.

Synchroconverter or current source inverter (CSI). This converter (Figure 15.10a) is typically used with a synchronous motor in ships other than icebreaker and dynamic positioning ships. As compared to the cycloconverter, its advantages are (1) much fewer components, hence simple and reliable; (2) lower volume and weight; (3) better power factor, around 0.85–0.90; (4) lower harmonics; and (5) improved overall system efficiency.

FIGURE 15.8 Interior view of 300-kW insulated gate bipolar transistor (IGBT) VFD for induction motor.

FIGURE 15.9 Exterior views of VFDs in various kilowatt ratings. (Courtesy of Converteam Inc., Pittsburgh, PA.)

PWM converter or voltage source inverter (*VSI*). It is often used with induction and permanent magnet motors, and sometimes with a synchronous motor. The voltage source drive (Figure 15.10b) can be designed with insulated gate bipolar transistor (IGBTs), gate turn-off thyristors (GTOs), or integrated gate commutated thyristors (GCTs) (IGCTs). The PWM drive can be designed to operate at a power factor near unity over the entire range of the motor speed compared to 0.85 power factor for CSI and 0.75 for the cycloconverter. The control voltage signal frequency (V_{control}) is compared to the triangular waveform, as shown in Figure 14.10. It reflects low harmonics on the ac source side, and lower-order harmonics are absent on the load side. It is typically used for azimuthing stern drive (in ships) with substantial weight savings compared to the conventional design.

Cycloconverter. This direct ac–ac frequency converter (Figure 15.10c) is used mainly with the synchronous motor, having a greater air gap as opposed to the induction motor. The cycloconverter typically uses many more power electronics devices, making it more expensive and less efficient. Also, it adds lag in the motor current due to the phase control modulation. Its typical power factor is around 0.75, which is at

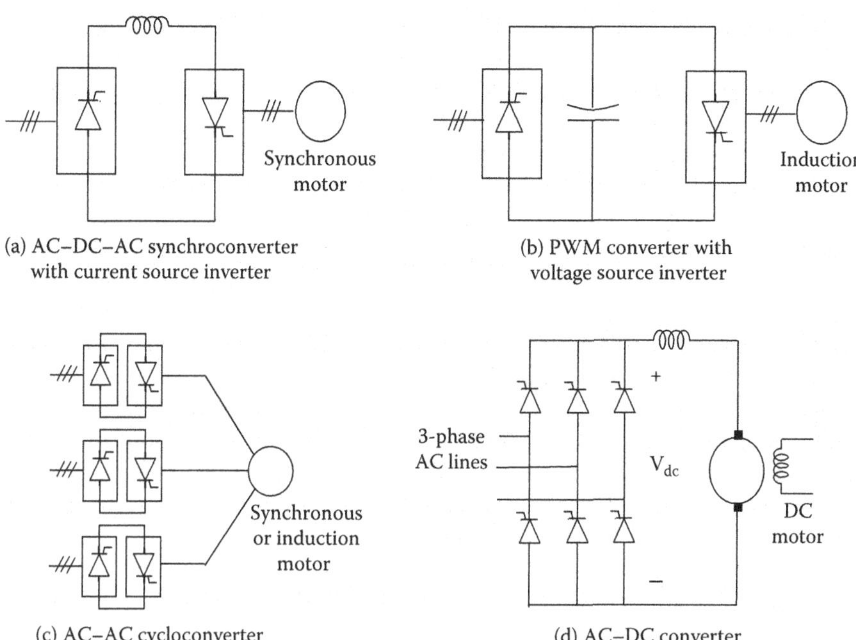

FIGURE 15.10 Alternative variable-speed drive (VSD) topologies.

the low end of the 0.7–0.9 range that many industry standards allow. The cycloconverter is an old converter technology, whose time had passed for a while but is now getting new attention for certain benefits. It works well in icebreakers and dynamic positioning ships requiring very high torque at low speed. However, its use is limited for output frequency up to 30%–40% of the supply frequency.

DC motor drive. The dc motor speed is controlled simply by changing the motor terminal voltage while maintaining the constant field current to maintain constant air gap flux. Unlike the ac motor drive, the dc motor drive (Figure 15.10d) has simple circuit configuration. It is discussed in further detail later in Section 15.11.

15.5 CONSTANT *V/F* RATIO OPERATION

All variable-frequency drives (VFDs) maintain the output voltage-to-frequency (V/f) ratio constant at all speeds for the reason that follows. The motor phase voltage V, frequency f, and the magnetic flux ϕ are related by Equation 15.4:

$$V = 4.444\,f N\phi_m \quad \text{or} \quad V/f = 4.444\,N\phi_m \tag{15.4}$$

where N = number of turns per phase. If we apply the same voltage at reduced frequency, the magnetic flux would increase and saturate the magnetic core, significantly distorting the motor performance. The magnetic saturation can be avoided only by keeping ϕ_m constant. Moreover, the motor torque is the product of the stator flux and the rotor current. The rated torque can be maintained at all speeds only by maintaining the flux constant at its rated value, which is done by keeping the voltage-to-frequency (*V/f*) ratio constant. That requires lowering the motor voltage in the same proportion as the frequency to avoid the magnetic saturation due to high flux or lower than rated torque due to low flux.

Maintaining the *V/f* ratio constant at all speeds requires controlling the rectifier firing delay angle in proportion to the desired speed. At very low speed, when the frequency and voltage are also low, the inductive reactance is negligible, and the voltage setting the air gap flux is $V - I_{stator} R_{stator}$. Therefore, for constant air gap flux, the $(V - I_{stator} R_{stator})/f$ ratio must be maintained constant. This requires a voltage boost of $I_{stator} R_{stator}$ to compensate for the voltage drop in the motor armature resistance. At low speeds, therefore, the motor terminal voltage must be boosted up by $I \times R$ drop in the motor armature as shown in Figure 15.11.

The exact VFD design with a constant *V/f* ratio is somewhat different for the induction motor than for the synchronous motor drive in order to optimize the system performance with the somewhat different nature of the two motors. In either case, controlling the motor terminal voltage requires controlling the rectifier firing delay angle in proportion to the desired speed as seen below.

With a three-phase, six-pulse rectifier with variable firing angle to vary V_{dc} and hence the VFDs ac output voltage, Equation 14.16 (with all ac voltages fundamental rms line-to-line values)

$$V_{dc} = 1.35 V_{LLline} \cos \alpha. \tag{15.5}$$

With a three-phase, six-pulse voltage source inverter (VSI), Equation 14.35 earlier gave the inverter output voltage at the motor terminal:

$$V_{LLmotor} = 0.78\ V_{dc} = 0.78 \times 1.35 V_{LLline} \cos \alpha = 1.05 V_{LLline} \cos \alpha. \tag{15.6}$$

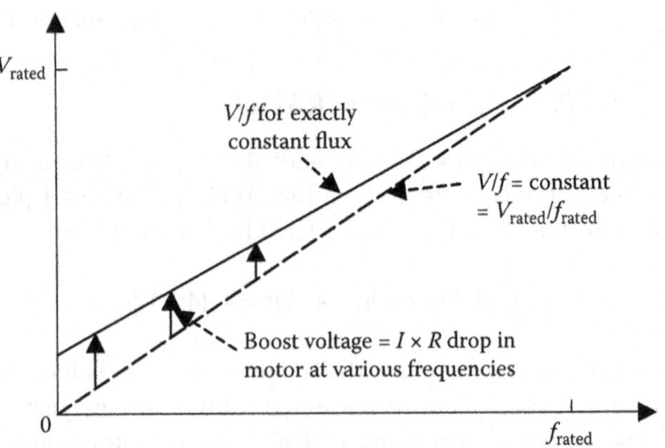

FIGURE 15.11 Voltage boost requirement in VFD at low speeds.

With a three-phase, six-pulse CSI, Equation 14.39 earlier gave the inverter output voltage at the motor terminal:

$$V_{\text{LLmotor}} = \frac{\sqrt{3}V_{\text{dc}}}{\sqrt{6}\cos\theta} = \frac{1}{\sqrt{2}\cos\theta} 1.35\ V_{\text{LLline}} \cos\alpha = \frac{0.955}{\cos\theta} V_{\text{LLline}} \cos\alpha \quad (15.7)$$

where $\cos\theta$ = motor power factor and α = rectifier firing delay angle. With a typical value of induction motor power factor of 0.9, Equations 15.6 and 15.7 give almost identical voltage at the motor terminals. It can further be shown that

$$\text{incoming line pf with CSI} = 0.955\cos\alpha = 0.955\left(\frac{\text{actual speed}}{\text{rated speed}}\right). \quad (15.8)$$

Also, Equation 15.6 gives the actual motor speed as follows:

$$\frac{\text{actual speed}}{\text{rated speed}} = \frac{f_{\text{motor}}}{f_{\text{rated}}} = \frac{V_{\text{LLmotor}}}{V_{\text{LLline}}} = 1.05\cos\alpha. \quad (15.9)$$

Thus, the motor speed is linearly related with $\cos\alpha$. When using the CSI with an induction motor, Equation 15.8 indicates that the incoming line power factor is good near the rated speed but degrades significantly at lower speeds due to the excessive delay in firing angle α that is necessary to reduce the voltage to a low value. In practice, this disadvantage can be overcome by controlling V_{dc} not by α but by step-down dc–dc buck converter.

Example 15.4

A 100-hp, three-phase, 60-Hz, four-pole, 1750-rpm, 0.85-pf induction motor operates with VFD connected to 460-V lines. The VFD design has CSI fed from a dc link. If the motor needs to run at 980 rpm, determine (1) the rectifier firing delay angle and (2) the VFD output voltage at the motor terminals.

Solution:

Using Equation 15.9, we have $\frac{980}{1750} = 1.05\cos\alpha$, from which we have $\cos\alpha$ = 0.533 and α = 57.77°.

Then, using Equation 15.7,

$$V_{\text{LLmotor}} = \frac{0.995}{0.85} \times 460 \times 0.533 = 287.2\ \text{V}.$$

Since the *V*/*f* ratio of the motor input is kept constant at reduced speed, the air gap flux remains constant, and the torque is linearly proportional to the armature current, which varies with the rotor slip rpm. For blade-type loads (fan, pump, and propeller) driven by the induction motor with constant flux, we have, with different values of constant k in different equations:

$$\text{torque} = k \times \text{current} = k \times \text{slip rpm}. \tag{15.10}$$

For blade-type loads, we know the following:

$$\text{torque} = k \times \text{speed}^2 = k \times \text{frequency}^2. \tag{15.11}$$

At stable operation, Equations 15.10 and 15.11 give slip rpm = $k \times \text{frequency}^2$, from which

$$\text{perunit slip} = \frac{\text{slip rpm}}{\text{synchronous speed}} = \frac{k \times \text{frequency}^2}{\text{frequency}} = k \times \text{frequency}. \tag{15.12}$$

Therefore, we can write the following:

$$\text{for blade-type loads, } \frac{\text{pu slip at new frequency}}{\text{pu slip at old frequency}} = \frac{\text{new frequency}}{\text{old frequency}}. \tag{15.13}$$

For constant torque loads—slow-moving loads such as conveyor belts, cranes, and winches that must overcome a constant friction—the above is not true. With constant flux at constant V/f ratio, the rotor current must remain constant, hence the slip. Therefore, for constant torque load below the rated frequency with constant flux, we have

$$\text{slip rpm} = \text{constant and } \frac{\text{hp at reduced frequency}}{\text{hp at rated frequency}} = \frac{\text{reduced frequency}}{\text{rated frequency}}. \tag{15.14}$$

For constant torque load above the rated frequency, the motor is operated at rated line voltage that produces lower flux, which in turn produces lower torque. The high speed and proportionately lower torque at high frequency result in a constant horsepower. Therefore, above rated frequency, we have

$$\text{HP} = \text{constant, } \frac{\text{torque at higher frequency}}{\text{torque at rated frequency}} = \frac{\text{rated frequency}}{\text{higher frequency}} \tag{15.15a}$$

and

$$\frac{\text{slip rpm at higher frequency}}{\text{slip rpm at rated frequency}} = \frac{\text{higher frequency}}{\text{rated frequency}}. \tag{15.15b}$$

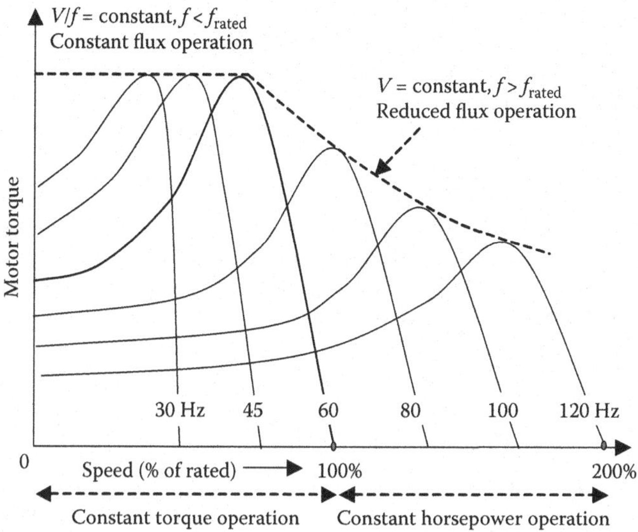

FIGURE 15.12 Induction motor torques at different frequencies below and above 60 Hz.

Figure 15.12 depicts the motor operation from 0% to 200% of the synchronous speed as the frequency varies from 0 to 2× rated value. The motor operates at constant torque at subsynchronous speeds and at constant horsepower at supersynchronous speeds.

Example 15.5

A 500-hp, 460-V, three-phase, 60-Hz, four-pole, 1728-rpm induction motor is driving a large ventilating fan. When it is running from a variable-frequency drive (VFD) output of 230 V, 30 Hz with a constant *V/f* ratio, determine (1) the approximate speed and (2) the exact speed.

Solution:

At 60 Hz, a four-pole ac motor has synchronous speed = 120 × 60/4 = 1800 rpm. The motor has a rated speed of 1728 rpm, at which the rotor slip = 1800 – 1728 = 72 rpm, which is 4% of the synchronous speed. The induction motor speed depends on the frequency, and so does the running speed.

(1) *Approximate estimate:* at 30-Hz operation of a 60-Hz motor, approximate speed = (30/60) × rated speed = 1/2 × 1728 = 864 rpm.

The synchronous speed at 30 Hz is 900 rpm, so the slip = 900 – 864 = 36 rpm, which is 4% of the new synchronous speed, the same percent slip as that at 60 Hz.

(2) *Exact solution:* the fan load torque varies with speed^2. With speed approximately 1/2, the load torque will be close to 1/4 of the rated torque.

Since the V/f ratio of the motor input is kept constant at reduced speed, its air gap flux remains constant. To produce 1/4 torque at the same flux, the rotor current must be 1/4 of the rated value. At constant flux, the rotor current linearly varies with the rotor slip rpm, which will become 1/4 × 72 = 18 rpm. This is 2% of the new synchronous speed, as we would have derived from Equation 15.13 also.

$$\text{Therefore, at 30 Hz, exact rotor speed} = 900 - 18 = 882 \text{ rpm.}$$

The exact speed (882 rpm) is about 2% higher than the approximate speed (864 rpm). This difference may matter in some applications and may not matter in others, such as in ventilating fans where only the approximate air flow rate within a few percent matters for the comfort.

Example 15.6

A three-phase, 60-Hz, 100-hp, 1750-rpm induction motor is driven by a variable-frequency drive. Determine (1) the power, torque and full load speed when operating at 20 Hz and (2) the power, torque, and full load speed when operating at 80 Hz.

Solution:

(1) We use the following relation that is valid for any machine delivering mechanical power—motor, diesel engine, turbine, etc.:

$$\text{horse power} = \frac{\text{torque}_{\text{lb.ft}} \times \text{speed in rpm}}{5252}.$$

Below rated frequency, the motor operates at a constant V/f ratio, that is, at constant flux, and hence will produce constant torque at rated current. Thus,

$$\text{torque at 20 Hz} = \text{torque at 60 Hz} = 100 \text{ hp} \times 5252 \div 1750 \text{ rpm} = 300 \text{ lb.-ft.}$$

The 60-Hz, 1750-rpm motor must be a four-pole motor with a synchronous speed of 1800 rpm and a rotor slip speed = (1800 – 1750) = 50 rpm. At 20 Hz, the motor will run at the same 50-rpm slip to produce the rated current to produce rated torque at rated flux. Thus, the speed at 20 Hz is (600 rpm synchronous speed – 50 rpm slip) = 550 rpm.

Therefore, 20-Hz hp = 550 × 300 ÷ 5252 = 31.42 hp. This reduced hp is approximately in the frequency ratio, that is, (20/60) × 100 = 33.33 hp.

(2) Above the rated frequency, the motor operates at constant hp since the flux is lower at the rated voltage and the current is kept at a rated value to limit the heating. At 80 Hz, the slip rpm can increase to (80/60) × 50 = 66.67 rpm. With the new synchronous speed of 2400 rpm, the rotor speed = 2400 – 66.67 = 2333.3 rpm and torque = 100 × 5252/2333.3 = 225 lb.-ft, which is the same as 300 × 60/80 = 225 lb.-ft that could have been derived from Equation 15.15a also.

15.6 COMMUTATION AND CONTROL METHODS

Forced-commutated VFD. With a lagging power factor induction motor load, the inverter thyristors need forced commutation with capacitors and diodes. The leakage inductance of the motor plays a significant role in the commutating circuit design for the inverters. Therefore, such VFD is designed for use with only a specific motor. On the positive side, it can be used for regenerative braking without additional circuits.

Load-commutated VFD. The load-commutated thyristors in a CSI are widely used with large multimegawatt synchronous motors for pumps, compressors, rolling mills, and ship propulsion. Figure 15.13 is such a VFD with a six-pulse rectifier and a six-pulse inverter. For high-power drives, 12-pulse converters are used to lower the harmonic content. The VFD is load-commutated by the leading power factor of the synchronous motor that can be achieved by overexciting the rotor field to obtain higher back-emf induced in the motor armature. The frequency and the phases of the three-phase currents are synchronized to the rotor position, and the current commutation in the inverter thyristor is accomplished by using the motor back-emf. Slow-speed thyristors can be used on both sides. This makes the inverter simple, cost effective, energy efficient, and reliable compared to the CSI with an induction motor.

At low speeds, however, the back voltage is not sufficient to achieve the force commutation, and the commutation is provided by using the thyristors in the rectifier operating at large α to make $I_{dc} = 0$, thus turning off the inverter thyristors. Also, since large synchronous motors have a few percent higher efficiency than the comparable size induction motors, the load-commutated VFD system with a synchronous motor offers a few percent efficiency advantage, which can make a significant positive difference in the energy consumption in multimegawatt systems running almost continuously over the year.

VFD feeding power back to the source. When the VFD is used to feed power back to fixed-frequency grid lines, such as in wind or photovoltaic energy farms, or during regenerative braking of a metro train, certain grid interface issues must

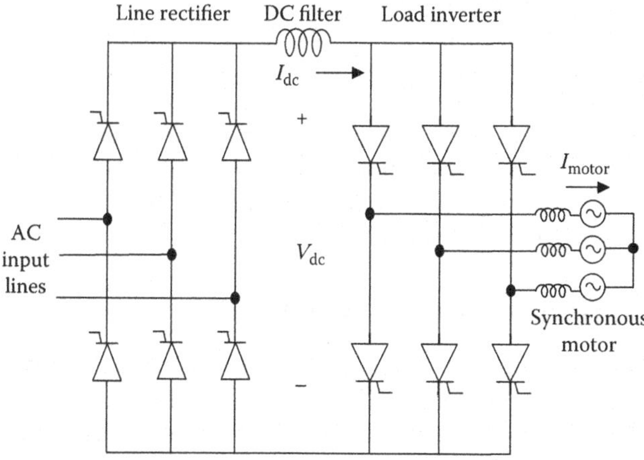

FIGURE 15.13 Load-commutated inverter drive for synchronous motor.

be addressed. At the grid interface, the power flow direction and the magnitude depend on the voltage magnitude and the phase relation of the inverter voltage with respect to the grid voltage. With the grid voltage being fixed, the inverter voltage must be controlled both in magnitude and in phase in order to feed power to the grid when available and to draw from the grid when needed. If the inverter is already included in the system for frequency conversion, the magnitude and phase control of the inverter output voltage is achieved with the same inverter with no additional hardware cost. Such controls are accomplished as follows.

- *Voltage control.* Since the magnitude of the ac voltage output from the inverter is proportional to the dc voltage input from the rectifier, the motor voltage can be controlled by operating the inverter with a variable dc-link voltage from a phase-controlled rectifier. However, there are two problems associated with this scheme. At a reduced output voltage, this method gives a poor distortion power factor and high harmonic content that require heavy filtering before feeding to the inverter. Also, in circuits deriving the load commutating current from the commutating capacitor voltage from the dc link, the commutating capability decreases when the output voltage is reduced. This could lead to an operational difficulty when the dc-link voltage varies over a wide range, such as in a motor drive controlling the speed in a ratio exceeding 4 to 1. In variable-speed wind power generation, such commutation difficulty is unlikely as the speed varies over a narrow range.
- *Frequency control.* The output frequency of the inverter solely depends on the rate at which the switching thyristors or transistors are triggered into conduction. The triggering rate is determined by the reference oscillator producing a continuous train of timing pulses, which are directed by logic circuits to the thyristor gating circuits. The timing pulse train is also used to control the commutating circuits. The frequency stability and accuracy requirements of the inverter dictate the selection of the reference oscillator. A simple temperature compensated R-C relaxation oscillator gives frequency stability within 0.02%. When a better stability is needed, a crystal-controlled oscillator and digital counters may be used, which can provide stability of .001% or better. The frequency control in a standalone power system is an open-loop system. The steady-state or transient load changes do not affect the frequency. This is a major advantage of the power electronics inverter over the old electromechanical means of frequency conversion using a motor–generator set.

15.7 OPEN-LOOP CONTROL SYSTEM

A simple open-loop volt/hertz control system for the induction motor drive using the pulse width modulated (PWM) voltage source inverter (VSI) is shown in Figure 15.14. It has no feedback and hence no instability concern in operation. Therefore, it is widely used in the industry. However, it has somewhat inferior performance. Its front end is a rectifier with a filter to get constant V_{dc}. The speed or frequency is the command signal, from which V_s^* is derived for the constant air gap flux. At low

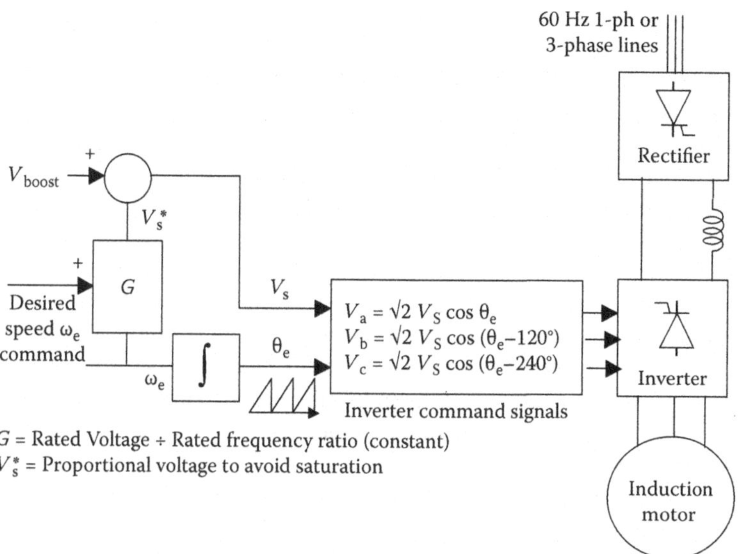

FIGURE 15.14 Open-loop *V/f* control schematic of induction motor with dc-link frequency converter.

frequency, the boost voltage V_{boost} is added to compensate for the large voltage drop in the stator armature resistance.

The three-phase sinusoidal voltages are derived from the voltage magnitude and angle command signals from the PWM inverter. The motor speed can be increased or decreased by slowly ramping up or down the speed command signal. While decelerating, the motor works as a generator, and the power is dissipated in the rotor resistance. When the command speed exceeds the base (full) speed, V_s saturates, the frequency proportionality is lost, the drive enters the field weakening region, and the developed torque is decreased due to the flux reduction. The motor delivers constant horsepower in such operation.

15.8 VECTOR CONTROL DRIVES

The vector control is often used for mooring winches and windlass on ships. Winches are designed to hold mooring lines to the pier in a constant tension mode. The operator can set the desired tension in tens of thousands of pounds by adjusting a potentiometer on the control panel. A windlass is a device that uses a rope or cable wound around a revolving drum to pull and/or lift things and to raise and lower the anchor. Typically a 50–100 hp induction motor is fed from the vector drive to operate the windlass. Some windlass failures with the vector drive at low-speed operations at low load have been reported in the past.

The induction motor draws lagging current having two components: (1) in phase with voltage that produces the real power and (2) lagging the voltage by 90° that produces the magnetic flux. The basic idea of the vector control is to control these two components separately in order to control the torque and the motor speed independently. For that reason, the induction motor with vector control behaves essentially like the dc motor. A true vector control requires information on the exact position of the rotor and the motor electrical parameters for a feedback control system, shown in Figure 15.15. The rotor position sensor is basically a permanent magnet on the rotor and a stationary sensor coil, the output of which is fed back in the speed and torque control loop. The three feedback loops are the output current, the motor speed, and the rotor position. The current loop prevents operation of the motor above rated current for a set time period. Above the base speed, the flux component is reduced for a constant-horsepower operation. Another key point is that the frequency of the voltage applied to the motor is calculated from the slip frequency and the motor speed. To simplify the design, some manufacturers have developed open-loop control software that, once tuned with the motor, can operate the motor in the open loop.

The vector-controlled inverter separates the motor current into the flux- and torque-producing components and maintains a 90° relationship between them, in what is also known as the field oriented control. The decoupled control of the flux and the torque offers the advantages of higher efficiency and full torque control. In order to separate the current into the flux and torque components, the drive must know the position of the motor magnetizing flux. An encoder attached to the motor shaft relays this information to the converter. The use of vector control produces the maximum torque from base speed down to zero speed. The other algorithm that could be used is scalar control, which would provide a constant volt/hertz output.

The vector control gives precise speed and position control over a wider operating range. If a variable-frequency drive (VFD) without vector control can control the motor speed in the 40:1 range, an open-loop vector control can control in the 120:1 speed range with speed regulation of 0.1%. A true vector control can provide the rated torque at zero speed, just like the dc motor, while the open-loop vector control

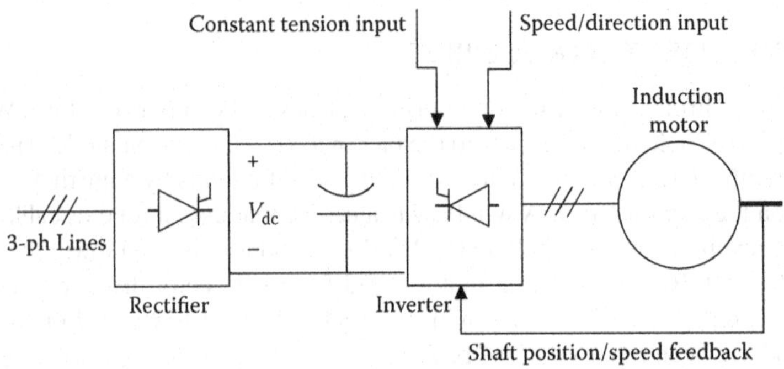

FIGURE 15.15 Vector control of induction motor for constant-torque applications.

can provide full torque up to about 15 rpm in a four-pole motor. Vector control also offers faster response to the torque and speed change command.

15.9 TWELVE-PULSE VFD DESIGN

Compared with the rarely-used three-phase, three-pulse converter, the three-phase, six-pulse converter produces low harmonic distortion. It is like a three-pulse converter using a three-phase center-tap transformer, where each of the six voltages is out of phase by 360/6 = 60°.

The harmonic content in the output ac can be further reduced by pulsing the inverters 12 times instead of six times. Both the 6-pulse and 12-pulse converters (often incorrectly called 6-phase and 12-phase converters) use 3-phase power mains. The 12-pulse inverter uses a three-winding transformer with one primary and two secondaries of equal ratings, one connected in Y and the other in Δ. The outputs of the two secondaries are 30° out of phase, pulsing the inverter's one block six times and the other block also six times, with total pulses in the combined output 12 times. The main advantage of the 12-pulse design is elimination of the 5th, 7th, 17th, 19th,... harmonics, thus improving the quality of the output power. Among the inverters commonly used for high-power applications, 12-pulse line-commutated full-wave topology prevails.

In large ships, the VFDs for large propulsion and thruster motors are often separate sets, as in Figure 15.16, which shows one pair of 12-pulse VFDs for large propulsion motors and one pair of 6-pulse VFDs for small thruster motors. In each VFD pair, one is for the port side and the other for the starboard side. Thus, in total, there are four separate VFDs. The 12-pulse VFDs for the propulsion motors are obtained by using the three-winding Δ–ΔY transformers from the high-voltage switchboard as explained above. The 12-pulse design is obtained by using the 60° phase difference in the output voltage of the Δ and Y secondaries, instead of 120° in the conventional 6-pulse converters for the thruster motors.

For further harmonic reduction, 18-pulse converters can be designed using a four-winding transformer—one primary and three secondaries—and 24-pulse converters can be designed using two sets of three-winding transformers in parallel with 15° phase shift from each other on the primary side.

15.10 SPECIAL VFD CABLES

Since the VFD output and input power contain high-frequency harmonics, the radiated EMI from the VFD cables can be of concern in systems with harmonic-sensitive equipment, especially in confined spaces in submarines. Moreover, additional harmonic heating and high dv/dt and di/dt stresses degrade the cable insulation at a higher rate than in the conventional sinusoidal equipment. The remedy is to use cables with higher-grade insulation and an enhanced EMI shield. Figure 15.17 is one such cable suitable for high-power applications in a high voltage class up to 15 kV. It is available in size American wire gage (AWG) 14 to 777 kcmil, in ampacity up to 472 A, and in temperature class up to 110°C. The resistance, inductance, and weight per 1000 ft. of 2-kV class VFD cable are listed in Table 15.1, those of 8-kV class cable are listed in Table 15.2, and those of 15-kV class cable are listed in Table 15.3.

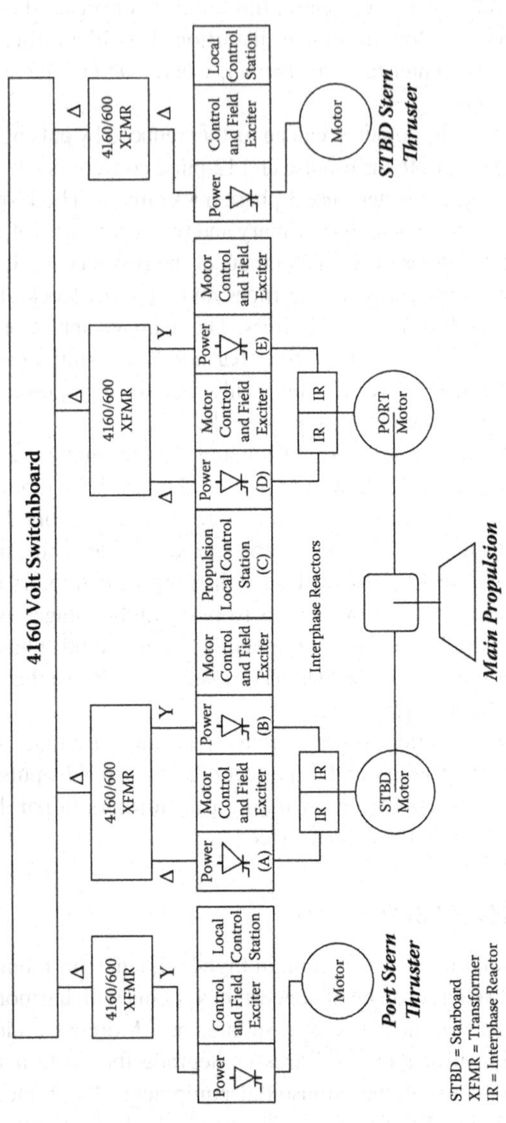

FIGURE 15.16 Electric ship propulsion power system with 12-pulse VFDs for main propulsion and 6-pulse VFDs for thrusters motors. (Courtesy of Dana Walker, U.S. Merchant Marine Academy.)

37-105VFD

Type MMV-VFD Power Cable

Three Conductor: 8 kV–15 kV • 133% Insulation Level • Rated 90°C

Oil & Gas Cables

Conductors (3)
Soft annealed flexible stranded tinned copper per IEEE 1580 Table 11.

Insulation
Extruded thermosetting 90°C Ethylene Propylene Rubber (EPR), meeting UL 1309 (Type E), IEEE 1580 (Type E) and UL 1072.

EMI Shield
Overall tinned copper braid plus aluminum/polyester tape providing 100% coverage

Insulation Shield
Composite shield consisting of 0.0126" tinned copper braided with nylon providing 60% copper shielded coverage meeting UL 1309, IEEE Std. 1580 and UL 1072. The nylon is colored for easy phase identification (three conductor = black, blue, red) without the need to remove the shield to find an underlying colored tape.

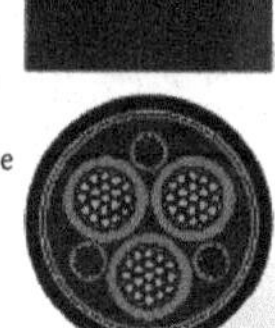

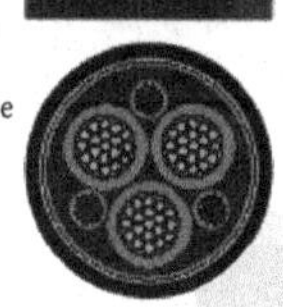

Conductor Shield
A combination of semi-conducting tape and exruded thermosetting semi-conducting material meeting UL 1309, IEEE 1580, and IL 1072.

Insulation Shield
Semi-conducting layer meeting UL 1309, IEEE 1580 and UL 1072.

Symmetrical Insulated Grounding Conductors
Three soft annealed flexible stranded tinned copper conductors per IEEE 1580 Table 11. Gexol insulation sized per Table 23.2 of UL 1072. Color: Green

Jacket
A black, arctic grade, flame retardant, oil, abrasion, chemical and sunlight resistant thermosetting compound meeting UL 1309/CSA 245, IEEE 1580, and UL 1072. Colored jackets for signifying different voltage levels are also available on special request (orange = 8 kV and red = 15 kV).

Armor (optional)
Basket weave wire armor per IEEE 1580 and UL 1309/CSA 245, Bronze standard. Tinned copper available by request.

Sheath (optional)
A black, arctic grade, flame retardant, oil, abrasion, chemical and sunlight resistant thermosetting compound meeting UL 1309/C SA 245, IEEE 1580, and UL 1072, Colored jackets for signifying different voltage levels are also available on special request (orange = 8 kV and red = 15 kV).

Low smoke halogen-free jacket avilable on request.

Termination Kits
AmerCable Systems offers per-sized and pre-formed termination kit packages specifically for VFD cable constructions

Ratings & Approvals

- UL Listed as Marine Shipboard Cable (E111461)
- American Bureau of Shipping (ABS)
- Det Norske Veritas (DNV) Pending
- Lloyd's Register of Shipping (LRS) Pending
- 90°C Temperature Rating
- Voltage Rating – 8 kV to 15 kV (25 kV available on request)

Applications

A flexible, braid and foil shielded, power cable specifically engineered for use in medium voltage variable frequency AC drive (VFD) applications.

Features

- Flexible stranded conductors, braided shields and a braided armor (when armored). Suitable for applications involving repeated flexing and high vibration.
- Small minimum bending radius (6 × OD for unarmored cables and 8 × OD for armored cables) for easy installation.
- Insulation resists the repetitive 3 × voltage spikes from VFDs and reduces drive over current trip problems due to cable charging current.
- Overall braid and foil shield provides 100% coverage containing VFD EMI emissions.
- Symmetrical insulated ground conductors reduce induced voltage imbalances and carry common mode noise back to the drive.
- High strand count conductors and braid shield design is much more flexible, easier to install and more resistant to vibration than type MC cable.
- Severe cold durability: exceeds CSA cold bend/cold impact (–40°C/–35°C).
- Flame retardant: IEC 332-3 Category A and IEEE 1202.
- Suitable for use in Class I, Division 1 and Zone 1 environments (armored and sheathed).

FIGURE 15.17 Three-conductor 8–15 kV class shielded, 8–15 kV class VFD cable with heavy insulation for propulsion motors. (From Gexol-insulated cable, a product of AmerCable Inc. With permission.)

TABLE 15.1
Three-Phase VFD Power Cable 2-kV, 110°C Class

		Unarmored		Armored		Armored and Sheathed (BS)			Ampacity				
Size AWG/kcmil	Part No. 37-102	Nominal Diameter (in.)[a]	Weight Per 1000 ft.	Nominal Diameter (in.)[a]	Weight Per 1000 ft.	Nominal Diameter (in.)[a]	Weight Per 1000 ft.	Green Insulated Grounding Conductor (x3) Size (AWG)	110°C	100°C	95°C	90°C	75°C
14	-508VFD	0.540	194	0.590	281	0.725	356	18	27	25	22	–	18
12	-516VFD	0.590	224	0.646	321	0.772	401	18	33	31	27	–	24
10	-308VFD	0.633	308	0.694	412	0.820	497	14	44	41	36	–	33
8	-309VFD	0.764	441	0.820	565	0.988	702	14	56	52	48	–	43
6	-310VFD	0.865	570	0.925	708	1.090	865	12	75	70	64	93	58
4	-312VFD	1.072	886	1.125	1061	1.295	1243	12	99	92	85	122	79
2	-314VFD	1.215	1421	1.271	1618	1.440	1822	10	131	122	113	159	105
1	-315VFD	1.340	1517	1.395	1743	1.560	1966	10	153	143	131	184	121
1/0	-316VFD	1.443	1803	1.493	2027	1.666	2327	10	176	164	152	211	145
2/0	-317VFD	1.572	2153	1.622	2399	1.854	2840	10	201	188	175	243	166
4/0	-319VFD	2.053	3463	2.103	3785	2.335	4347	8	270	252	235	321	223
262	-320VFD	2.193	4175	2.243	4522	2.475	5120	6	315	294	267	365	254
313	-321VFD	2.370	4727	2.420	5104	2.652	5747	6	344	321	299	408	287
373	-322VFD	2.501	5415	2.551	5809	2.845	6674	6	387	361	334	451	315
444	-323VFD	2.670	6707	2.721	7141	3.014	8059	6	440	411	372	499	350
535	-324VFD	2.972	7483	3.022	2966	3.316	8981	6	498	443	418	–	390
646	-326VFD	3.164	8916	3.214	9428	3.508	10504	4	553	516	470	–	431
777	-327VFD	3.388	10395	3.438	10940	3.732	12088	4	602	562	529		473

Source: AmerCable, Inc.
[a] Cable diameters are subject to a ±5% manufacturing tolerance.

TABLE 15.2

Three-Phase VFD Power Cable 8-kV, 90°C, 133% Insulation Class

Three-Conductor Type MMV-VFD Marine Medium Voltage—8 kV • 133% Insulation Level														
		Unarmored			Armored and Sheathed (BS)			Ampacity						
Size (AWG/ kcmil)	mm²	Part No. 37-105	Nominal Diameter (in.)	Weight (lb./1000 ft.)	Part No. 37-105	Nominal Diameter (in.)	Weight (lb./1000 ft.)	In Free Air (A)	Single Banked in Trays (A)	DC Resistance at 25°C (Ω/1000 ft.)	AC Resistance (Ω/1000 ft.)	Inductive Reactance (Ω/1000 ft.)	Voltage Drop (V/A/ 1000 ft.)	Green Insulated Grounding Conductor (3×) Size (AWG)
6	12.5	-332VFD	1.541	1349	-332BSVFD	1.879	2048	88	75	0.445	0.556	0.048	0.820	10
4	21	-333VFD	1.728	1770	-333BSVFD	2.069	2548	116	99	0.300	0.376	0.043	0.564	10
2	34	-334VFD	1.939	2335	-334BSVFD	2.283	3201	152	129	0.184	0.230	0.040	0.359	10
1	43	-335VFD	2.031	2664	-335BSVFD	2.378	3570	175	149	0.147	0.184	0.038	0.294	8
1/0	54	-336VFD	2.133	3065	-336BSVFD	2.481	4014	201	171	0.117	0.147	0.037	0.242	8
2/0	70	-337VFD	2.269	3593	-337BSVFD	2.621	4600	232	197	0.093	0.117	0.036	0.199	8
3/0	86	-338VFD	2.370	4064	-338BSVFD	2.723	5113	266	226	0.074	0.094	0.035	0.166	6
4/0	109	-339VFD	2.511	4770	-339BSVFD	2.866	5878	306	260	0.058	0.075	0.033	0.139	6
262	132	-340VFD	2.691	5544	-340BSVFD	3.119	6936	348	296	0.048	0.063	0.032	0.121	6
313	159	-341VFD	2.841	6340	-341SBVFD	3.273	7805	386	328	0.040	0.053	0.032	0.106	6
373	189	-342VFD	3.058	7435	-342BSVFD	3.494	9007	429	365	0.034	0.045	0.031	0.094	4
444	227	-343VFD	3.218	8596	-343BSVFD	3.650	10250	455	387	0.028	0.039	0.030	0.085	4
535	273	-344VFD	3.403	9900	-344BSVFD	3.836	1164	528	449	0.024	0.033	0.030	0.076	4

Source: AmerCable, Inc.

TABLE 15.3
Three-Phase VFD Power Cable 15-kV, 90°C, 133% Insulation Class

Three-Conductor Type MMV-VFD Marine Medium Voltage—15 kV • 133% Insulation Level

		Unarmored			Armored and Sheathed (BS)			Ampacity						
Size (AWG/kcmil)	mm²	Part No. 37-105	Nominal Diameter (in.)	Weight (lbs./1000 ft.)	Part No. 37-105	Nominal Diameter (in.)	Weight (lb./1000 ft.)	In Free Air (A)	Single Banked in Trays (A)	DC Resistance at 25°C (Ω/1000 ft.)	AC Resistance at 90°C, 60 Hz (Ω/1000 ft.)	Inductive Reactance (Ω/1000 ft.)	Voltage Drop (V/A/1000 ft.)	Green Insulated Grounding Conductor (3x) Size (AWG)
2	34	-357VFD	2.474	3375	-357BSVFD	2.829	4468	156	133	0.184	0.230	0.0440	0.364	10
1	43	-358VFD	2.561	3748	-358BSVFD	2.917	4876	178	151	0.147	0.184	.0430	0.299	8
1/0	54	-359VFD	2.663	4184	-359BSVFD	3.090	5562	205	174	0.117	0.147	.041	0.246	8
2/0	70	-360VFD	2.795	4764	-360BSVFD	3.226	6208	234	199	0.093	0.117	0.0390	0.203	8
3/0	86	-361VFD	3.013	5622	-361BSVFD	3.447	7171	269	229	0.074	0.094	.038	0.170	6
4/0	109	-362VFD	3.155	6404	-362BSVFD	3.591	8024	309	263	0.058	0.075	0.037	0.142	6
262	132	-363VFD	3.168	6925	-363BSVFD	3.607	8457	352	299	0.048	0.063	0.035	0.124	6
313	159	-364VFD	3.276	7549	-364BSVFD	3.666	9211	389	331	0.040	0.053	0.034	0.109	6
373	189	-365VFD	3.396	8529	-365BSVFD	3.832	10271	432	367	0.034	0.045	0.034	0.097	4
444	227	-366VFD	3.548	9589	-366BSVFD	3.985	11405	456	388	0.028	0.039	0.033	0.08	4

Source: AmerCable, Inc.

15.11 VARIABLE-VOLTAGE DC MOTOR DRIVE

The term *variable-frequency drive* applies to the ac motor drive that requires a variable-frequency power source, whereas the term *variable-speed drive* (VSD) applies to the dc motor drive that requires only a variable-voltage power source. The dc motor speed is linearly related to the applied voltage and inversely with the flux (field current). The VSD is, therefore, basically a dc–dc full-wave converter with the firing delay angle control to vary the output voltage that is applied to the motor. The electric propulsion system using VSD may be found on old ships with a dc motor driving fixed pitch propellers, where the thrust is controlled by the motor speed.

In the VSD, a full-wave thyristor rectifier with variable-voltage output feeds the dc motor. The motor voltage is varied by the thyristor firing delay angle α between 0° and 180°. The field winding is excited with a constant regulated field current. Therefore, the speed of the motor is proportional to the dc armature voltage. Such a drive is generally limited to 2–3 MW power ratings at 600 V_{ac} or 750 V_{dc}. Its disadvantages for high-power applications are that (1) the maintenance cost of the dc motor commutator surface and brushes is high and (2) the presence of high EMI radiated from the motor commutator can interfere with other equipment (EMI-sensitive equipment, computers, etc.).

Many speed control strategies are used in the industry. For speed below the rated speed, the dc-link current that determines the motor current is varied with torque load at a given speed, while simultaneously controlling the field current to keep the air gap flux and the torque constant at the rated values. The horsepower output, therefore, decreases with decreasing speed. Above the rated speed, the motor torque capability degrades as a result of the armature current limitation due to heating, but the motor can supply constant rated horsepower, which is the product of the torque and the speed, as shown in Figure 15.18.

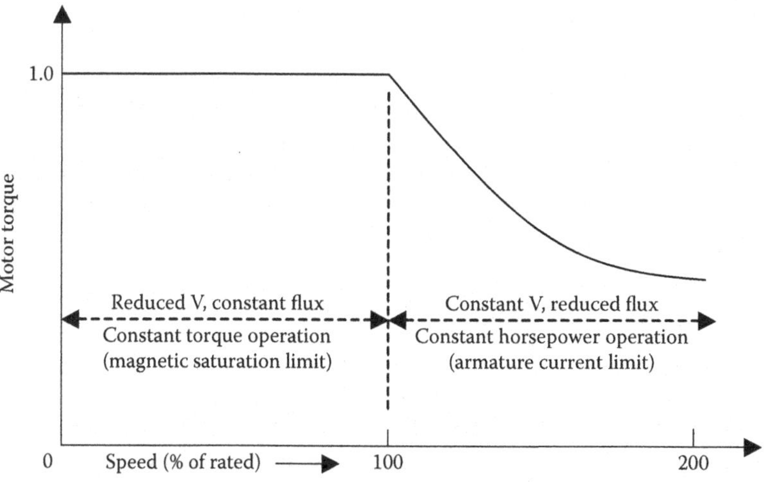

FIGURE 15.18 DC motor speed control regions with VSD.

At present, most drives in the industry are ac, whereas dc drives are used selectively for applications where the system offers special benefits. The dc motor speed can be changed over a wide range by a combination of voltage and field control. As stated earlier, the drive delivers constant horsepower above the rated speed and constant torque below rated speed all the way down to zero speed. Thus, it is possible for the dc machine to deliver the full rated torque at zero speed, that is, producing a holding torque for dynamic positioning. Such operation is possible with an induction motor only when driven by vector control, which is complex and costly in design and in operation.

15.12 VSD IN METRO TRAINS

The modern railway traction incorporates the advances made in motor and power electronics technologies. In large city metro lines, the power is typically dc, which is fed by a trolley wire or a third rail typically at 600 or 750 V dc. Some new installations have used 1500 V_{dc} and even higher voltages up to 3000 V_{dc}. In metro trains with frequent stops and starts, the drive must have a four-quadrant operating capability for regenerative braking in both forward and reverse rotations. The train dc drives in the past have used thyristor-based designs, but the newer metro train dc drives use GTOs, and most recent designs use IGBTs for lower cost and higher efficiency. For 750-V dc drives, 1200-V series-connected IGBTs or 1700-V IGBTs have been found adequate in a simple two-level inverter design. For 1500-V dc drives, 3.3-kV IGBTs have been available since the early 2000s. Even for higher 3000-V_{dc} traction systems, 6.6-kV IGBTs have been available for 10 years for two-level inverter designs.

The bullet trains in Japan are successful high-speed trains implemented in the mid-1960s. In their early version, the trains used dc drives from single-phase ac lines through tap-changing transformers and diode rectifiers. This was later replaced by thyristor phase-controlled rectifier dc drives. Since the 1990s, the trains have used GTO-based pulse width modulated (PWM) rectifiers with two-level inverters with induction motor drives. In some parts of Kolkata, India, and Santiago, Chile, rheostat-controlled dc motors have been used in some sections where the station distances are greater than a mile or so, but are not used anymore due to their poor efficiency.

In both dc and ac cases, dc motors have been used in the past, but they are being replaced with induction motors or ac series motors. For example, many long-distance trains in India use 440-V ac series motors specially designed for high torque needed in railway traction applications. In some new long-distance trains, 25–35 kV main frequency ac power is used.

15.13 VFD AS LARGE MOTOR STARTER

Small induction motors can be started directly from the lines. However, the starting inrush current in a large motor with high inertia can be high and last for a prolonged time, causing the motor to overheat. Moreover, lights may flicker under excessive voltage dip in the lines. The starting current can be reduced by starting the motor with reduced voltage using a Y–Δ starter, an autotransformer, or a phase-control voltage step-down converter. The latter, however, is the least preferred method due to its unacceptably high harmonic current. All these methods result in low starting torque proportional to the applied voltage squared and are not really suitable for starting a large motor with high initial torque load. The wound rotor induction motor can give high starting torque with an external rotor resistance temporarily introduced through the slip rings. However, the cost of a wound rotor can be high and the reliability low due to the slip rings and brushes. The squirrel cage induction motor—the most widely used ac motor—can be most effectively started with full rated torque by using a variable-frequency starter. Such a scheme incurs additional cost but has the following advantages:

- It gives high starting torque compared to other methods.
- Line current can be made sinusoidal with displacement power factor unity with a pulse width modulated (PWM) rectifier.
- The same converter can be used on a time-sharing mode for other motors.
- The converter can be bypassed after starting the motor to eliminate the converter losses.
- At light load running, the voltage can be adjusted to improve the running efficiency.

The variable-frequency starters—very similar to variable-frequency drives (VFDs) in design—are gradually becoming widely used due to falling power electronics prices and stringent power quality standards on the line harmonics imposed on the variable-frequency motor drives.

Shared motor starter. The motor takes less than a minute to start from zero speed and come up to full speed. Therefore, many motors can share a variable-frequency starter by staggering the starting operations. The rating of the power electronics motor starter in such shared applications must correspond to the largest motor in the group. Each motor can be brought to full speed in an open-loop control mode with a constant V/f ratio. The motor can be induction or wound rotor synchronous or permanent magnet synchronous.

During soft starting of a motor, the frequency and voltage start from a small value and ramp up to a full rated value in a linear ramp. However, near zero speed and frequency at starting, the inductive reactance is negligible, and the voltage setting the air gap flux is $V - I_{stator} R_{stator}$. For constant air gap flux, then, $(V - I_{stator} R_{stator})/f$ must be maintained constant. As discussed earlier, this requires a voltage boost of $I_{stator} R_{stator}$ during starting. The required boost decays as the motor speeds up under increasing frequency. This results in a V/f ratio linearly ramping up, as was shown in Figure 15.11. The rate at which V and f are ramped up depends on the load inertia. The ramping is done over a longer time for high inertia loads than for low inertia loads.

Synchronous motor with starter. The variable-frequency power electronics starter for synchronous motor is possible in theory but is hardly used in the synchronous motor mode. Instead, the synchronous motor is normally started as the induction motor using the damper bars on the pole faces. A small permanent magnet synchronous motor can be started as the induction motor directly online using the cage bars, but a large motor may be started softly by using a variable-frequency power electronics starter. A large synchronous motor can also be started using a pony motor on the shaft. Either way, the synchronous motor is first started and brought near the synchronous speed, often at no load or light load to keep the rating of the starting device low. The field is then excited, which brings the motor to the full synchronous speed, and then the load is applied after bypassing the starting device and connecting the motor to the main supply lines.

15.14 CONVERTER TOPOLOGIES COMPARED

The synchronous motor is generally more efficient than the induction motor. It is often used in very large ratings that run for more numbers of hours per year, so that the energy savings add up to a large sum at the end of the year.

As for the drives, the synchronous motor generally uses the load commutated inverter (LCI), whereas the large induction motor generally uses the CSI. In construction, a typical CSI uses 6 thyristors + 6 diodes + 6 capacitors. Since the induction motor is a lagging power factor load, the load commutation from one phase of the thyristor to another phase requires a forced commutation circuit made of diodes and capacitors, tuned to the specific motor leakage inductance. This makes it difficult to design a general-purpose drive for multiple motor ratings. Therefore, the CSI is used only with large induction motors and is designed specifically for a given motor.

The LCI typically uses only six thyristors. The load commutation is done using the synchronous motor's back-emf. However, at speeds below 10%, the back-emf is not enough for the current commutation in the load converter. The commutation is, therefore, provided by the line converter by going into the inverter mode and forcing I_{dc} to become zero to commutate the load inverter. The LCI control scheme requires sensing the rotor position for commutating the load current, making the control more complex.

The LCI with a synchronous motor and the CSI with an induction motor compare as follows:

- LCI has low cost and high reliability since it requires fewer devices.
- High efficiencies of both the inverter and synchronous motor make the LCI a much more efficient drive system.
- LCI is not a motor-specific design, whereas the CSI must match with the induction motor's leakage inductance.
- LCI requires a complex control scheme.
- Both drives are capable of regenerative braking, if desired.

The CSI with a forced-commutated converter has the following advantages:

- Independent control of active and reactive powers
- Absence of commutation failure
- Lower harmonic distortion factor
- Lower adverse impact of large wind energy farms on weaker grid
- Ability of start-up from zero load (i.e., from cold network)

The cycloconverter and CSI and LCI converters have seen a recent decline in use with large motors, where most of the drives being installed at present tend to be the

VSI-PWM type. Advanced high-voltage motors optimized with the power electronics drives are being developed that do not require a transformer for voltage matching.

Regarding the Load Commutated Inverter (LCI) versus VSI versus any other power electronics converter, the debate will perhaps continue on the most optimum design, more so when the total system with the motor is considered. In standalone systems such as on ships, designing the transformer-less system is possible by using a 12-phase or 15-phase generator and all the converters designed accordingly.

15.15 NOTES ON VFDs

High electrical stresses on motor insulation induced by power electronics. In many applications using fast-switching insulated gate bipolar transistor (IGBT) in PWM inverters, the motor winding insulation experiences high dv/dt stress due to steep wave front of the PWM wave. Capacitive ground leakage current flowing in the winding insulation under high-frequency harmonics causes additional heating. The phase-to-phase, phase-to-ground, and turn-to-turn insulations all see high dielectric stress at high temperature, causing accelerated aging and premature insulation failure. With long cables between the inverter and the motor, overvoltages can also result from the reflected waves. The common-mode stray current in the bearings is another deteriorating effect. A soft-switched inverter or a low-pass filter can minimize such problems.

Torsional stress in a motor shaft in a variable-speed motor drive. The variable-speed motor drives were rapidly adopted in the oil refineries due to significant energy-saving potentials in their large and continuous oil-pumping requirements. The early drives had high harmonic content due to lack of industry standard limiting the harmonics. This resulted in a high pulsating torque superimposed on the steady load torque on the motor shaft. The torsional vibrations in the shaft due to such torque pulsations shortened the fatigue life, leading to premature mechanical failures of the shaft. The failures were often seen particularly at the shaft key slots where the mechanical stress concentration was high due to sharp corners. The problem gradually became less severe as the harmonic content was reduced in newer designs, and the shafts were redesigned for a longer fatigue life by reducing the stress concentration.

Induction motor under single phasing. The power electronics device may fail open or short, but it often fails by internal short circuit and then opening up the converter phase by the fuse clearing. This results in a single-phase supply to the three-phase motor between two lines. With three-phase induction motor drive, the motor will continue to run as a single-phase motor but will introduce a high degree of pulsating torque. At full frequency of 60 or 50 Hz, the torque pulsation has a double frequency (100 or 120 Hz) component superimposed on the average steady torque. Pulsation frequency is high enough to filter out the speed pulsations due to the mechanical inertia. However, at low frequency (i.e., at low motor speed), the pulsations may be excessive and may be felt in the load.

VFD in air-conditioning. In conventional air-conditioning, the single-phase induction motor is used at a fixed supply frequency. The motor runs essentially at constant speed, and the temperature is controlled by a thermostat. The efficiency of a constant-speed air-conditioner is poor. The load-proportional VFD can be more efficient. In countries like Japan, where the electrical power cost is high, over 90% of home air-conditioners use VFDs with a permanent magnet synchronous motor, which gives even better efficiency compared with the induction motor drive. Refrigerators and washing machines can also use such drives for improving the energy efficiency.

Conducting band around a permanent magnet rotor. In the permanent magnet synchronous motor, a metal band is often used around the permanent magnets to brace the rotor parts against the centrifugal force. Such a band made of a conducting metal works like a cage winding and may provide a starting torque and a damping torque during transient oscillations of the rotor following a step load change. It has no effect on the synchronous operation since no current is induced in it at synchronous speed. However, in a motor with a VFD feeding the harmonic voltages or harmonic currents, the band carries harmonic frequency currents and generates high power losses and heating on the rotor. Nonconducting bands, such as a fiber–epoxy composite band, in such cases are preferred.

Bearing currents and heating. The currents induced in the motor bearings are due to dv/dt across the stray capacitance of the motor air gap. The air gap capacitance varies inversely with length of the air gap. Therefore, the bearing currents can be of concern in the induction motor where the air gap is typically very small, compared with the synchronous motor (permanent magnet or wound rotor).

PROBLEMS

1. A water pump driven by a 1800-hp induction motor pumps 200,000 m^3/day at rated speed. If the motor were run at slower speed using VFD to reduce the flow rate to 60% over a longer time to pump the same quantity of water per day, determine the percent savings in kilowatt-hour energy consumed per day.
2. A ship was designed to travel at 15 knots with a dedicated 4.6-kV, 30-MW_e propulsion power plant consisting of the generators, transformers, VFDs, and motors rated accordingly. A new line of ships is being designed with essentially the same hull, except for a higher speed of 20 knots. Determine the rating of the new propulsion power plant and the recommended voltage level.
3. A ship travels 3000 nautical mi. at 20 knots speed using 400,000 gal. of fuel oil. If the speed was reduced by 15%, determine the percent change in total gallons of fuel consumption.
4. A 200-hp, three-phase, 60-Hz, six-pole, 1150-rpm, 0.90-pf induction motor operates with VFD connected to 480 V lines. The VFD design has a current source inverter (CSI) fed from a dc link. If the motor needs to run at 1/2 × rated speed, determine (1) the rectifier firing delay angle and (2) the VFD output voltage at the motor terminals.
5. A 300-hp, 480-V, three-phase, 50-Hz, four-pole, 1446-rpm induction motor is driving a large ventilating fan. When it is running from a VFD output of 25 Hz with a constant V/f ratio, determine (1) the approximate speed and (2) the exact speed.

6. A VFD drives a three-phase, 60-Hz, 100-hp, 1760-rpm induction motor. Determine (1) the power, torque, and full load speed when operating at 40 Hz and (2) the power, torque, and full load speed when operating at 80 Hz.
7. A small ship designed to travel at 10 knots needs 2000-kW_e propulsion power and 500 kW_e for housekeeping loads. If it is desired to redesign the power plant for a higher speed of 15 knots, determine the total (1) kilovolt-ampere rating of the new electrical generators, (2) horsepower rating of the new diesel engines, and (3) diesel fuel consumption in U.S. gallons for a 1000-nautical-mi. round trip at 15 knots. Assume that (i) the housekeeping load remains the same, (ii) the new generators have 90% power factor and 92% efficiency, and (iii) the thermodynamic cycle efficiency of the engines is 50%.

QUESTIONS

1. In an unthrottled fluid pipeline, explain why you should expect a higher percentage of reduction in the motor input power than in the flow output rate.
2. List key types of inverters that are used in VFDs.
3. What is the primary reason for maintaining a constant V/f ratio at the motor terminal at all speeds?
4. Why is a voltage boost needed in the VFD output voltage when starting and at low motor speeds?
5. With VFD, explain how the motor torque varies below and above the rated speed.
6. With VFD, explain how the motor horsepower varies below and above the rated speed.
7. How does the variable-speed drive (VSD) for a dc motor differ for the VFD for an ac motor in construction and in operation?
8. Explain how the VFD can also be used for soft starting a large motor and the benefits of doing so.
9. Why might metal bands on the permanent magnet synchronous motor rotor pose overheating risk? What is the remedy?
10. Explain the construction and working of the special transformer used to build a 12-pulse VFD.

FURTHER READING

Barnes, M. 2003. *Practical Variable Speed Drives and Power Electronics*. Burlington: Elsevier.

Bose, B. 2006. *Power Electronics and Motor Drives*. Burlington: Academy Press.

Mohan, N., Undeland, T.M., and Robbins, W. 2003. *Power Electronics Converters, Applications, and Design*. New York: John Wiley & Sons.

Murphy, M. 2001. Variable speed drives for marine electric propulsion, *Trans IMarE*, 108, Part 2, 97–107.

Wildi, T. 2002. *Electrical Machines, Drives, and Power Systems*. Upper Saddle River: Prentice Hall.

Wu, Bin. et al. 2006. *High-Power Converters and AC Drives*. New York: John Wiley & Sons.

16 Quality of Power

Although the quality of power has recently become a part of the industry standards for electrical power that utility companies and other power providers are expected to deliver at a user's terminals, the term *quality of power* has no single universally accepted definition and measurement at present. However, organizations such as the Institute of Electrical and Electronics Engineers (IEEE), the International Electrotechnical Commission (IEC), and the North American Reliability Council have developed working definitions that include the following key performance measures:

- Transient voltage deviations during system faults and disturbances
- Periodic dips in voltage that may cause light flickers
- Transient change in voltage due to large step loading or motor starting
- Steady-state voltage regulation (voltage rise at light load)
- Harmonic distortion (primarily generated by power electronics converters)

The discussion of power quality actually addresses the compatibility between the power delivered by the bus and the ability of the connected load equipment to perform as specified. Any gap in the compatibility between the two has two alternative solutions: (1) to clean up the bus power or (2) to make the equipment tougher to deliver full performance even with a poor-quality power input.

16.1 POWER QUALITY TERMINOLOGY

The ac bus voltage should ideally be sinusoidal with the amplitude and frequency set by the national standards for the utility mains or by the system specifications for a standalone bus that is independent of the mains. The bus voltage may deviate from normal in many ways. The commonly used terms in describing the bus voltage deviations are as follows:

Sag (American term) or dip (British term): the rms voltage falling below a rated value by 10% to 90% for a duration lasting for 0.5 cycle to 1 min.

Swell: the rms voltage rising above a rated value by 10% to 90% for a time duration lasting for 0.5 cycle to 1 min.

Undervoltage: the voltage dropping by more than 10% of the rated value for more than 1 min (also known as brownout). This term is commonly used to describe a reduction in system voltage that requires the utility or system operator to decrease the load demand or to increase the system operating margin.

Overvoltage: the voltage rising by more than 10% of the rated value for more than 1 min. Some lights may burn out, electromagnetic equipment may increase hum due to magnetic saturation, and many equipment may overheat due to increased power losses.

Transients voltage deviation: the voltage may deviate within the acceptable band defined by voltage versus time ($v - t$) limits on both sides of the nominal system voltage. For example, the American National Standard Institute (ANSI) requires the system voltage maintained within the $v - t$ envelope shown in Figure 16.1. The narrow right-hand side of the band is set primarily from the steady-state performance considerations of the motor and transformer-like loads. The middle tapering portion of the band comes from visible light flicker annoyance considerations, and the broad left-hand side of the band comes from electronics load susceptibility considerations. In general, the deviation that can be tolerated by user equipment depends on its magnitude and the time duration. A smaller deviation can be tolerated for a longer time than a larger deviation. The ANSI requires the steady-state voltage of the utility source to be within 5% and the short-time frequency deviations to be less than 0.1 Hz.

The computer and business equipment using microelectronics circuits are more susceptible to voltage transients than rugged power equipment like motors and transformers. They require a narrower band than that shown in Figure 16.1, which allows deviations only for microseconds based on the power supply's volt-second or flux limitation in the magnetic core to avoid saturation (since $v \times dt = N \times d\varphi$ by Faraday's law).

Flicker: the random or repetitive variations in the rms voltage between 90% and 110% of the nominal value can cause flickers in lights, especially in fluorescent lights. Light flickers cause unsteadiness of visual sensation on

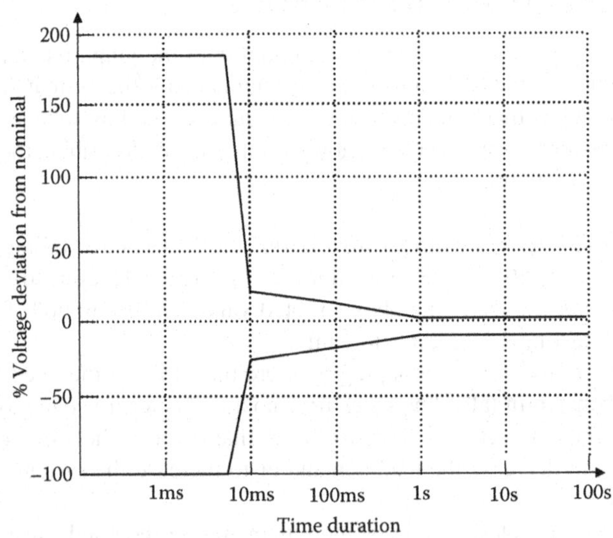

FIGURE 16.1 Allowable voltage deviation versus time duration in utility power (ANSI).

the human eye and can be a source of annoyance. Flickers are often caused by frequently switching on and off large loads. In small standalone systems, switching a relatively large load may result in a large voltage flicker that may impact other neighboring equipment and distract people working in the area. In order to minimize flickers in the neighboring equipment, the grid company may restrict large loads from switching on and off to no more than three to four times per hour. Flicker limits are specified in IEEE Standards 141 and 519, which have served the industry well for many years. Cooperative efforts between the International Electrotechnical Commission (IEC), International Union for Electricity Applications (UIE), Electric Power Research Institute (EPRI), and IEEE have resulted in updated standards as documented in IEC Standard 61000-3.

Spike, impulse, or surge: abrupt but very brief (in microseconds) increase in voltage, generally caused by a large inductive load being turned off within the system or by the lightning surge coming from outside via cables. The surge suppressor at a power outlet used by many small users to protect computer-type sensitive equipment does not eliminate spikes or solve the power quality problem. It merely diverts large voltage spikes to ground, which would otherwise enter the equipment, possibly causing damage.

Frequency deviation: the frequency deviation from a rated value is caused mainly due to the prime mover speed regulation or other reasons.

Harmonics: the deviation from pure sinusoidal waveshape that contains high-frequency components superimposed on the fundamental frequency voltage or current.

Bus, source, or line resistance: the Thevenin equivalent series resistance behind the bus terminals. It causes the steady-state voltage to drop when the load current increases.

Bus, source, or line inductance: the Thevenin equivalent series inductance behind the bus terminals. It causes a transient dip or spike in the voltage when the load current suddenly changes in one step.

Each of these power quality problems has a different cause and solution. Some problems result from using a shared generator, transformer, or line feeder for multiple users. A problem on one piece of user equipment may cause the bus voltage transient to affect other users on the same bus. Other problems, such as harmonics, arise within the user's own installation and may or may not propagate onto the bus to affect other users. Harmonic distortion can be minimized by a combination of good practice in the harmonic reduction at the source and the harmonic filter design.

16.2 ELECTRICAL BUS MODEL

Some power quality issues at the bus terminals can be better understood in terms of the Thevenin model of the bus as a single generator, although multiple generators could be working in parallel behind the bus in the power grid. The Thevenin equivalent source model consists of a single source voltage V_s with single internal impedance Z_s in series as shown in Figure 16.2a, where

$$\text{Thevenin source voltage } V_s = \text{open-circuit voltage at load terminals 1 and 2} \tag{16.1a}$$

$$\text{Thevenin source impedance } Z_s = \text{voltage drop per ampere of load current.} \tag{16.1b}$$

If Z_s is derived under a steady-state power frequency test, then it is called the steady-state or static bus impedance. Under steady load current I, the internal bus impedance Z_s causes the bus voltage to drop by $I \times Z_s$ as shown in Figure 16.2b. In absence of any feedback control on the bus voltage, this drop will remain steady. A source with low Z_s will have a low internal voltage drop but will result in a high short-circuit current, making the circuit breaker rating unmanageable. The design engineer must strike a balance between these two conflicting requirements.

The bus voltage drops under load current due to the internal voltage drop in the Thevenin source impedance Z_s in series with the voltage source. If $Z_s = R + j\,X$ Ω/phase, then the voltage drop under current I at power factor pf = cos θ is given by the following formula that is fairly accurate for all practical power factors:

$$V_{\text{drop}} = I \times (R\cos\theta + X\sin\theta) = I\left\{R \times \text{pf} + X \times \sqrt{1-\text{pf}^2}\right\} \text{ volts/phase.} \tag{16.2}$$

Under a sudden step load increase of ΔI, shown in Figure 16.3, the voltage momentarily drops by $\Delta V_d = \Delta I \times Z_d$ and settles to $\Delta I \times Z_s$ after several transient oscillations of the R-L-C circuit. Here, $Z_d = \Delta V/\Delta I$ is called the *dynamic bus impedance*, which

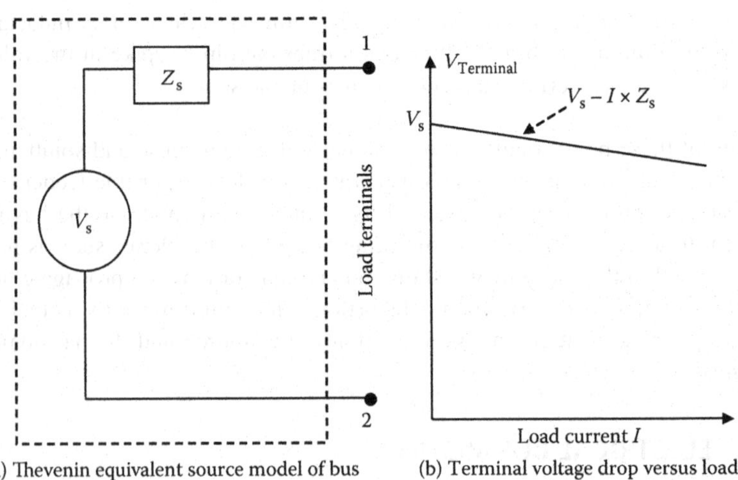

(a) Thevenin equivalent source model of bus (b) Terminal voltage drop versus load

FIGURE 16.2 Thevenin equivalent source model and bus terminal voltage drop versus load current.

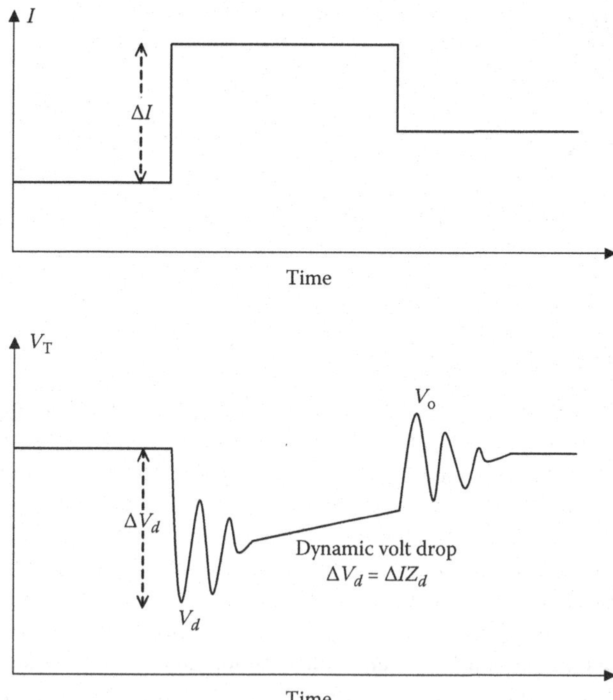

FIGURE 16.3 Voltage deviation ΔV after step change ΔI in load current.

is quite different from the static bus impedance Z_s. If a synchronous generator is the dominant contributor to the bus impedance Z_d, then the generator d-axis subtransient reactance X_d'' contributes in Z_d, and the d-axis synchronous reactance X_d contributes in Z_s. The voltage dip V_d falling below a certain value at the lower end may cause a light flicker. In a system with feedback voltage control, the bus voltage slowly moves toward the rated value within the controller's regulation limit. Under a step load decrease, the opposite happens; the voltage rises, oscillates, and then moves toward the rated value. Such oscillations are often called *ringing*.

Example 16.1

A three-phase 1-MW bus has a Thevenin source voltage of 520 V and a Thevenin source impedance of $5 + j35$ mΩ/phase. Determine the bus voltage when delivering the rated current at 0.85 pf lagging.

Solution:

Assuming a *Y*-connected source (generator):

rated current in each line (phase) = 1,000,000 ÷ (√3 × 480) = 1203 A.

Using Equation 16.2, the voltage drop in an impedance is given fairly accurately at all practical power factors by $R\cos\theta + X\sin\theta$ volts per phase per ampere.

For this machine,

$$V_{drop} = 0.005 \times 0.85 + 0.035 \times \sqrt{1 - 0.85^2} = 0.0227 \text{ V/phase per ampere.}$$

Therefore, voltage drop = 1203 × 0.0227 = 27.3 V/phase or $\sqrt{3}$ × 27.3 = 47.3 V_{LL}.

Terminal line voltage = 520 – 47.3 = 472.7 V.

16.3 HARMONICS

The utility power source is primarily sinusoidal, with one dominant frequency called the fundamental frequency, which is 60 or 50 Hz. Harmonics is the term used to describe higher-frequency sinusoids superimposed on the fundamental wave. The power electronics converter is the most common source of harmonics in the electrical power system. The magnetic saturation in power equipment also generates harmonics. The generator and transformer cores are normally designed to carry flux near the magnetic saturation limit where the magnetizing current is slightly nonlinear and nonsinusoidal, peaking more than the normal peak value of $\sqrt{2}$ × rms value.

As discussed in Section 14.1.3, any nonsinusoidal alternating current (or voltage) can be decomposed into a Fourier series of sinusoidal current I_1 of the fundamental frequency and sinusoidal harmonic current I_h of higher frequencies, where h = 3, 5, 7, 9,.... All even harmonics are absent in practical power systems. The frequency of the hth harmonic current = h × fundamental frequency. For example, the frequency of the 9th harmonic in 60-Hz power system = 9 × 60 = 540 Hz.

If a three-phase load is fed by a Δ-connected transformer, all triplen (multiples of three) harmonics are absent in the line current, that is, $I_h = 0$ for h = 3, 9, 15, and so on. The six-pulse full-wave inverter circuit contains harmonics of the order $h = 6k \pm 1$, where k = 1, 2, 3,.... Therefore, major harmonics present in a six-pulse inverter are 5, 7, 11, 13, 17, 19, 23, and 25. The 12-pulse full-wave inverter circuit contains harmonics of the order $h = 12k \pm 1$, where k = 1, 2, 3, ... Therefore, major harmonics present in a 12-pulse inverter are 11, 13, 23, and 25, which are fewer than in a six-pulse inverter. In both cases, the harmonic current magnitude is found to be inversely proportional to the harmonic order h, that is,

$$I_h = \frac{I_1}{h} \tag{16.3a}$$

where I_1 is the fundamental sinusoidal current. This formula gives approximate harmonic contents in 6- and 12-pulse inverters, as given in the second and third columns of Table 16.1. It clearly shows the benefits of using the 12-pulse converter.

The harmonics can be determined either by the circuit calculations leading to the converter output wave and then going through the Fourier series analysis or by measurements using a harmonic spectrum analyzer or a power quality analyzer—equipment

TABLE 16.1
Harmonic Currents in 6- and 12-Pulse Converters (% of Fundamental)

	Theoretical Value		Actual Value
Harmonic Order h	6-Pulse Converter (Equation 16.2)	12-Pulse Converter (Equation 16.2)	3- and 6-Pulse Converters (IEEE Std 519)
5	20	–	18.5
7	14.5	–	11.1
11	9.1	9.1	4.5
13	7.7	7.7	2.9
17	5.9	5.9	1.5
19	5.3	5.3	1.0

that displays and prints percentage of harmonics of the voltage or current fed to it. The IEEE Standard 519 gives the actually measured current harmonics in typical three-pulse and six-pulse converters as listed in the last column of Table 16.1, which are lower than those given by Equation 16.3a and listed in the second column. The harmonic current induces harmonic voltage on the bus, which is given by

$$V_h = I_h \cdot Z_h \tag{16.3b}$$

where Z_h = harmonic impedance of the bus for the hth harmonic frequency. It is similar to the dynamic impedance Z_d and can be determined from tests at the harmonic frequency of interest. The skin effect makes R_h higher than R_1. The generator d-axis synchronous reactance X_d contributes to X_1, and the subtransient reactance X''_d contributes to X_h for high-frequency harmonics. The direct way of measuring Z_h is by injecting I_h amperes of harmonic current in the bus from an outside source, as shown in Figure 16.4, and measuring the harmonic voltage rise V_h at the bus, which

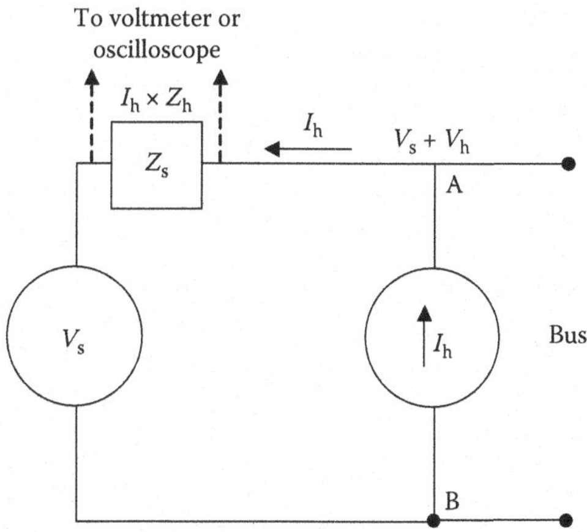

FIGURE 16.4 Harmonic impedance determination by injecting harmonic current source.

is given by $I_h \times Z_h$. Then, $Z_h = V_h/I_h$. The presence of harmonics in a waveform is often indicated by the crest factor (CF), defined as

$$\text{crest factor} = \frac{\text{peak of waveform}}{\text{rms value of waveform}}. \tag{16.4}$$

For an ideal sine wave, CF = $\sqrt{2}$ = 1.414. The CF deviating away from this value would indicate the presence of harmonics. For an ideal square wave, CF = 1.0.

16.3.1 Harmonic Power

The performance of a system driven by a source voltage with multiple harmonics is determined by superimposing the system performance under each harmonic separately. For such an analysis, the system is represented by a series of equivalent circuits, each for the corresponding harmonic frequency. For example, when a motor is driven by a voltage containing multiple harmonics as shown in Figure 16.5a, we first determine the motor performance from the equivalent circuit model for each source

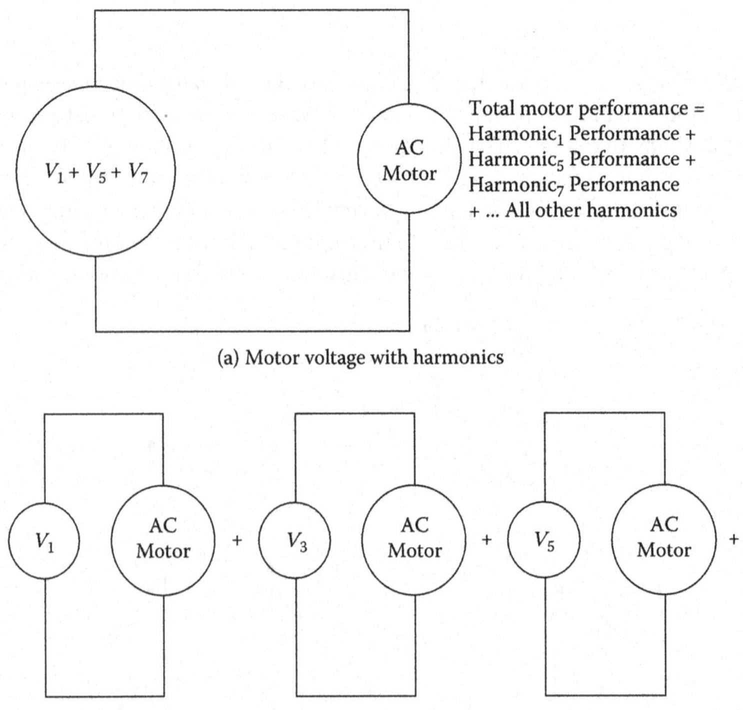

FIGURE 16.5 Superposition model of motor powered by voltage with multiple harmonics.

frequency separately as shown in Figure 16.5b, one at a time, and then superimpose all of them to get the total performance, that is,

$$\text{total performance} = \text{performance}_1 + \sum \text{performance}_h \tag{16.5}$$

where performance_1 = fundamental frequency performance and performance_h = performance under the hth harmonic voltage, h = 5, 7, 11, 13,.... Most motors are Δ-connected, in which the triplen harmonics are zero as explained later in Section 16.5.2.

To analyze the harmonic power in any equipment, we write the bus voltage and current as the sum of their respective fundamentals plus the harmonics, that is,

$$V(t) = V_1 + \sum_{\text{all},h} V_h \text{ and } I(t) = I_1 + \sum_{\text{all},h} I_h. \tag{16.6}$$

The power is the product of voltage and current, that is, $p(t) = v(t) \cdot i(t)$. Using the rms values in Equation 16.6, we write

$$P(t) = V_1 \times I_1 + \sum_{\text{all},h} V_h \times I_h + (V_1 \times I_h + V_h \times I_1). \tag{16.7}$$

On the right-hand side of Equation 16.7, the first term $V_1 \times I_1$ is the real active fundamental power. The third and fourth terms in the parentheses each amounts to zero average power since V and I are of different frequency (known as the orthogonal functions in Fourier series). Their multiplication has a nonzero instantaneous value but has a zero average value. It works like reactive power, adding into volt-ampere reactive that is unproductive and undesirable. The second term $V_h I_h$ is the harmonic power due to each harmonic with average value ≠ 0, which adds undesirable high-frequency pulsations in the equipment performance. For example, it would produce high-frequency pulsations in the motor torque and resulting torsional vibrations in the shaft that may lead to early fatigue failure.

All harmonic currents produce additional I^2R power loss in the conductor. The harmonic power loss can be significantly high due to skin and proximity effects since the high-frequency current concentrates near the conductor surface. The effective skin depth δ_h of the conductor for the hth harmonic current is given by

$$\delta_h = \frac{\delta_1}{\sqrt{f_h}}. \tag{16.8}$$

For a copper conductor, δ_1 for 60-Hz fundamental frequency current is about 9 mm, δ_7 for the 7th harmonic (420 Hz) current is 3.4 mm, and δ_{25} for 1500-Hz current is 1.8 mm. The thin skin effectively reduces the conductor's current carrying area, making the high-frequency resistance significantly higher than the 60-Hz value. Therefore, the harmonic power loss can be much higher than the main frequency power loss in thick conductors, particularly in heavy bus bars made with solid conductors.

Thus, harmonics produce pulsating power, reactive power (kVAR), and additional real power loss in the conductor; all of them are undesirable. Therefore, poor quality of power also results in poor efficiency of the system.

16.3.2 THD and Power Factor

In order to compare the harmonic content in various power buses, we recall the results from Section 14.1.4. The total harmonic distortion (THD) factor in current is defined as

$$\mathrm{THD_i} = \frac{\text{total harmonic current crms}}{\text{fundamental current } I_1 \text{ rms}} = \frac{I_{\mathrm{Hrms}}}{I_{\mathrm{1rms}}}. \tag{16.9}$$

Also, we recall that the total rms value is obtained by the *root sum square* of the component rms values, that is,

$$\text{total rms current } I_{\mathrm{Total}} = \sqrt{I_1^2 + I_3^2 + I_5^2 + \ldots.} = \sqrt{I_1^2 + I_{\mathrm{H}}^2} \tag{16.10}$$

where the total harmonic current

$$I_{\mathrm{H}} = \sqrt{I_3^2 + I_5^2 + I_7^2 + \ldots..} \tag{16.11}$$

The TDH factor in the current waveform, using all rms values, is given by

$$\mathrm{THD_i} = \frac{I_{\mathrm{H}}}{I_1} = \sqrt{\frac{I_{\mathrm{H}}^2}{I_1^2}} = \sqrt{\frac{I_3^2}{I_1^2} + \frac{I_5^2}{I_1^2} + \frac{I_7^2}{I_1^2} \ldots} = \sqrt{\left(\frac{I_3}{I_1}\right)^2 + \left(\frac{I_5}{I_1}\right)^2 + \left(\frac{I_7}{I_1}\right)^2 \ldots} \tag{16.12}$$

The $\mathrm{TDH_i}$ can be easily derived from the harmonic current ratios often given to the power engineer by the converter manufacturer. Similar calculations apply to the total harmonic distortion factor $\mathrm{THD_v}$ in the voltage waveform. The THD is useful in comparing the quality of ac power at various locations. In a pure sine-wave ac source, THD = 0. A higher value of THD indicates a more distorted sine wave, resulting in lower efficiency of the system.

Example 16.2

The actual values of the harmonic currents in a practical 6-pulse converter are given in the I_h/I_1 ratios below, which are less than the theoretical values of I_1/I_h.

Fundamental	5	7	11	13	17	19	23	25
I_h/I_1 ratio	0.173	0.108	0.041	0.032	0.021	0.015	0.011	0.009

Determine the $\mathrm{THD_i}$ with the actual and theoretical harmonic currents.

Solution:

Recall that the total effective rms value is the root-sum-squared (rss) value of the (fundamental + harmonics), that is,

$$I_T = \sqrt{I_1^2 + I_3^2 + I_5^2 + I_7^2 + \text{....25th}}.$$

The table above gives the actual harmonic current ratios, from which we obtain using Equation 16.12:

$$\text{THD}_i = \sqrt{0.173^2 + 0.108^2 + 0.041^2 + 0.032^2 + 0.021^2 + 0.015^2 + 0.011^2 + 0.009^2}$$
$$= 0.212 \text{ pu or } 21.2\%.$$

Similar calculations with theoretical values of $I_h = I_1/h$ would give $\text{THD}_i = 28.7\%$, which is significantly higher than the actual THD_i found in practical converters.

As seen earlier, the harmonic current I_h drawn by any nonlinear load on the bus causes the harmonic distortion in the bus voltage that is given by $V_h = I_h \times Z_h$. V_h in turn causes the harmonic current to flow even in a purely linear resistive load connected to the same bus, called the victim load. If the power plant is relatively small, a nonlinear power electronics load may cause significant distortion on the bus voltage, which then supplies distorted current to the linear resistance-type loads. The IEEE Standard 519 limits the THD_v for the utility-grade voltage to less than 5%. A THD_v above 5% is considered unacceptable, and above 10% needs major correction.

As mentioned earlier, a rough measure of THD_v is the CF—ratio of the peak to rms voltage—measured by the true rms voltmeter. In a pure sine wave, this ratio is $\sqrt{2} = 1.414$. Most acceptable bus voltages will have this ratio in the range of 1.35 to 1.45, which can be used as a quick approximate check on the THD_v at any location in the system. True (or total) rms meter is another way of identifying the power quality. The true rms value of a voltage with harmonic contents is the total rms value given by $V_{\text{Trms}} = \left(V_1^2 + V_3^2 + V_5^2 + \ldots + V_n^2\right)^{1/2}$, where V_1, V_3, V_5,…, V_n are the rms values of the fundamental and all harmonic voltages present. In terms of THD_v, we can write

$$V_{\text{Trms}} = V_{\text{1rms}} \sqrt{1 + \text{THD}_v^2}. \quad (16.13)$$

High harmonic content distorts (degrades) the load power factor also since only the fundamental component out of the total is doing the useful work. True (total) power factor = fundamental power factor × distortion power factor, where the distortion power factor is defined as

$$\text{distortion pf} = \frac{V_{\text{1rms}}}{V_{\text{Trms}}} = \frac{1}{\sqrt{1 + \text{THD}_v^2}}. \quad (16.14)$$

$$\text{Therefore, true pf} = \frac{\text{fundamental pf}}{\sqrt{1+\text{THD}^2}}. \tag{16.15}$$

The Canadian Standards Association and the IEC define the harmonic distortion factor based on the total rms current $I_{T.rms}$, that is, $\text{THD}_{True} = I_{Hrms}/I_{T.rms}$. The IEEE and other U.S. standards, on the other hand, define the THD based on the fundamental current, that is, $\text{THD} = I_{Hrms}/I_{1rms}$, which is somewhat higher than THD_{True}. The two are related as follows:

$$\text{THD}_{True} = \frac{\text{THD}}{\sqrt{1+\text{THD}^2}} \tag{16.16}$$

where THD is as defined by the U.S. standards. The above can also be written as

$$\text{THD}_{international} = \frac{\text{THD}_{USA}}{\sqrt{1+\text{THD}_{USA}^2}}. \tag{16.17}$$

Although the technical definition of THD is the same in the international and U.S. standards, the bases are different. This may create a legal tangle when the THD is specified in the contract without clear base. The contract should have a clear definition of THD, specifically defining the base that can be the fundamental or the total (true) rms value.

Example 16.3

A three-phase, *Y*-connected load draws current containing the following Fourier harmonics (all rms amperes) in each line:

Order h	1	3	5	7	9	11	13	15
$I_{h\,rms}$ (A)	50	30	20	15	11	9	7	6

Determine (1) the total (true) rms value of the current, (2) the THD based on the fundamental current, (3) the true THD_T as defined in some international standards, and (4) the current in the neutral conductor.

Solution:

1. Total (true) rms $I_{Trms} = \sqrt{[50^2 + 30^2 + \ldots + 6^2]} = 65.67$ A.
2. The fundamental current $I_{1rms} = 50$ A and

total harmonic current $I_{Hrms} = \sqrt{[30^2 + \ldots + 6^2]} = 42.57$ A.

Therefore, based on the fundamental current, THD = 42.57/50 = 0.8514 or 85.14%.

(3) Based on the total rms current, $THD_T = 42.57/65.67 = 0.648 = 64.8\%$, which is much lower than the THD of 85.14%. The THD_T could have been derived from Equation 16.17 also. The difference between THD and THD_T must be well understood and clarified particularly in international contracts to avoid any litigation.

(4) All harmonics other than triplen have 120° phase difference between line currents; hence, they cancel out, requiring no return conductor. The triplen harmonics (i.e., 3, 9, and 15 in this case), on the other hand, are all in phase in three lines and require the neutral conductor to carry the sum that is equal to 3 times the rss value of all triple harmonics in each phase, that is,

$$I_{neutral} = 3 \times \sqrt{[30^2 + 11^2 + 6^2]} = 97.53 \text{ A rms}.$$

This neutral current is much higher than the fundamental current of 50 A rms in each phase line.

So, the neutral conductor for this load has to be much heavier than the line conductor. This is not uncommon in systems containing harmonic currents.

16.3.3 *K*-Rated Transformer

As seen in Section 16.3.1, harmonics of different orders—being orthogonal functions in the Fourier series—do not contribute in delivering real average power but produce additional I^2R heating due to skin and proximity effects. Such additional heating in generators, motors, and transformers causes higher temperature rise due to their limited cooling areas, as opposed to long running cables. For this reason, the National Electrical Code (NEC) requires all distribution transformers to state the *K*-rating on a permanent nameplate. This is useful in buying a proper transformer for harmonic-rich power electronics loads. The *K*-rated transformer does not eliminate line harmonics. The *K*-rating merely represents the transformer's ability to tolerate harmonics within the design temperature limit of the conductor.

Since the magnetic power loss in the core and the eddy current loss in the conductor vary with frequency squared, the harmonic losses also increase with frequency squared. Therefore, the higher-order harmonic currents produce additional power losses in the transformer, leading to higher temperature rise. In order to limit such temperature rise, the transformer catering to high harmonic currents is designed with thicker conductors and thinner core laminations, both of which cost more. The transformer designed this way is designated by its *K*-rating, which is the indication of the transformer's ability to deliver power to a power electronics-type harmonic-rich load without exceeding its rated operating temperature. The *K*-rating is defined as

$$K = \frac{\sum h^2 \times I_h^2}{I_{rms}^2} = \sum_{\text{all } h} h^2 \left(\frac{I_h}{I_{rms}}\right)^2 \tag{16.18}$$

for $h = 1, 3, 5, 7, \ldots$ (includes fundamental).

Underwriters Laboratories (UL) has designated *K*-factor as a means of rating a transformer's ability to handle loads that generate harmonic currents. The UL recognizes *K*-factor values of 4, 9, 13, 20, 30, 40, and 50, which are based on information contained in ANSI/IEEE C57.110-1986, *Recommended Practice for Establishing Capability When Supplying Nonsinusoidal Load Currents*. The *K*-factor number tells us how much a transformer must be derated to handle a definite nonlinear load or, conversely, how much it must be oversized to handle the same load.

The transformer designed to power a sinusoidal load from a sinusoidal source requires a *K*-rating of 1.0. Typical transformer *K*-ratings required for power electronics load are *K*-4, *K*-9, *K*-13, *K*-20,..., *K*-50. A *K*-9 rated transformer, for example, can withstand 9 × 60-Hz eddy current loss without exceeding the design temperature. The *K*-rated transformer is used anywhere nonlinear loads are present. Typically a *K*-13 rated transformer is sufficient in most power electronics systems. The following rules are generally found acceptable in selecting the *K*-rating of the transformer:

- Use a standard transformer where the harmonic-producing equipment kilovolt-ampere rating is less than 15% of the source rating.
- A *K*-4 rated transformer is needed where the power electronics equipment represent 30%–40% of the total load.
- Use a *K*-13 rated transformer where the power electronics equipment represent 70%–80% of the total load.
- A *K*-20 rated transformer is needed where supplying 100% power electronics load, such as the variable frequency motor drive.
- Higher *K*-factor ratings are generally reserved for specific pieces of equipment where the harmonic spectrum of the load is known to be high.

If it is desired to use an old existing transformer without a *K*-rating (i.e., with *K*-1 rating), the transformer must be derated so as not to exceed the design temperature. The advantage of using a *K*-rated transformer is that it is usually more economical than using a derated, oversized transformer.

Example 16.4

A power electronics converter draws the fundamental and harmonic currents from a transformer as listed below.

Fundamental	5th h	7th h	11th h	13th h	17th h	19th h
20 A	5 A	3 A	2 A	1.5 A	1 A	1 A

Determine (1) the total rms current in the transformer output and (2) the *K*-rating of the transformer needed to power this converter.

Solution:

The total rms value of the transformer output current is given by the rss value of the fundamental plus all harmonic currents, that is,

$$I_{T.rms} = \sqrt{20^2 + 5^2 + 3^2 + 2^2 + 1.5^2 + 1^2 + 1^2} = 21.03 \text{ A}.$$

The K-rating of the transformer derived from Equation 16.18 is

$$K = 1^2\left(\frac{20}{21.03}\right)^2 + 5^2\left(\frac{5}{21.03}\right)^2 + 7^2\left(\frac{3}{21.03}\right)^2 + 11^2\left(\frac{2}{21.03}\right)^2$$
$$+13^2\left(\frac{1.5}{21.03}\right)^2 + 17^2\left(\frac{1}{21.03}\right)^2 + 19^2\left(\frac{1}{21.03}\right)^2.$$

Therefore, $K = 6.72$; hence, the transformer must have a K-rating of 7 or higher.

16.3.4 Motor Torque Pulsations

In the induction motor driven by variable-frequency drive (VFD), the 3rd harmonic currents in three phases have 3 × 120° = 360° or 0° phase difference from each other, that is, they are in phase, and so are the 6th, 9th,... (all triplen) harmonics. The in-phase triplen harmonic currents cannot flow in Δ-connected motor lines since there is no return conductor to and from the motor terminals. Therefore, the largest current harmonics going in the motor are 5th and 7th. However, the actual torque harmonics on the motor shaft are at 6× fundamental frequency, as explained below.

The fundamental three-phase currents are, say, $I_{a1} = 100 \cos \omega t$, $I_{b1} = 100 \cos (\omega t - 120)$, and $I_{c1} = 100 \cos (\omega t - 240)$, where all angles are in degrees. Such currents produce a magnetic flux rotating at synchronous speed, driving the motor in that direction also, say, the forward direction.

The 5th harmonic three-phase currents are $I_{a5} = 100 \cos (5\omega t)$, $I_{b5} = 100 (5\omega t - 600) = 100 \cos (5\omega t + 120)$, and $I_{c5} = 100 \cos (5\omega t - 1200) = 100 \cos (5\omega t + 240)$. These currents produce a flux *rotating backward* at 5× synchronous speed.

The 7th harmonic three-phase currents are $I_{a7} = 100 \cos (7\omega t)$, $I_{b7} = 100 (7\omega t - 840) = 100 \cos (7\omega t - 120)$, and $I_{c7} = 100 \cos (7\omega t - 1680) = 100 \cos (7\omega t - 240)$. These currents produce a flux *rotating forward* at 7× synchronous speed. The speed and rotation of the stator flux created by various harmonic currents are given in Table 16.2.

The three-phase induction motor runs near the synchronous speed in the forward direction. The 5th harmonic currents set up a magnetic field rotating backward at 5× synchronous speed with respect to the stator and at about 6× synchronous speed with respect to the rotor. Therefore, the rotor would have 6th harmonic currents, producing the 6th harmonic torque. The 7th harmonic currents set up a magnetic field rotating forward at 7× synch speed with respect to the stator and at about 6× synchronous speed with respect to the rotor. Therefore, the rotor currents would have 6th harmonic currents, producing the 6th harmonic torque.

So, both the 5th and 7th harmonics currents produce the 6th harmonic (6 × 60 = 300 Hz) torque pulsations that add up to cause significant shaft jitters (Figure 16.6)

TABLE 16.2
Speed and Direction of Rotation of Magnetic Flux Created by Various Harmonic Currents

Harmonic Order	Frequency in 60-Hz System (Hz)	Synchronous Speed in Four-Pole, 60-Hz Motor (rpm)	Direction of Rotation
1 (fundamental)	60	1800	Forward
5	300	5 × 1800	Backward
7	420	7 × 1800	Forward
11	660	11 × 1800	Backward
13	780	13 × 1800	Forward
17	1020	17 × 1800	Backward
19	1140	19 × 1800	Forward
23	1380	23 × 1800	Backward
25	1500	25 × 1800 (45,000 rpm)	Forward

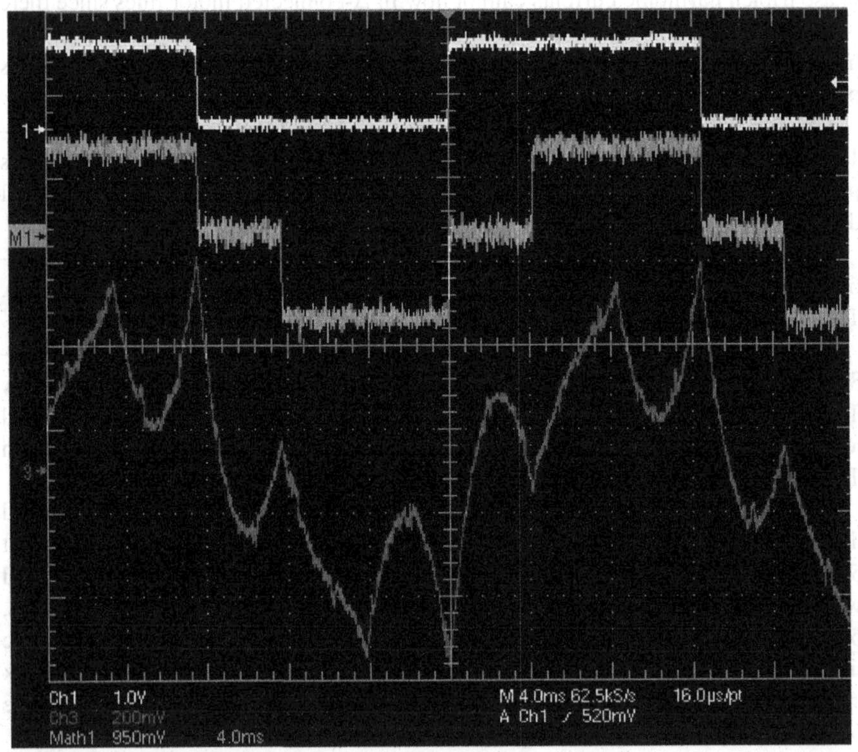

FIGURE 16.6 Sixth harmonic motor torque pulsations produced by 5th and 7th harmonic currents. (Courtesy of Bill Veit, U.S. Merchant Marine Academy.)

and other performance degradations. The total torque has 300-Hz pulsations, giving the speed pulsations with ripple amplitude as follows:

$$\text{speed ripple amplitude} = \text{constant} \times \frac{\text{torque ripple amplitude}}{\text{ripple frequency} \times \text{mass inertia}}. \quad (16.19)$$

The impedance of a rotor circuit to each harmonic is different due to much different slips, such as approximate slip = 6 pu for the 5th and 7th harmonics, slip = 12 pu for the 11th and 13th harmonics, etc. This is because the synchronous speed of the harmonic flux is much greater than the actual rotor speed. Therefore, the harmonic impedance is primarily due to the leakage reactance at the harmonic frequency, with the ratio (harmonic resistance ÷ harmonic slip) approaching zero. The harmonic current, however, produces high I^2R loss due to skin effect at high frequency, and the harmonic flux produces high loss in the magnetic core. Thus, the total increase in power loss in the motor driven by VFD is significant and must be considered in selecting the motor type and the rating. When operating at lower than rated speed, the low-frequency operation of the inverter produces relatively higher harmonics. Up to a certain speed, the constant *V/f* ratio maintains rated flux and rated torque. Below that speed, the torque loading on a given motor should be proportionately reduced as shown in Figure 16.7 to limit the harmonic-related heating and to allow for the reduced self-cooling at lower speed. At speed and frequency above rated values, the torque is inversely proportional to the speed, and the motor delivers constant horsepower.

For fan and pump motors, which generally operate over a limited speed range of 2:1 or 3:1, the horsepower output varies as the speed cubed. So, the motor load naturally

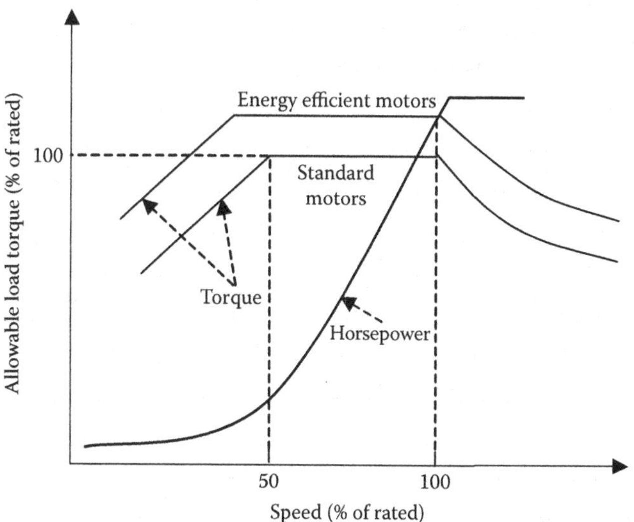

FIGURE 16.7 Motor load torque derating versus speed with VFD harmonics.

falls off considerably at lower speed, and derating from the rated horsepower does not require any action from the user.

16.3.5 Harmonic-Sensitive Loads

In small standalone power systems, such as in navy and commercial ships with electric propulsion, the power electronics motor drives shown as nonlinear load-2 in Figure 16.8 draw a substantial percentage of the total power system capacity. This makes the harmonic problems much more severe than typically seen in large land-based power systems. The harmonic distortion in the main bus voltage, even with clean generator voltage, could be pronounced and could inflict on clean linear load-1. Such a situation is also true in land-based systems in the neighborhood of a large nonlinear load, such as steel mills and other industrial plants using power electronics converters for process controls. Referring to Figure 16.8, if the clean load-1 draws I_{L1} and the power electronics load-2 draws $I_{L2} + I_h$, the total current drawn for the generator is $I_{L1} + I_{L2} + I_h$. The voltage at the generator bus is the clean generated voltage minus the voltage drop in the internal source impedance, that is,

$$V_{AB} = V_{gen.clean} - (I_{L1} + I_{L2} + I_h) \times Z_s = \{V_{gen.clean} - (I_{L1} + I_{L2}) \times Z_s\} - I_h \times Z_s. \quad (16.20)$$

The term on the right-hand side inside { } is clean voltage, whereas $I_h \times Z_s$ is harmonic voltage at the bus output that will cause harmonic currents in the clean load-1 as well, making it the victim load.

In the marine industry, a special harmonic-related problem may occur in fishing ships. The fishery electronics on such ships works on a hydroacoustics principle by sending out surveying signals of high-frequency power in the frequency range of 20 to 800 kHz. Each fish species echoes differently, from which the information on the type of fish present in the survey area is determined. The very-high-order harmonics in the power source can distort the signals going out, and the echo readings on the type of the fish present may also be distorted.

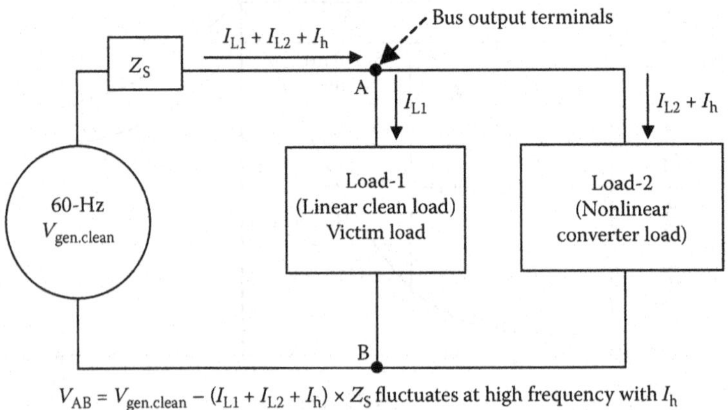

FIGURE 16.8 Bus voltage contamination by power electronics load current harmonics.

16.4 POWER QUALITY STUDIES

Power engineers can use the following means to identify poor power quality:

- A true rms meter can identify the total harmonic distortion factor when compared with the fundamental rms value. A ratio of true to fundamental rms values close to 1.0 indicates good power quality.
- The ratio of the peak to rms value can be a quick indication of the power quality. The peak/rms ratio of $\sqrt{2}$ = 1.414 indicates the perfect sinusoidal power quality. A generally acceptable quality would have the ratio in the 1.35–1.45 range, and a poor unacceptable quality would have the ratio outside this range.
- A current sensor in the neutral wire of a three-phase power system showing a significant value indicates the presence of triplen harmonic currents that are in phase and return through the neutral wire.
- Oscilloscope display of the voltage or current waveshape that has many high-frequency ripples would surely indicate poor power quality.
- A harmonic analyzer, often available in a handheld version, gives a variety of useful information, such as real, reactive, and apparent power; CF; true power factor; total harmonic distortion (THD) as percentage of the fundamental or true rms; and the entire harmonic spectrum. The Fluke F41 model (Figure 16.9) has a capability of downloading information to a serial port of a personal computer.

FIGURE 16.9 Harmonic analyzer Fluke F41 handheld model.

- Power quality analyzers not only offer the harmonic analysis described above but also monitor and record transient events for days at a time. They are available in a three-phase version in larger bench-type or handheld units, such as the one shown in Figure 16.10, which is AEMC Instruments' PowerPad-3945 model.

AEMC 3945B PowerPad
3-Phase Power Quality Analyzer

- True RMS single, dual and three phase measurements at 256 samples/cycle plus DC
- Real time color display of waveforms including transients
- Easy on-screen graphics for setup
- True RMS voltage and current
- Display voltage, current and power harmonics to the 50th order, including direction, in real time
- Phasor diagram display
- Nominal frequency from 40 to 70 Hz
- VA, VAR and W per phase and total
- KVAh, VARh and kWh
- Neutral current
- Crest factors for current & voltage
- K-factor for transformer
- Power factor, displacement PF and tangent
- Captures up to 50 transients
- Short-term flicker and voltage
- Phase unbalance
- Harmonic distortion (Total and individual)
- Includes DataView configuration and analysis software
- 3945 has a CAT III safety rating
- 3945-B has a CAT IV safety rating

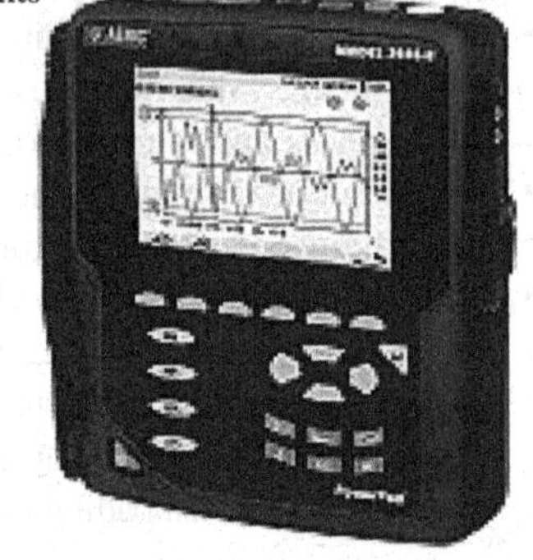

AEMC 3945B Shown

Product Information	
Datasheet	PDF 1.4 MB
Interactive Demo	Flash

AEMC 3945 PowerPad is a hand-held three phase power and power quality meter with a large easy-to-read graphical color display. Measurements are displayed numerically and graphically with colored waveforms. The meter is menu driven with pop-up functions that are activated at the push of a button. All necessary measurements are available for a comprehensive power system check or analysis to 830Vrms; 6500 Arms.

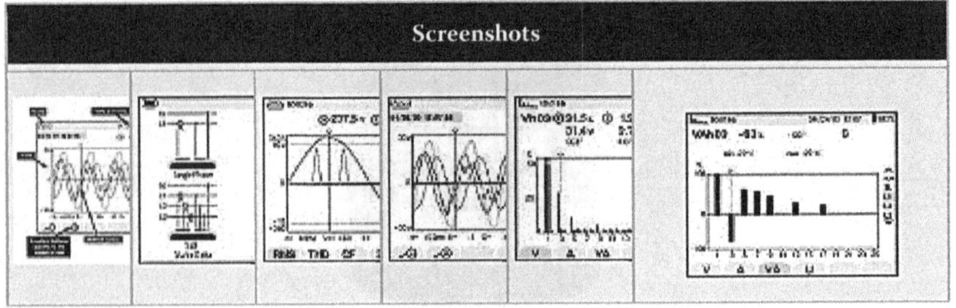

FIGURE 16.10 Three-phase power quality analyzer. (Courtesy of Chauvin Arnoux Inc./AEMC Instruments, Dover, NH.)

16.5 HARMONIC REDUCTION

Harmonic currents can propagate from one piece of equipment to another through solid conductors or by electromagnetic coupling. Equipment not connected by conductors cannot inflict each other by conduction. However, high-frequency harmonic currents can inflict other equipment in the proximity by the electromagnetic interference (EMI) through the magnetic coupling via the leakage flux in air. Therefore, in addition to filtering out major harmonics currents, we need to protect harmonic-sensitive loads both from the conducted harmonics by not connecting through wire and from the radiated EMI by minimizing the leakage flux. Some of the design and operational methods used to minimize the harmonic problems in power systems are described below.

16.5.1 Harmonic Filter Design

The bus voltage harmonics induced by the load current harmonics can be eliminated (filtered out) by a series-resonant *L–C* circuit that is called the harmonic filter. A single-phase version of the filter is shown by a one-line diagram in Figure 16.11. The harmonic current is then supplied or absorbed by the filter, rather than by the bus, thus improving the quality of power at the bus. The *L* and *C* values are tuned to the harmonic frequency desired to be filtered out. The series resistance *R* (often internal to *L*) is kept small by design. If $R \neq 0$, then a small I_h will be drawn from the bus, which is determined by the current sharing rule between the load impedance Z_{load} and the harmonic filter resistance *R*. If we ignore *R* for simplicity here, then the values of *L* and *C* should be such that they are in series resonance at the harmonic frequency f_h, such that

$$2\pi f_h L - \frac{1}{2\pi f_h C} = 0 \quad \text{or} \quad f_h = \frac{1}{2\pi\sqrt{L \times C}} \text{ Hz}. \tag{16.21}$$

Then, the harmonic impedance of the filter $Z_h = 0$, which behaves like a shorted wire. All the harmonic current I_h will pass through the filter and bypass the bus. The bus that was delivering $I_{\text{load}} + I_h$ current before to the load will now see only the pure sine-wave I_{load}. Thus, the harmonic filter does not eliminate the harmonic current; it

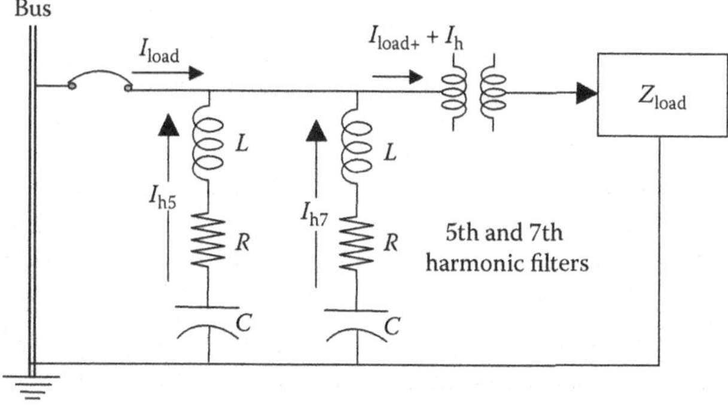

FIGURE 16.11 Load harmonic filter keeps bus free from harmonic currents.

merely bypasses it, not reaching the bus, hence keeping the bus voltage clean. If we wish to filter out multiple harmonics, there must be one such filter tuned to each harmonic frequency. Or, for cost considerations, two dominant harmonics of adjacent orders can be combined into one filter design at the midrange frequency.

In designing a bus harmonic filter for harmonic current I_h, the capacitor voltage rating must be at least equal to the bus voltage, the kVAR rating per phase must be equal to $V_{bus.LN} \times I_h \div 1000$, and the value of $C = 1000\text{kVAR}_{cap} \div \left(2\pi f_{bus} \times V^2_{bus.LN}\right)$ farads per phase. Then, the inductor coil is designed to have inductance value L derived from Equation 16.21 that is capable of carrying the harmonic current I_h continuously without overheating.

The harmonic filter can be placed near the bus on the transformer primary side or near the load on the transformer secondary side. Both have certain advantages. Placing the filter near or on the HV bus is preferred when many small power electronics loads without individual harmonic filters are connected to a common bus where one central filter can be cost effective. On the other hand, placing the filter next to the load prevents harmonics going into the load-side transformer and cables as well, eliminating the overheating in both.

In designing a harmonic filter, the series resonance frequency of the filter is kept about 5% lower than the harmonic frequency. For a typical six-pulse converter with a dominant 5th harmonic (300 Hz) current, the filter series resonance frequency is kept around 285 Hz. Since the filter now supplies the harmonic current (instead of coming from the bus), its L and C components must be designed to carry the harmonic current, which can be calculated by the current divider rule applied on the source impedance in parallel with the filter impedance looking from the harmonic load point. If the bus-frequency capacitor's kilovolt and kVAR ratings are known, then we derive its capacitance value as follows (the formula remains the same with single- or three-phase line ratings):

$$C = \frac{\text{kVAR}_{cap}}{1000 \times \text{kV}^2_{cap} \times 2\pi f_{bus}} \text{ farads/phase.} \tag{16.22}$$

Then, Equation 16.21 gives the required inductance value in the harmonic filter:

$$L_h = \frac{1000\text{kV}^2_{cap} f_{bus}}{\text{kVAR}_{cap} 2\pi f_h^2} \text{ henrys/phase.} \tag{16.23}$$

When a large power electronics load draws power from the utility grid, there are certain interface issues that must be addressed to maintain the quality of power for the rest of the users drawing power from the same bus or the same substation of the grid. On ships, the large propulsion motor power electronics drive drawing power from the source (generators) can contaminate the quality of power for all other users by introducing high content of harmonics. For this reason, the bus itself may require a central harmonic filter to benefit all loads connected to the bus. Figure 16.12 shows a harmonic filter placed on the bus.

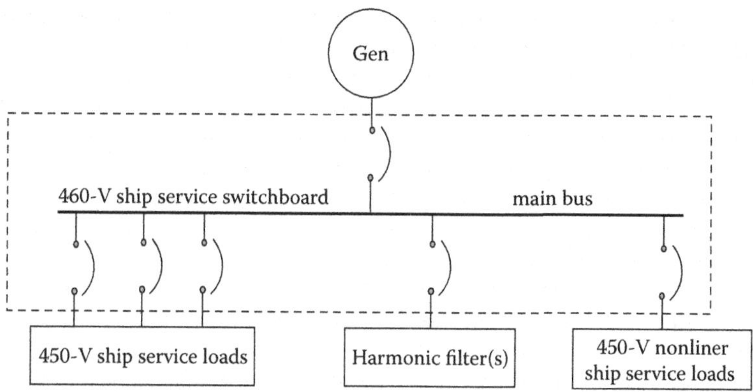

FIGURE 16.12 Power distribution with harmonic filter on main bus.

Example 16.5

A 480-V bus in an industrial plant supplying power to several feeders has a Thevenin source impedance behind the bus equal to 0.003 + *j*0.015 Ω/phase. A power factor correction capacitor rated 600 $kVAR_{3ph}$ is connected to the bus as shown in the figure below. Determine the parallel resonance frequency and identify the precaution needed for connecting a power electronics load to this bus.

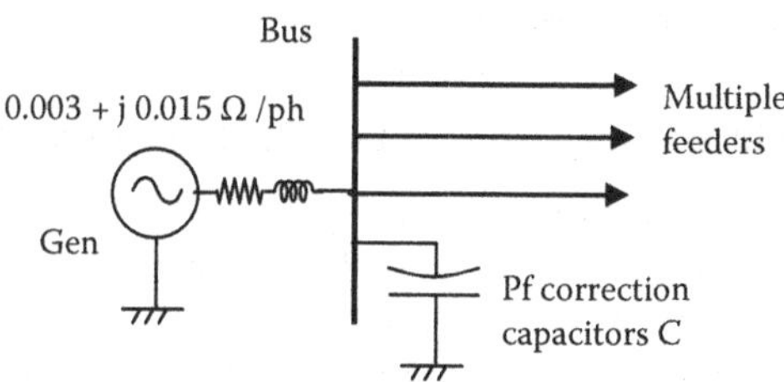

Solution:

Using Equation 16.22, we obtain the capacitance of 600 $kVAR_{3ph}$ capacitor bank:

$$C = \frac{600}{1000 \times 0.480^2 \times 2p \times 60} = 6.91 \times 10^{-3} \text{ F/phase.}$$

The inductance behind the bus is given by

$$L = \frac{0.015}{2\pi \times 60} = 39.5 \times 10^{-6} \text{ H/phase.}$$

From Equation 16.21, which applies for parallel resonance also, the undamped parallel resonance frequency (ignoring resistance) is given by

$$f_o = \frac{1}{2\pi\sqrt{39.5\times10^{-6}\times6.91\times10^{-3}}} = 305\,\text{Hz},$$

which is 5.08 × 60 Hz.

Precaution:

Since the parallel resonance frequency of 305 Hz is essentially equal to the 5th harmonic frequency in a 60-Hz system, any power electronics load that draws significant 5th harmonic current from the bus would resonate and severely distort the bus voltage. The next example determines such distortion.

Example 16.6

The bus in Example 16.5 supplies a 200-kVA, 480-V power electronics load that has the harmonic spectrum shown in Table E5.1 below, which also shows the harmonic impedances behind the bus. Determine (1) the rss and THD of the source current without the 600-kVAR capacitor bank and (2) the rss and THD of the bus voltage with a 600-kVAR capacitor connected to the bus.

TABLE E5.1
Bus Impedances and Voltage Drops at Bus due to Harmonic Currents

Harmonic Order	Frequency (Hz)	Line current I_h (A)	R_h (Ω/ph)	X_h (Ω/ph)	Z_h (Ω/ph)	Volt Drop/ph $V_h = I_h Z_h$
5	300	50	0.003	0.075	0.0751	3.76
7	420	30	0.003	0.105	0.105	3.15
11	660	15	0.003	0.165	0.165	2.48
13	780	7	0.003	0.195	0.195	1.37

Solution:

The fundamental impedance behind the bus, as given in Example 16.5, is Z_1 = 0.003 + j0.015 Ω/phase. In the above table, the harmonic resistance is constant (ignoring the skin effect), and $X_h = h \times X_1$, which rises with the harmonic frequency.

Fundamental load current = 200,000 ÷ (√3 × 480) = 240.6 A.

The total rms current is given by the root sum square (rss) of all harmonic rms values, and the THD of the source current and the bus voltage are

$$I_{rss} = \sqrt{240.6^2 + 50^2 + 30^2 + 15^2 + 7^2} = 248 \text{ A}.$$

$$\text{Therefore, THD}_i = \frac{\sqrt{50^2 + 30^2 + 15^2 + 7^2}}{240.6} = 0.252 \text{ or } 25.2\%$$

as defined by the U.S. standards.

Since the harmonic voltages are volts per phase, THD_v calculations must use the phase voltage of the bus, which is $480/\sqrt{3} = 277$ V.

Therefore, $\text{THD}_v = \dfrac{\sqrt{3.76^2 + 3.15^2 + 2.48^2 + 1.37^2}}{277} = 0.0204$ or 2.04% of the bus voltage.

With a 600-kVAR capacitor bank connected to the bus, the Thevenin source impedance behind the bus is in parallel with the capacitor impedance on the bus. The total equivalent impedance of the parallel combination is then calculated for each harmonic, which gives the following results.

Frequency (Hz)	$Z_{Total.h}$ (Ω/ph)	I_h (A)	V_h (V/ph)
300	1.58	50	79
420	0.116	30	3.48
660	0.045	15	0.675
780	0.035	7	0.245

The rss value and THD of the bus voltage are

$$V_{bus.rss} = \sqrt{277^2 + 79^2 + 3.48^2 + 0.675^2 + 0.245^2} = 288\,V$$

$$\text{THD}_v = \frac{\sqrt{79^2 + 3.48^2 + 0.675^2 + 0.245^2}}{277} = 0.285 \text{ or } 28.5\% \text{ of the bus voltage.}$$

Thus, the harmonic distortion in the bus voltage with the pf correction capacitor is much higher—28.5% versus 2.04% without the capacitor—due to the parallel resonance between the bus impedance and the capacitor at one of the power electronics load harmonics, which is the 5th harmonic in this example. The amplification of the 5th harmonic voltage as seen in the above table has resulted in such unacceptable harmonic distortion on the bus voltage. The system study should identify such possibility and avoid it to maintain a good quality of power.

Example 16.7

Determine the value of filter inductance required to filter out the 11th harmonic from a 60-Hz bus voltage that supplies a 12-pulse converter with a 100-kVAR, 4160-V bus capacitor.

Solution:

As generally practiced by power engineers, we tune the harmonic filter at 5% lower than the harmonic frequency, that is, $f_h = 0.95 \times 11 \times 60 = 627$ Hz for the 11th harmonic. Then using Equation 16.23, we get

$$L_{11} = \frac{1000 \times 4.160^2 \times 60}{100 \times 2\pi \times 627^2} = 0.0042\text{H or } 4.2 \text{ mH/phase.}$$

16.5.2 Very Clean Power Bus

High-power electronics converters switching on and off high currents produce high harmonic distortion on the bus that can adversely impact the performance of harmonic-sensitive loads, such as science instruments, computers in data centers, navigation instruments and radar on ships, etc. In such applications, very clean harmonic-free electrically isolated power is derived by a dedicated motor–generator set, shown in Figure 16.13. Since there is no electrical or magnetic connection between the generator and the motor, the motor–generator set output is very clean, that is, free of all harmonics.

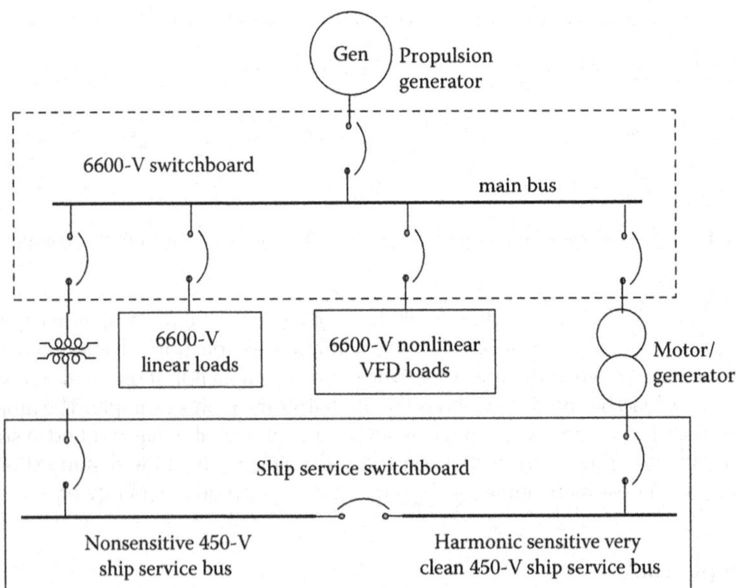

FIGURE 16.13 Electric propulsion bus (6600 V) with 460-V motor–generator set for very clean power for harmonic-sensitive loads.

16.5.3 Δ-Connected Transformer

If the load draws fundamental plus triplen and other harmonic currents, the triplen harmonics can be effectively eliminated from the bus lines by using a Δ–*Y* transformer with neutral, as shown in Figure 16.14. It works as follows, where all angles are in degrees.

The balanced three-phase fundamental currents are 120° out of phase from each other, that is,

$$I_{a1} = 100 \cos \omega t \quad I_{b1} = 100 \cos (\omega t - 120) \quad I_{c1} = 100 (\omega t - 240). \qquad (16.24)$$

The neutral current is the phasor sum of three phase currents, that is, $\tilde{I}_{N1} = \tilde{I}_{a1} + \tilde{I}_{b1} + \tilde{I}_{c1} = 0$; that is, no fundamental neutral current flows in the neutral wire. The triplen harmonic currents are a different matter. If we have 40% 3rd harmonic current in each phase, then the 3rd harmonic current in three phases are

$$\begin{aligned} I_{a3} &= 40 \cos \omega t \\ I_{b3} &= 40 \cos 3(\omega t - 120) = 40 \cos(3\omega t - 360) = 40 \cos 3\omega t \\ I_{c3} &= 40 \cos 3(\omega t - 240) = 40 \cos(3\omega t - 720) = 40 \cos 3\omega t. \end{aligned} \qquad (16.25)$$

Equation 16.25 shows that the 3rd harmonic currents in three lines are in phase, and their sum in the return wire is $I_{N3} = 120 \cos 3\omega t$, which has an amplitude of 120 A. So, in the four-wire *Y*-connected system, the neutral wire will carry 120 A, 20% higher than the 100-A fundamental current in the main lines in this typical example.

The triplen harmonic currents in the Δ-connected primary coils come from the *Y*-connected secondary coils by the transformer turn ratio. However, these currents are absorbed in the Δ-windings and do not propagate further to the main lines on the source side, since there is no 4th wire for the return current. However, the triplen harmonic currents in the windings increase the internal power loss and the operating temperature. They reduce the transformer's load capability and hence are important for selecting the transformer's *K*-rating.

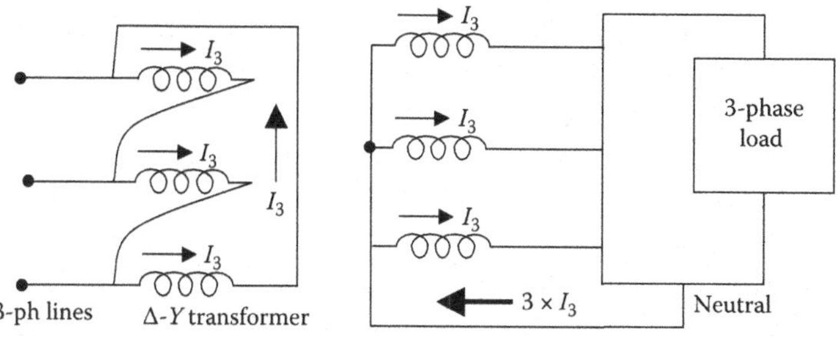

FIGURE 16.14 Triplen harmonics elimination by using Δ–*Y* transformer.

If the neutral wire were not provided in the *Y*-connected system, then also $I_{N3} = 0$. Thus, in a three-wire *Y*- or Δ-connected system, all triplen harmonics are zero (filtered out from the lines), although they circulate within the phase coils of Δ. With all triplen harmonics filtered, the remaining harmonics are $h = 6k \pm 1$, where $k = 1, 2, 3, \ldots$, that is, $h = 5, 7, 11, 13, 17, 19, 23$, and 25 in practical systems. For special 12-pulse connections of rectifiers and inverters using a three-winding transformer with Δ/Δ–*Y* connections, the harmonics orders are $h = 12k \pm 1$, where $k = 1, 2, 3, \ldots$, that is, $h = 11, 13, 23$, and 25. The harmonic current magnitudes $I_h = I_1/h$, so the higher-order harmonics have inherently lower magnitudes. In most applications, the 23rd and higher-order harmonics can be ignored without significant loss of accuracy.

16.5.4 Cable Shielding and Twisting

The EMI—sometimes called the high-frequency or radio-frequency electrical noise—is generally caused by the leakage flux coming from one electrical equipment (culprit) linking to another equipment (victim) in the same proximity, as shown in Figure 16.15a. The leakage flux—although small but of high frequency—induces voltage in the victim equipment as per Faraday's law. This can distort the rated performance of the victim equipment, which may pose a severe problem in cases wherein many pieces of equipment are in a small confined space, such as in a submarine with high-power electronics loads (motor drives). Delicate low-power equipment in close proximity of high-current cables may be impacted more severely.

The leakage flux from the power cables carrying harmonic currents can also inflict serious ill effect on the nearby control signal wires carrying telephone communication or data. Higher-order harmonics having higher frequencies pose a greater problem due to higher interference voltage induced in the signal wires. For this reason,

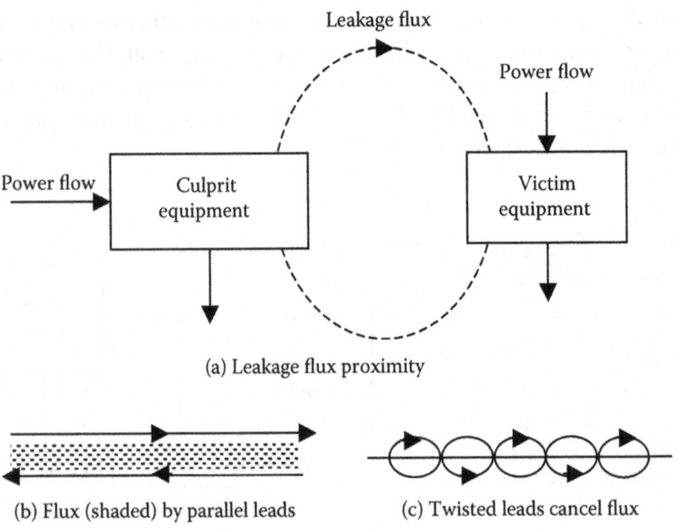

FIGURE 16.15 EMI from culprit equipment inflicting victim equipment.

many systems require the power cables to have a metallic shield that will prevent the leakage flux from escaping the Faraday cage formed by the shield. Additionally, the data and signal cables are required to be in separate metal trays placed with sufficient physical distance between them.

The amount of leakage flux radiated from or received by wire loops depends on the wire loop area, which is large for lead and return wires placed far apart. It can be minimized by placing the lead and return wires as close as possible and using twisted wires in both culprit and victim equipment, as shown in Figure 16.15c.

16.5.5 Isolation Transformer

An isolation transformer can inhibit the propagation of conducted EMI. Having no direct conduction of current by a metallic wire, it provides electrical isolation, hence the name. However, some energy transfer from the input to the output can still take place either by the inductive or capacitive coupling between the coils. The circuit mode of the transformer for this purpose is shown in Figure 16.16, where L_{12} = leakage inductance and C_{12} = capacitance between the primary and secondary windings, C_{1G} = capacitance of the primary coil to the ground, and C_{2G} = capacitance of the secondary coil to the ground. The ground is usually the static shield foil placed between the primary and secondary coils.

At high frequency, the high value of $X_L = \omega L$ blocks the high-frequency current from going further, and the low value of $X_c = 1/\omega C$ diverts the high-frequency current to the ground. Thus, both X_L and X_C jointly block the high-frequency current and voltage spikes going from one side to the other side of the transformer. The isolation transformer has three purposes: (1) safety of the personnel working on the low-voltage side, (2) safety of the equipment from switching or accidental overvoltages in wires on the high-voltage side, and (3) filtering of the high-frequency magnetic and electrical noise.

As per the often-used definition, the isolation transformer has two separate coils with a 1:1 turn ratio. Its sole purpose is to isolate two sides without changing the voltage. However, many power engineers use the term for any two-winding transformer with any turn ratio. Such a transformer does isolate, but it also steps up or down the

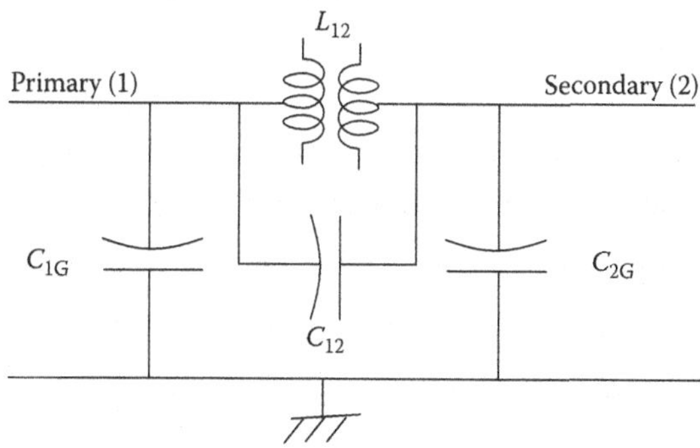

FIGURE 16.16 Transformer model for EMI studies.

voltage on the other side. The auto transformer with metallic connection between the primary and secondary coils is not an isolation transformer.

16.6 IEEE STANDARD 519

The earlier version of IEEE Standard 519 specified the total harmonic distortion (THD) limits on the voltage provided by the utility company but said nothing about the customer's load current causing the voltage distortion. The new version recognizes that both the utility and the customer share the responsibility of maintaining the quality of power in the area being served. We recall that the customer may draw a sinusoidal current as in a transformer, or a chopping current as in a power electronics converter, that would cause the bus voltage harmonics for other area customers. On that basis, the IEEE standard now defines the harmonics current limits listed in Table 16.3 for individual customers. Because a large customer can cause more distortion on the system voltage than a small customer, the standard allows lower harmonics to large customers than small ones. The customer size is measured by the *short-circuit ratio* (SCR), which is defined as

$$\text{short-circuit ratio} = \frac{\text{short-circuit kVA at service point}}{\text{Customer's maximum kVA demand}} \tag{16.26}$$

where

$$\begin{aligned}\text{short-circuit kVA} &= \text{line voltage} \times \text{short-circuit current at service point} \div 1000\\ &= [(\text{line voltage})^2 \div \text{source impedance at service point}] \div 1000.\end{aligned} \tag{16.27}$$

The source impedance is also known as the Thevenin impedance or the internal bus impedance. The maximum demand kilovolt-ampere is known from the monthly utility bill of the medium and large power user. Large customers have low SCR since they demand relatively higher kilovolt-amperes compared to the system capacity.

TABLE 16.3
IEEE Standard 519 Limits on THD_i in Individual Customer Load Current

SCR	$h < 11$	$h = 11 - 15$	$h = 17 - 21$	% THD
<20	4.0	2.0	1.5	5.0
20–50	7.0	3.5	2.5	8.0
50–100	10.0	4.5	4.0	12.0
100–1000	12.0	5.5	5.0	15.0
>1000	15.0	7.0	6.0	20.0

They can cause more distortion to other area customers and hence are imposed with a smaller total harmonic distortion (THD) limit in their load current. A small customer can hardly distort the system voltage even by continuously drawing high harmonics in small load current; hence, they do not draw much attention.

16.7 INTERNATIONAL STANDARDS

The generally acceptable line voltage variations at a medium voltage distribution point in four countries are listed in Table 16.4. The permissible variations are higher in developing countries like China, India, and many Eastern European and South American countries. The allowable step changes in voltage customers can cause on loading or unloading are listed in Table 16.5. The number of voltage disturbances of various durations in a typical European land-based power distribution system (Table 16.6) shows what is expected; the small voltage deviations are more frequent, and vice versa.

TABLE 16.4
Acceptable Line Voltage Variation at Medium-Voltage Distribution Points

Country	Acceptable Range (%)[a]
United States	± 5
France	± 5
United Kingdom	± 6
Spain	± 7

[a] Low voltage consumers may see wider variations.

TABLE 16.5
Allowable Step Change in Voltages Customer Can Cause by Step Loading or Unloading

Country	Allowable Range (%)
France	± 5
United Kingdom	± 3
Germany	± 2
Spain	± 2 for grid-connected renewable systems ± 5 for standalone power systems

TABLE 16.6
Voltage Dips, Interrupts, and Variations on European Power Distribution Networks

	Number of Events per Year for Various Durations of Disturbances			
Voltage Drop	**10 to 100 ms**	**100 ms to 0.5 s**	**0.5 to 1 s**	**1 to 3 s**
10% to 30%	61	66	12	6
30% to 60%	8	36	4	1
60% to <100%	2	17	3	2
100%	0	12	24	5

Source: Adapted from Lutz, M. and Nicholas, W. *Conformity Magazine,* November 2004, p. 12.

PROBLEMS

1. A 50-Hz 800-kW bus has a Thevenin source voltage of 480 V and a Thevenin source impedance of $3 + j25$ mΩ/phase. Determine the bus voltage when delivering rated current at 0.90 pf lagging.
2. The values of the harmonic currents measured in a six-pulse converter are given in the I_h/I_1 ratios below, which are less than the theoretical values of I_1/I_h.

Harmonic	5	7	11	13	17	19	23	25
I_h/I_1 ratio	0.18	0.12	0.05	0.04	0.02	0.015	0.01	0.01

Determine the THD_i with the actual and theoretical harmonic currents.

3. A three-phase, *Y*-connected load draws current containing the following Fourier harmonics (all rms amperes) in each line:

Order h	1	3	5	7	9	11	13	15	17
I_h rms (A)	80	48	32	25	17	15	10	9	8

Determine (1) the total (true) rms value of the current, (2) the THD based on the fundamental current, (3) the true THD_T as used in some international standards, and (4) the current in the neutral conductor.

4. A power electronics converter draws the fundamental and harmonic currents from a transformer as listed below.

Fundamental	5th h	7th h	11th h	13th h	17th h	19th h	23rd h	25th h
30 A	7.5 A	4.5 A	3 A	2 A	1.5 A	1.3 A	1.1 A	1.0 A

Determine (1) the total rms current in the transformer output and (2) the *K*-rating of the transformer needed to power this converter.

5. A 460-V bus in a factory supplying power to several feeders has a Thevenin source impedance behind the bus equal to $0.002 + j\,0.01$ Ω/phase. A power factor–correction capacitor rated 500 $\text{kVAR}_{3\text{ph}}$ is connected to the bus as shown in the following figure. Determine the parallel resonance frequency with the bus and identify the precaution needed for connecting a power electronics load to this bus.

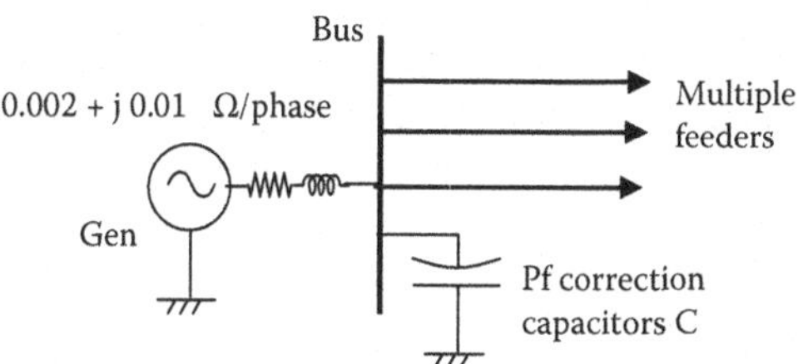

6. It is desired to filter out the 5th harmonic from the 60-Hz supply lines that powers a six-pulse converter in an industrial power distribution system. Design the harmonic filter capacitor using a 100-mh inductor coil that is readily and economically available for use.
7. For a 60-Hz 4160-V bus that supplies power to 12-pulse converter, determine two separate harmonic filter inductance values (1) to filter out the 11th harmonic with a 100-kVAR capacitor and (2) to filter out the 13th harmonic with a 120-kVAR capacitor.
8. The product of the fundamental voltage and the fundamental current results in a net average power delivered to the loads. On the other hand, the product of the fundamental voltage and a harmonic current results in zero net average power over one fundamental cycle. Using Excel, plot the product of a 100-V_{rms} 60-Hz fundamental voltage and an 80-A_{rms} 3rd harmonic current in 15° increments over one fundamental cycle, and verify that their product averages out to be zero over one cycle.
9. A three-phase 460-V transformer is to be procured to power a motor drive that will draw a line current containing the following harmonics as per the VFD supplier:

Harmonic order *h*	1	5	7	11	13	17	19	23	25
Harmonic amps rms	25	5	4	2.5	2	1.5	1.5	1	1

Determine the three-phase kilovolt-ampere rating and the *K*-factor of the transformer that must be specified in the procurement contract.

10. A 120-V power source with no feedback voltage regulator has the static source impedance of $0.2 + j\,0.6$ Ω. Determine its steady-state voltage drop at its terminals under an increase of 8 A in the load current.

11. The utility line voltage displayed on a harmonic spectrum analyzer shows the harmonic content of 10%, 7%, 5%, 3%, and 1% of the fundamental—percentages based on the fundamental magnitude—for the 5th, 7th, 11th, 13th, and 17th harmonic, respectively. Determine the THD_v factor for this line voltage as per the U.S. standards and also as per some international standards.

QUESTIONS

1. In a power distribution system, the line voltage often drops by 50% for 10 ms. Does it meet the ANSI standard on quality of power?
2. What is the major cause of harmonic currents in the electrical power system?
3. Explain the difference between the static source impedance and the dynamic source impedance.
4. Explain the term *ringing* in the bus voltage, and when does it occur?
5. Where and how do we use the superposition theorem when the source voltage has many harmonics?
6. What is the transformer *K*-rating, and why does the higher *K*-rated transformer cost more?
7. Identify equipment around your work area that may severely suffer in performance due to poor quality of power.
8. Identify major ill effects of poor quality of power on the performance of a large induction motor.
9. How does one identify poor quality of a power anywhere (at home, office, industry, etc.)?
10. What are the triplen harmonic currents, and where are they able and unable to flow?
11. Why can the triplen harmonics currents not flow in the Δ-connected system?
12. Explain how a very clean harmonic-free power is obtained on navy ships.
13. Explain how cable shielding and twisting significantly reduce the electromagnetic interference (EMI).

FURTHER READING

Baggini, A. 2010. *Handbook of Power Quality*. New York: John Wiley and Sons.

Fuchs, E. and Masoum, M.A. 2008. *Power Quality in Power Systems and Electrical Machines*. Burlington: Elsevier Academic Press.

Santoso, S. 2010. *Fundamentals of Electric Power Quality*. Austin: University of Texas.

Vedan, R.S. and Sarma, M.S. 2009. *Power Quality*. Boca Raton: CRC Press/Taylor & Francis.

17 Power Converter Cooling

The power rating of equipment, electrical or mechanical, is the maximum continuous output power it can deliver while keeping the steady-state operating temperature below the allowable limit in normal use. The cooling is, therefore, an integral part of the equipment design for limiting the temperature rise under rated load. The heat is generated inside the equipment by internal power loss, which must be carried outside and eventually dissipated in the ambient air. We recall that higher-than-rated operating temperature significantly reduces the equipment life. The 10°C rule for half-life for electrical power equipment holds approximately true also for power electronics devices. Other adverse effects of high temperature on power electronics components are summarized in Table 17.1.

Conventional power equipment, such as motors, generators, and dry-type transformers, are cooled primarily by air circulated inside either by natural convection or by forced air from cooling fans. In power electronics devices, the heat is generated way inside a small semiconducting device having small wafer-thin volume around the junction area. This results in a high thermal gradient from the heat-generating junction to the ambient air where the heat is finally dissipated. Therefore, limiting the semiconductor junction temperature is much more challenging. It generally requires a heat sink (finned metal) with a large surface area around the semiconducting device to keep the junction temperature below the allowable limit. The heat sinks, commonly made of aluminum, remove heat from the inside junction to the outside surface by conduction and then to the ambient air by convection and radiation. Figure 17.1 shows a small power electronics subassembly mounted on a heat sink.

17.1 HEAT TRANSFER BY CONDUCTION

The heat transfer rate by conduction from one surface to another in Btu/hour or joules/second (watts) is given by

$$\text{watts} = \frac{\sigma \cdot \Delta T}{d} = \frac{\Delta T}{d/\sigma \cdot A} = \frac{\Delta T}{R_{\text{Th}}} \text{ where } R_{\text{Th}} = \frac{d}{\sigma \cdot A} \quad (17.1)$$

where σ = thermal conductivity of medium between two surfaces; A = area available for heat transfer between two surfaces; ΔT = temperature difference (gradient) between two surfaces; d = distance between two heat transfer surfaces; and R_{Th} = thermal resistance between two surfaces (°C rise/W transferred or *thermal ohms*).

Recalling the electrical Ohm's law, $I = V/R$, Equation 17.1 can be viewed as Ohm's law in heat transfer by conduction. The analogy between the electrical and thermal Ohm's laws is given in Table 17.2. Once this analogy is understood, the heat

TABLE 17.1
Adverse Effects of High Temperature in Power Electronics Equipment Performance

Semiconductor Devices	Capacitors	Magnetic Components
• Unequal power sharing in parallel or series devices. • Reduction in breakdown voltage in some devices. • Increase in leakage currents. • Increase in switching times.	• Electrolyte evaporation rate increases significantly with temperature increases and thus shortens lifetime.	• Losses increase above 100°C even at constant power input. • Winding insulation (lacquer or varnish) degrades above 100°C.

FIGURE 17.1 Heat sink of thin aluminum fins with power electronics subassembly.

TABLE 17.2
Analogy between Electrical and Thermal Circuits

	Electrical	Thermal
Ohm's law	$I = V/R$	$P = \Delta T/R_{Th}$
Flow	Current I (A or C/s)	Heat P (W or J/s)
Driver	Potential difference (voltage V)	Temperature gradient ΔT (°C)
Resistance	Electrical resistance $R = \frac{\text{length}}{\sigma A}$ (Ω)	Thermal resistance $R_{Th} = \frac{\text{distance}}{\sigma A}$ (thermal Ω)

conduction in a complex system can be broken down in various series–parallel paths, which can be reduced to one equivalent thermal resistance by using the same formulas as for the electrical resistances in series–parallel combinations.

17.2 MULTIPLE CONDUCTION PATHS

Once the heat sink conducts the heat from the inside junction of a power electronics device to the outside surface, the heat is then dissipated by natural radiation and convection in the ambient air. Over the operating temperature range of power electronics equipment, the radiation and convections can be combined into one expression similar to Equation 17.1, that is,

$$\text{watts} = \frac{\theta}{R_{\theta sa}} \tag{17.2}$$

where θ = temperature gradient ΔT (we use θ for ΔT for simplicity in writing equations from hereunder), and $R_{\theta sa}$ = heat sink surface-to-air thermal resistance.

From the junction to outside air, there are three thermal resistances in series as shown in Figure 17.2, where

$R_{\theta jc}$ = thermal resistance (in conduction mode) from device junction to case
$R_{\theta cs}$ = thermal resistance (in conduction mode) from device case to heat sink
$R_{\theta sa}$ = thermal resistance (in convection and radiation modes) from heat sink to ambient air.

The $R_{\theta jc}$ value is given by the power electronics device vendor, and the $R_{\theta cs}$ value depends on the materials and methods used by the design engineer to mount the device on the heat sink. The value of $R_{\theta sa}$ comes from the heat sink manufacturer's technical product data sheet. Representative data for selected heat sinks from one manufacturer are given in Table 17.3.

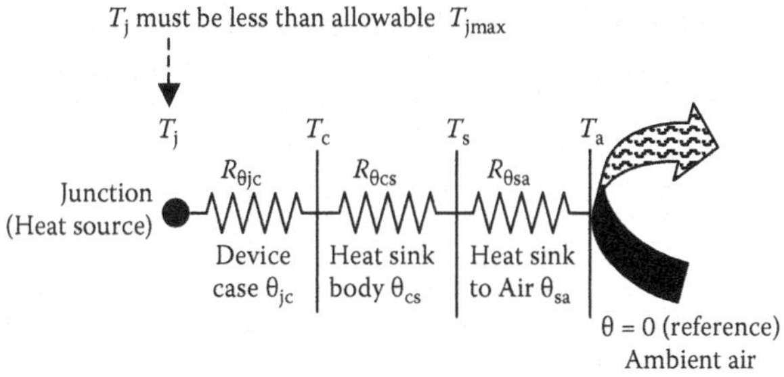

FIGURE 17.2 Temperature gradients from power electronics device junction to ambient air.

TABLE 17.3
Thermal Resistance of Selected Heat Sinks

Heat Sink No.	Thermal Resistance $R_{\theta sa}$ (°C/watt)	Heat Sink Volume (cm³)
1	3.2	75
3	2.2	180
6	1.7	300
9	1.25	600
12	0.65	1300

The three temperature rises (gradients) in Figure 17.2 are determined by the power loss multiplied by the respective thermal resistance. The actual operating temperatures are then given by

$$
\begin{aligned}
&\text{operating temperature of the heat sink surface} \quad && T_s = T_a + \theta_{sa} \\
&\text{operating temperature of the devices case} \quad && T_c = T_s + \theta_{cs} \\
&\text{operating temperature of the device junction} \quad && T_j = T_c + \theta_{jc}.
\end{aligned}
\tag{17.3}
$$

The total thermal resistance from junction to air is the sum of all three resistances in series, that is,

$$\text{total resistance from junction to air,} \quad R_{\theta ja} = R_{\theta jc} + R_{\theta cs} + R_{\theta sa}. \tag{17.4}$$

The junction temperature rise above the ambient air is then given by

$$\theta_{ja} = \frac{\text{power loss in watts}}{R_{\theta ja}}. \tag{17.5}$$

Knowing the maximum power loss under permissible overload and the maximum possible ambient air temperature, the heat sink must be selected to keep

$$R_{\theta ja} < \frac{T_{jmax} - T_{amax}}{\text{maximum watt loss}}, \tag{17.6}$$

The analysis of steady-state heat transfer by conduction through solid metal can show that the optimum heat sink shape is a cone with the heat-generating devices at the base, as shown in Figure 17.3. Therefore, many heat sinks that are solid (i.e., not made of sheet metal) are approximately triangular in shape.

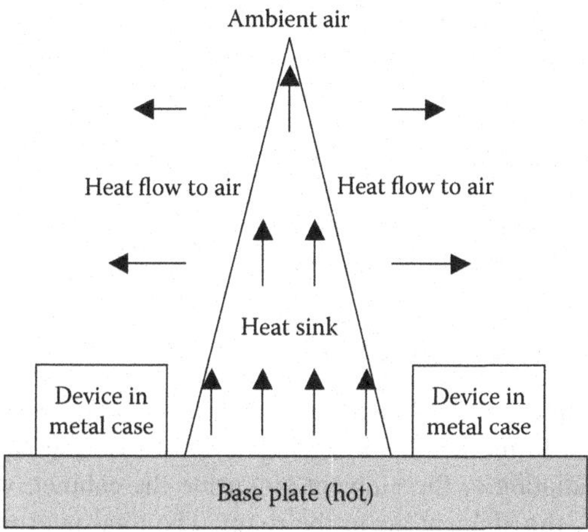

FIGURE 17.3 Optimum shape of solid metal heat sink is approximately conical.

Example 17.1

A power electronics device has an on-state power loss of 40 W and switching loss in watts equal to 1.1 × switching frequency in kilohertz. The junction to case thermal resistance is 1.85°C/W, and the maximum allowable junction temperature T_{jmax} is 150°C. If the device case is mounted on a heat sink to limit the case temperature to 50°C, determine the maximum allowable switching frequency for this device on this heat sink.

Solution:

With switching frequency f_{sw} in kilohertz, the total power loss = 40 + 1.1f_{sw}. The maximum allowable thermal gradient from the device case to the junction is 150 – 50 = 100°C.

Therefore, the maximum power loss this device can withstand = 100°C ÷ 1.85°C/W = 54.05 W, which should be equal to 40 + 1.1 × f_{sw}.

Equating the two, we get 54.05 = 40 + 1.1f_{sw} or f_{sw} = 12.78 kHz, which is the maximum allowable switching frequency for this device.

Example 17.2

A transistor junction has a total power loss of 25 W, and the junction to case thermal resistance is 0.9°C/W. The case is mounted on heat sink #9 with a 50-μm-thick mica insulation with thermal grease resulting in the case to heat sink thermal resistance of 0.5°C/W. The air temperature inside the converter cabinet can be as high as 55°C. Determine the maximum junction temperature in this device.

Solution:

We read from Table 17.3 that the thermal resistance from the heat sink surface to air for heat sink #9 is 1.25°C/W.

Therefore, the total thermal resistance of three elements in series = 0.9 + 0.5 + 1.25 = 2.65°C/W.

The junction temperature rise above cabinet air = 25 × 2.65 = 66.25°C, and the maximum total junction temperature = 66.25 + 55 = 121.15°C.

17.3 CONVECTION AND RADIATION

Conduction is the only mode of heat transfer from the interior of the semiconducting device junction to the heat sink surface. From there, the heat is dissipated by convection and radiation to the ambient air inside the cabinet, which also rises in temperature above the ambient air in the room. The total internal power loss must eventually be dissipated from the equipment's outer surface to the ambient air. The heat gets transferred from a heated cabinet to the surrounding air by three modes: (1) conduction to solids in contact that can be ignored, since most cabinets stand on thin legs with negligible contact with the ground or other solid surfaces; (2) convection in air; and (3) radiation in space. If θ = temperature difference (rise) between the heated body and the cooler surrounding air, then the total heat dissipation in typical operating temperature range can be expressed with some approximation as

$$\text{watts dissipated} = K_1 \cdot \theta + K_2 \cdot \theta^{\alpha} + K_3 \cdot \theta^{\beta} \tag{17.7}$$

where K_1, K_2, and K_3 are the conduction, convection, and radiation constants, respectively, which depend on the areas and thermal conductivities of the materials involved in the cooling path; α and β are the convection and radiation exponents.

The temperature rise is determined by the surface area available for dissipating the internal power loss that heats the equipment body. Air-cooled equipment dissipate internal heat mostly by convection, some by radiation, and negligible by conduction. A heated surface with the same exposed area for convection and radiation in the normal operating range (around 100°C) dissipates approximately equal heat by convection and radiation, and its temperature rise above the ambient air is given by an empirical relation:

$$\theta_{\text{rise°C}} = \frac{540 \times H_{\text{cm}}^{0.1} \times (\text{watts})^{0.8}}{\text{cm}^2}, \tag{17.8}$$

where watts = power loss to be dissipated, cm^2 = surface area (both horizontal and vertical) exposed to ambient air, and H_{cm} = height in centimeters.

Greater height results in greater θ_{rise} because of reduced convection cooling at upper height, where the heated air is close to the surface temperature, thereby reducing the convective heat transfer rate. Having determined the temperature rise from Equation 17.8, the enclosure operating temperature of the heat dissipating body is then

$$T_{\text{operating}} = \theta_{\text{rise}} + T_{\text{room}}. \tag{17.9}$$

The equipment cabinet may be sealed if it is working in an area with dust or combustible vapor. In this case, the cooling calculations are done in two stages, one from the equipment to the cabinet inside air, and then from the cabinet surface to the room air. Industry standards require that the enclosure surface temperature be less than the specified limit considered safe for human touch in a given application.

Example 17.3

The power train devices in a power electronics converter are mounted on a metal base plate 10 cm high × 20 cm wide. The total switching power loss in the devices is 40 W. Determine the temperature rise of the plate due to convection and radiation cooling from both sides of the plate. The ambient air circulating inside the converter cabinet is at 50°C. Also, determine the effective thermal resistance of the plate in units of degrees Celsius per watt.

Solution:

The plate surface area on both sides = 2 × 10 × 30 = 600 cm².

Using Equation 17.8, the plate temperature rise is given by

$$\theta_{\text{rise}} = \frac{540 \times 10^{0.1} \times 40^{0.8}}{600} = 21.7°\text{C}.$$

Therefore, effective thermal resistance = 21.7/40 = 0.54°C/W, and surface temperature of the plate = 21.7 + 50 = 71.7°C.

The junction temperature of the devices will be higher than 71.7°C by the conductive thermal gradients from the plate surface to the device case and then from the device case to the junction.

17.4 THERMAL TRANSIENT

Upon turning on the equipment, the transient temperature rise θ above the ambient air at any time can be determined from the following differential equation:

$$\frac{d\theta}{dt} = \frac{(P - SK\theta)}{GC_p} = \frac{\dfrac{P}{SK} - \theta}{\dfrac{GC_p}{SK}} = \frac{(\theta_m - \theta)}{\tau}, \tag{17.10}$$

where P = power loss in equipment at rated load, S = dissipating surface area, K = dissipation rate per unit area per degree Celsius, G = equipment mass, C_p = specific heat of mass (average), $\tau = \frac{GC_p}{SK}$ = time constant, and $\theta_m = P/(SK)$ = final temperature rise that the equipment will reach.

For equipment starting from initial room temperature, the particular solution to Equation 17.10 is

$$\theta = \theta_m \left[1 - e^{-\frac{t}{\tau}}\right], \tag{17.11}$$

which is in the same form as the capacitor voltage or inductor current rising during the charging period (engineering fields have many such similarities). In the reverse, when the equipment is turned off, we can show the body temperature above the air decays as

$$\theta = \theta_m e^{-\frac{t}{\tau}}. \tag{17.12}$$

Equation 17.12 gives an exponential decay rate $\frac{d\theta}{dt} = -\frac{1}{\tau}\theta_m e^{-\frac{t}{\tau}}$ at any time t, and the initial rate of decay at $t = 0$ equal to $-\theta_m/\tau$. If the body continues to cool at this initial rate of temperature decay, it will reach the room temperature (i.e., $\theta = 0$) in one time constant τ. This leads to an alternative useful way of defining the thermal time constant as follows:

thermal time constant τ = time taken by body under cooling to reach room temperature if the initial cooling rate continues until the end. (17.13)

From the electrical circuit theory, we know that the capacitor and inductor charging and discharging take five time constants ($5 \times \tau$) for the transient to reach the final value (99.67%). That applies to the thermal heating and cooling as well. A simple and very practical use of this definition is illustrated in the example below.

Example: Say we wish to turn off a piece of equipment and start an urgent fix as soon as it reaches the room temperature. To estimate how long we must wait, we can measure the initial surface temperature of the equipment body at $t = 0$ (say, it is 75°C) and then measure it again at $t = 1$ min after it is turned off (say, it is 73°C). This gives a decay rate of 2°C per minute. If the room temperature is 25°C, the surface temperature has to fall a total of 75 − 25 = 50°C, which gives a time constant of τ = (50°C total decay ÷ 2°C per minute initial decay rate) = 25 min. The final room temperature will then be reached in 5 × 25 = 125 min, when we can start working on the repairs.

In the reverse, after turning on, we can predict the time to reach the final temperature of the equipment by measuring the temperature rise in the first minute of turning on at full load. If the rated temperature rise is given on the equipment nameplate (it usually is), then dividing it by the first minute rise gives the time constant τ of the equipment enclosure, and the time to reach the final operating temperature will then be $5 \times \tau$.

17.5 WATER COOLING

Modern pieces of power electronics equipment handle high power and yet are very compact in volume. Although the efficiency is high, a 10-MW, 95% efficient converter would generate $10{,}000 \times \{1/0.95 - 1\} = 526.3$ kW heat internally in a small volume around the device junctions. Since cooling such high-power equipment by natural air convection can be a design challenge, they are often cooled by internally circulating water through copper tubes. Water is a much more effective coolant than air because of its high density and specific heat compared to air. Large water-cooled generators have stator coils wound with hollow copper conductors that carry both the electrical current and the cooling water. Figure 17.4 shows the cross section of a water-cooled stator conductor of a large three-phase, 1200-MVA (1100-MW), 30-kV synchronous generator for a large power plant. It has 6 × 8 cm overall dimensions and 23,100-A_{rms} rated current.

The heat transfer from the wall of a thermally conductive metal tube carrying circulating water depends on many fluid dynamic variables, for example, the Nusselt, Reynolds, and Prandtl numbers, whether the flow is streamline or turbulent, etc. However, we take a simplified view of the fluid flow and heat transfer to the circulating water in conductor tubes running through the heat-generating equipment (power electronics converter or electrical machine). The basic equation that determines the water temperature rise is

$$\text{heat transferred to water} = \text{water mass} \times \text{specific heat} \times \text{temperature rise}$$

which leads to

$$\text{water temperature rise} = \frac{\text{watts transferred to water}}{\text{water flow rate kg per second} \times \text{specific heat}}. \quad (17.14)$$

FIGURE 17.4 Cross section of hollow water-cooled stator conductor of three-phase, 1200-MW 30-kV generator (6 × 8 cm overall dimensions and 23,100 A_{rms} rated current).

Measuring the temperature rise of water between inlet and outlet in degrees Celsius, the above equation reduces to a simplified relation between the power loss and the water quantity flow rate in two systems of units as follows:

$$\text{liters/minute} = \frac{14.3 \times \text{kW power loss}}{\text{water temperature rise in °C}}$$

$$\text{U.S. gallons per minute} = \frac{3.75 \times \text{kW power loss}}{\text{water temperature rise in °C}} \tag{17.15}$$

Equation 17.15 gives the water flow rate for a given piece of equipment assuming that all water is in contact with the cooling tube. In a large-diameter tube with laminar flow, only a fraction of the total water flow is in contact with the tube wall. Therefore, even with the required quantity of flow rate passing through the tubes, the *heat may not be removed* unless the tube is narrow enough to have high-velocity turbulent flow with all water in contact with the tube. On the other hand, narrow tubes result in high pressure drop. The design engineer may take several iterations before arriving at a satisfactory design that optimizes the heat transfer and the pressure drop.

17.5.1 Cooling Tube Design

The cooling water progressively heats up while passing through copper tubes in the heat-generating equipment. The water enters one end at room temperature (10°C to 25°C) and leaves the other end about 30°C–40°C hotter. Therefore, the temperature range of the cooling water in the equipment could be 10°C–65°C. The properties of water (density, specific heat, etc.) would change with temperature but by negligible amount over this temperature range. Therefore, we assume the water density and specific heat constant to simplify the temperature rise estimate. Over the practical temperature range, we use constant values for water density, 1000 kg/m^3, and specific heat, 4180 J/(kg °C). With these values, a fluid flow and heat transfer analysis will lead to the following relation for the cooling water path parameters shown in Figure 17.5.

$$\text{watts}/(L\,\theta) = 148.2\,(V \cdot D)^{0.8} \tag{17.16}$$

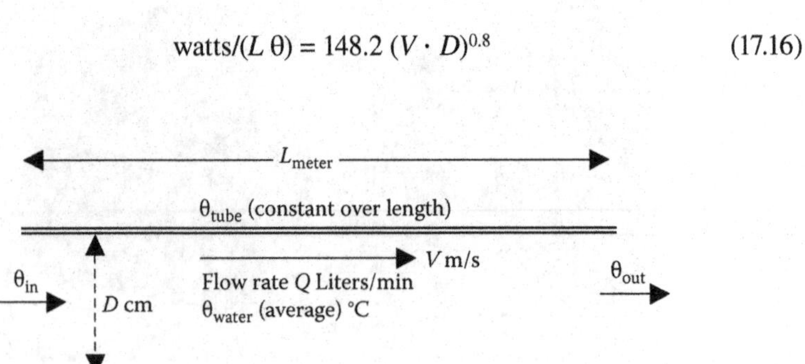

FIGURE 17.5 Water cooling tube parameters for thermal analysis.

where watts = equipment power loss transferred to water; L = total length of cooling tube in one series path (m); θ = mean temperature difference, $\theta_{tube(constant)} - \theta_{water(average)}$ (°C); V = velocity of water in tube (m/s); and D = inside diameter of cooling tube (cm).

In terms of the water quantity flow rate Q in liters/minute,

$$Q = \frac{\pi(D/2)^2 100V}{1000} \times 60 \quad \text{or} \quad V = \frac{Q}{1.5\pi D^2}. \tag{17.17}$$

Using Equation 17.17 in 17.16, we obtain

$$\frac{\text{Watts}}{L \times \theta} = 148.2\left(\frac{Q}{1.5 \times \pi \times D}\right)^{0.8} = 42.9\left(\frac{Q}{D}\right)^{0.8}$$

$$\text{Therefore, } Q = \frac{\text{liters}}{\text{minute}} = D\left(\frac{\text{watts}}{42.9L \times \theta}\right)^{1.25} \tag{17.18}$$

$$\text{water velocity } V_{m/s} = 0.212\frac{\text{liters/minute}}{D_{cm}^2}. \tag{17.19}$$

The left-hand side of Equation 17.18 suggests that for the same power loss, tube length, and temperature difference, the Q/D ratio must be maintained constant. If we wish to have fewer liters of water circulated, then the diameter of the tube must be reduced. It would, however, increase both the water temperature and the tube wall temperature. It would also increase the pressure drop through the tube running through the equipment, which would require a larger pump and higher pumping power. These are the design trades the engineer makes to meet the overall system requirements.

As the cooling water circulates through the tube, its temperature rises by $\Delta\theta_{water} = \theta_{in} - \theta_{out}$, where θ_{in} = inlet temperature of water, and θ_{out} = outlet temperature of water. If θ_{tube} = tube temperature (assumed constant), then the mean temperature difference between tube and water is given by

$$\theta = \frac{1}{2}\{(\theta_{tube} - \theta_{in}) + (\theta_{tube} - \theta_{out})\} = \theta_{tube} - \frac{1}{2}(\theta_{in} + \theta_{out}). \tag{17.20}$$

These equations are accurate within 5% for $\theta_{in}/\theta_{out} < 2$.

17.5.2 Pressure Drop

Determining the water pump rating requires calculating the pressure difference (drop) between the water inlet and outlet. Various sources of pressure drop in a fluid flow through a series of cooling tubes are frictions at (1) the tube wall along its length, (2) the tube entrance and exit, and (3) each bend in the tube routing path. The water is generally drawn from a large header (or reservoir) and is discharged into another large header (or reservoir). The entry and exit velocities are, therefore, extremely low and can be ignored. The pressure drop is then approximately given by the following:

$$\Delta P \text{ in pascals} = 251\frac{5.55L \times V^{1.8}}{D_{\text{cm}}^{1.2}} + (750 + 50 \times \text{number of } 90^\circ \text{ bends})V^2. \quad (17.21)$$

In Equation 17.21, the multiplier 50 with the number of 90° bends assumes that the ratio of the bend radius to the tube diameter is greater than three (soft bends). The multiplier will be 75 for less soft bends with a ratio of two and 100 for a ratio of one (sharp bends). For 180° bends (U-turns), the number of 90° bends to use in Equation 17.21 = 2 × 180° bends.

The unit of ΔP in SI unit is pascal (newtons per square meter). In practice, it is often expressed in head of water in meters, which is

$$\Delta H \text{ in meters of water head} = \frac{\Delta P \text{ in pascals}}{9.81 \times 1000}. \quad (17.22)$$

Parallel cooling paths. Multiple parallel paths may be required for limiting the pressure drop in a long tube. If N = number of identical parallel paths, then the above formulas apply to each path with its own kilowatt power loss, length, and water quantity per path. However, the pressure drop ΔP is the same for all parallel paths. For nonidentical parallel paths, the kilowatt power loss, water flow rate Q, and water velocity will be different in each path, but ΔP would be the same.

Noncircular tubes. The cooling water tubes are often round copper or aluminum tubes in most applications. When they are of another shape (rectangular, square, oblong, or triangular tubes, or even two parallel plates with narrow gap), their equivalent diameter is used in the above formulas. It is derived such as to give the same water flow rate for a given ΔP:

$$\text{equivalent tube diameter} = \frac{4 \times \text{cross section of the water flow}}{\text{wetted perimeter of the water flow path}}. \quad (17.23)$$

The power electronics converter subassemblies are often mounted on a solid metal plate—called base plates or the cold plate—with a labyrinth of water tubes, shown in Figure 17.6. Parallel plate heat exchangers are also used in many places, such as in large power converters and lubricating oil cooling systems. Figure 17.7 shows one such parallel plate heat exchanger.

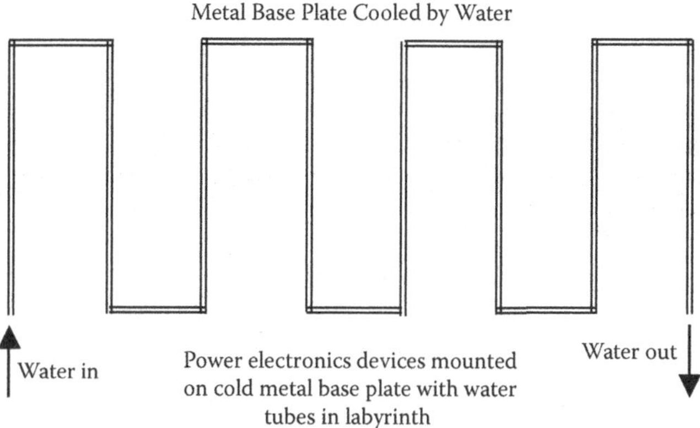

FIGURE 17.6 Labyrinth metal tubes for water-cooled power electronics converter assembly.

FIGURE 17.7 Parallel plate heat exchanger for main lubricating oil cooler. (Courtesy of Raul P. Osigian, U.S. Merchant Marine Academy.)

Example 17.4

A power converter assembly with 800-W power loss is mounted on a 20-cm square base plate cooled by water. The water tubes are of 2-mm internal diameter placed on a 15-cm square with four 90° bends as shown below. If the inlet and outlet water temperature difference is to be limited to 25°C, determine the water flow rate in liters per minute.

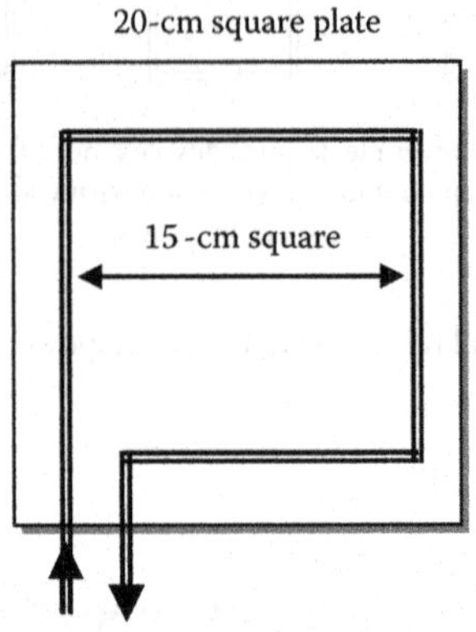

Solution:

For the cooling tube length $L = 4 \times 15 \div 100 = 0.6$ m, $D = 0.2$ cm, and $\theta = 25$°C, Equation 17.15 gives the cooling water flow rate as follows:

$$\text{liters/minute} = 14.3 \times 0.80 \div 25 = 0.458.$$

From Equation 17.19, water velocity in tubes = $0.212 \times 0.458 \div 0.2^2 = 2.43$ m/s. From Equations 17.21 and 17.22

$$\text{pressure drop} = 251\frac{5.55\times 0.6\times 2.43^{1.8}}{0.2^{1.2}} + (750+50\times 4)\times 2.43^2$$

$$= 28{,}614 + 5{,}610 = 34{,}224 \text{ Pa}$$

$$= \frac{34.224}{1000\times 9.81} = 3.49 \text{ meters of water head.}$$

Example 17.5

A piece of electrical equipment is cooled by a labyrinth (Figure 17.6) of 1.0-cm ID water tubes made of 15 straight sections, each 0.6 m long joined at the ends by 90° connectors and 0.1-m straight bridging pieces. The cooling water enters the labyrinth from a large header and discharges to another large header such that the velocity at both ends is negligible. If the water velocity in the tubes is 0.8 m/s, determine the pressure drop and heat removal capacity of the labyrinth with 30°C water temperature rise.

Solution:

Total tube length = (15 × 0.6) + (14 × 0.1) = 10.4 m

number of bends = 14 × 2 = 28.

Equation 17.21 gives the pressure drop as follows:

$$P = 251\frac{5.55\times 10.4\times 0.8^{1.8}}{1.0^{1.2}} + (750+50\times 28)\times 0.8^2 = 9695+1376 = 11{,}071\,\text{Pa}.$$

$$\text{Pressure drop in meters of water head} = \frac{11{,}071}{9.81\times 100} = 1.13\,\text{m} = 1.6\,\text{psi (small)}.$$

From Equation 17.19, we get liters/minute = 0.8 × 1.0²/0.212 = 3.774. Then, from Equation 17.15, we obtain the heat removal capacity:

kW = 3.774 × 30/14.3 = 7.917.

Example 17.6

A vertical heat exchanger dissipates 1 MJ/h per pair of cooling plates that are 25 cm wide, 50 cm high, and separated by a 1.5-mm water gap. Determine the water flow rate and pressure drop to limit the water temperature rise to 20°C.

Solution:

Heat exchange rate = 1 × 10⁶ ÷ 3600 = 278 J/s = 0.278 kW.

From Equation 17.15, water flow rate = 14.3 × 0.278/20 = 0.2 L/min.

Actual water velocity between plates = 0.2 × 1000/(25 × 0.15) = 53.3 cm/s = 0.533 m/s.

From Equation 17.23,

$$\text{equivalent diameter of flow cross section} = \frac{4\times(25\times 0.15)}{2\times(25+0.15)} = 0.298\,\text{cm}.$$

Using Equation 17.19, effective velocity = 0.212 × 0.2 ÷ 0.298² = 0.480 m/s (close match with the actual 0.533 m/s).

Length of the water flow = 50 cm = 0.5 m. Then, Equation 17.21 gives

$$\text{Pressure drops } \Delta P = 251\frac{5.55 \times 0.50 \times 0.48^{1.8}}{0.298^{1.2}} + 0 \text{ for no bends} = 793 \text{ Pa}$$

$$= \frac{793}{9.81 \times 1000} = 0.08 \text{ m of water head.}$$

The pressure drop here is negligible, as expected in a plate heat exchanger. The pump rating is primarily determined by the pressure drops in other plumbing that brings the water to the plates.

17.5.3 Cooling Water Quality

The electrically live hollow stator conductors (Figure 17.4) in large utility-scale synchronous generators are cooled with deionized distilled water that gives the equipment a life of 25 to 30 years. For a reasonably long life of the cooling tubes that are not electrically live, the circulating water inside must be at least clean fresh water. Equipment dissipating the heat ultimately in a lake, river, or ocean may require an intermediate freshwater or deionized-water heat exchanger. For example, the United States Coast Guard (USCG) Healy icebreaker cycloconverter is water-cooled with deionized water circulating through the thyristor heat sinks. Deionized water is used for many power electronics converter cooling processes where the heat sinks are electrically live. The deionized water is passed through stainless-steel coils cast into aluminum body. Such a water-cooling technique clearly enables a compact thyristor module assembly. The ultimate heat rejection in the sea is achieved by water flowing between parallel plates.

Water cooling is used in many high-power VFDs for large motors on ships. It is generally configured in a closed loop consisting of primary and secondary sides separated by a plate-type heat exchanger. The primary side has a pump, heat exchanger, plumbing, and distribution manifolds for each side of the propulsion switchboard. The secondary side also has a pump, plumbing, three-way heat exchanger bypass valve, and the ultimate cooling medium (seawater in this case). The cooling water circuit is monitored with differential pressure flow meters in the primary pump discharge pipes, pressure sensors in distribution manifolds, and temperature sensors in both water circuits. Such complex monitoring is installed to alert the operator of cooling system problems. It also minimizes the pumping power and maintains the temperature by varying primary pump motor speed as required to meet the VFD load demand. The scheme often uses pulse width modulated (PWM) drive for the centrifugal pump motor. The monitoring circuit also has shutdown functions to prevent damage due to out-of-range faults. In addition, leak detection is installed within VFDs under the cooling plates, which alters operators of either plate leakage or excessive condensation due to low temperature.

The typical cooling water is a 20%/80% mixture of propylene glycol and deionized water, respectively. Although propylene glycol has a lower specific heat and higher viscosity than plain water, and therefore reduced heat transfer performance, it raises the boiling point, lowers the freezing point, and contains many anticorrosive properties, which extend the service life of components and reduce the maintenance.

The mixture is cooled by seawater circulating through a brazed plate heat exchanger. All plumbing components are stainless-steel tubing.

With a range of seawater temperature typically encountered in practice, monitoring the VFD operating temperature and corresponding load adjustment is required. High water temperature would result in high drive temperature, poor performance, and possible damage if the drive load is not reduced. On the other hand, too-low water temperature results in condensation in the VFD itself and possible damage. The recommended water temperature in the secondary-side heat exchanger is +23°C, and the recommended minimum temperature difference between the primary and secondary sides of the heat exchanger is 5°C. Therefore, the primary-side water is below 30°C. A three-way valve can bypass the cooling water around the heat exchanger to maintain the proper temperature range of the secondary cooling medium.

PROBLEMS

1. A power electronics device has an on-state power loss of 30 W and switching loss in watts equal to 1.5 × switching frequency in kilohertz. The junction to case thermal resistance is 2.0°C/W, and the maximum allowable junction temperature T_{jmax} is 180°C. If the device case is mounted on a heat sink to limit the case temperature to 60°C, determine the maximum allowable switching frequency for this device.
2. A transistor junction has a power loss of 45 W and a junction to case thermal resistance of 1.1°C/W. The case is mounted on the heat sink #6 with 35-μm-thick mica insulation with thermal grease resulting in the case to heat sink thermal resistance of 0.7°C/W. The air temperature inside the converter cabinet can be as high as 65°C. Determine the maximum junction temperature in this device.
3. The power electronics devices in a converter assembly are mounted on a metal base plate 15 cm high × 25 cm wide. The total switching power loss in the devices is 95 W. Determine the temperature rise of the plate due to convection and radiation cooling from both sides of the plate. The ambient air circulating inside the converter cabinet is at 55°C. Also, determine the effective thermal resistance of the plate in units of degrees Celsius per watt.
4. A power converter assembly with 1.2-kW power loss is mounted on a 30-cm square base plate cooled by water. The water tubes are of 2-mm internal diameter placed on a 25-cm square frame with four 90° bends. If the inlet and outlet water temperature difference is to be limited to 30°C, determine the water flow rate in liters per minute.
5. A piece of electrical equipment is cooled by a labyrinth of 1.0-cm ID water tubes made of 12 straight sections, each 80 cm long joined at ends by 90° connectors and 12-cm straight bridging pieces. The cooling water enters the labyrinth from a large header and discharges to another large header such that the velocity at both ends is negligible. If the water velocity is 1.5 m/s, determine the pressure drop and heat removal capacity of the labyrinth with 40°C water temperature rise.

6. A large vertical heat exchanger dissipates 10 kW per pair of cooling plates that are 30 cm wide, 95 cm high, and separated by 10-mm water gap. Determine the water flow rate and pressure drop to limit the water temperature rise to 30°C.
7. A 10-MW water-cooled power electronics converter is 97% efficient. Determine the water requirement in gallons per minute if the available cooling water inlet temperature is 30°C and the outlet temperature must be limited to 60°C.
8. A three-phase, three-cm outer diameter cable in air has a resistance of 2 mΩ/phase per meter and carries 100-A/phase load. Determine its outside surface temperature (rise + ambient) if the ambient air is 40°C.

QUESTIONS

1. Briefly summarize the mathematical analogy between the heat transfer and the flow of electrical current.
2. How would you determine the thermal time constant of a piece of equipment merely by using two readings of a thermometer that are 1 min apart?
3. Why is passing the required quantity of cooling water not sufficient to limit the equipment temperature rise?
4. Passing less water through a thinner tube can remove the same heat with a given length of tube. What is the downside of it?
5. How can you reduce the pressure drop in the water cooling path into one-half?
6. Discuss your experience (if any) in water cooling of the electrical or power electronics equipment.

FURTHER READING

Ellison, G. 2010. *Thermal Computations for Electronics.* Boca Raton: CRC Press/Taylor & Francis.

Shabani, Y. 2010. *Heart Transfer, Thermal Management of Electronics.* Boca Raton: CRC Press/Taylor & Francis.

Smith, S. 1985. *Magnetic Components.* New York: Van Nostrand Reinhold.

Appendix: Symmetrical Components

The unsymmetrical operation of a three-phase ac system—in steady state or during short circuit—is analyzed by resolving the unsymmetrical applied voltages and resulting currents into symmetrical components that are orthogonal in a mathematical sense (not physical space). The method was first developed by Fortesque and later fully expanded by Wagner, Evans, and Clarke for applications in practical power systems. It is analogous to resolving any space vector force into three orthogonal components. It involves algebras of complex numbers and matrices. We study here a brief theory and its applications in power system analyses with unbalanced voltages applied in operation or during unsymmetrical short-circuit faults. The L-G, L-L, and L-L-G short circuits are called *unsymmetrical faults* since they do not involve all three lines symmetrically.

A.1 THEORY OF SYMMETRICAL COMPONENTS

The method of symmetrical components uses the operator $a = 1\angle 120°$ similar to the operator $j = 1\angle 90°$ we routinely use in ac circuits. We recall that the operator j was a shorthand notation for 90° phase shift in the positive (counterclockwise) direction. Similarly, we use in this Appendix, often the operator α as a shorthand notation for 120° phase shift in the positive direction. Although often written as Greek letter α in the literature, we will hereafter write it as the Latin letter a without confusion. With $a = 1\angle 120°$, we have $a^2 = 1\angle 240°$ and $a^3 = 1\angle 360° = 1\angle 0° = 1$. In exponential form, $a = e^{j2\pi/3}$. Obviously, the phasor sum of $1 + a + a^2 = 0$.

We can resolve any three unbalanced phase currents $\tilde{I}_a$, $\tilde{I}_b$, and $\tilde{I}_c$ into three symmetrical sets of balanced three-phase currents $\tilde{I}_0$, $\tilde{I}_1$, and $\tilde{I}_2$ as follows:

$$
\begin{aligned}
\tilde{I}_a &= \tilde{I}_0 + \tilde{I}_1 + \tilde{I}_2 \\
\tilde{I}_b &= \tilde{I}_0 + a^2\tilde{I}_1 + a\tilde{I}_2 \\
\tilde{I}_c &= \tilde{I}_0 + a\tilde{I}_1 + a^2\tilde{I}_2
\end{aligned}
\tag{A.1}
$$

where $\tilde{I}_1$ = phase-a value of the positive sequence set of three-phase currents; $\tilde{I}_2$ = phase-a value of the negative sequence set of three-phase currents; and $\tilde{I}_0$ = phase-a value of the zero sequence (in phase) set of three-phase currents.

Figure A.1a and b depicts three unsymmetrical currents and their three symmetrical components, respectively. The symmetrical sets of balanced three-phase zero, positive, and negative sequence currents $\tilde{I}_0$, $\tilde{I}_1$, and $\tilde{I}_2$ are also denoted by I^0, I^+, and I^-, respectively, in some books and literature.

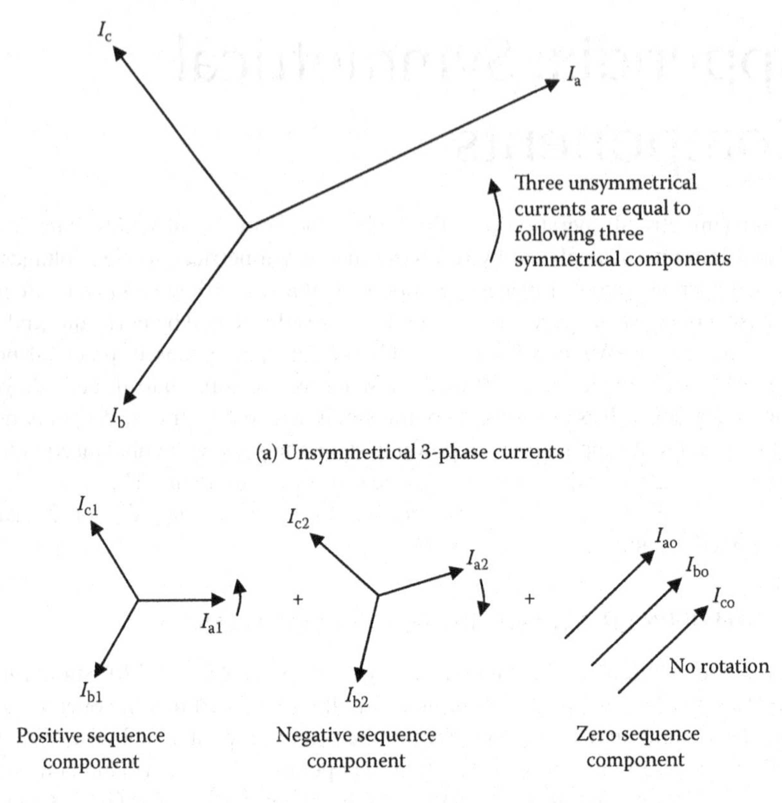

FIGURE A.1 Symmetrical components of three-phase unbalanced currents.

The argument Fortesque made was as follows. A vector has two degrees of freedom—in magnitude and in direction—so it can be resolved in two orthogonal components on x–y axes. A three-dimensional vector has three degrees of freedom—magnitude, angle θ in the x–y plane, and angle φ in the z-plane—so it can be resolved into three orthogonal components on x–y–z axes. A set of three-phase balanced currents has *two* degrees of freedom—magnitude and phase angle of any one phase current. The magnitudes and phase angles of the other two phase currents have no additional freedom, as they are fixed in magnitude and 120° phase shifts. However, a set of three unbalanced (unsymmetrical) currents has six degrees of freedom—two for each of the three phase currents—so it can be resolved into three balanced (symmetrical) sets (called symmetrical components), each having two degrees of freedom, making a total of six degrees of freedom. This is shown in Figure A.1, where the upper set depicts unbalanced three-phase currents, and the three lower sets show the symmetrical components, which are three-phase balanced sets.

In resolving any unsymmetrical three-phase currents into three symmetrical components as in Equation A.1, we recognize the following. Like all ac quantities, the components $\tilde{I}_0$, $\tilde{I}_1$, and $\tilde{I}_2$ are also phasors. By writing $\tilde{I}_a$ as in Equation A.1, we imply that the component phase angles are with respect to $\tilde{I}_a$. The correct way of writing Equation A.1 is $\tilde{I}_a = \tilde{I}_{0a} + \tilde{I}_{1a} + \tilde{I}_{2a}$. Instead of writing such long notations repeatedly, we just imply that $\tilde{I}_0$, $\tilde{I}_1$, and $\tilde{I}_2$ component phase angles are with respect to $\tilde{I}_a$, that is, $\tilde{I}_0 = \tilde{I}_{0a}$, $\tilde{I}_1 = \tilde{I}_{1a}$, and $\tilde{I}_2 = \tilde{I}_{2a}$.

The set of Equation A.1 can be written in the matrix form:

$$\begin{pmatrix} I_a \\ I_b \\ I_c \end{pmatrix} = \begin{pmatrix} 1 & 1 & 1 \\ 1 & a^2 & a \\ 1 & a & a^2 \end{pmatrix} \begin{pmatrix} I_o \\ I_1 \\ I_2 \end{pmatrix} = [A] \cdot \begin{pmatrix} I_o \\ I_1 \\ I_2 \end{pmatrix} \tag{A.2}$$

where $[A] = \begin{pmatrix} 1 & 1 & 1 \\ 1 & a^2 & a \\ 1 & a & a^2 \end{pmatrix}$ is called the operator matrix.

The matrix in Equation A.2 can be solved for $\tilde{I}_0$, $\tilde{I}_1$, and $\tilde{I}_2$ by using the matrix inversion software installed in many computers and even in advanced calculators. The solution for the symmetrical components can then be written in terms of the inverse of matrix $[A]$, that is,

$$\begin{pmatrix} I_o \\ I_1 \\ I_2 \end{pmatrix} = \frac{1}{[A]} \cdot \begin{pmatrix} I_a \\ I_b \\ I_c \end{pmatrix} = [A]^{-1} \cdot \begin{pmatrix} I_a \\ I_b \\ I_c \end{pmatrix} = \frac{1}{3} \begin{pmatrix} 1 & 1 & 1 \\ 1 & a & a^2 \\ 1 & a^2 & a \end{pmatrix} \begin{pmatrix} I_a \\ I_b \\ I_c \end{pmatrix} \tag{A.3}$$

Thus, the inverse of the operator matrix $[A]$ multiplied by the phase currents gives the following set of component currents:

$$\begin{aligned} \tilde{I}_0 &= \frac{1}{3}\left(\tilde{I}_a + \tilde{I}_b + \tilde{I}_c\right) \\ \tilde{I}_1 &= \frac{1}{3}\left(\tilde{I}_a + a\tilde{I}_b + a^2\tilde{I}_c\right) \\ \tilde{I}_2 &= \frac{1}{3}\left(\tilde{I}_a + a^2\tilde{I}_b + a\tilde{I}_c\right). \end{aligned} \tag{A.4}$$

In a Y-connected system with neutral, the neutral current must be the phasor sum of three phase currents. In a Δ-connected system, it must be zero as there is no neutral wire. Thus, the neutral current $\tilde{I}_n$ is related to $\tilde{I}_0$ as follows.

In a four-wire Y-n system,

$$\tilde{I}_n = -(\tilde{I}_a + \tilde{I}_b + \tilde{I}_c) = -3\tilde{I}_0, \text{ which gives } \tilde{I}_0 = -\frac{1}{3}\tilde{I}_n. \tag{A.5}$$

In a Δ system, absence of neutral wire makes

$$\tilde{I}_n = 0, \text{ which gives } \tilde{I}_0 = 0. \tag{A.6}$$

The symmetrical components of the three-phase unbalanced voltages are found from relations exactly similar to Equations A.1 through A.4. It can be verified via actual plots of the symmetrical component phasors that the current $\tilde{I}_1$ in three phases has positive phase sequence (i.e., a–b–c, the same as $\tilde{I}_a$, $\tilde{I}_b$, and $\tilde{I}_c$), $\tilde{I}_2$ in three phases has negative phase sequence (i.e., a–c–b), and $\tilde{I}_0$ in three phases is all in phase (zero phase sequence). Since all three zero sequence currents are in phase, they must have a return path through neutral wire. Therefore, in absence of the neutral wire in three-wire ungrounded *Y*- and Δ-connected systems, $\tilde{I}_n = \tilde{I}_0 = 0$.

A.2 SEQUENCE IMPEDANCES

Each symmetrical component current set sees different impedance in rotating machines due to different rotating direction of the resultant flux. They are called positive, negative, and zero sequence impedances Z_1, Z_2, and Z_0, respectively. The positive sequence impedance is what we normally deal with in balanced three-phase systems. For example, the positive sequence three-phase currents in a generator set up flux in the positive direction (same as the rotor direction), and the positive sequence impedance Z_1 = (synchronous reactance + armature resistance). On the other hand, the negative sequence three-phase currents in a generator set up flux rotating backward, and the rotor has slip of –2.0 with respect to the negative sequence flux. The rotor, therefore, works like an induction motor with rotor slip –2.0, offering very low impedance that will be similar to the subtransient impedance we discussed in Section 8.5.3. Since the negative sequence flux alternately sweeps the *d*- and *q*-axes of the rotor, $Z_2 = 1/2\ (Z''_d + Z''_q)$. In static equipment like transformers and cables, Z_1 and Z_2 are equal since the flux rotation sequence does not matter in the equipment performance. As for Z_0, it is different than Z_1 and Z_2 in both the generator and the transformer depending on the flux pattern of the zero sequence current.

In general, $Z_2 \ll Z_1$ in the rotating machines, $Z_2 = Z_1$ in transformers and cables, and Z_0 depends on the equipment connection and neutral grounding method. When the phasor sum of 3-ph currents cannot return back, such as in ungrounded *Y*- and Δ-connected systems, I_0 cannot flow, and hence, $Z_0 = \infty$ (effectively open circuit). The values of $\tilde{Z}_1$, $\tilde{Z}_2$, and $\tilde{Z}_0$ can be derived from the machine configuration. In many cases where the sequence impedances cannot be calculated, they are best derived from tests. First, a purely positive sequence voltage $\tilde{V}_1$ is applied to the equipment, and the resulting positive sequence current $\tilde{I}_1$ is measured. Then, the positive sequence impedance of the equipment $\tilde{Z}_1 = \tilde{V}_1/\tilde{I}_1$. Similarly, the negative and zero sequence impedances are determined separately. In practice, they are usually determined by applying simulated unbalanced faults at low voltages and measuring the currents. For the system analysis under unsymmetrical conditions, the sequence components are added to make the total sequence impedance of the system.

The symmetrical component analysis is often used to determine the fault currents in unsymmetrical faults for circuit breaker sizing and for designing the protective relaying schemes. In such analyses, the system's equivalent Thevenin network up to the fault location is established for each sequence voltage with the sequence impedance. The three sequence networks are then connected as dictated by the nature of the unsymmetrical fault and the correspondingly imposed boundary conditions. The classical circuit solution then gives the sequence currents, from which the actual phase currents are determined.

With this introduction to the theory of symmetrical components, we now examine its applications in the fault current analysis, starting with our familiar symmetrical fault and then moving into unsymmetrical faults.

A.3 SYMMETRICAL FAULT CURRENT

As such, the symmetrical component analysis is not required for symmetrical faults since the fault current calculations follow the analysis as we covered in Chapter 8. However, we apply the symmetrical components theory to see what we already know about symmetrical faults.

Equation A.3 for three-phase symmetrical fault current gives

$$\tilde{I}_0 = \frac{1}{3}\left(\tilde{I}_a + \tilde{I}_b + \tilde{I}_c\right) = 0$$

$$\tilde{I}_1 = \frac{1}{3}\left(\tilde{I}_a + a\tilde{I}_b + a^2\tilde{I}_c\right) = \frac{1}{3}\left(\tilde{I}_a + \tilde{I}_a + \tilde{I}_a\right) = \tilde{I}_a \tag{A.7}$$

$$\tilde{I}_2 = \frac{1}{3}\left(\tilde{I}_a + a^2\tilde{I}_b + a\tilde{I}_c\right) = \frac{1}{3}\left(\tilde{I}_a + a\tilde{I}_a + a^2\tilde{I}_a\right) = 0.$$

Similarly, assuming symmetrical 3-ph voltages before the fault, only positive sequence is involved, that is, $\tilde{V}_1 = \tilde{V}_a$, and $\tilde{V}_1 = \tilde{V}_2 = 0$. Then,

$$\tilde{I}_1 = \tilde{V}_1/\tilde{Z}_1, \tilde{I}_2 = 0, \tilde{I}_0 = 0. \tag{A.8}$$

Also, since $\tilde{I}_a$, $\tilde{I}_b$, and $\tilde{I}_c$ are all balanced,

$$\tilde{I}_a = \tilde{I}_1, \tilde{I}_b = a^2\,\tilde{I}_1, \tilde{I}_c = a\tilde{I}_1, \text{ and } \tilde{I}_n = 0. \tag{A.9}$$

This gives only the positive sequence current with the other two components zero, as expected.

A.4 L-G FAULT CURRENT

This is the most frequent fault in practical power systems, and it can be analyzed only using the theory of symmetrical components. For a grounded system, shown in Figure A.2, $\tilde{Z}_f$ = ground fault impedance (called soft fault or arcing fault impedance as opposed to hard dead fault of zero impedance), $\tilde{Z}_g$ = actual earth impedance in the return path, and $\tilde{Z}_n$ = impedance intentionally placed in the neutral. For a ground fault on any one phase, say phase A, we know the boundary conditions imposed by the circuit, which are, by inspection, $\tilde{I}_a \neq 0$ and $\tilde{I}_b = \tilde{I}_c = 0$.

Resolving the phase currents into their symmetrical components leads to $\tilde{I}_1 = \tilde{I}_2 = \tilde{I}_0 = 1/3\,\tilde{I}_a$. The equal sequence currents indicate that all sequence networks are in series. Moreover, at the point of fault, $\tilde{V}_a = \tilde{I}_a\left(\tilde{Z}_f + \tilde{Z}_g + \tilde{Z}_n\right) = 1/3\,\tilde{I}_a\left(3\tilde{Z}_f + 3\tilde{Z}_g + 3\tilde{Z}_n\right)$. This indicates that the total $(3\tilde{Z}_f + 3\tilde{Z}_g + 3\tilde{Z}_n)$ is in series with the sequence currents.

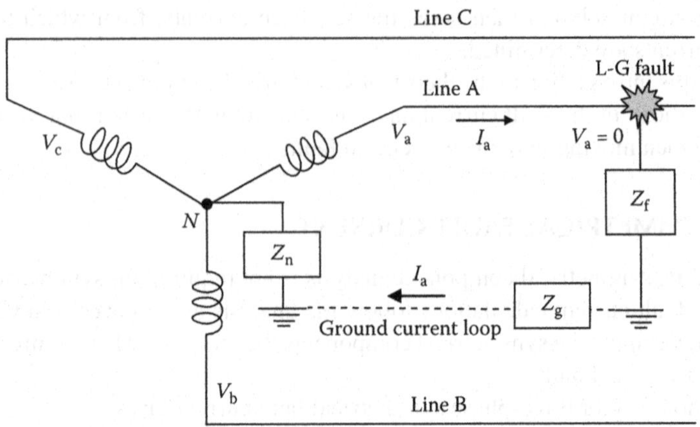

FIGURE A.2 Single-line to ground fault in three-phase power system.

These two indications lead us to the sequence network shown in Figure A.3. For this circuit, therefore, the total driving voltage divided by the total impedance gives the current. The total driving voltage is the sum of three sequence voltages (i.e., $\tilde{V}_1 + \tilde{V}_2 + \tilde{V}_0$), and the total impedance is $\tilde{Z}_1 + \tilde{Z}_2 + \tilde{Z}_0 + \left(3\tilde{Z}_f + 3\tilde{Z}_g + 3\tilde{Z}_n\right)$.

$$\text{Therefore, } \tilde{I}_1 = \tilde{I}_2 = \tilde{I}_0 = \left(\tilde{V}_1 + \tilde{V}_2 + \tilde{V}_0\right) \div \left(\tilde{Z}_1 + \tilde{Z}_2 + \tilde{Z}_0 + 3\tilde{Z}_f + 3\tilde{Z}_g + 3\tilde{Z}_n\right). \tag{A.10}$$

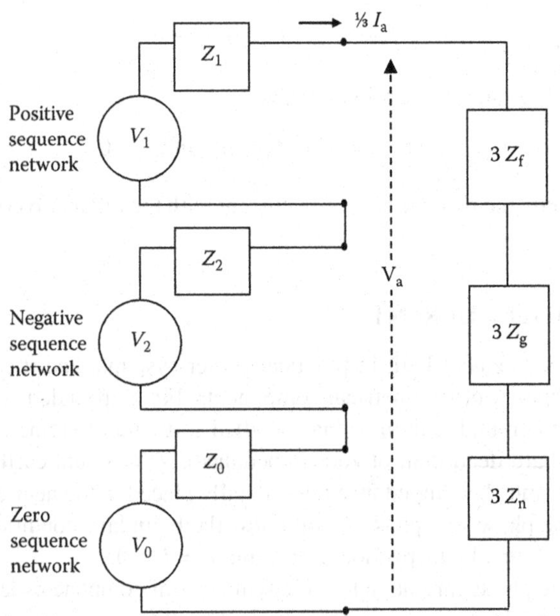

FIGURE A.3 Sequence networks connected in series for L-G fault (depends on type of fault).

Assuming that the prefault voltages are balance voltages, the sequence voltages are

$$\tilde{V}_1 = \tilde{V}_a \quad \text{and} \quad \tilde{V}_2 = \tilde{V}_0 = 0 \tag{A.11}$$

$$I_1 = I_2 = I_3 = \frac{V_1}{Z_1 + Z_2 + Z_0 + 3Z_f + 3Z_g + 3Z_n} \tag{A.12}$$

$$\text{Therefore, } I_a = 3I_1 = \frac{V_a}{\frac{1}{3}(Z_1 + Z_2 + Z_0) + Z_f + Z_g + Z_n}. \tag{A.13}$$

For this fault, we already know that $\tilde{I}_b = \tilde{I}_c = 0$ and $\tilde{I}_n = -\tilde{I}_a$.

For special cases of dead fault with a solidly grounded neutral, $\tilde{Z}_f = \tilde{Z}_g = \tilde{Z}_n = 0$ in Equation A.13.

A.5 L-L-G FAULT

For L-L-G fault from phase B to C to solid ground ($\tilde{Z}_f = \tilde{Z}_g = \tilde{Z}_n = 0$), we know the boundary conditions, which are $\tilde{V}_b = \tilde{V}_c = 0$ and $\tilde{V}_a \neq 0$. For these phase voltages, we derive the sequence voltages, which lead to $\tilde{V}_0 = \tilde{V}_1 = \tilde{V}_2 = 1/3\,\tilde{V}_a$, indicating that the three sequence networks are connected in parallel, from which we derive the sequence currents.

For this fault, a similar analysis as we discussed for L-G fault would lead to the sequence currents, where the // sign indicates the parallel connected impedances:

$$\tilde{I}_1 = \tilde{V}_1 \div \left(\tilde{Z}_1 + \tilde{Z}_2 // \tilde{Z}_0\right), \quad \tilde{I}_2 = -\tilde{I}_1 \times \tilde{Z}_0 \div \left(\tilde{Z}_2 + \tilde{Z}_0\right), \quad \text{and}$$
$$\tilde{I}_0 = -\tilde{I}_1 \times \tilde{Z}_2 \div \left(\tilde{Z}_2 + \tilde{Z}_0\right). \tag{A.14}$$

From Equation A.14 values of $\tilde{I}_1$, $\tilde{I}_2$, and $\tilde{I}_0$, we can derive $\tilde{I}_b$ and $\tilde{I}_c$ using Equation A.1, which are

$$\tilde{I}_b = \tilde{I}_0 + a^2\tilde{I}_1 + a\tilde{I}_2 \quad \text{and} \quad \tilde{I}_c = \tilde{I}_0 + a\tilde{I}_1 + a^2\tilde{I}_2. \tag{A.15}$$

We know that $\tilde{I}_a = 0$, so the neutral current is given by

$$\tilde{I}_n = -(\tilde{I}_b + \tilde{I}_c) = -3\tilde{I}_0. \tag{A.16}$$

A.6 L-L FAULT

For L-L fault from phase B to C, the boundary conditions are $\tilde{V}_b = \tilde{V}_c$ and $\tilde{I}_a = 0$ applying the same methods as above, except that now $\tilde{I}_b$ must return via phase C, that is, $\tilde{I}_b = -\tilde{I}_c$ and $\tilde{I}_a = 0$, which gives $3\tilde{I}_0 = \tilde{I}_n = 0$. Since $\tilde{I}_0 = 0$, we deduce that the zero sequence network is not involved at all in the circuit. We would then get the positive and negative sequence currents equal to

$$I_1 = -I_2 = \frac{V_1}{Z_1 + Z_2}. \tag{A.17}$$

Using these values of $\tilde{I}_1$, $\tilde{I}_2$, and $\tilde{I}_0$ in Equation A.1 would give us $\tilde{I}_b$, and then $\tilde{I}_c = -\tilde{I}_b$. We already know that $\tilde{I}_a = 0$ and $\tilde{I}_0 = 0$. These results can also be derived from a simple viewpoint that L-L fault (no ground involved) is a special case of L-L-G fault with $\tilde{Z}_0$ = infinity (open) since $\tilde{I}_0 = 0$.

Then, $\tilde{I}_1 = -\tilde{I}_2 = \tilde{V}_1 \div \left(\tilde{Z}_1 + \tilde{Z}_2\right)$, which is the same result as in Equation A.17.

FURTHER READING

Clarke, E. 1943. *Circuit Analysis of AC Power Systems, Volume I, Symmetrical and Related Components*. New York: John Wiley & Sons Inc.

Fortecque, C.L. 1918. Method of symmetrical components applied to the solutions of polyphone networks. *AIEE* 37, 1027.

Wagner, C.F. and Evans, R.D. 1933. *Symmetrical Components*. New York: McGraw Hill.

Index

C

D

E

H

I

J

R

W

Z

Printed in the United States
by Baker & Taylor Publisher Services